Springer Series in Surface Sciences

Volume 56

For further volumes:
http://www.springer.com/series/409

This series covers the whole spectrum of surface sciences, including structure and dynamics of clean and adsorbate-covered surfaces, thin films, basic surface effects, analytical methods and also the physics and chemistry of interfaces. Written by leading researchers in the field, the books are intended primarily for researchers in academia and industry and for graduate students.

Andrea Cusano · Marco Consales
Alessio Crescitelli · Armando Ricciardi
Editors

Lab-on-Fiber Technology

Springer

Editors
Andrea Cusano
Marco Consales
Alessio Crescitelli
Armando Ricciardi
Department of Engineering
University of Sannio
Benevento
Italy

ISSN 0931-5195
ISBN 978-3-319-06997-5 ISBN 978-3-319-06998-2 (eBook)
DOI 10.1007/978-3-319-06998-2
Springer Cham Heidelberg New York Dordrecht London

Library of Congress Control Number: 2014944338

Printed on acid-free paper

Springer is part of Springer Science+Business Media (www.springer.com)

Foreword

Almost half a century ago—in 1967—the first fiber optic sensor concepts were published. The five decades since that time have seen considerable change. Optical fiber technology has transformed the communications sector and established itself as a cornerstone of the Internet. Fiber sensors have also made modest but impressive inroads by making new measurement modalities become practical. The very special features of fiber sensing, particularly the capability for distributed measurements over ranges of a 100 km and the capability to address complex multiplexed networks with total electrical interference immunity, have facilitated an immense range of novel and intriguing measurements. Meanwhile, the original 1967 invention remains available as the Fotonic probe for surface finish assessment.

However, examining this a little more closely reveals the perhaps surprising feature that by far the vast majority of this progress has been in the context of physical measurements, predominately strain (and the associated acoustic/vibrational fields) and temperature. Optical fiber-based chemical measurements have made but modest impact and closer examination shows that this is concentrated in applications that involve providing a convenient light source, be it laser, light emitting diode, or even occasionally incandescent lamp, coupled into an optical fiber to facilitate illumination at a remote location. This feeds into a conventional spectroscopic cell containing typically a gas or a liquid sample whose optical characteristics (or those of some intermediate compound) are monitored through some sort of spectroscopic interaction either via absorption or through inelastic processes such as fluorescence and Raman or Brillouin scatter. The remote optic source has many benefits including, for example, precise spectral control of the optical source, physical separation of the source and the interaction cell, and a cold, electrically passive, illumination system. The optical interaction processes in the cell are derived from conventional chemistry.

Perhaps then, chemical measurements using fiber sensor approaches should more closely emulate the essential feature of physical sensors, namely that the interaction between the parameter to be measured and the light takes place either within the fiber itself or within a structure that is totally integrated with the fiber and of a scale comparable to the fiber core diameter. Then possibly, the Lab on the Fiber could follow the example of the "ab on the chip" through which fiber coupled spectroscopic analysis could be realized within the fiber geometry itself.

For this to become practical two basic enablers need to be present. The first is to scale the chemistry, whether through intermediate optically active compounds or for direct spectroscopy, into a technology compatible with the fiber dimensions. The second is to ensure that samples can be effectively and efficiently handled with the necessary spatial precision.

The past half century through which fiber sensors have made their mark has also seen the emergence of new precision tools. In the 1970s, photolithography relied on very sharp knives and rubylith masks. Now we have the electron beam. This has facilitated achievable light guiding to evolve from diffraction through free space through ever-more complex waveguide structures into the region of nanoprobes and plasmonics. The last two of these may be simplistically viewed as the optical equivalent of an electrical wire enabling subwavelength precision in spatial sampling—exactly as in the Lab on the Chip. This has provided the optical sensor researcher with an entirely new toolbox through which to facilitate innovative measurements.

This book explores this rapidly evolving and extremely intriguing toolbox for innovative fiber optic-based chemical sensing in considerable and very authoritative detail. The following chapters embrace not only the new tools themselves but also a variety of ways through which they may be realized. The fibers themselves can be fabricated in an immense variety of microstructured "photonic crystal" geometries comprising one or many material structures. This potentially enables chemical activity to take place along the fiber length to modify the transmission and/or backscattering spectral properties in response to chemical stimuli. The fiber ends provide a platform on which precision nanolithography gives the capability to realize complex structures into which interesting chemistry can be built. These structures can either be fabricated on the end of the fiber or attached using nanoscale bonding techniques. The materials that are attached can range from silicon photonic crystals to polymer films. The near-field taper—as used in the scanning near field optical microscope—can also be modified as a tool for micro and nanoscale chemistry. Then there is plasmonics—optical wires—which precisely locate highly concentrated optical fields and facilitate the interaction of these fields with local chemical variations. There is also a plethora of techniques that can be used to fabricate these nanoscale structures ranging from imprinting to e-beam lithography. Add to this the many new approaches to realizing optical tweezers to position the samples and the toolbox is complete.

Within these pages you will encounter all of these and many other evolving innovations that promise to stimulate research, development, and exploitation in fiber-based chemical sensing over the coming couple of decades or so. The need for this flexible and highly responsive fiber optic chemical analysis probe will also consolidate through the coming years as more and more must become known about the environment in which we live and its influence on our overall well-being. This book brings together the essentials of this new toolbox for the fiber optic chemical sensing community. Its contents make stimulating reading and will enthuse future generations of young researchers while providing a technological backdrop for those who guide and encourage the young enthusiasts. The book promises to make an important and effective contribution to what will become an essential measurement enabler in future communities.

Glasgow Brian Culshaw

Preface

Since its discovery, the fiber optics field has undergone a tremendous growth and advancement over the past several decades. As optical fibers cemented their position in the telecommunications industry and its commercial markets matured, parallel efforts were carried out by a number of different research groups around the world to exploit them also in the sensors field. Nowadays, a variety of commercial fiber optic sensors have widespread use for structural sensing and monitoring applications in civil engineering, aerospace, marine, oil and gas, smart structures, and bio-medical devices, to name a few.

Although fiber optic sensors have, in many cases, completely replaced and outperformed their counterparts based on more conventional technologies, it is out of the question that optical fibers are still mainly conceived as communication medium. So far, the fiber optic industry focused its efforts on the development of devices and components whose functionalities are mainly related to the silica glass properties. For this reason, in the development of photonic systems both for communication and sensing applications, several out-of-fiber optical components, ranging from light sources, modulators, polarizers, up to photo-detectors, are currently employed. This is why there is currently an increasing market demand for highly integrated and multifunctional devices with advanced performances and unrivaled features. A significant technological breakthrough would be the development of these components and devices "all in fiber" through the integration of advanced functional materials at micro and nanoscale. All the above considerations lead to envision the "Lab-on-Fiber" concept as a concrete technological solution to actual market demand.

In this context, the "Lab-on-Fiber" technology is an emerging research field which envisions a novel class of advanced and multifunctional photonic devices and components arising from the integration onto optical fibers of different materials at micro and nanoscale with suitable physical, chemical, and biological properties. This fascinating and intriguing research topic is essentially aimed at the development of novel technological platforms where functionalized materials, devices, and components are constructed, embedded all together in a single optical fiber providing the necessary physical connections and light–matter interaction, usefully exploitable and useful in many strategic sectors such as optical processing, environment, life science, safety, and security. The addition of new features and functionalities to optical fibers involves the concurrent approach of different disciplines ranging from nanotechnology to photonics engineering passing through material science.

This technological innovation is opening the way for the creation of fiber-based multifunction sensing and actuating systems, with enhanced performances and functionalities with respect to conventional technologies. For example, novel fiber optic nanoprobes could be judiciously integrated with microfluidics components to provide new Lab-on-chip implementations, taking advantage from the easier connection of the optical chain to complex lighting systems as well as sophisticated remote interrogation units.

"Lab-on-Fiber Technology" is becoming a key and enabling technology behind many devices, components, and systems found in the modern home, factory, and research lab as well as in many strategic industrial sectors. The enormous potentialities of this new intriguing technological world could envisage a primary role in what we can safely label as the "Photonic Century," especially if main issues, concerning the identification and definition of viable fabrication methodologies, routes, and strategies, enabling the integration of a large set of materials onto non-conventional substrates as the case of optical fibers are judiciously addressed.

Many research efforts have been carried out aimed to translate the vision in a technological reality, and according to the methodologies proposed so far, the book has been divided into the following sections, comprising a series of commissioned chapters from leading experts in the Lab-on-fiber technology field:

Section 1: The *macro to micro/nano approach,* enabling the thermal scaling of a macroscopic pre-form down to the microscopic and nanoscopic scale, by means of the fiber drawing technique.

In this context, Chap. 1 provides an overview of multimaterial fibers, starting from a discussion on material constrains, fiber drawing, and pre-form fabrication up to the introduction of photonic and optoelectronic multimaterial fibers with novel optical and sophisticated electronics. The fabrication and the applications of optical microfibers and nanofibers, i.e., optical fibers with diameters close to or smaller than the wavelength of the guided light, are presented in Chap. 2 as a valuable technological platform for Lab-on-Fiber with strong near-field interaction.

Section 2: The *transferring method*, essentially consisting of the previous fabrication of dielectric and metallic structures onto planar substrates (by using standard nanofabrication techniques) followed by their successive transfer onto optical fiber substrates.

By following this approach, with the aim of combining the good performances of photonic biosensors on chip with the advantages offered by optical fibers, in Chap. 3 an optical fiber probe for label-free biosensing is presented, based on a high Q silicon-on-insulator ring resonators transferred on the fiber tip. Similarly, Chap. 4 deals with a fiber device for both refractive index and temperature measurements based on the integration of a photonic crystal slab fabricated on standard Si wafers and transferred onto optical fiber facet. In addition to the transferring of silicon membranes, both hybrid and double transfer UV-curing nanoimprint-soft lithography techniques (discussed in Chap. 5) enable pattern transfer on highly curved surfaces (e.g., the sidewall of an optical fiber) with a resolution down to a few nanometers. In the same line of argument, Chap. 6 discusses the fabrication strategies to apply on the fiber tip a flexible, ultrathin polymeric membrane supporting metallic nano-features, which acts as a guided mode resonance filter.

Section 3: The *direct-writing method*, relying on the use of nanofabrication techniques suitably adapted to directly operate on the optical fiber. Both top-down, using Electron-Beam Lithography (EBL) or Focused-Ion Beam (FIB), and bottom-up approaches, by means of self-assembly, have been so far demonstrated.

In Chap. 7, a fabrication process involving EBL on the optical fiber tip is discussed, for creating multifunctional optical probes for both label-free chemical and acoustic sensing. By using FIB milling and two-photon lithography techniques, the realization of single-fiber optical tweezers able to create a purely optical three-dimensional trap is provided in Chap. 8.

The FIB milling technique is also demonstrated in Chap. 9 to enable the development of innovative fiber optic hydrogen sensors based on C-shaped nano-apertures patterned onto the tip of a Pd coated fiber.

By using bottom-up self-assembly approaches, as discussed in Chap. 10, it has been shown that 2-D structures made of metallic nanoparticles can be coated on D-shaped fibers. Moreover, a fabrication process based on the so-called "breath-figure technique" can be exploited to realize metallo-dielectric crystals onto optical fiber tip, as presented in Chap. 11.

Section 4: *Integration of functional materials onto conventional and unconventional optical fibers,* opening up interesting possibilities for adding new functionalities and developing advanced optical fiber probes for sensing applications.

For example, techniques for manipulating liquids on micro- and nanoscale are presented in Chap. 12, providing possibility of functionalizing sensing area of lab-in-fiber devices. Chapter 13 deals with the integration of optical fibers with different types of nanoscale structured materials (such as multilayer-based nanostructures, sol-gel-derived materials, molecularly imprinted polymers, and metallic thin films and nanoparticles) that so far significantly contributed to give a push to the continuous development of new fiber optic sensors.

The biological functionalization of grating-based fiber optic transducers by means of T4 bacteriophages, enabling the realization of a label-free biosensor for the detection of *E. coli,* is discussed in Chap. 14. Concerning unconventional optical fibers, Chap. 15 reviews the state-of-the-art advances of photonic crystal fibers as inherent lab-in-fiber optofluidics platforms (thanks to the easy access of the fiber air channels for surface functionalization) for monitoring important chemical and biological events. Finally, Chap. 16 provides an overview of micro- and nano-structured optical fiber sensors, directly associated with surface plasmon resonance sensing.

The book thus highlights the main achievements of the Lab-on-Fiber technology roadmap, providing an exhaustive overview of the main fabrication approaches including first demonstrations of optical devices and components and their use in practical applications. The text aims to make the readers aware of the enormous opportunities offered by this innovative technology that is leading it to be one of the hottest topics in the photonics community. As this topic is still a subject of dynamic research, this book can represent a reference and at the same time a source of inspiration for new ideas and concepts.

In conclusion, the editors hope that this book, reflecting ongoing and latest research advances in this promising area, could bring readers with very broad competencies ranging from chemistry, biology, electronics, photonics, material science, nanotechnologies, and other disciplines to seek and promote multidisciplinary and synergic cooperations for reaching further innovations in this area.

Benevento, Italy

Andrea Cusano
Marco Consales
Alessio Crescitelli
Armando Ricciardi

Contents

Contributors

Ayman F. Abouraddy CREOL, The College of Optics and Photonics, University of Central Florida, Orlando, FL, USA

Francisco J. Arregui Department of Electric and Electronic Engineering, Public University of Navarre, Madrid, Spain

Peter Bienstman Photonics Research Group (INTEC), Ghent University, Ghent, Belgium

Wojtek J. Bock University of Quebec in Outaouais, Gatineau, QC, Canada

Francesca Bragheri Dipartimento di Ingegneria Industriale e dell'Informazione, Universita' di Pavia, Pavia, Italy

Patrizio Candeloro BioNEM Lab, Dipartimento di Medicina Sperimentale e Clinica, Universita' Magna Graecia di Catanzaro, Catanzaro, Italy

John Canning Interdisciplinary Photonics Laboratories (IPL), The School of Chemistry, The University of Sydney, Sydney, NSW, Australia

Taerin Chung School of Electrical Engineering, Seoul National University, Seoul, Korea

Tom Claes Photonics Research Group (INTEC), Ghent University, Ghent, Belgium

Gheorghe Cojoc Max Planck Institute of Molecular Cell Biology and Genetics MPI-CBG, Dresden, Germany

M. Consales Optoelectronic Division, Department of Engineering, University of Sannio, Benevento, Italy

Sara Coppola INO—CNR, Pozzuoli, NA, Italy

A. Crescitelli Optoelectronic Division, Department of Engineering, University of Sannio, Benevento, Italy; Institute for Microelectronics and Microsystems (IMM), National Council of Research, NA, Italy

Ilaria Cristiani Dipartimento di Ingegneria Industriale e dell'Informazione, Universita' di Pavia, Pavia, Italy

A. Cusano Optoelectronic Division, Department of Engineering, University of Sannio, Benevento, Italy

Katrien De Vos Photonics Research Group (INTEC), Ghent University, Ghent, Belgium

Henry Du Department of Chemical Engineering and Materials Science, Stevens Institute of Technology, Hoboken, NJ, USA

Cesar Elosua Department of Electric and Electronic Engineering, Public University of Navarre, Madrid, Spain

E. Esposito Institute for Microelectronics and Microsystems (IMM), National Council of Research, NA, Italy

Enzo Di Fabrizio PSE and BESE divisions, King Abdullah University of Science and Technology (KAUST), Jeddah, Saudi Arabia; BioNEM Lab, Dipartimento di Medicina Sperimentale e Clinica, Universita' Magna Graecia di Catanzaro, Catanzaro, Italy

Andrea Di Falco SUPA, School of Physics and Astronomy, University of St Andrews, St Andrews, UK

Lorenzo Ferrara Nanostructures, Istituto Italiano di Tecnologia, Genoa, Italy

Pietro Ferraro INO—CNR, Pozzuoli, NA, Italy

Yoel Fink Research Laboratory of Electronics, Massachusetts Institute of Technology, Cambridge, MA, USA

Francesco Galeotti Istituto per lo Studio delle Macromolecole, Consiglio Nazionale delle Ricerche, Milan, Italy

Haixiong Ge Department of Materials Science and Engineering, College of Engineering and Applied Sciences, National Laboratory of Solid State Microstructures, Nanjing University, Nanjing, People's Republic of China

Oriella Gennari INO—CNR, Pozzuoli, NA, Italy

Michele Giordano Institute for Composite and Biomedical Materials, Consiglio Nazionale delle Ricerche, Portici, NA, Italy

Lynford L. Goddard University of Illinois at Urbana-Champaign, Urbana, IL, USA

Simonetta Grilli INO—CNR, Pozzuoli, NA, Italy

Miguel Hernaez Department of Electric and Electronic Engineering, Public University of Navarre, Madrid, Spain

Agostino Iadicicco Department of Engineering, University of Naples "Parthenope", NA, Italy

Katarzyna Komorowska Photonics Research Group (INTEC), Ghent University, Ghent, Belgium

Byoungho Lee School of Electrical Engineering, Seoul National University, Seoul, Korea

Cristina Lerma Arce Photonics Research Group (INTEC), Ghent University, Ghent, Belgium

Wen-Di Li Department of Mechanical Engineering, The University of Hong Kong, Hong Kong, People's Republic of China

Carlo Liberale Nanostructures, Istituto Italiano di Tecnologia, Genoa, Italy

Ignacio R. Matias Department of Electric and Electronic Engineering, Public University of Navarre, Madrid, Spain

Eric Mazur School of Engineering and Applied Sciences, Harvard University, Cambridge, MA, USA

Steven J. McKeown University of Illinois at Urbana-Champaign, Urbana, IL, USA

Francesco Merola INO—CNR, Pozzuoli, NA, Italy

Lisa Miccio INO—CNR, Pozzuoli, NA, Italy

Paolo Minzioni Dipartimento di Ingegneria Industriale e dell'Informazione, Universita' di Pavia, Pavia, Italy

Bryan Park E. L. Ginzton Laboratory, Department of Electrical Engineering, Stanford University, Stanford, CA, USA

Melania Paturzo INO—CNR, Pozzuoli, NA, Italy

Gerardo Perozziello BioNEM Lab, Dipartimento di Medicina Sperimentale e Clinica, Universita' Magna Graecia di Catanzaro, Catanzaro, Italy

Marco Pisco Department of Optoelectronic Division—Engineering, University of Sannio, Benevento, Italy

Giuseppe Quero Optoelectronic Division, Department of Engineering, University of Sannio, Benevento, Italy

Vijayakumar Rajamanickam Nanostructures, Istituto Italiano di Tecnologia, Genoa, Italy

Peter Reader-Harris SUPA, School of Physics and Astronomy, University of St Andrews, St Andrews, UK

Armando Ricciardi Department of Engineering, University of Sannio, Benevento, Italy

Mateusz Smietana Warsaw University of Technology, Warsaw, Poland

Olav Solgaard E. L. Ginzton Laboratory, Department of Electrical Engineering, Stanford University, Stanford, CA, USA

Alexander M. Stolyarov Research Laboratory of Electronics, Massachusetts Institute of Technology, Cambridge, MA, USA

Svetlana Sukhishvili Department of Chemical Engineering and Materials Science, Stevens Institute of Technology, Hoboken, NJ, USA

Guangming Tao CREOL, The College of Optics and Photonics, University of Central Florida, Orlando, FL, USA

Fei Tian Department of Chemical Engineering and Materials Science, Stevens Institute of Technology, Hoboken, NJ, USA

Limin Tong State Key Laboratory of Modern Optical Instrumentation, Department of Optical Engineering, Zhejiang University, Hangzhou, China

Saurabh Mani Tripathi Indian Institute of Technology Kanpur, Kanpur, Uttar Pradesh, India

Veronica Vespini INO—CNR, Pozzuoli, NA, Italy

Xiaoqin Wu State Key Laboratory of Modern Optical Instrumentation, Department of Optical Engineering, Zhejiang University, Hangzhou, China

Wei Wu Department of Electrical Engineering, University of Southern California, Los Angeles, CA, USA

Chapter 1
Multimaterial Fibers

Guangming Tao, Ayman F. Abouraddy, Alexander M. Stolyarov and Yoel Fink

Abstract In recent years, new materials processing approaches have emerged that enable the realization of fiber devices with unique photonic, optoelectronic, and acoustic functionalities. At the heart of this achievement is the identification of materials and processing conditions that mitigate surface energy effects, allowing for materials with disparate optical, electronic, and thermo-mechanical properties to be monolithically drawn from a preform into kilometer-long fibers with complex micro- and nano-structured cross-sectional features. We review this nascent but rapidly growing field and highlight future research directions.

1.1 Introduction

Optical fibers have enhanced the quality of life throughout the world and have fundamentally transformed the human condition in both obvious and imperceptible ways. Today, the worldwide delivery of the internet, and most telecommunications, is achieved through optical fibers [1, 2]. Additionally, optical fibers are used in a multitude of medical and industrial applications ranging from noninvasive

G. Tao · A. F. Abouraddy
CREOL, The College of Optics and Photonics, University of Central Florida,
Orlando, FL, USA
e-mail: guangmingtao@gmail.com

A. F. Abouraddy
e-mail: raddy@creol.ucf.edu

A. M. Stolyarov · Y. Fink (✉)
Research Laboratory of Electronics, Massachusetts Institute of Technology,
Cambridge, MA, USA
e-mail: yoel@mit.edu

A. M. Stolyarov
e-mail: sashok@mit.edu

A. Cusano et al. (eds.), *Lab-on-Fiber Technology*,
Springer Series in Surface Sciences 56, DOI: 10.1007/978-3-319-06998-2_1,

medical surgery [3, 4] to monitoring the structural integrity of bridges and oil pipelines [5, 6], and fiber lasers are finding applications in materials processing and manufacturing [7]. Interestingly, despite the many complex technological applications of optical fibers, the fibers themselves are remarkably simple from the perspective of materials composition. A single material, silica glass, is used to fabricate the majority of optical fibers in use today. While optical fibers have also been made from other glasses or polymers, silica glass remains the dominant material in producing optical fibers.

With the new physics of photonic band gaps (PBG's) [8], a novel class of optical fibers emerged in the mid 1990s, photonic crystal fibers (PCF's) and photonic band gap fibers [9]. These fibers presented a significant step in overcoming perceived limitations of the structures that could be produced by the traditional process of thermal fiber drawing. In fact, the initial proposal for fabricating silica PBG fibers was dismissed as unfeasible by seasoned practitioners [10] even though no new material was combined with silica, only air holes. Nevertheless, the rapid success of PBG fibers has had unexpected consequences for the process of fiber fabrication itself. One such consequence was the introduction of the concept of 'multimaterial fibers' over the past decade. This new class of fibers leverages the capabilities of traditional fiber fabrication, but aims at developing new fiber structures, functionalities, and applications that stem from altering the materials composition of the fiber. An all-encompassing and conclusive definition to this emerging concept is difficult to provide. Here, as in our recent review article on the subject [11], we define multimaterial fibers to be *high-aspect-ratio structures that comprise multiple distinct materials, typically produced by thermal drawing from a macroscopic scaled-up model called a 'preform'*. Materials with different optical, electronic, thermo-mechanical, and acoustic properties have now been incorporated into the fiber form-factor [12]. The set of fabrication approaches developed in this emerging field are enabling new functionalities that are not usually associated with optical fibers. Examples include fibers that produce an electrical signal when light is incident on the fiber external surface [13–15] or when the temperature of the surrounding environment changes [16]; self-monitoring fibers that measure their own structural integrity [17]; flexible, lightweight fiber arrays that can image the surrounding environment [14, 18]; fibers that can emit or detect acoustic signals [19, 20]; and fibers that incorporate microfluidics for actively tuning their optical properties [21, 22]. Fiber drawing is therefore morphing into a fabrication route for producing photonic and optoelectronic fiber devices that may be potentially incorporated or woven into fabrics, thereby endowing them with new and sophisticated functionalities [12].

Along with the evolution of multimaterial fibers come new ways of thinking about the fiber drawing process itself. Through dimensional reduction, nanostructures such as nanowires with few-nanometer diameters and unprecedented lengths have been produced [23–26]. The fiber cladding may be viewed as a crucible for chemical synthesis and the thermal fiber draw as a catalyst for enabling chemical reactions to take place between reactive species placed in the preform [27–29]. Multimaterial fibers are also a new playground for the controllable study of fluid dynamics in confined space and over a wide range of length scales [30]. These developments have recently led to a scalable, top-down, in-fiber fabrication process capable of producing complex structured particles over an unprecedented range of diameters spanning both micro- and nano-scales [31].

There are two aspects of fiber production that have been appropriated by the emerging field of multimaterial fibers. First, the process typically starts by preparing a macroscopic preform. Since the preform is structured on the centimeter scale, it is straightforward to create a complex transverse cross section with controllable placement of multiple materials. Second, thermal fiber drawing is an inherently scalable manufacturing process, producing kilometers of fiber with accurate control over size and axial uniformity [32]. In order for the fiber to maintain the complex transverse structure of the multimaterial preform and have all the materials to flow together at the same temperature, restrictions are placed on the allowable materials combinations compatible with this fabrication process. There are some exceptions to this overall approach. For example, multimaterial fibers may be produced by starting with a single-material fiber or wire as a scaffold and proceeding with additive manufacturing. This can include approaches such as dip coating on the outer surface of extended fiber lengths [33] or vapor deposition inside hollow enclaves in short lengths of a pre-existing fiber [34].

Multimaterial fiber research is at the crossroads of many disciples, including optics, materials science, device physics, nanotechnology, and fluid dynamics. While this field is still in its initial stages of development, there has been tremendous recent progress. This chapter consequently aims at providing only a brief and selective overview of some of the main achievements based on our definition of multimaterial fibers provided above. We first discuss in Sect. 1.2 some general constraints on the materials that may be codrawn in the same fiber, and in Sect. 1.3 we describe several general approaches that–to date–have been utilized to produce multimaterial preforms. Section 1.4 focuses on multimaterial fiber devices with novel optical functionalities, such as transverse omnidirectional emitting fiber lasers. In Sect. 1.5 we describe multimaterial fibers endowed with sophisticated electronic, optoelectronic, and even piezoelectric functionalities, in addition to recent work on producing fibers with traditional crystalline semiconductor cores, such as silicon. In the Sect. 1.6, we highlight novel approaches that utilize fiber drawing as a route to synthesize new chemical compounds. We end the chapter by describing some alternative fabrication approaches (Sect. 1.7) and future possibilities (Sect. 1.8).

1.2 Material Constraints and Fiber Drawing

The preform-to-fiber approach for producing optical fibers relies on thermally drawing the fiber from a macroscopic preform in a fiber draw tower [35]. By feeding the preform into a furnace that softens the material, the molten gob at the preform tip is 'pulled' by gravity or an external force and is stretched into a thin fiber strand, after which polymer coatings are added to the fiber surface for mechanical protection if necessary. Despite the relative simplicity of this fabrication methodology, its efficiency and robustness are behind the thousands of kilometers of optical fiber that span the globe today enabling worldwide communications networks [1, 2]. An alternative approach, the so-called double-crucible method [36], produces optical fibers from glass melts of the core and cladding materials that are pushed through a nozzle. To date, this process has not been used to produce multimaterial fibers.

The transition from traditional single-material fibers to multimaterial fibers dictates materials constraints that stem from the nature of the preform-to-fiber process. In the case of a single-material preform, the parameters of the draw (e.g., the drawing speed) are determined by the material viscosity at the drawing temperature. On the other hand, since multimaterial preforms may contain materials that are incompatible with thermal drawing when taken separately, such as crystalline semiconductors or metals, thermal drawing imposes some constraints on the materials combinations that are compatible with this fabrication strategy. One may gain insight into the feasibility of various materials combinations by examining the viscosity of silica glass, silicon, and gold as a function of temperature, presented in Fig. 1.1a. These materials represent three distinct classes from the perspective of electronic properties: an amorphous insulator, a crystalline semiconductor, and a metal, respectively. Additionally, these three materials have different optical and mechanical properties. Silica is amenable to thermal drawing over a broad range of conditions since its softening temperature ranges from 1400 to 2350 °C. On the other hand, crystalline materials such as silicon and gold are characterized by an abrupt drop in viscosity above the melting temperature T_m where a phase transition takes place. While this physical feature excludes the use of thermal drawing to produce a fiber from a single-material preform made of silicon or gold, nevertheless a multimaterial-preform approach enables the use of such materials in fiber drawing. By making use of a crystalline material such as silicon or gold as a 'core' embedded in an amorphous 'cladding' such as silica, this multimaterial preform may be thermally drawn above the core T_m. In this scenario, the cladding (silica) acts as a supporting scaffold that contains and restricts the flow of the low-viscosity core material (silicon or gold).

This example enables us to outline some general constraints on the construction of multimaterial preforms and the ensuing thermal drawing conditions. First, at least one material should be amorphous and resist devitrification during thermal drawing, typically a glass or polymer. This amorphous 'backbone' material constitutes an outer cladding that supports the other materials during the draw process, thereby maintaining the fiber cross-sectional structure. The amorphous material constituents must be chosen to have overlapping softening temperatures, while the crystalline constituents must have T_m below the drawing temperature. The drawing temperature must be lower than the boiling temperature of the core material. Additionally, care must be taken to avoid fluid instabilities that may occur when the viscosity of the materials is lowered and the transverse dimensions reduced from preform to fiber (see [23, 24, 30, 31] for further details on this issue). Finally, the materials should also have relatively similar thermal expansion coefficients in

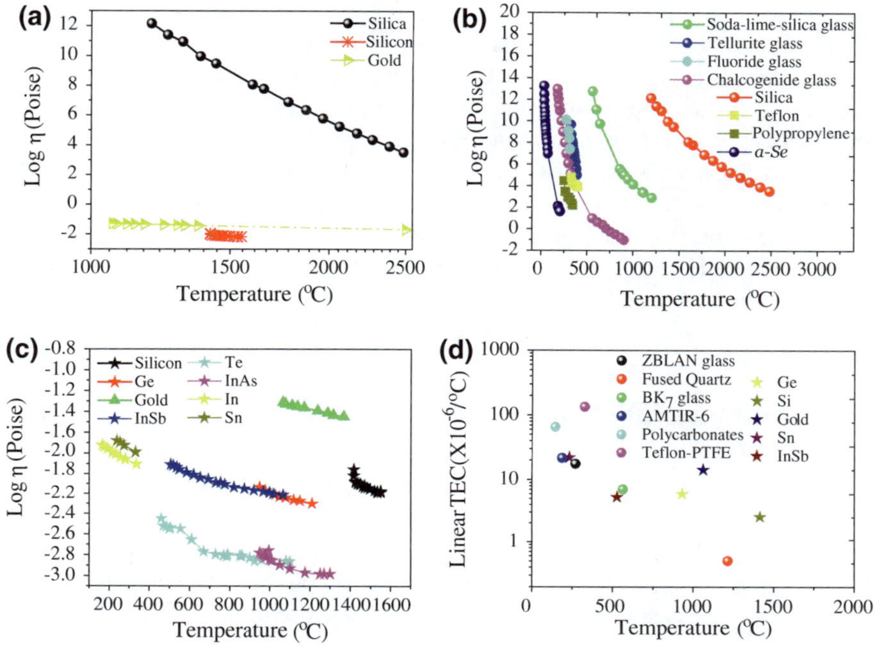

Fig. 1.1 Dynamic viscosity (logarithm of viscosity η in poise) of selected materials versus temperature. **a** Viscosity for silica [37], silicon [38] and gold [39]. The viscosity for silicon and gold are measured above their melting temperatures. **b** Same as (**a**), showing the viscosity for silica [37], soda-lime-silica glass [40], fluoride glass [41], tellurite glass [42], chalcogenide glass (As_2S_3) [43], Teflon®PTFE-6 polymer [44], polypropylene (PP) polymer [45], and amorphous selenium [46]. **c** Same as (**a**), showing the viscosity for silicon [38], germanium [47], indium antimonide [48], tellurium [47], indium arsenide [47], indium [49], tin [49], and gold [39]. **d** Linear thermal expansion coefficient (TEC) at room temperature for selected materials plotted against the melting temperature (T_m) for metals and semiconductors (*solid stars*) and the glass transition temperature (T_g) for the amorphous materials (*solid dots*). References: Ge: T_m [50], TEC [51]; Si: T_m [50], TEC [52]; Gold: T_m [53], TEC [52]; Sn: T_m [53], TEC [54]; InSb: T_m [55], TEC [56]; Teflon® PTFE: T_g [57]; TEC [58]; Fused quartz: T_g [58], TEC [59]; BK7 glass: T_g [58], TEC [58]; AMTIR-6: T_g [60], TEC [60]; Polycarbonates: T_g [61], TEC [59]; ZBLAN glass: T_g [58], TEC [59]

the temperature range extending to the drawing temperature, to avoid fractures resulting from thermo-mechanical mismatches.

The wide range of possibilities for constructing multimaterial preforms within the above-prescribed constraints may be appreciated from Fig. 1.1b where we present the viscosity of a wide range of *amorphous* materials. While some glasses, such as soda-lime-silica, have a broad temperature range suitable for thermal drawing, others (such as fluorides, chalcogenides, and tellurites) have a relatively narrower temperature window. Next, we plot in Fig. 1.1c the viscosity of some typical *crystalline* materials above their T_m. It is possible to choose potential pairs of amorphous-crystalline materials that may be combined in a preform and co-drawn into a multimaterial fiber by consulting Fig. 1.1b, c together. For example, it is possible to draw a fiber containing a core of Si, Ge, or gold by using a cladding made of silica glass; InSb clad with soda-lime-silica glass glass; or Sn or Se clad with fluoride or chalcogenide glass, or even a polymer. Furthermore, we identify in Fig. 1.1d the glass transition temperature T_g of the amorphous materials and the melting temperature T_m of crystalline materials used in Fig. 1.1a–c versus the linear thermal expansion coefficient (TEC) for completeness. Finally, we note that it is possible for more than two materials to be chosen according to the above criteria and thus be incorporated in the same multimaterial fiber. Indeed, this important possibility enables the construction of in-fiber photonic, electronic and, optoelectronic devices, as described in Sects. 1.4 and 1.5.

1.3 Multimaterial Preform Fabrication

We describe here four general classes of approaches to the preparation of multimaterial preforms: (a) the rod-in-tube approach [62–64], (b) extrusion [65–67], (c) the stack-and-draw approach [9, 68], and (d) thin-film-rolling technology [12]. The choice of any specific approach, or combinations thereof, is dictated both by the various materials involved and the transverse structure targeted.

1.3.1 Rod-in-Tube Approach

The rod-in-tube approach, depicted in Fig. 1.2a, relies on inserting a solid rod of one material (the core) into a tube of another material (the cladding) to form a preform with a core-cladding structure. In a variation on this procedure, the solid core rod may be replaced with a powder that is placed in the tube. The preform is then sealed and thermally drawn. If T_m of the core powder is lower than the drawing temperature, we designate the process molten-core-in-tube method, a process that enables a wide range of materials to be incorporated as a core. The recognition of the usefulness of this method for multimaterial fibers may be traced back to the pioneering work of E. Snitzer in 1989 [69], where selective volatization combined with the rod-in-tube method was used to produce a silica-clad fiber containing a soft-glass core.

1.3.2 Extrusion

Extrusion is a well-known process used to create axially symmetric objects with a fixed complex cross-sectional profile by 'pushing' a soft material through a die under pressure. The first extrusion process was patented by J. Bramah in 1802 for producing lead pipes [70]. Subsequently, E. Roeder [71–73] extended this approach to soda-lime

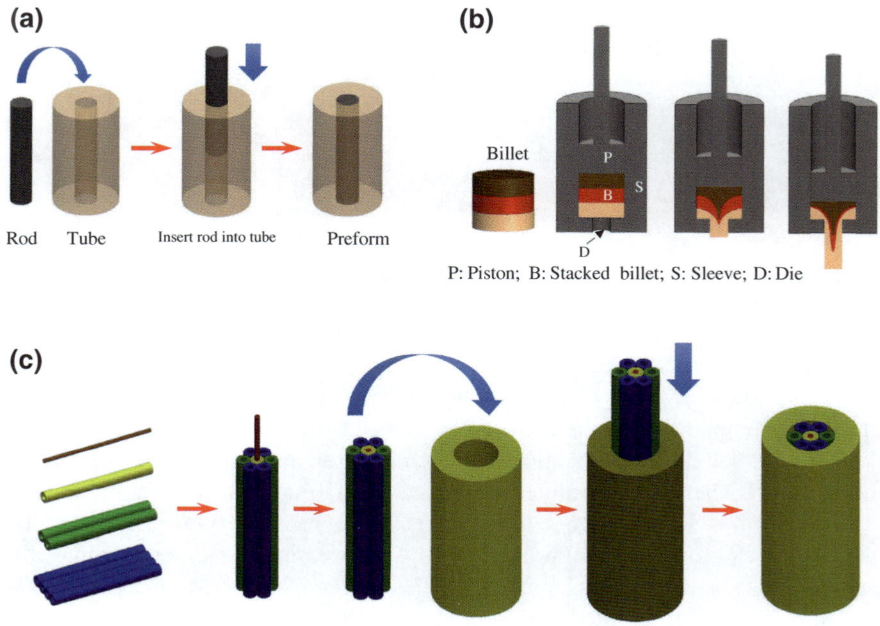

Fig. 1.2 General methodologies for multimaterial fiber preform fabrication. **a** Rod-in-tube, **b** extrusion, and **c** stack-and-draw methods

silica, lead silicate, calcium aluminate and boric oxide glass in the 1970s. Extrusion makes use of material in the form of a rod, typically called a *billet*, which is placed in a sleeve held inside a furnace (a vertically stacked billet is shown in Fig. 1.2b for illustration). By heating the billet to the softening temperature of the incorporated materials, applied pressure pushes the material through a die that imparts shape to the extruded rod, which may constitute a preform that is subsequently drawn into a fiber. Using a multimaterial billet consisting, for example, of vertically stacked discs, the transverse structure of the extruded rod may be rationally engineered [65, 67].

1.3.3 Stack-and-Draw Approach

An alternative procedure that has been used extensively in preparing the preforms drawn into microstructured fibers, PCFs, and PBG fibers [9] is the stack-and-draw approach. It is instructive that the first prescient demonstration of the stack-and-draw method to produce an optical fiber may be traced to Bell Labs in 1974 [74], at the dawn of the development of silica fibers. In that effort, a fiber containing a hanging core surrounded by air was produced. This procedure offers flexibility in the choice of geometry of the domains for each material by starting from rods, tubes, and/or plates from a single or multiple materials that are assembled into a preform–Fig. 1.2c–with dimensions determined by the targeted fiber structure.

Furthermore, recursive application of multiple stack-and-draw steps enables one to reach the required dimensions (that may not be achievable in a single thermal-drawing step) and attain complex transverse structures.

1.3.4 Thin-Film Rolling

Any of the above three approaches may be used to incorporate polymers into a multimaterial preform. Additionally, a unique process exists to incorporate a polymer in a preform made possible by the availability of polymers in the form of extended thin films. In this technique, a thin polymer film is 'rolled' around a pre-existing rod, followed by thermal consolidation under vacuum above the glass transition temperature of the constituent materials until the individual films fuse. Figure 1.3a details the fabrication approach towards making a multimaterial PBG fiber [75–79], thereby illustrating schematically a particular instance of this process. The thin-film-rolling technique has also been used recursively to realize preforms with complex cross-sections through multiple processing steps. For example, after consolidating a preform constructed in this fashion, voids or enclaves may be introduced that can be filled, for example, by conductors and encapsulated in the preform, followed by rolling additional thin films and reconsolidation. These multistep preform preparation processes form the basis for many recently developed photonic and optoelectronic multimaterial fiber devices [12, 18, 21, 22].

1.4 Photonic Multimaterial Fibers

This section provides an overview of multimaterial fibers with both axial and transverse optical functionality.

1.4.1 Multimaterial PBG Fibers

The MIT fiber group led by Y. Fink demonstrated the first successful draw of multimaterial PBG optical fibers with a hollow core and an all-solid cladding [75]. This was made possible by combining and co-drawing two materials having a large refractive-index contrast, yet compatible thermomechanical properties. Below we describe this and several other examples of photonic multimaterial fiber devices.

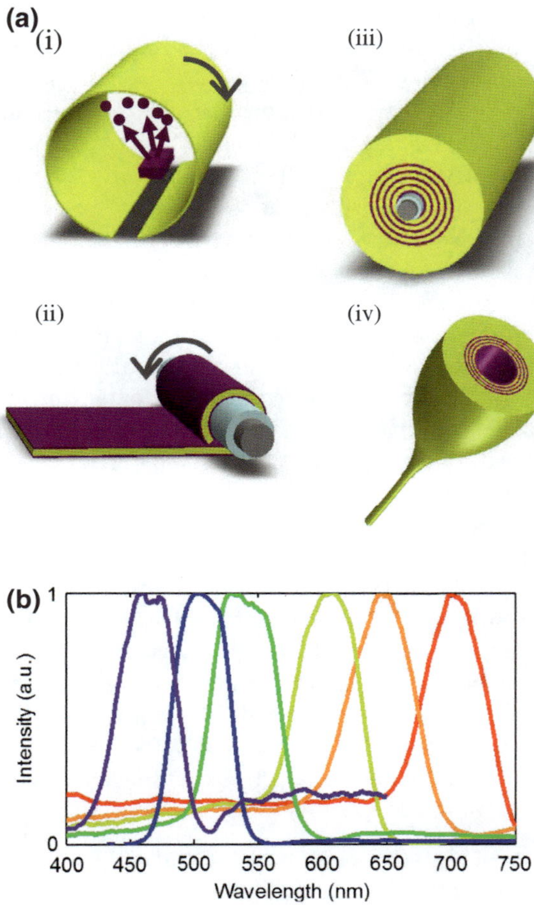

Fig. 1.3 PBG fiber fabrication and characterization **a** PBG fiber fabrication flow. (*i*) The chalcogenide glass is thermally evaporated onto both sides of the polymer film. (*ii*) This multilayer film is then rolled onto a Teflon-lined mandrel and additional polymer cladding layers are rolled for mechanical support. (*iii*) The entire structure is thermally consolidated under vacuum until the materials fuse together into one solid preform. (*iv*) The preform is then thermally drawn into 100 m of fiber by applying uniaxial tension. The ratio of the preform down-feed speed and fiber draw speed dictates the final layer thicknesses. **b** Transmission band gap spectra for hollow-core PBG fibers. The fiber structure in each case is the same. The only difference is that the length of the period in the multilayer structure lining the hollow core is reduced (from *right to left*) resulting in the shift of the transmission PBG spectrum

1.4.1.1 Hollow-Core PBG Fibers

A one-dimensional (1D) periodic stack of alternating high- and low- refractive index materials with a sufficiently large refractive-index contrast can be used to construct an omnidirectional mirror [80, 81], one that reflects light incident at any angle and polarization within a given spectral window. If this type of a mirror lines the inner surface of a hollow-core fiber, light can be guided axially through the air core [82]. Ideas surrounding such a light-guiding mechanism were considered in the 1970s [83], but it was only in 2002 that the standard preform-to-fiber drawing technique was used to fabricate a multimaterial hollow-core fiber, which realized the first waveguide of this type [75]. The challenge in fabricating these types of multimaterial fibers lies in discovering (or creating) two different thermally compatible materials as well as maintaining the geometric structure during the fiber drawing process. Specifically, the alternating materials must have compatible thermomechanical properties (such that their viscosities overlap at the drawing temperature), and simultaneously they must have a high refractive-index contrast. Furthermore, in order to create a low-loss fiber, the multilayer structure needs to be maintained with low-scattering interfaces down to the micro- and nanoscale. To date, several pairs of materials have been identified which yield low-loss multimaterial optical transmission fibers under appropriate fabrication conditions. Typically, the pair consists of a low-refractive-index polymer and a high-refractive-index chalcogenide glass, which are both amorphous and thermomechanically compatible for fiber drawing [75–77]. The fabrication process for creating hollow-core multimaterial fibers with PBG's is outlined in Fig. 1.3a. The same technique can be used to produce fibers that guide light from the infrared down to ultraviolet. Precise control over the layer thicknesses at the nanometer scale is confirmed by the measured transmission spectra spanning the visible range; see Fig. 1.3b. During the thermal draw, simply increasing the draw speed leads to cross-sectional feature reduction and consequently blue-shifting of the PBG. Such hollow-core, multimaterial PBG fibers have been used in a variety of applications which we describe below

(1) A multiplicity of eigenmodes is supported by the fibers [84] [85]. The ability to precisely excite, propagate, and reconstruct specific fiber modes and their superpositions [78, 86, 87] provides a means to sculpt and tune the vectorial field distribution along the axis of the fiber. These capabilities are particularly interesting for the study of light-matter interaction over extended lengths. Moreover, as the core size decreases, the spiral nature of the multilayer structure— arising from the fabrication approach in Fig. 1.3a—imparts unique properties to the optical modes, such as asymmetric wave propagation [88].
(2) Numerous medical and industrial applications rely on the delivery of high-peak-power laser pulses in well-controlled spatial modes. However, the use of traditional solid-core fibers for handling high peak powers is limited due to material absorption, which sets the damage threshold limit. Multimaterial hollow-core PBG fibers have been shown to transmit peak powers of 11.4 MW at 1.55 μm with 97 % of the fiber output in the fundamental mode, a record for any fiber at this wavelength. Additionally, PBG fibers designed for the transmission of high-power continuous-wave CO_2 laser light at 10.6 μm are used as optical scalpels in minimally invasive medical procedures [75, 89].

(3) Chemical vapor sensing constitutes another application for hollow-core PBG fibers [79, 90]. For example, a chemiluminescent dye that emits light in reaction with peroxide vapor can be introduced into the fiber core. In this way, the fiber provides confinement for the analyte flow and simultaneously serves as an optical waveguide for transmitting the chemiluminescent signal to an optical detector mounted at the distal fiber end [79].
(4) Hollow-core PBG fibers can be used as refractive index sensors since the transmission bandgap spectral position and linewidth change with varying core refractive indices [91, 92].

1.4.1.2 Radial Lasing and Azimuthal Intensity Control

The hollow core of a multimaterial PBG fiber can host a gain medium, enabling the creation of radially emitting fiber lasers [22, 93]. This laser design stands in clear contrast to all other fiber lasers, which emit light axially from their end facet. Two configurations of radially emitting fiber lasers have been explored thus far. One such configuration incorporates an organic dye dissolved in a solid host while another makes use of a liquid host. In both cases, the gain medium plug is introduced in the fiber core and is pumped axially as shown in Fig. 1.4a. In the radial fiber laser architecture, the multilayer structure plays a dual role. The reflection at glancing incidence angles delivers the pump beam to the gain medium plug. The reflection at normal incidence to the fiber reflects the fluorescence back into the core. This radial feedback facilitates laser action in the transverse plane. As shown in Fig. 1.4a, the radiation pattern emitted by the solid plug is anisotropic (with directionality dictated by the pump beam polarization), while the emission from the liquid is rotationally symmetric. This contrast arises from the differences in the relaxation dynamics of excited molecules in the solid and liquid phases.

The rotational symmetry and specific polarization of the radially emitting liquid laser forms the basis for a fiber device with controllable directional emission [22]. Multiple electrically controllable and individually addressable liquid-crystal-based light modulators [22] can be integrated in an annular fashion around the laser cavity, as shown in Fig. 1.4b. Each of the liquid-crystal-filled channels serves to modulate the polarized wavefront emanating from the fiber core and passing through it as it exits the fiber. By controlling multiple microchannels simultaneously, a laser with a dynamically controlled intensity distribution spanning the full azimuthal angular range can be achieved. Figure 1.4c depicts the laser intensity control through one of the channels as a function of the applied voltage. This new capability, implemented monolithically within a single fiber, presents opportunities ranging from flexible multidirectional displays to minimally invasive directed light delivery systems for medical applications.

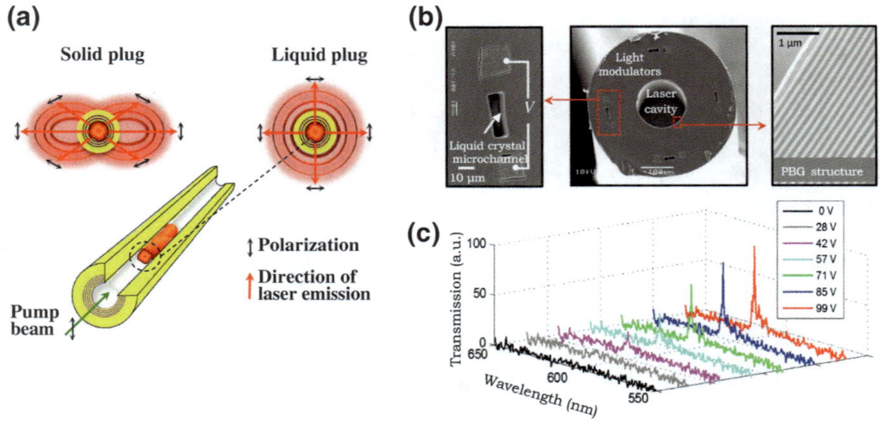

Fig. 1.4 Radial fiber lasers (**a**, *center*) Schematic of a radially emitting fiber laser. The *red cylinder* in the core corresponds to an organic-dye doped plug, which can be a liquid or a solid. (**a**, *left*) Emission patterns from the solid plug and (**a**, *right*) a liquid plug. The *red arrows* denote the direction of laser emission. The direction of the lobes for the solid plug is orthogonal to the pump beam polarization. **b** Scanning electron micrograph of the hybrid fiber containing a cylindrical PBG structure encircled by four hollow microchannels, each flanked by a pair of conductive electrodes. **c** Laser spectra measured for light transmitted through one of the liquid-crystal-filled channels for various driving voltages

1.4.1.3 External Reflection Fibers

Instead of lining the hollow core, the multilayer structure can be introduced at the external surface of the fiber, thereby creating omnidirectional externally reflective fibers [94]. While a purely periodic structure leads to PBG's that reflect light over a broad wavelength range, the intentional deviation from periodicity by introducing a defect layer facilities the existence of localized modes, which leads to narrow band transmission dips within the PBG [8]. These narrowband cavity resonances can be tuned dynamically either by applying an axial strain to the fiber [95] or by optically modulating the defect layer via a transient photodarkening effect [96]. Incorporated into fabrics, such fibers could be used for spectral identification, as radiation barriers, or as large-area tunable optical filters.

1.5 Optoelectronic Fibers

1.5.1 Metal- Insulator-Semiconductor Fibers

The creation of electronic and optoelectonic devices requires the assembly of metals (M), insulators (I), and semiconductors (S), in prescribed geometries with low-scattering interfaces and nanoscale feature sizes. While traditional wafer-based approaches have been applied to this problem, they are limited to primarily rigid and low-aspect ratios structures. However, if all the constituent device components could be integrated into a fiber, the complex functionality associated with modern-day chip-based devices could extend into the fabrics that we wear.

While the insulating and semiconducting domains can both be amorphous and thus can be drawn in a highly viscous state, a particular challenge involves the incorporation of a crystalline metallic domain into the fiber. Since the metal undergoes a phase transition into the liquid state during the draw, it is susceptible to capillary breakup and can easily disrupt the structural integrity of the fiber. These challenges can be addressed by confining the metallic electrodes within highly viscous boundaries, which can help to suppress capillary instabilities and direct the flow of the metal during the draw. Several M–I–S fiber structures have been created thus far using this approach. Figure 1.5a depicts a M–I–S preform being elongated into a fiber and the corresponding scanning electron microscope (SEM) image of the cross section is shown in Fig. 1.5b [15]. The core of the fiber is an amorphous photoconductive semiconductor (a semiconducting chalcogenide glass). The electrodes which contact the semiconductor can be connected to an external circuit post draw, and this fiber can then be used as a photodetector, which produces a photocurrent signal in proportion to the optical power incident onto its surface. Two advantages are gained by reducing the bulk semiconductor cylinder to a thin-film structure such as shown in Fig. 1.5c [13]. First, the dark current, which sets the noise level for optoelectronic applications, is reduced by more than an order of magnitude. Second, the creation of thin-film structures enables higher device-density per fiber (Fig. 1.5d) as well as the creation of more elaborate geometries (Fig. 1.5e) [18]. All of these structures are axially symmetric, with the electrodes forming intimate contacts with the semiconducting chalcogenide glass layers along the entire fiber length.

Both the composition of the semiconducting chalcogenide glass and the geometry of other structural elements comprising the preform can be tuned to target particular applications.

(1) By using a photoconductive glass, photodetecting fibers are created with the ability of detecting illumination along their entire length [13, 15, 18, 97], which has applications spanning chemical vapor sensing [98] to lensless imaging [14, 18]. Each pair of adjacent electrodes can act as an independent photodetector, thereby imparting increasing functionality. For example, a semiconductor ring structure (Fig. 1.5c) can be used to determine the angle of incidence of an incoming beam in the plane perpendicular to the fiber axis. Moreover, the semiconductor's wavelength-dependent absorbance may be leveraged using a duel ring structure (Fig. 1.5d) to reconstruct the wavelength of an incident beam by measuring the ratio of photocurrents in the inner and outer rings.

(2) Photoconductive glasses typically respond over a broad wavelength range. In order to impart spectral selection to a photosensitive fiber, a multilayer reflector containing a defect cavity can be combined with a photodetecting

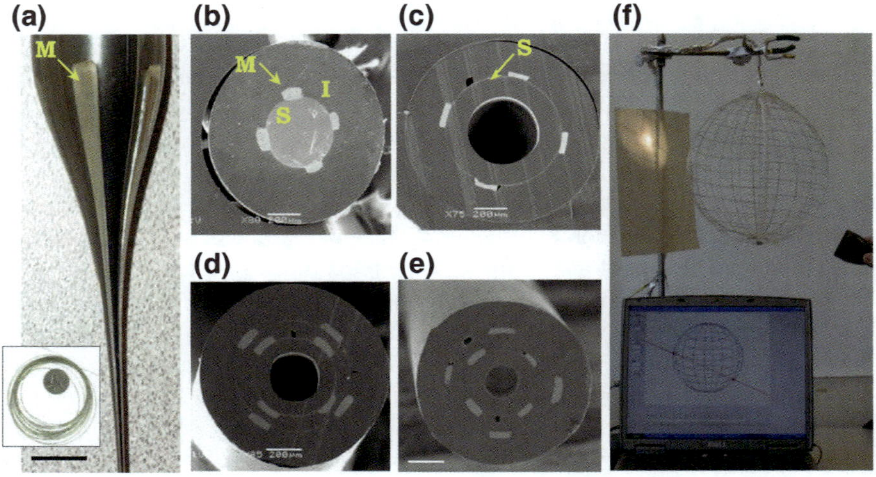

Fig. 1.5 Metal (*M*)—Insulator (*I*)—Semiconductor (*S*) Fibers **a** Picture of an M-I-S preform scaling down into a fiber. (Scale bar is 1 cm). Inset shows several meters of coiled fiber. **b–e** SEM micrographs of various M-I–S fiber architectures. (Scale bar for **b**, **c**, **d**, **e**: 200, 200, 200, 100 μm). **f** A spherical fiber web is an omnidirectional photodetector which can detect the direction on an optical beam incident from any angle. The reconstructed path of the flashlight beam passing through the sphere is seen on the computer monitor

structure [15]. The thickness of the defect cavity determines the narrowband transmission window within the PBG, allowing the engineering of spectroscopic photosensitive fibers tuned to respond at specific wavelengths.

(3) A semiconductor with a temperature-dependent resistivity can be drawn into a fiber with the identical geometry as shown in Fig. 1.5b, but with the ability to detect changes in temperature along its entire length [16]. Integrated into fabrics, such fibers can be used for temperature sensing over large areas. Moreover, a thin film ring of the same semiconductor can be integrated into a PBG transmission fiber to predict the onset of fiber failure [17].

(4) The electronic properties of the semiconductor can also be significantly improved through crystallization post-draw. This method has been used for realizing in-fiber field-effect transistors [99] and rewritable memory devices reminiscent of the ovonic switches [100].

Meters of fiber result from a single draw, thus facilitating the construction of large area assemblies and fabrics with optoelectronic, thermal, and acoustic functionalities. For example, a transparent sphere made of photodetecting fibers (Fig. 1.5f) can reconstruct the direction of an optical beam incident from any angle. Furthermore, the wavelength-reconstruction property of the dual-ring fiber structures (Fig. 1.5d) enable a two-dimensional fiber grid to perform lensless imaging [18], thus opening the possibility for fabrics that can *see*.

In addition to optoelectronic functionality, multimaterial fibers can be endowed with acoustic functionality through the piezoelectric effect. This can be achieved by drawing a copolymer, P(VDF- TrFE), that has a stable beta-phase at room temperature, and therefore will directly crystallize from the melt into the piezoelectric phase. Rectangular and cylindrical piezoelectric fibers have been drawn, both with the capability to emit and detect acoustic radiation over a broad frequency range spanning the kHz to MHz [19]. The ability to change the fiber cross-sectional geometry is particularly interesting for piezoelectric fibers, as this provides the means to tune the acoustic radiation pattern emanating from the fiber. Arrays of piezoelectric fibers can thus be constructed, leading to sophisticated functionalities such as coherent acoustic wave interference and beam steering [20], and large-area acoustic fiber webs can pave the way towards novel applications ranging from large-area pressure-wave monitoring to communications.

1.5.2 Crystalline-Semiconductor-Core Fibers

The mainstay of the microelectronic industry is crystalline semiconductors such as silicon (Si) and germanium (Ge), which have superior electronic properties compared to amorphous semiconductors. Therefore, to expand the performance of in-fiber electronic devices, there is a growing interest in developing methods to thermally draw these more traditional semiconductor materials. The optical properties of Si and Ge are also attractive: they are transparent in the midinfrared, have high optical damage threshold [101, 102], and are also highly nonlinear [103, 104]. For example, the Raman gain coefficient in Si is $\sim 10^4$ times greater than that of silica. Furthermore, the high thermal conductivity of Si and Ge offers opportunities in better removal of dissipated heat. The extension from planar substrates to optical fibers [34, 62] is therefore a significant complement to the emerging field of silicon photonics [105–108].

There has been significant recent progress in developing fibers containing traditional crystalline semiconductors. The first thermally drawn Si-core optical fiber was demonstrated using the molten-rod-in-tube method by J. Ballato's group in 2008 [62] (Fig. 1.6a). The melting temperature of Si ($\sim$1416 °C) overlaps with the softening temperature of silica (Fig. 1.1a); therefore, this materials combination in a multimaterial fiber is feasible from a thermal compatibility standpoint. A Si rod (3 mm diameter, 40 mm length) is inserted into a thick-walled silica tube, and the assembled preform is drawn into tens of meters of fiber with core diameters of approximately 60–120 μm. This approach has been extended to several other semiconductors such as Ge [109, 110] (Fig. 1.6b) and InSb [111], as well as to more complex materials systems, such as sapphire-derived high-alumina-content optical fibres [112].

Although this fiber is extremely simple from an elemental-analysis perspective (Si and O are the only two elements involved), this glass/crystalline-semiconductor materials system has nevertheless proven very fruitful in fundamental studies of

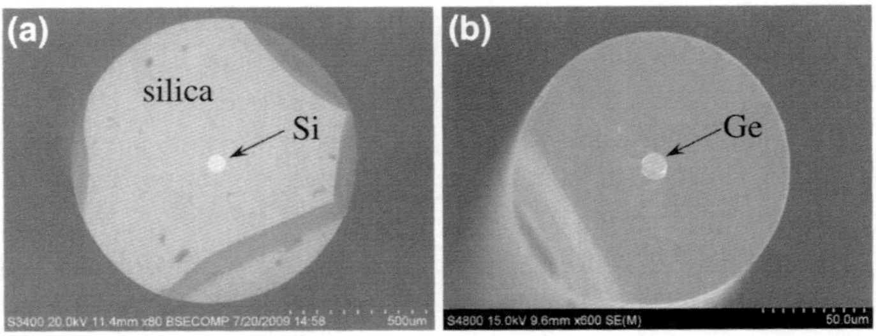

Fig. 1.6 Crystalline semiconductor core fibers. **a** Scanning electron micrograph of the core region of the silicon core, silica-clad optical fiber in [62]. **b** Electron micrographs of a crystalline Ge core optical fiber from [109]

diffusion and crystallization in confined volumes. Extensive work by J. Ballato's group, using X-ray diffraction, Raman spectroscopy, and spatially resolved crystallographic orientation studies [110, 113], have elucidated the polycrystalline domain structure of the core along the fiber. Several approaches have been exploited to increase the axial length of the single-crystal domains, including fiber-annealing by rapid photothermal processing (RPP) and controlling the core geometry [114], which have led to larger single-crystal domains. It is expected that such approaches will yield fibers with improved structural, electrical, optical, and mechanical properties.

Another research effort to produce Si-core fibers using the powder-in-tube method is reported in [63]. Silicon powder is packed into a silica tube which is then evacuated in order to limit the oxidation of Si. Outer diameters in the range of several hundred microns and core diameters in the range of tens of microns of n-type Si optical fibers have been demonstrated by this method [63]. High optical losses have been observed, which are attributed to irregularities along the fiber, such as thermal-expansion-induced micro-cracks at the core/cladding boundary and the polycrystallinity of the Si core.

This area of research is expected to blossom in the next few years with more research groups entering this field. While the focus to date has been on the materials science aspect of this class of multimaterial fibers, it is anticipated that the maturation of this field will lead to the development of fibers endowed with electronic and optoelectronic functionalities having potential applications in functional clothing and solar energy harvesting.

1.6 In-Fiber Synthesis

A recent surprising development in multimaterial fibers has resulted from the observation that the high-temperatures associated with thermal fiber drawing combined with the potential of using core and cladding materials with very different chemical reactivity offer the possibility of exploiting fiber drawing in *chemical synthesis*. On this view, the preform cladding acts as a crucible in which a chemical synthesis process is confined in the core. First, physical changes take place during fiber drawing, namely a reduction in the core and cladding viscosity, and potentially melting or volatization of the core, in addition to the geometric constriction of the core material(s). Crucially, in the in-fiber synthesis approach, such *physical* changes are then accompanied by *chemical* changes. For example, new compounds may be produced in the core, either via reactions between pre-existing core compounds, or through diffusion of elements or compounds from the cladding into the core. This methodology constitutes one example of a re-imagining of the fiber drawing process itself, where the elongation at high temperature of multimaterial fibers is used to achieve a new goal: controlled chemical reactions along an extended length in a precisely controlled, confined geometry.

The concept of multimaterial in-fiber synthesis may be traced back to the work by E. Snitzer and R. Tumminelli in 1989 [69] (see Sect. 1.3). The fiber drawing temperature in that realization was higher than T_m of the core material (a soft-glass) and its higher vapor pressure in the liquid phase led to the volatization of some compounds in the hot neck-down region resulting in a different residual core composition. This early work may also be viewed as a first instance of co-drawing two families of materials having incompatible thermal properties.

Another example drawn from more recent work demonstrated the possibility of synthesizing a material—whose melting temperature far exceeds the draw temperature—starting from low-temperature precursors. Specifically, low-temperature materials $Se_{97}S_3$ and $Sn_{85}Zn_{15}$ were co-drawn and a ZnSe compound was produced in the drawn fiber [27]. The synthesized compound—ZnSe here—has a much higher melting temperature than the drawing temperature. Additionally, J. Ballato et al. [28] have used crystalline Bi_2O_3-rich ($Bi_2O_3 + GeO_2$) and $Bi_{12}GeO_{20}$ powders (core) placed in borosilicate glass tube (cladding) to produce fibers where an in situ chemical reaction occurs during the thermal drawing in both amorphous and crystalline cores. Furthermore, in an attempt to reduce the diffusion of oxygen from the silica cladding to the Si core, silicon carbide (SiC) was introduced into the core to provide an in situ reactive oxygen getter during the drawing process [115]. This novel processing route can be used potentially to produce fiber cores from materials that are difficult to fabricate or machine into a rod using conventional methods among many other possible applications.

1.7 Other Approaches

In this chapter, we have described various multimaterial fibers that have been produced using the general methodology of preform-to-fiber fabrication. Recently, alternative strategies have emerged that also produce multimaterial fibers but do not rely on first constructing a macroscopic preform. We briefly describe these interesting exceptions in this section for completeness. First, an approach developed by

Konarka, Inc., starts from a long steel wire that is used in successive steps of dip-coating in organic solutions [33] to produce an organic photovoltaic fiber. The resulting fiber formed of a multilayer coating along its whole length constitutes an organic photovoltaic cell that demonstrated ~3% efficiency. A second approach uses a silica fiber with hollow enclaves (either a hollow-core fiber or a PCF) as a scaffold for vapor deposition of traditional crystalline semiconductors, a process typically called high-pressure microfluidic chemical deposition (HPMCD) [34] that extends the material phases that may be incorporated into the fiber to *single-crystal* semiconductors [116] and polycrystalline elemental or compound semiconductors [111, 117]. This process requires high-pressure flow (2–1000 MPa) in silica microstructured optical fibers (MOF's) to overcome mass-transport constraints, resulting in uniform annular deposition onto hollow pore walls. Two well-developed fabrication methodologies are combined in this approach: chemical vapor deposition (CVD) [118] and silica fiber drawing. The basic idea is shown in Fig. 1.7a, which involves heating small quantities of a high-pressure precursor within the interior of a MOF that decomposes upon heating to deposit on the walls in an amorphous state to ensure that it bonds smoothly, followed by annealing for crystallization. Figure 1.7b is an example of deposited doped semiconductor layers and metals [119]. As the annular deposited film grows thicker, the central holes from which depositions occurs becomes smaller until it is completely plugged and flow is extinguished. The empty pores in the MOFs are thus treated as micro- or nano-scale reaction chambers. A fiber-based device fabricated using this approach enabled all-optical modulation of 1.55 μm guided light via free-carrier absorption mediated by a 532 nm pump pulse [120]. Other examples include producing a ZnSe-core fiber [117], in-fiber Si and Ge wires and tubes used as field effect transistors [121], and in-fiber crystalline Si p–n homojunctions and Pt/n-Si Schottky heterojunctions [119]. The HPCVD technique can accommodate different capillary core dimensions and may also be used to fill a large number of micro- and nano-scale pores in MOF's. The main drawback of this technology is the limited lengths of fiber devices produced compared to those resulting from fiber drawing.

1.8 Conclusions and Outlook

We have presented an outline of the wide range of exciting ideas currently under active investigation in the emerging field of multimaterial fibers. Over the past decade, this field has witnessed rapid growth and it is impossible to cover all the ideas being explored in this chapter. For example, we did not describe interesting new work on producing THz multimaterial-fiber metamaterials [122–126]. Metamaterials are synthetic photonic structures that may potentially enable exotic optical phenomena such as 'cloaking' [127]. Some of the necessary features of a metamaterial, such as control over the electrical and magnetic resonances experienced by THz radiation transmitted through the transverse fiber cross sections, have been observed using polymer fibers containing arrays of metallic structures.

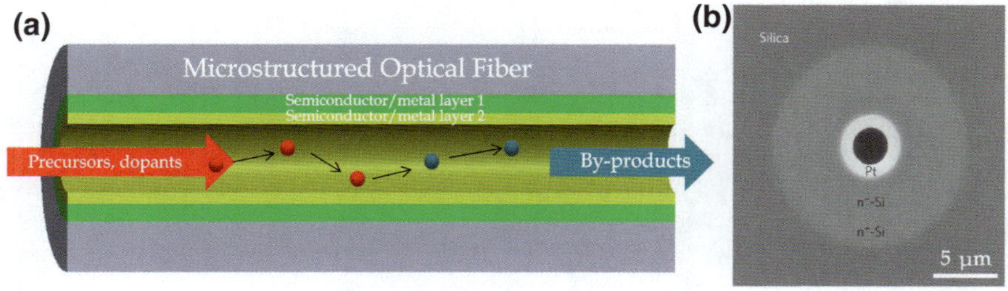

Fig. 1.7 Integration of semiconductor junctions in MOFs. **a** Illustration of HPCVD in a MOF pore. **b** A Pt/n^--Si Schottky junction formed by sequential deposition of phosphorous doped n^+-Si, n^--Si and platinum layers [119]

Extending this approach to optical wavelengths will require further reduction of the feature sizes in the multimaterial fibers. Other examples include the use of gold in silica fibers for plasmonic studies [128–133], the use of polymers to produce robust infrared chalcogenide fibers and nano-tapers [67, 134, 135], and the use of soft glasses inside silica fibers for mid-infrared nonlinear applications [136, 137].

An important avenue for further research is increasing the density of optical, electronic and optoelectronic devices incorporated in a single multimaterial fiber. Another opportunity is the production of fibers endowed with such functionalities in a form that may be directly woven into fabrics, which would constitute a milestone for this field, leading to a marriage of optics, electronics, and textiles. Another recent avenue has focused on using in-fiber capillary instabilities to produce three-dimensional macroscopic synthetic photonic structures with nanoscale control over the material distribution [30, 31], which may enable unprecedented control over the behavior of light and sound. A recent example of this can be found in [138] where the ability to produce in-fiber, electrically rectifying p-n composite micro-particles from the controlled fluidic breakup of two adjacent p- and n-doped silicon cores embedded in a silica cladding was demonstrated. Many surprises, no doubt, lie ahead, and it is thus safe to predict that the progress in multimaterial fibers reviewed here is pointing to a renaissance in fiber fabrication that promises to continue for a long time to come.

Acknowledgments The authors would like to thank John D. Joannopoulos for his encouragement, support, and vision. We also thank John Badding, John Ballato, Daosheng S. Deng, Steven G. Johnson, Guifang Li, and Siddharth Ramachandran for useful and illuminating discussions. We also thank Yan Zhang, Yifan Liu, Tiansi Wang, Fang Chen, Cen Xia, and Zhiyong Yang for assistance in preparing this review. This work was supported by the US National Science Foundation under award number ECCS-1002295, the NSF Materials Research Science and Engineering Program under award number DMR-0819762, and in part by the US Air Force Office of Scientific Research (AFOSR) under contract FA-9550-12-1-0148.

References

1. G.P. Agrawal, *Fiber-Optic Communication Systems*, 4th edn. (Wiley, Hoboken, 2010)
2. R. Ramaswami, K.N. Sivarajan, G.H. Sasaki, *Optical Networks: A Practical Perspective*, 3rd edn. (Morgan Kaufmann Publishers Inc., San Francisco, 2009)
3. J.I. Peterson, G.G. Vurek, Fiber-optic sensors for biomedical applications. Science **13**(224), 123–127 (1984)
4. A.G. Mignani, F. Baldini, Biomedical sensors using optical fibres. Rep. Prog. Phys. **59**, 1–28 (1996)
5. Z. Zhou, J. He, M. Huang, J. He, and G. Chen, Casing Pipe Damage Detection with Optical Fiber Sensors: A Case Study in Oil Well Constructions, Adv. Civil Eng. 638967, 9 (2010)
6. N. Mohamed, I. Jawhar, J. Al-Jaroodi, L. Zhang, Sensor network architectures for monitoring underwater pipelines. Sensors **11**(11), 10738–10764 (2011)
7. T. Webber, in *Advances in Fiber Lasers for the Materials Processing Market, Quantum Electronics and Laser Science Conference (QELS) on Advances in High-Power Lasers and their Applications III: Processing (JTh4I)*, San Jose, California, USA, 6 May 2012
8. J.D. Joannopoulos, S.G. Johnson, J.N. Winn, R.D. Meade, *Photonic Crystals. Molding the Flow of Light*, 2nd edn. (Princeton University Press, Princeton, 2008)
9. P. Russell, Photonic crystal fibers. Science **299**, 358–362 (2003)
10. P. Russell, Photonic crystal fibers: a historical account. IEEE LEOS Newsletters **21**(5), 11–15 (2007)
11. G. Tao, A.M. Stolyarov, A.F. Abouraddy, Multimaterial fibers. I. J. Appl. Glass Sci. **3**(4), 349–368 (2012)
12. A.F. Abouraddy, M. Bayindir, G. Benoit, S.D. Hart, K. Kuriki, N. Orf, O. Shapira, F. Sorin, B. Temelkuran, Y. Fink, Towards multimaterial multifunctional fibres that see, hear, sense and communicate. Nat. Mater. **6**, 336–347 (2007)
13. F. Sorin, A.F. Abouraddy, N. Orf, O. Shapira, J. Viens, J. Arnold, J.D. Joannopoulos, Y. Fink, Multimaterial photo detecting fibers: a geometric and structural study. Adv. Mat. **19**, 3872–3877 (2007)
14. A.F. Abouraddy, O. Shapira, M. Bayindir, J. Arnold, F. Sorin, D. Saygin-Hinczewski, J.D. Joannopoulos, Y. Fink, Large-scale optical-field measurements with geometric fibre constructs. Nat. Mater. **5**, 532–536 (2006)
15. M. Bayindir, F. Sorin, S. Hart, O. Shapira, J.D. Joannopoulos, Y. Fink, Metal-insulator-semiconductor optoelectronic fibres. Nature **431**, 826–829 (2004)
16. M. Bayindir, A.F. Abouraddy, J. Arnold, J.D. Joannopoulos, Y. Fink, Thermal-sensing fiber devices by multimaterial codrawing. Adv. Mat. **18**, 845–849 (2006)
17. M. Bayindir, O. Shapira, D. Saygin-Hinczewski, J. Viens, A.F. Abouraddy, J.D. Joannopoulos, Y. Fink, Integrated fibers for self monitored optical transport. Nat. Mater. **4**, 820–824 (2005)
18. F. Sorin, O. Shapira, A.F. Abouraddy, M. Spencer, N.D. Orf, J.D. Joannopoulos, Y. Fink, Exploiting collective effects of multiple optoelectronic devices integrated in a single fiber. Nano Lett. **9**, 2630–2635 (2009)
19. S. Egusa, Z. Wang, N. Chocat, Z.M. Ruff, A.M. Stolyarov, D. Shemuly, F. Sorin, P.T. Rakich, J.D. Joannopoulos, Y. Fink, Multimaterial piezoelectric fibres. Nat. Mater. **9**, 643–648 (2010)
20. N. Chocat, G. Lestoquoy, Z. Wang, D. M. Rodgers, J. D. Joannopoulos, Y. Fink, Piezoelectric fibers for conformal acoustics, Adv. Mat. doi: 10.1002/adma.201201355. (2012)
21. A.M. Stolyarov, L. Wei, F. Sorin, G. Lestoquoy, J.D. Joannopoulos, Y. Fink, Fabrication and characterization of fibers with built-in liquid crystal channels and electrodes for transverse incident-light modulation. Appl. Phys. Lett. **101**, 011108 (2012)

22. A.M. Stolyarov, L. Wei, O. Shapira, F. Sorin, S.L. Chua, J.D. Joannopoulos, Y. Fink, Microfluidic directional emission control of an azimuthally polarized radial fibre laser. Nat. Photonics **4**, 229–233 (2012)
23. D.S. Deng, N.D. Orf, A.F. Abouraddy, A.M. Stolyarov, J.D. Joannopoulos, H.A. Stone, Y. Fink, In-fiber semiconductor filament arrays. Nano Lett. **8**, 4265–4269 (2008)
24. D.S. Deng, N.D. Orf, S. Danto, A.F. Abouraddy, J.D. Joannopoulos, Y. Fink, Processing and properties of centimeter-long, in-fiber, crystalline-selenium filaments. Appl. Phys. Lett. **96**, 023102 (2010)
25. M. Yaman, T. Khudiyev, E. Ozgur, M. Kanik, O. Aktas, E.O. Ozgur, H. Deniz, E. Korkut, M. Bayindir, Arrays of indefinitely long uniform nanowires and nanotubes. Nat. Mater. **10**, 494–501 (2011)
26. J.J. Kaufman, G. Tao, S. Shabahang, D.S. Deng, Y. Fink, A.F. Abouraddy, Thermal drawing of high-density macroscopic arrays of well-ordered sub-5-nm-diameter nanowires. Nano Lett. **11**, 4768–4773 (2011)
27. N.D. Orf, O. Shapira, F. Sorin, S. Danto, M.A. Baldo, J.D. Joannopoulos, Y. Fink, Fiber draw synthesis. P. Natl. Acad. Sci. USA **108**(12), 4743–4747 (2011)
28. J. Ballato, C. McMillen, T. Hawkins, P. Foy, R. Stolen, R. Rice, L. Zhu, O. Stafsudd, Reactive molten core fabrication of glass-clad amorphous and crystalline oxide optical fibers. Opt. Mater. Express **2**(2), 153–160 (2012)
29. C. Hou, X. Jia, L. Wei, A.M. Stolyarov, O. Shapira, J.D. Joannopoulos, Y. Fink, Direct atomic-level observation and chemical analysis of ZnSe synthesized by in situ high-throughput reactive fiber drawing. Nano Lett. **13**(3), 975–979 (2013)
30. S. Shabahang, J.J. Kaufman, D.S. Deng, A.F. Abouraddy, Observation of the plateau-rayleigh capillary instability in multi-material optical fibers. Appl. Phys. Lett. **99**, 161909 (2011)
31. J.J. Kaufman, G. Tao, S. Shabahang, E.-H. Banaei, D.S. Deng, X. Liang, S.G. Johnson, Y. Fink, A.F. Abouraddy, Structured spheres generated by an in-fibre fluid instability. Nature **487**, 463–467 (2012)
32. M. Bayindir, A.F. Abouraddy, O. Shapira, J. Viens, D. Saygin-Hinczewski, F. Sorin, J. Arnold, J. Joannopoulos, Y. Fink, Kilometer-long ordered nanophotonic devices by preform-to-fiber fabrication. IEEE J. Sel. Top. Quant. **12**(6), 1202–1213 (2006)
33. M.R. Lee, R.D. Eckert, K. Forberich, G. Dennler, C.J. Brabec, R.A. Gaudiana, Solar power wires based on organic photovoltaic materials. Science **324**, 232–235 (2009)
34. P.J.A. Sazio, A. Amezcua-Correa, C.E. Finlayson, J.R. Hayes, T.J. Scheidemantel, N.F. Baril, B.R. Jackson, D.-J. Won, F. Zhang, E.R. Margine, V. Gopalan, V.H. Crespi, J.V. Badding, Microstructured optical fibers as high-pressure microfluidic reactors. Science **311**, 1583–1586 (2006)
35. T. Li (ed.), *Optical fiber communications: fiber fabrication* (Academic Press, Waltham, 1985)
36. H. Tokiwa, Y. Mimura, T. Nakai, O. Shinbori, Fabrication of long single-mode and multimode fluoride glass fibres by the double-crucible technique. Electronics Lett. **21**(24), 1131–1132 (1985)
37. G. Urbain, Y. Bottinga, P. Richet, Viscosity of liquid silica, silicates and alumino-silicates. Geochim. Cosmochim. Acta. **46**, 1061–1072 (1982)
38. K. Kakimoto, M. Eguchi, H. Watanabe, T. Hibiya, Natural and forced convection of molten silicon during czochralski single crystal growth. J. Cryst. Growth **94**(2), 412–420 (1989)
39. D. Ofte, The viscosities of liquid uranium, gold and lead. J. Nucl. Mater. **22**(1), 28–32 (1967)
40. A. Napolitano, E.G. Hawkins, Viscosity of a standard soda-lime-silica glass. J. Res. NBS A Phys. Ch. **68A**(5), 439–448 (1964)
41. M. Braglia, C. Bruschi, D. Cavalli, G. Cocito, D. Guojun, J. Kraus, S. Mosso, Rheology of fluoride glasses. J. Crystal. Solid **213–214**(12), 325–329 (1997)
42. A. Belwalkar, W.Z. Misiolek, J. Toulouse, Viscosity study of the optical tellurite glass: $75TeO_2$-$20ZnO$-$5Na_2O$. J. Non-Crystal. Solid **356**(1), 1354–1358 (2010)

43. A.S. Tverjanovich, Temperature dependence of the viscosity of chalcogenide glass-forming melts. Glass Phys. Chem. **29**(6), 532–536 (2003)
44. B. Chu, K. Linliu, Viscosity characterization of poly(tetrafluoroethylene) by centrifuge ball viscosimetry. Macromolecules **28**(8), 2723–2727 (1995)
45. B. Collins, J. Shields, K. Butler, M. Seck, T.J. Ohlemiller, Exploring the Role of Polymer Melt Viscosity in Melt Flow and Flammability Behavior, National Institute of Standards and Technology (BFRL), Gaithersburg, MD., ProductType: Technical report, NTIS Order Number: PB2007-105069, p 29 (2000)
46. P. Koštál, J. Málek, Viscosity of selenium melt. J. Non-Crystal. Solid **356**(50–51), 2803–2806 (2010)
47. V.M. Glazov, S.N. Chizhevskaya, N.N. Glagoleva, *Liquid Semiconductors* (Plenum Press, New York, 1969)
48. Y. Sato, T. Nishizuka, T. Takamizawa, T. Yamamura, Y. Waseda, Viscosity of molten GaSb and InSb. Int. J. Thermophys. **23**, 235–243 (2002)
49. M.F. Culpin, The viscosity of liquid indium and liquid tin. Proc. Phys. Soc. B **70**(11), 1069–1078 (1957)
50. A. Dargys, J. Kundrotas, *Handbook on physical properties of Ge, Si, GaAs and InP* (The Science and Encyclopaedia Publishing Centre, Vilnius, 1994)
51. http://www.owlnet.rice.edu/~msci301/ThermalExpansion.pdf
52. ASM Handbook: Properties and Selection: Nonferrous Alloys and Special-Purpose Materials (ASM Handbook) vol. 2, 10th edn, ASM International, Metals Park, OH, 1990, 704–705 (Au), 1154–1156 (Si)
53. http://www.engineersedge.com/properties_of_metals.htm
54. F.C. Nix, D. MacNair, The thermal expansion of pure metals: copper, gold, aluminum, nickel, and iron. Phys. Rev. **60**, 597–605 (1941)
55. http://www.ioffe.ru/SVA/NSM/Semicond/InSb/thermal.html
56. D.F. Gibbons, Thermal expansion of some crystals with the diamond structure. Phys. Rev. **112**, 136–140 (1958)
57. http://www2.dupont.com/Teflon_Industrial/en_US/tech_info/techinfo_compare.html
58. M.J. Weber, *Optics Handbook of Optical Materials* (CRC PRESS, Boca Raton 2003), section 3.4 (Teflon-PTFE), section 2.4 (fused silica), section 2.2.4 (BK7) and section 2.5.1 (ZBLAN)
59. M. Bass, E.W. Van Stryland, D.R. Williams, W.L. Wolfe, *Handbook of Optics Volume II Devices, Measurements, and Properties*, 2nd edn. (McGraw-Hill INC, New York, 1995), p33.55 (fused Silica and ZBLAN) p7.9 (Polycarbonate)
60. www.amorphousmaterials.com
61. http://www.polymerprocessing.com/polymers/PC.html
62. J. Ballato, T. Hawkins, P. Foy, R. Stolen, B. Kokuoz, M. Ellison, C. McMillen, J. Reppert, A.M. Rao, M. Daw, S.R. Sharma, R. Shori, O. Stafsudd, R.R. Rice, D.R. Powers, *Silicon Optical Fiber*. Opt. Express **16**(23), 18675–18683 (2008)
63. B.L. Scott, K. Wang, G. Pickrell, Fabrication of n-type silicon optical fiber. IEEE Photon. Technol. Lett. **21**(24), 1798–1800 (2009)
64. H.K. Tyagi, H.W. Lee, P. Uebel, M.A. Schmidt, N. Joly, M. Scharrer, P.St.J. Russell, Plasmon Resonances on Gold Nanowires Directly Drawn in a Step-index Fiber, Opt. Lett. **35**(15) 2573–2575 (2010)
65. D.J. Gibson, J.A. Harrington, Extrusion of hollow waveguide preforms with a one-dimensional photonic bandgap structure. J. Appl. Phys. **95**(8), 3895–3900 (2004)
66. X. Feng, T.M. Monro, P. Petropoulos, V. Finazzi, D.J. Richardson, Extruded single-mode high-index-core one-dimensional microstructured optical fiber with high index-contrast for highly nonlinear optical devices. Appl. Phys. Lett. **87**(8), 081110 (2005)
67. G. Tao, S. Shabahang, E.-H. Banaei, J.J. Kaufman, A.F. Abouraddy, Multimaterial preform coextrusion for robust chalcogenide optical fibers and tapers. Opt. Lett. **37**(13), 2751–2753 (2012)

68. M. Liao, C. Chaudhari, G. Qin, X. Yan, C. Kito, T. Suzuki, Y. Ohishi, M. Matsumoto, T. Misumi, Fabrication and characterization of a chalcogenide-tellurite composite microstructure fiber with high nonlinearity. Opt. Express **17**(24), 21608–21614 (2009)
69. E. Snitzer, R. Tumminelli, SiO_2-clad fibers with selectively volatilized soft-glass cores. Opt. Lett. **14**(14), 757–759 (1989)
70. H.L. Blackmore, *A Dictionary of London Gunmakers* (Phaidon—Christie's Limited, Oxford, 1986), p. 59
71. E. Roeder, Extrusion of glass. J. Non-Cryst. Solids **5**(5), 377–388 (1971)
72. E. Roeder, Flow behaviour of glass during extrusion. J. Non-Cryst. Solids **7**(2), 203–220 (1972)
73. W. Egel-Hess, E. Roeder, Extrusion of glass melts-influence of wall friction effects on the die swell phenomenon. Glasstech. Ber. **62**(8), 279–284 (1989)
74. P. Kaiser, H.W. Astle, Low-loss single-material fibers made from pure fused silica. AT&T Tech. J. **53**(6), 1021–1039 (1974)
75. B. Temelkuran, S.D. Hart, G. Benoit, J.D. Joannopoulos, Y. Fink, Wavelength-scalable hollow optical fibres with large photonic bandgaps for co_2 laser transmission. Nature **420**, 650–653 (2002)
76. K. Kuriki, O. Shapira, S. Hart, G. Benoit, Y. Kuriki, J. Viens, M. Bayindir, J. Joannopoulos, Y. Fink, Hollow multilayer photonic bandgap fibers for nir applications. Opt. Express **12**(8), 1510–1517 (2004)
77. Z. Ruff, D. Shemuly, X. Peng, O. Shapira, Z. Wang, Y. Fink, Polymer-composite fibers for transmitting high peak power pulses at 1.55 microns. Opt. Express **18**(15), 15697–15703 (2010)
78. D. Shemuly, A.M. Stolyarov, Z.M. Ruff, L. Wei, Y. Fink, O. Shapira, Preparation and transmission of low-loss azimuthally polarized pure single mode in multimode photonic band gap fibers. Opt. Express **20**(6), 6029–6035 (2012)
79. A.M. Stolyarov, A. Gumennik, W. McDaniel, O. Shapira, B. Schell, F. Sorin, K. Kuriki, G. Benoit, A. Rose, J.D. Joannopoulos, Y. Fink, Enhanced chemiluminescent detection scheme for trace vapor sensing in pneumatically-tuned hollow core photonic bandgap fibers. Opt. Express **20**(11), 12407–12415 (2012)
80. J.N. Winn, Y. Fink, S. Fan, J.D. Joannopoulos, Omnidirectional reflection from a one-dimensional photonic crystal. Opt. Lett. **23**, 1573–1575 (1998)
81. Y. Fink, J.N. Winn, S. Fan, C. Chen, J. Michel, J.D. Joannopoulos, E.L. Thomas, A Dielectric Omnidirectional Reflector. Science **282**, 1679–1682 (1998)
82. Y. Fink, D.J. Ripin, S. Fan, C. Chen, J.D. Joannopoulos, E.L. Thomas, Guiding optical light in air using an all-dielectric structure. J. Lightwave Technol. **17**(11), 2039–2041 (1999)
83. P. Yeh, A. Yariv, E. Marom, Theory of bragg fiber. J. Opt. Soc. Am. **68**, 1196–1201 (1978)
84. S. Johnson, M. Ibanescu, M. Skorobogatiy, O. Weisberg, T. Engeness, M. Soljacic, S. Jacobs, J. Joannopoulos, Y. Fink, Low-loss asymptotically single-mode propagation in large-core omniguide fibers. Opt. Express **9**(13), 748–779 (2001)
85. M. Ibanescu, S.G. Johnson, M. Soljacic, J.D. Joannopoulos, Y. Fink, Analysis of mode structure in hollow dielectric waveguide fibers. Phys. Rev. E **67**, 046608 (2003)
86. O. Shapira, A.F. Abouraddy, J.D. Joannopoulos, Y. Fink, Complete modal decomposition for optical waveguides. Phys. Rev. Lett. **94**, 143902 (2005)
87. O. Shapira, A.F. Abouraddy, Q. Hu, D. Shemuly, J.D. Joannopoulos, Y. Fink, Enabling coherent superpositions of iso-frequency optical states in multimode fibers. Opt. Express **18**(12), 12622–12629 (2010)
88. D. Shemuly, Z.M. Ruff, A.M. Stolyarov, G. Spektor, S.G. Johnson, Y. Fink, O. Shapira, Asymmetric wave propagation in planar chiral fibers, **21**(2), 1465–1472 (2012)
89. http://www.omni-guide.com/
90. A. Yildirim, M. Vural, M. Yaman, M. Bayindir, Bioinspired optoelectronic nose with nanostructured wavelength-scalable hollow-core infrared fibers. Adv. Mat. **23**, 1263–1267 (2011)

91. H. Qu, M. Skorobogatiya, Liquid-core low-refractive-index-contrast bragg fiber sensor. Appl. Phys. Lett. **98**, 201114 (2011)
92. K.J. Rowland, S. Afshar, A. Stolyarov, Y. Fink, T.M. Monro, Bragg waveguides with low-index liquid cores. Opt. Express **20**(1), 48–62 (2012)
93. O. Shapira, K. Kuriki, N.D. Orf, A.F. Abouraddy, G. Benoit, J.F. Viens, A. Rodriguez, M. Ibanescu, J.D. Joannopoulos, Y. Fink, M.M. Brewster, Surface-emitting fiber lasers. Opt. Express **14**(9), 3929–3935 (2006)
94. S.D. Hart, G.R. Maskaly, B. Temelkuran, P.H. Prideaux, J.D. Joannopoulos, External reflection from omnidirectional dielectric mirror fibers. Science **296**, 510–513 (2002)
95. G. Benoit, S.D. Hart, B. Temelkuran, J.D. Joannopoulos, Y. Fink, Static and dynamic properties of optical microcavities in photonic bandgap yarns. Adv. Mater. **15**(24), 2053–2056 (2003)
96. G. Benoit, K. Kuriki, J.F. Viens, J.D. Joannopoulos, Y. Fink, Dynamic all-optical tuning of transverse resonant cavity modes in photonic bandgap fibers. Opt. Lett. **30**(13), 1620–1622 (2005)
97. F. Sorin, G. Lestoquoy, S. Danto, J.D. Joannopoulos, Y. Fink, Resolving optical illumination distributions along an axially symmetric photodetecting fiber. Opt. Express **18**(23), 24264–24275 (2010)
98. A. Gumennik, A.M. Stolyarov, B. Schell, C. Hou, G. Lestoquoy, F. Sorin, W. McDaniel, A. Rose, J.D. Joannopoulos, Y. Fink, All-in-fiber chemical sensing. Adv. Mater. adma201203053 (in press, 2012)
99. S. Danto, F. Sorin, N.D. Orf, Z. Weng, S.A. Speakman, J.D. Joannopoulos, Y. Fink, Fiber field-effect device via in situ channel crystallization. Adv. Mater. **22**, 4162–4166 (2010)
100. S. Danto, Z. Ruff, Z. Wang, J.D. Joannopoulos, Y. Fink, Ovonic memory switching in multimaterial fibers. Adv. Funct. Mat. **21**(6), 1095–1101 (2011)
101. R.M. Wood, S.K. Sharam, P. Waite, in *Variation of Laser Induced Damage Threshold with Laser Pulse Repetition Frequency, Laser Induced Damage in Optical Material, 1982*. Ed. by H.E. Bennett, A.H. Guenther, D. Milam, B.E. Newnam, Proceedings of a symposium sponsored by National Bureau of Standards, Boulder, Colorado, 44–49, 1984
102. B.M. Cowan, in *Optical Damage Threshold of Silicon for Ultrafast Infrared Pulses, Laser-Induced Damage in Optical Materials: 2007*. ed by G.J. Exarhos, A.H. Guenther, K.L. Lewis, D. Ristau, M.J Soileau, C.J Stolz, Proceedings of the SPIE, 6720, 67201 M, 2008
103. B. Jalali, V. Raghunathan, D. Dimitropoulos, O. Boyraz, Raman-based silicon photonics. IEEE J. Sel. Top. Quantum Electron. **12**(3), 412–421 (2006)
104. V. Raghunathan, D. Borlaug, R. Rice, B. Jalali, Demonstration of a mid-infrared silicon raman amplifier. Opt. Express **15**, 14355–14362 (2007)
105. M. Lipson, Guiding, modulating, and emitting light on silicon—challenges and opportunities. J. Lightwave Technol. **23**(12), 4222–4238 (2005)
106. G.T. Reed, A.P. Knights, *Silicon Photonics: An Introduction* (wiley, Chichester, 2004)
107. J. Leuthold, C. Koos, W. Freude, Nonlinear silicon photonics. Nat. Photonics **4**, 535–544 (2010)
108. J. Michel, J. Liu, L.C. Kimerling, High-performance ge-on-si photodetectors. Nat. Photonics **4**, 527–534 (2010)
109. J. Ballato, T. Hawkins, P. Foy, B. Yazgan-Kokuoz, R. Stolen, C. McMillen, N.K. Hon, B. Jalali, R. Rice, Glass-clad single-crystal germanium optical fiber. Opt. Express **17**(10), 8029–8035 (2009)
110. C. McMillen, T. Hawkins, P. Foy, D. Mulwee, J. Kolis, R. Stolen, R. Rice, J. Ballato, On crystallographic orientation in crystal core optical fibers. Opt. Mater. **32**(9), 862–867 (2010)
111. J. Ballato, T. Hawkins, P. Foy, C. McMillen, L. Burka, J. Reppert, R. Podila, A. Rao, R. Rice, Binary iii–v core semiconductor optical fiber. Opt. Express **18**(5), 4972–4979 (2009)
112. P. Dragic, T. Hawkins, P. Foy, S. Morris, J. Ballato, Sapphire-derived all-glass optical fibres. Nat. Photonics **6**, 627–633 (2012)

113. S. Morris, C. McMillen, T. Hawkins, P. Foy, R. Stolen, J. Ballato, R. Rice, The influence of core geometry on the crystallography of silicon optical fiber. J. Cryst. Growth **352**(1), 53–58 (2011)
114. N. Gupta, C. McMillen, R. Singh, R. Podila, A.M. Rao, T. Hawkins, P. Foy, S. Morris, R. Rice, K.F. Poole, L. Zhu, J. Ballato, Annealing of silicon optical fibers. J. Appl. Phys. **110**, 093107 (2011)
115. S. Morris, T. Hawkins, P. Foy, C. McMillen, J. Fan, L. Zhu, R. Stolen, R. Rice, J. Ballato, Reactive molten core fabrication of silicon optical fiber. Opt. Mater. Express **1**(6), 1141–1149 (2011)
116. B.R. Jackson, P.J.A. Sazio, J.V. Badding, Single-crystal semiconductor wires intetrated into microstructured optical fibers. Adv. Mat. **20**, 1135–1140 (2008)
117. J.R. Sparks, R. He, N. Healy, M. Krishnamurthi, A.C. Peacock, P.J.A. Sazio, V. Gopalan, J.V. Badding, Zinc selenide optical fibers. Adv. Mat. **23**, 1647–1651 (2011)
118. D.M. Dobkin, M.K. Zuraw (eds.), *Principles of Chemical Vapor Deposition* (Kluwer, Dordrecht, 2003)
119. R. He, P.J.A. Sazio, A.C. Peacock, N. Healy, J.R. Sparks, M. Krishnamurthi, V. Gopalan, J.V. Badding, Integration of gigahertz-bandwidth semiconductor devices inside microstructured optical fibres. Nat. Photonics **6**, 174–179 (2012)
120. D.-J. Won, M.O. Ramirez, H. Kang, V. Gopalan, N.F. Baril, J. Calkins, J.V. Badding, P.J.A. Sazio, All-optical modulation of laser light in amorphous silicon-filled microstructured optical fibers. Appl. Phys. Lett. **91**, 161112 (2007)
121. C.E. Finlayson, A. Amezcua-Correa, P.J.A. Sazio, N.F. Baril, J.V. Badding, Electrical and raman characterization of silicon and germanium-filled microstructed optical fibers. Appl. Phys. Lett. **90**, 132110 (2007)
122. A. Tuniz, B.T. Kuhlmey, R. Lwin, A. Wang, J. Anthony, R. Leonhardt, S.C. Fleming, Drawn metamaterials with plasmonic response at terahertz frequencies. Appl. Phys. Lett. **96**, 191101 (2010)
123. E.J. Smith, Z. Liu, Y. Mei, O.G. Schmidt, Combined surface plasmon and classical waveguiding through metamaterial fiber design. Nano Lett. **10**, 1–5 (2010)
124. A. Wang, A. Tuniz, P.G. Hunt, E.M. Pogson, R.A. Lewis, A. Bendavid, S.C. Fleming, B.T. Kuhlmey, M.C.J. Large, Fiber metamaterials with negative magnetic permeability in the terahertz. Opt. Mater. Express **1**(1), 115–120 (2011)
125. A. Tuniz, R. Lwin, A. Argyros, S.C. Fleming, E.M. Pogson, E. Constable, R.A. Lewis, B.T. Kuhlmey, Stacked-and-drawn metamaterials with magnetic resonances in the terahertz range. Opt. Express **19**(17), 16480–16490 (2011)
126. A. Tuniz, B. Pope, A. Wang, M.C.J. Large, S. Atakaramians, S.-S. Min, E.M. Pogson, R.A. Lewis, A. Bendavid, A. Argyros, S.C. Fleming, B.T. Kuhlmey, Spatial dispersion in three-dimensional drawn magnetic metamaterials. Opt. Express **20**(11), 11924–11935 (2012)
127. R.C. McPhedran, I.V. Shadrivov, B.T. Kuhlmey, Y.S. Kivsha, Metamaterials and metaoptics. NPG Asia Mater. **3**, 100–108 (2011)
128. H.K. Tyagi, H.W. Lee, P. Uebel, M.A. Schmidt, N. Joly, M. Scharrer, P.St.J. Russell, Plasmon resonances on gold nanowires directly drawn in a step-index fiber. Opt. Lett. **35**(15) 2573–2575 (2010)
129. J. Hou, D. Bird, A. George, S. Maier, B.T. Kuhlmey, J.C. Knight, Metallic mode confinement in microstructured fibres. Opt. Express **16**(9), 5983–5990 (2008)
130. C.G. Poulton, M.A. Schmidt, G.J. Pearce, G. Kakarantzas, P.St.J. Russell, Numerical study of guided modes in arrays of metallic nanowires, Opt. Lett. **32**(12), 1647–1649 (2007)
131. M.A. Schmidt, L.N.P. Sempere, H.K. Tyagi, C.G. Poulton, P.St.J. Russell, Waveguiding and plasmon resonances in two-dimensional photonic lattices of gold and silver nanowires. Phys. Rev. B **77**, 033417 (2008)
132. H.W. Lee, M.A. Schmidt, H.K. Tyagi, L.N. Prill Sempere, P.St.J. Russell, Polarization-dependent coupling to plasmon modes on submicron gold wire in photonic crystal fiber, Appl. Phys. Lett. **93**(11), 111102 (2008)

133. H.W. Lee, M.A. Schmidt, R.F. Russell, N.Y. Joly, H.K. Tyagi, P. Uebel, P.St.J. Russell, Pressure-assisted melt-filling and optical characterization of au nano-wires in microstructured fibers, Opt. Express **19**(13), 12180–12189 (2011)
134. S. Shabahang, M.P. Marquez, G. Tao, J.J. Kaufman, M.U. Piracha, D. Nguyen, P.J. Delfyett, A.F. Abouraddy, Octave-spanning infrared supercontinuum generation in robust chalcogenide fiber nano-tapers using picosecond pulses. Opt. Lett. **37**(22), 4639–4641 (2012)
135. S. Shabahang, G. Tao, J.J Kaufman, A.F. Abouraddy, Dispersion characterization of chalcogenide bulk glass, composite fibers, and robust nano-tapers, J. Opt. Soc. Am. B, in press (2012)
136. N. Granzow, S.P. Stark, M.A. Schmidt, A.S. Tverjanovich, L. Wondraczek, P.St.J. Russell, Supercontinuum generation in chalcogenide-silica step-index fibers. Opt. Express **19**(21), 21003–21010 (2011)
137. N. Granzow, P. Uebel, M.A. Schmidt, A.S. Tverjanovich, L. Wondraczek, P.St.J. Russell, Bandgap guidance in hybrid chalcogenide–silica photonic crystal fibers. Opt. Lett. **36**(13), 2432–2434 (2011)
138. A. Gumennik, L. Wei, G. Lestoquoy, A.M. Stolyarov, X. Jia, P.H. Rekemeyer, M.J. Smith, X. Liang, S.G. Johnson, S. Gradečak, A.F. Abouraddy, J.D. Joannopoulos, Y. Fink, Silicon-in-Silica spheres via axial thermal gradient in-fibre capillary instabilities. Nat. Commun. **4**, 2216 (2013). doi:10.1038/ncomms3216

Chapter 2
Optical Micro/Nanofiber as Valuable Technological Platform for Lab on Fiber

Xiaoqin Wu, Limin Tong and Eric Mazur

Abstract In this Chapter, we introduce optical micro/nanofibers (MNFs) as valuable technological platform for lab on fiber. Starting from the fabrication and micromanipulation of individual MNFs, we put emphases on optical properties and technological applications of MNFs, including both passive components and active devices. A brief outlook regarding future possibilities is also presented.

Optical microfibers and nanofibers (MNFs) are optical fibers with diameters close to or smaller than the wavelength of the guided light. A number of names, such as "subwavelength-diameter optical fiber/wire", "optical microfiber", "optical nanofiber", "fiber taper", "tapered fiber", "fiber nanowire", "photonic/optical nanowire", "submicron-diameter fiber" and "ultrathin fiber", have been used to mention these one-dimensional (1D) micro/nanoscale waveguides in various literatures. For simplicity, here we use the name "MNF" to cover these tiny optical fibers with diameters from tens of nanometers to several micrometers.

When its size goes down to the wavelength of the guided light, an optical MNF exhibits interesting properties such as tight optical confinement, strong near-field interaction, steep field gradient, abnormal dispersion, enhanced surface effects and excellent mechanical flexibilities [1–7], which opens opportunities for combining waveguide optics with near-field optics, nonlinear optics, plasmonics, quantum optics and optomechanics on the nanoscale [8–12], and intrigues a variety of novel nanophotonic components/devices including nanolasers and nanosensors that may serve as a valuable technological platform for Lab on fiber. In this section, we

X. Wu · L. Tong (✉)
State Key Laboratory of Modern Optical Instrumentation, Department of Optical Engineering, Zhejiang University, Hangzhou 310027, China
e-mail: phytong@zju.edu.cn

E. Mazur
School of Engineering and Applied Sciences, Harvard University, Cambridge, MA 02138, USA
e-mail: mazur@physics.harvard.edu

A. Cusano et al. (eds.), *Lab-on-Fiber Technology*,
Springer Series in Surface Sciences 56, DOI: 10.1007/978-3-319-06998-2_2,

introduce the principles and technological applications of optical nanofibers and nanowires. Future applications of these 1D building blocks for optical communication and sensing are also discussed.

2.1 Fabrication

So far, MNFs are mostly fabricated from amorphous materials including glasses and polymers. For these materials, the top-down physical drawing is the optimal technique to yield MNFs with circular cross-section, uniform diameter, smooth surface, large length and low cost. Usually, taper drawing is the most commonly used technique for drawing MNFs directly from a glass fiber [1, 3, 13–19] or bulk glass [20]. For facilitating fabrication and light launching, as well as the availability of high-purity glasses, standard glass fibers are routinely used as the starting material, with a flame [1, 3, 13–19]-/laser [3, 21–23] -heated or electrically heated [17, 24–26] taper drawing configuration.

Besides the glass, polymer is another excellent material for drawing optical MNFs with special advantages including perm-selective nature, biocompatibility and hospitability to a wide range of functional dopants. Typical techniques including electrospinning [27–30] and physical drawing [31–38] from polymer solutions have been developed for the fabrication of polymer MNFs.

To bestow as-fabricated MNFs with greater versatilities, a number of post-fabrication techniques including micromanipulation [5, 32], plastic bend [5], coating [39, 40], embedding [41–46] and fusion splicing [23, 47, 48] of MNFs have been investigated in the past years. Especially, coating or embedding MNFs with low-index polymers have been proved a feasible approach to protecting the MNF from environmental contamination and mechanical perturbation [41–44]. In addition, novel structures, such as MNFs with elliptical cross-sections for high birefringence [19], panda-core for polarization maintaining [49] or suspended cores for isolating surface contamination and contact leakage [4, 50–52] have also been reported. Figure 2.1 shows microscope images of typical MNFs fabricated with physical drawing technique, demonstrating the great diversity and flexibility of MNFs drawn from glasses or polymers.

2.2 Micromanipulation

To integrate as-drawn MNFs into nanophotonic circuits or devices, micromanipulation techniques must be developed to finely tailor and assemble these tiny building blocks into functional structures or geometries. Using precisely controlled tungsten or tapered fiber probes with tip sizes of tens to hundreds of nanometers, MNFs can be cut, picked up, transferred, bent and shaped under an optical microscope.

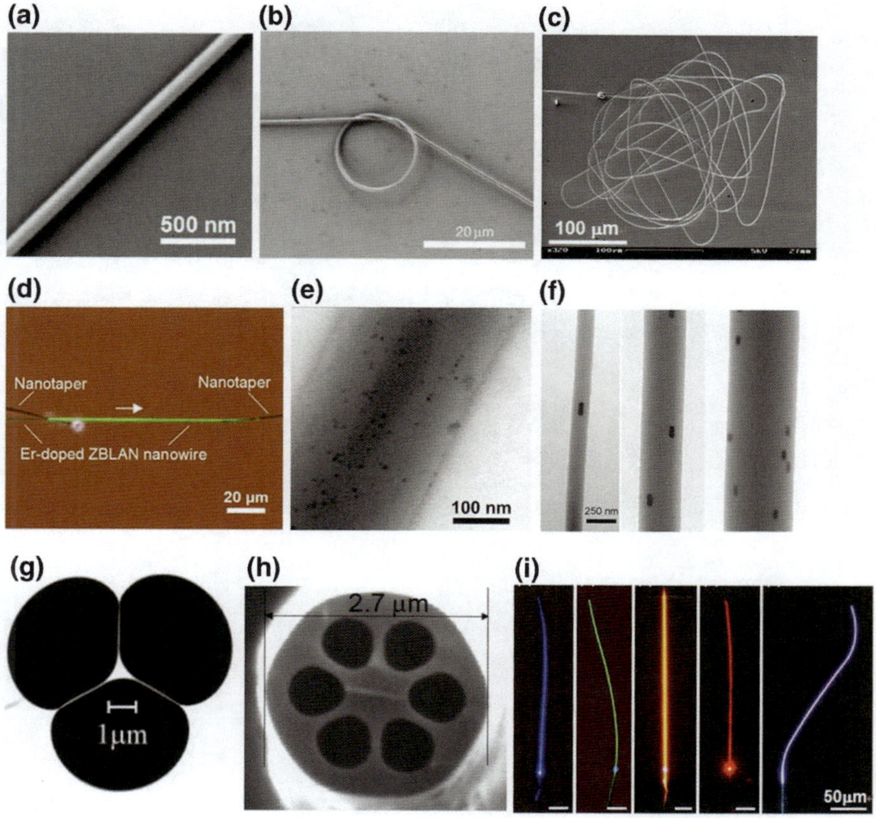

Fig. 2.1 Microscope images of typical glass or polymer MNFs [12]. **a** A 100-nm-diameter tellurite glass MNF [20]. **b** A 520-nm-diameter silica MNF tied into a knot [1]. **c** A long 260-nm-diameter silica MNF with total length of about 4 mm [1]. **d** A 320-nm-diameter Er-doped fluoride (ZBLAN) glass MNF emits up-conversion green light when launched by a 980-nm-wavelength light [20]. **e** A 280-nm-diameter polystyrene MNF doped with CdSe quantum dots [36]. **f** Three Polyacrylamide MNFs doped with aligned Au nanorods [37]. **g** A suspended-core MNF with core size of about 1 μm [51]. **h** A microstructured MNF with diameter of about 2.7 μm [50]. **i** Typical light emitting polymer MNFs excited by a 355-nm-wavelength light. The MNFs are doped with different fluorescent dye molecules to emit different colors of light. Scale bars 50 μm [35]

As shown in Fig. 2.2, to intercept a section of polymer MNF with a desired length from a polymer MNF, an electrochemically sharpened tungsten probe that is mounted on a three-dimensional translation stage, is used to cut the polymer MNF at the desired point. The intercepted polymer MNF is then picked up (Fig. 2.2b), transferred and deposited (Fig. 2.2c) on a certain substrate (e.g., low-index MgF_2 wafer or silica aerogel) using a tapered fiber probe (drawn from a standard glass optical fiber). When deposited on a certain substrate with a smooth surface, polymer MNFs can be firmly attracted in position by van der Waals and electrostatic forces on the substrate. Using directional pushing or dragging operation on the substrate

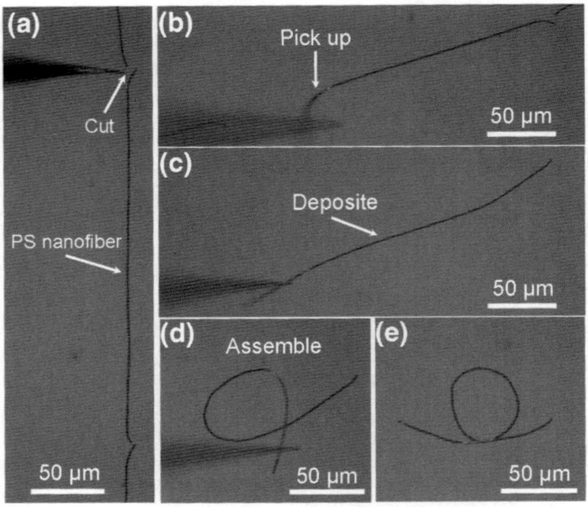

Fig. 2.2 Micromanipulation of single nanofibers [70]. **a** Cutting a PS nanofiber using a tungsten probe. **b** Picking up the cut PS nanofiber using a tapered fiber. **c** Depositing the PS nanofiber on an MgF2 wafer. **d, e** Assembling the PS nanofiber into a ring structure

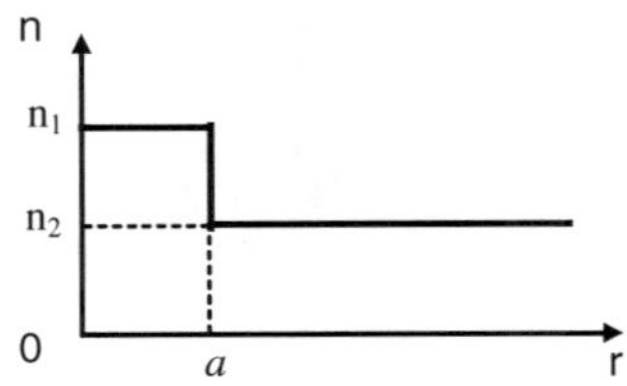

Fig. 2.3 Index profile of an optical MNF [2]

surface against the friction force using micromanipulation probes, polymer MNFs can be bent and assembled to desired structures or patterns (Fig. 2.2d, e). Using micromanipulation, a variety of MNF-based photonic components or devices, e.g., micro-couplers [5, 20, 53, 54], interferometers [55–60] and resonators [41, 42, 47, 58, 61–69] have been reported.

2.3 Optical Properties

2.3.1 Waveguiding Modes in MNFs

The mathematic model of an MNF is shown in Fig. 2.3, in which the refractive indices of the MNF and the cladding material are assumed to be n_1 and n_2, respectively, and the radius of the MNF is a. The index profile of the waveguiding system is then expressed as

$$n(r)\begin{cases} n_1, & 0<r<a, \\ n_2, & a<r<\infty \end{cases} \tag{2.1}$$

For non-absorptive MNF materials, the waveguiding properties of a MNF can be obtained by analytically solving the Helmholtz equations

$$\begin{aligned}(\nabla^2 + n^2k^2 - \beta^2)\,\overline{e} = 0,\\ (\nabla^2 + n^2k^2 - \beta^2)\,\overline{h} = 0\end{aligned} \tag{2.2}$$

where $k = 2\pi/\lambda$, λ is the wavelength of the light in vacuum, and β is the propagation constant.

Benefitted from the circular cross section, (2.2) can be analytically solved in cylindrical coordinate [2], with eigenvalue equations

$$HE_{\nu m} \text{ and } EH_{\nu m} \text{ modes } \left\{\frac{J'_\nu(U)}{UJ_\nu(U)} + \frac{K'_\nu(W)}{WK_\nu(W)}\right\}\left\{\frac{J'_\nu(U)}{UJ_\nu(U)} + \frac{n_2^2K'_\nu(W)}{n_1^2WK_\nu(W)}\right\} = \left(\frac{\nu\beta}{kn_1}\right)^2\left(\frac{V}{UW}\right)^4 \tag{2.3}$$

$$TE_{0m} \text{ modes } \quad \frac{J_1(U)}{UJ_0(U)} + \frac{K_1(W)}{WK_0(W)} = 0 \tag{2.4}$$

$$TE_{0m} \text{ modes } \quad \frac{n_1^2J_1(U)}{UJ_0(U)} + \frac{n_2^2K_1(W)}{WK_0(W)} = 0 \tag{2.5}$$

where J_ν is the Bessel function of the first kind, and K_ν is the modified Bessel function of the second kind, and $U = a(k^2n_1^2 - \beta^2)^{1/2}$, $W = a(\beta^2 - k^2n_2^2)^{1/2,}$ $V = k{\cdot}a(n_1^2 - n_2^2)^{1/2}$.

By numerically solving (2.3–2.5), propagation constants (β) of guiding modes supported by the MNFs can be obtained. For reference, Fig. 2.4a gives diameter-dependent β of air-cladding silica MNFs guiding a 633-nm-wavelength light ($D = 2a$, $n_1 = 1.46$ at $\lambda = 633$ nm, $n_2 = 1$, and $V = k{\cdot}D(n_1 - n_2^2)^{1/2}/2)$. It is clear that, when the wire diameter reduced to a certain value (denoted as dashed vertical line in Fig. 2.4a, at $V = 2.405$, $D = 457$ nm), only the HE_{11} mode exists, corresponding to the single mode operation. Figure 2.4b–d show mode profiles of the first 4 modes (HE_{11}, TE_{01}, TM_{01} and HE_{21}) of a 600-nm-diameter MNF at 633-nm wavelength.

Generally, when its diameter goes close to or smaller than the wavelength of the guided light (Fig. 2.5), an MNF with a low-index clad (e.g., vacuum, air or water) offers unusual properties such as tight optical confinement, high fractional evanescent fields (Fig. 2.5c) and tailorable waveguide dispersions, which intrigue new opportunities for manipulating light on the micro/nano-scale. As one can see, while a 800-nm-diameter silica MNF confines major energy inside the wire (Fig. 2.5a), a 200-nm-diameter MNF leaves a large amount of light guided outside as evanescent waves (Fig. 2.5c).

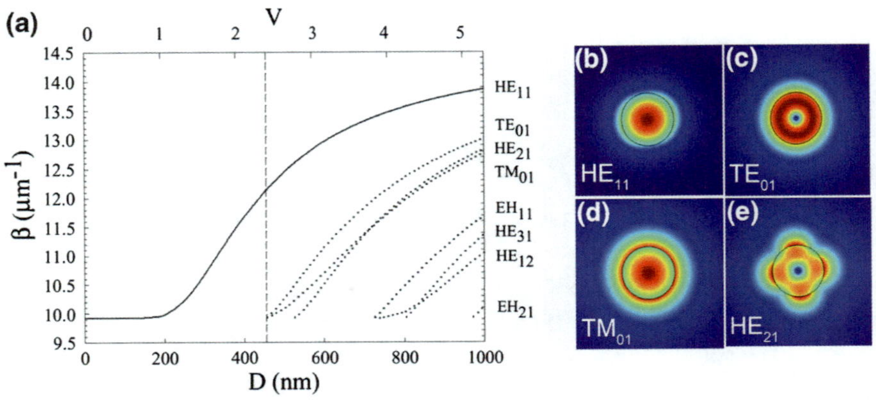

Fig. 2.4 Waveguiding properties of MNFs. **a** Numerical solutions of propagation constant (β) of air-clad silica MNFs at 633-nm wavelength. Solid line, fundamental mode. *Dotted lines*, high-order modes. *Dashed vertical line*, critical diameter for single-mode operation (V = 2.405, corresponding to D = 457 nm) [2]. **b–e** Power distributions (Poynting vectors) of the first four modes at the transverse cross plane of a 600-nm-diameter silica MNF

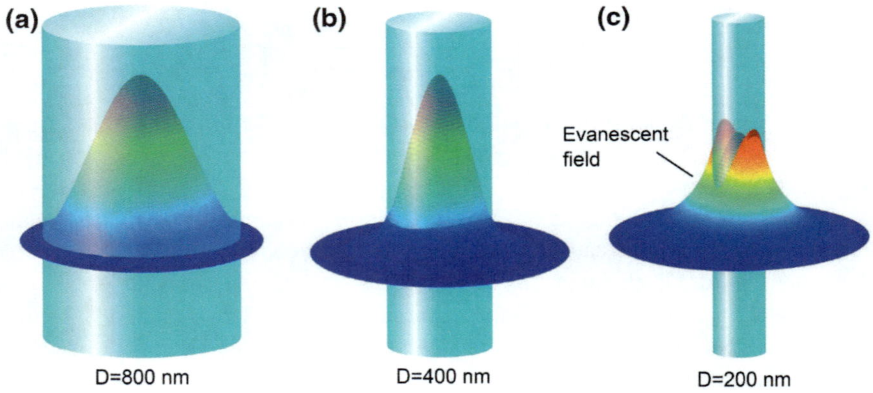

Fig. 2.5 Z-direction Poynting vectors of silica MNFs at 633-nm wavelength with diameters of **a** 800 nm, **b** 400 nm, and **c** 200 nm

2.3.2 Evanescent Coupling

Evanescent coupling between two parallel MNFs is one of the most important parameters to be considered when designing near-field coupling-based functional components like couplers [5, 20, 53, 54], modulators [71], resonators [41, 42, 47, 58, 61–69], lasers [34, 72–77], and sensors [43, 78–85]. Generally, evanescent coupling between adjacent weakly guiding waveguides (low-index-contrast waveguides) with a certain space (e.g., a few hundred nanometers) can be described by the perturbation theory [86]. However, when two MNFs are closely contacted, they are

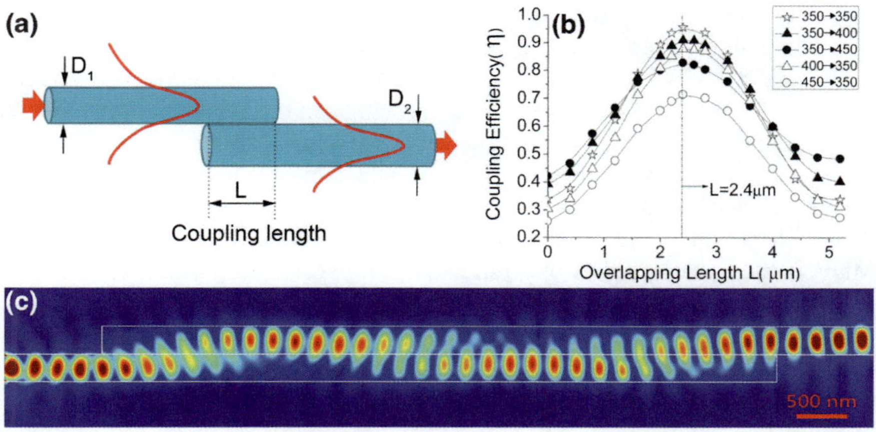

Fig. 2.6 Coupling between closely contacted parallel MNFs [87]. **a** Mathematic model for the coupling of two parallel MNFs. **b** Overlapping-length-dependent coupling efficiency of two silica nanowires with a z-polarized 633 nm wavelength source. Diameters of the nanowires are denoted as x → y, in which x and y stand for the diameters of the input and output nanowires, respectively [6]. **c** Power maps of evanescent coupling between two parallel 350 nm diameter silica MNFs with overlapping length of 7.2 μm. The source is z polarized with wave-length of 633 nm

no longer weak coupling system, in which perturbation theory cannot be applied. Numerical calculation proves to be a simple and direct way to investigate the mode coupling between the closely contacted MNFs. In 2007, using the finite-difference time-domain (FDTD) simulations, K. J. Huang et al. calculated the evanescent-coupling efficiency between two air-clad parallel MNFs [6] (Fig. 2.6a). For reference, Fig. 2.6b gives a simulation result of two parallel 350-nm-diameter silica MNFs. As one can see, the minimum transfer length (2.4 μm) for energy exchange is much shorter than that in weakly coupled waveguides [86]. Also, the coupling efficiency shows an oscillating behavior depending on the overlapping length (Fig. 2.6b, c), with a minimum considerably higher than zero (e.g. 34 % for two 350-nm-diameter silica MNFs) owing to the strong coupling and a maximum lower than 100 % (e.g., 96 % for 350-nm-diameter silica MNFs). Moreover, for two MNFs with different diameter, the coupling efficiency is direction dependent: coupling light from a thinner MNF to a thicker one (e.g., 350 → 400 or 350 → 450) shows higher efficiency (e.g., 91 % for 350 → 400 or 82 % for 350 → 450) than in the opposite direction (e.g., 88 % for 400 → 350 or 71 % for 450 → 350), which may be explained as thinner MNF have stronger evanescent field (as depicted in Fig. 2.6).

2.3.3 Bending Loss

Bent MNFs are important building blocks in micro- or nano-scale photonic components or devices such as couplers [5, 20, 53, 54], interferometers [55–60], resonators [41, 42, 47, 58, 61–69], and lasers [34, 72–77]. Usually, bending losses of conventional fibers can be calculated using weakly guiding or adiabatic approximation. However, these approximation are not valid for sharply bent (e.g., a few micrometers) MNFs, which are usually high-index-contrast waveguides. Based on FDTD method, H. K. Yu et al. investigated the bending losses of MNFs with circular 90° bends, with an acceptable value of 1 dB/90° for bending radii down to micrometer level (e.g., with a minimum allowable bending radius of 5 μm for an 350-nm-diameter silica MNF) [7]. For reference, based on the mathematical model shown in Fig. 2.7a, numerical simulations of a 450-nm-diameter silica MNF with a bending radius of 5 μm and 1 μm are shown in Fig. 2.7b–e. As one can see, there is virtually no power leakage (with calculated bending loss of 0.14 dB/90°) for the 5 μm bent silica MNF, owing to its strong optical confinement ability. When the bending radius decreases to 1 μm, obvious energy leakage occurs around the bending region (Fig. 2.7d), with evident lateral shift of the modal field (Fig. 2.7e) and a calculated bending loss of about 4.8 dB/90°. Bending-radius-dependent bending losses in MNFs of some typical materials at 633-nm wavelength are also investigated (Fig. 2.7f). As one can see, with increasing bending radius or index of the MNF (from silica, to polystyrene (PS) or ZnO wire), the bending loss reduces as a result of the increasing optical confinement.

2.4 Platform for Lab on Fiber

Inspired by the great potential and curiosity of guiding light on a lower dimension, the research on optical MNFs have been attracting more and more attentions. A variety of works, ranging from theoretical designs/simulations, experimental fabrication and optical properties to photonic applications (e.g., MNF-based effects, techniques, components and devices), have been reported. As summarized in the "MNF tree" in Fig. 2.8, the research on MNFs, mainly in the last 10 years, has brought numerous opportunities in renewing and expanding the fiber optics and technology on micro/nanoscale [88–196]. Briefly, these recent advances are categorized into 5 areas based on the optics behind: waveguide and near-field optics, nonlinear optics, quantum and atom optics, plasmonics, and optomechanics.

2.4.1 MNF-Based Passive Components and Devices

Benefitting from their strong evanescent fields, two MNFs can be efficiently coupled within a short interaction length, which can be used to derive a series of MNF-based components and devices with high compactness. This section introduces three types of evanescent-coupling-based components:

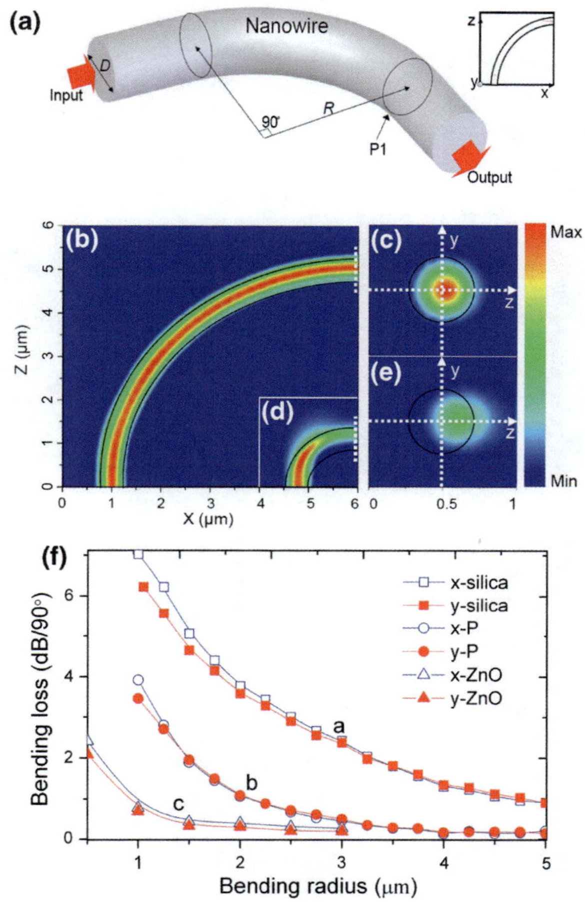

Fig. 2.7 Bending loss of MNFs. **a** Mathematical model for 3D-FDTD simulation of a circular 90° bent MNF. Inset is a topography profile of the bent MNF. **b–e** Electric field intensity distributions in x–z plane (y = 0) of **b** 5 μm and **d** 1 μm bent MNFs. The wavelength of the quasi-x-polarized light is 633 nm and the diameter of the MNFs is 450 nm. The output mode profiles of **b** 5 μm and **d** 1 μm bent MNFs at the P1 transverse cross planes as are located at *black dashed lines* are shown in (**c**) and (**e**), respectively. The *black solid lines* map the topography profile of the MNFs. **f** Bending losses of 350-nm-diameter silica MNF (*a-line*, *squares*), 350-nm-diameter PS nanowire (*b-line*, *circles*), and 270-nm-diameter ZnO nanowire (*c-line*, *triangles*) at 633-nm wavelength (quasi-*x* and quasi-*y* polarizations) with respect to the bending radius [7]

(1) MNF couplers

A MNF coupler is a most basic element in evanescent-coupling-based components. In 2005, using a silica aerogel as the substrate, Tong et al. reported an X-shape structure assembled from two 420-nm-diameter MNFs. With an overlap less than 5 μm, the structure worked as a 3-dB splitter with an excess loss of less than 0.5 dB [5]. In 2006, using MNFs drawn from high-index materials, a more robust coupler can be formed on the surface of a solid substrate [20] (silicate glass

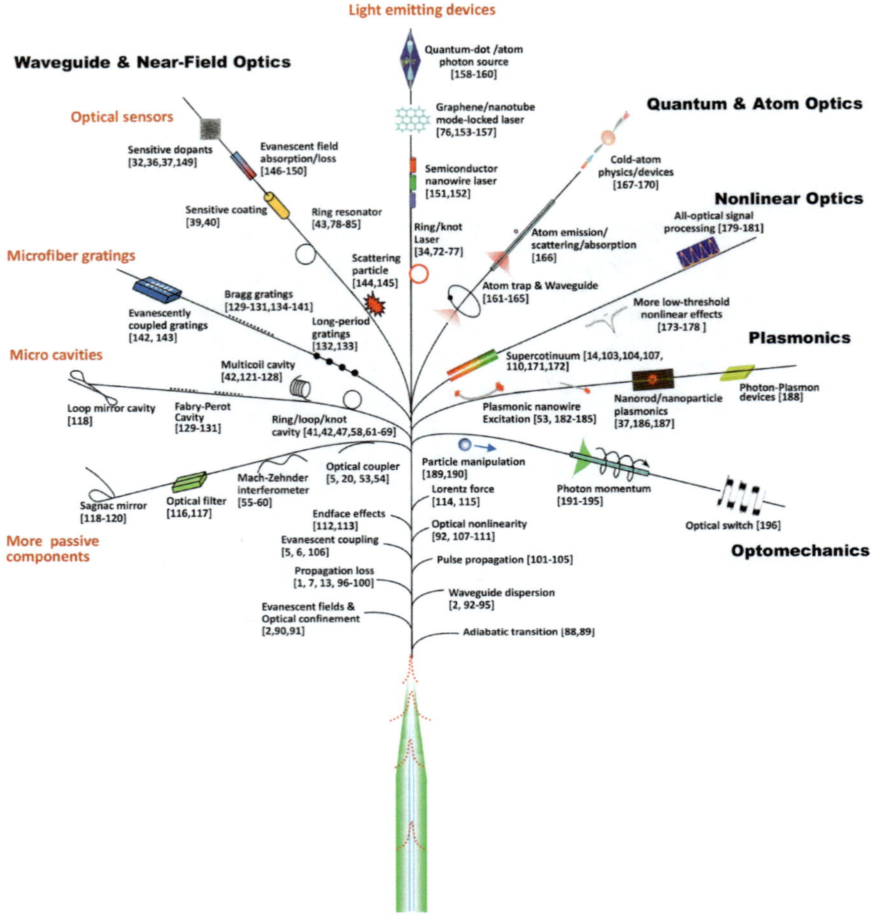

Fig. 2.8 Tree plot of typical applications of optical MNFs (adapted from [12])

or MgF_2 crystal). Figure 2.9 shows an optical coupler assembled using two tellurite glass MNFs (with diameters of 350 and 450 nm, respectively) supported by a silica substrate with a refractive index of 1.46. When 633-nm-wavelength light is launched into the bottom left arm, the coupler splits the flow of light in two, with an interaction length of less than 4 μm and virtually no excess loss (no scattering is observed around the coupling area). Also, the MNFs can be integrated with semiconductor or metal nanowires through near-field coupling, to fabricate various hybrid nanophotonic component [53, 54].

(2) MNF-based Mach-Zehnder interferometer

As a typical functional structure for phase-sensitive measurement, the MZI has been attracting broad attentions for various applications ranging from optical sensors to modulators. Owing to their compact sizes and high sensitivity,

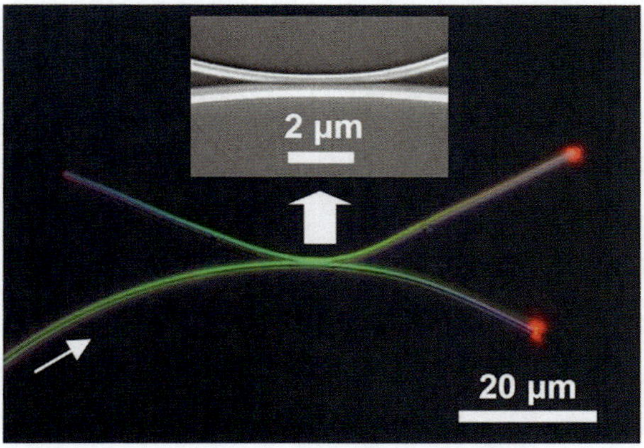

Fig. 2.9 MNF-based couplers. Optical micrograph of an optical coupler assembled using two tellurite glass MNFs (350 and 450 nm in diameter respectively) on the surface of a silicate glass. The coupler splits the 633-nm-wavelength light equally [20]

MNF-based MZIs have been investigated in recent years. In 2008, Li et al. reported a MNF MZI assembled from two tellurite glass MNFs, with footprints down to about 50 micrometers (Fig. 2.10a) [55]. As-assembled MNF MZI showed evident interference fringes with extinction ratios of 10 dB, and the free-spectral range (FSR) could be readily tuned by micromanipulation. In 2010, Chen et al. proposed a novel hybrid MNF MZI using a knot resonator as an alternative, achieving a higher Q-factor of ~15000 and extinction ratio of ~15 dB compared to the single knot resonator and single MZI [56]. To enhance the device stability and robustness, Sulaiman et al. investigated embedding the MNF MZI in low index polymer, which exhibited a larger FSR and a slight degradation in transmission loss and extinction ratio [59]. In 2012, Hu et al. demonstrated splicing microfibers via polymer nanowires in the contact regions to enhance robustness and as-processed MNF MZI was also evaluated in air and water, both showing good performance [58]. Very recently, by integrating Ag nanowires with MNFs (Fig. 2.10b, c), Li et al. demonstrated a hybrid photon-plasmon MZI [54] with Q-factor of 6×10^6 and extinction ratio up to 30 dB (Fig. 2.10d). Using this MZI, fiber-compatible plasmonic sensing was also demonstrated with high sensitivity and low optical power.

(3) MNF resonators

Relying on strong near-field coupling, tying a MNF into a loop or knot proves to be an efficient approach to micro-ring resonators [41, 42, 47, 58, 61–69]. Figure 2.11a, for example, shows a 15-μm-diameter micro-knot tied with a 520-nm-diameter silica MNF [1]. In 2006, Jiang et al. demonstrated a 850-μm-diameter MNF knot resonator using a 1.73-μm-diameter MNF with Q factor as high as 57 000 and finesse of 22 (Fig. 2.11b) in free space [62]. Based on micro-ring resonators, several functional components or devices, including filters [65], lasers [34, 72–77] and sensors [43, 78–85], have been reported. Figure 2.11c, d shows an all-fiber add-drop filter composed of a microfiber knot (working as a resonator) and a fiber taper (working as a dropping fiber) [65]. The 308-μm-diameter knot, assembled with a 2.7-μm-diameter silica MNF, shows a Q factor of about 13,000

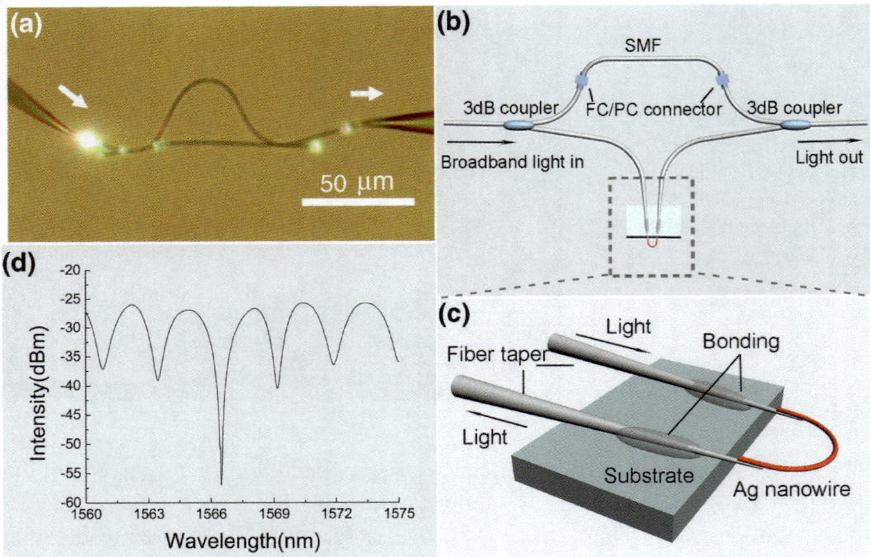

Fig. 2.10 MNF-based MZI. **a** Optical microscope image of a MZI assembled with two 480 nm diameter tellurite MNFs. White light from a supercontinuum source is launched into and picked up from the MZI by two silica fiber tapers. The *white arrows* indicate the direction of light propagation [55]. **b** Schematic of a hybrid photonic-plasmonic MZI. The structure in the dashed box represents the in-fiber return-signal plasmonic probe. A closed-up view of the plasmonic probe is shown in (**c**). **d** Typical transmission spectrum of the hybrid MZI [54]

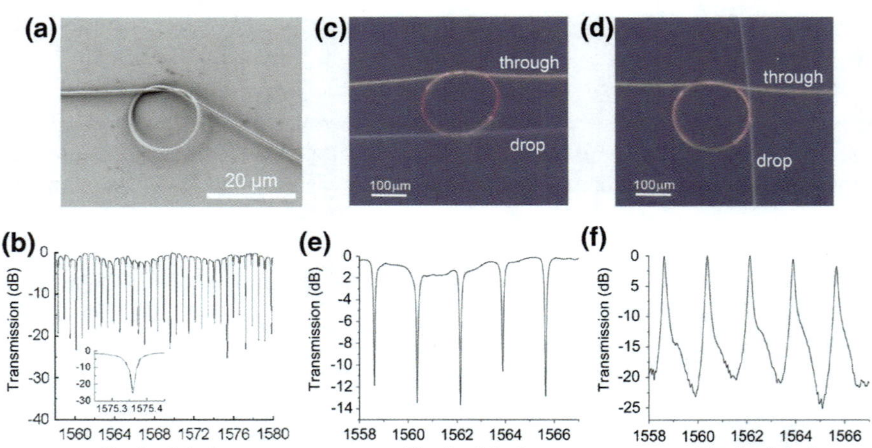

Fig. 2.11 Optical MNF resonators. **a** SEM image of knot resonator using a 520-nm-diameter silica MNF [1]. **b** Transmission spectra of a 850-µm-diameter MNF knot assembled using a 1.73-µm-diameter MNF. The inset shows a single resonance peak [62]. **c**, **d** Optical microscope image of a 200-µm-diameter microknot add–drop filter with the dropping taper placed **c** parallel or **d** perpendicular to the input port of the filter. **e**, **f** Transmission spectra of a 308-µm-diameter add-drop filter (using a 2.7-µm-diameter MNF) with the dropping taper placed parallel to the input port of the filter at the **e** through port and **f** drop port [65]

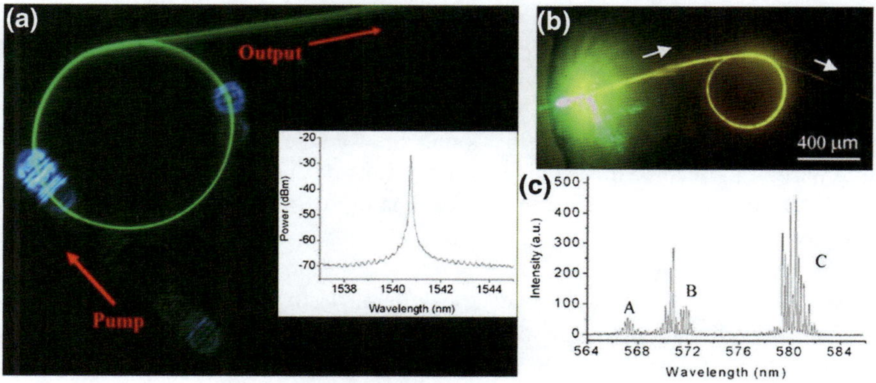

Fig. 2.12 MNF lasers. **a** Optical microscope image of an Er:Yb-doped phosphate glass microfiber knot pumped at a wavelength of 975 nm. The *green* upconverted photoluminescence is clearly seen. Inset is laser emission spectrum of a 2-mm-diameter knot assembled with a 3.8-μm-diameter microfiber [72]. **b** Optical microscope image of a typical 450-μm-diameter knot immersed in a 5 mM/l rhodamine 6G dye solution. The dye is evanescently pumped by 532 nm wavelength light guided along the knot. Strong *yellow* photoluminescence is clearly seen along the microfiber knot. **c** Laser emission from a 350-μm-diameter microfiber knot assembled with a 3.9-μm-diameter microfiber [74]

and a finesse of 14.6 with a FSR of 1.8 nm. Figure 2.11e, f shows the transmission spectra of the add-drop filter with the dropping taper placed at the through port and drop port.

2.4.2 MNF Lasers

Similar to the a fiber laser, an MNF laser can be constructed by incorporating active materials (e.g., dye molecules and rare earth ions) with an MNF-based optical cavity(e.g., an MNF loop or knot). The first MNF laser was reported by Jiang et al. in 2006, in which the MNF knot was assembled using a 3.8-μm-diameter Er:Yb-doped phosphate glass MNF [72] (Fig. 2.12a). With a doping concentration of 1.25 mol % in Er^{3+} ions, single-longitudinal mode emission around 1.5 μm wavelength was obtained in a 2-mm-diameter knot, with a threshold of about 5 mW (for pumping light of CW 975-nm light) and a maximum output of 8 μW. Theoretically, Li et al. predicted that, by resonating of pump light and the signal light along the MNF knot, it was possible to significantly reduce the threshold of the rare-earth-doped MNF laser to μW level, and subsequently increase the external quantum efficiency [73]. In 2009, Song et al. reported a knot laser assembled using dye-doped polymer MNF. When optically pumped by pulsed 532-nm-wavelength light, lasing emission around 625-nm wavelength was obtained with a linewidth of 0.07 nm [34]. In 2011, by fusion splicing a rare-earth-ion-doped phosphate glass MNFs into a 3.5-mm-diameter closed-ring, lasing emission around 1.5-μm wavelength was observed with a linewidth of about 0.05 nm [75].

Besides the above-mention approaches that dope gain materials inside the fiber core, the strong evanescent field around the surface of the MNF provided an alternative approach to dope active materials outside but in the optical near fields of the MNF. In 2007, by immersing a pure silica MNF knot into a rhodamine 6G dye solution, Jiang et al. realized a miniaturized MNF knot dye laser via evanescent-wave coupled gain [74] (Fig. 2.12b). Under optical pumping (532-nm wavelength laser pulses), the 350-um-diameter knot offered lasing emission around 570 and 580 nm wavelength, providing possibilities for integrating MNF lasers with opto-fluidic systems (Fig. 2.12c). Similarly, in 2009, by attaching zinc oxide (ZnO) nanowires to a silica MNF knot cavity, Yang et al. reported a hybrid structure laser with ZnO-nanowire serving as gain [152]. Benefitting from the high-quality cavity of the MNF knot, the lasing threshold of the ZnO nanowire was as low as 0.2 μJ/pulse, which is much lower than that of a freestanding ZnO nanowire. It is also noticeable that, by attaching three distinct semiconductor nanowires (CdSe, CdS and ZnO) to a silica MNF, Ding et al. demonstrated a compact hybrid-structure red-green-ultraviolet three-color laser in a single MNF [151].

Recently, by incorporating saturable absorbers such as carbon nanotubes or graphene films, several groups have reported a variety of mode-locked MNF lasers, providing an opportunity to generating femtosecond or picosecond pulses from compact MNF lasers [76, 153–157].

2.4.3 MNF Sensors

As has been described in Sect. 2.3.1, subwavelength MNFs exhibit large fractional evanescent fields outside the fiber core, bestowing the MNFs with high sensitivity to environmental changes, which will consequently lead to instant changes in optical intensity, phase or spectral components for optical sensing. According to the sensing mechanisms, MNF sensors can be summarized into two categories spectral/intensity detection, and phase sensitive detection.

(1) Spectral/intensity detection

Spectral/intensity detection is one of the most commonly used schemes in MNF optical sensors. These sensors probe the environment by monitoring the changes of transmission properties in response to the variation of the ambient medium. Owing to the large fractional evanescent fields of a waveguiding MNF, the intensity-dependent MNF sensor usually offers high sensitivity in either air or liquid solutions. In 2005, Polynkin et al. reported a MNF optical sensor for measuring the refractive index of liquids in microfluidic channels [147]. By measuring the refractive-index-dependent leakage loss of 1.5-um-wavelength light guided in a 700-nm-diameter silica MNF, they realized an accuracy of refractive-index

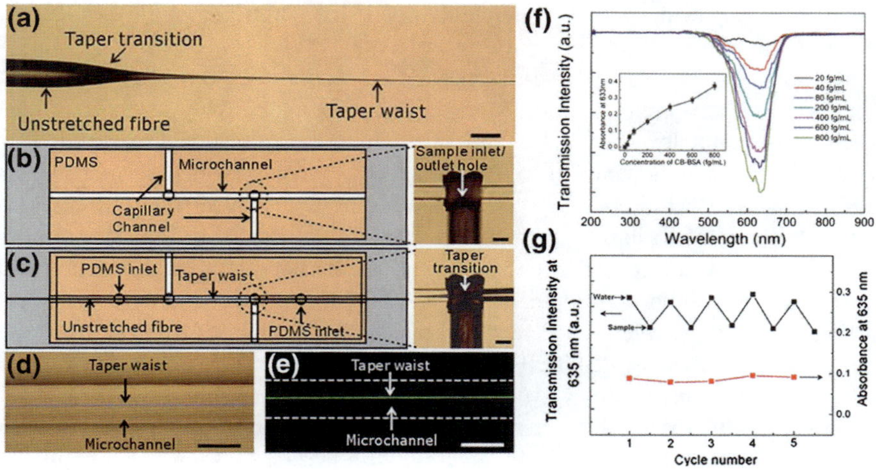

Fig. 2.13 Silica MNF sensors. **a** Biconical tapered fibre with a 900 nm diameter waist (MNF). Scale bars 125 μm. **b**, **c** Cartoon and optical micrographs of microfluidic chip based MNF sensor fabrication procedures. **d** Optical micrograph of a 1.5-μm-diameter MNF guiding a laser with a wavelength of 473 nm embedded in a microchannel. **e** Optical micrograph of the fluorescence excited by evanescent field outside a 1.5-μm-diameter MNF. Scale bars, 125 μm. **f** Transmission spectra of different BSA concentrations for the 900 nm diameter MNF. *Inset* absorbance at 633 nm wavelength versus BSA concentrations. The y-errors are determined from 3 repeated measures. **g** Cycling measurement with 500 pM MB solutions for a 900 nm diameter MNF [150]

measurement of 5.3×10^{-4}. Around the same time, Villatoro et al. demonstrated a miniature hydrogen sensor that consists of a 1.3-um-diameter silica MNF coated with an ultra-thin palladium film [39]. Relying on the hydrogen-concentration-dependent transmission intensity of the MNF at 1,550-nm wavelength, they successfully achieved a fast-response (~10 s) hydrogen sensor with low detection limit. In 2007, based on nanoparticle-induced Rayleigh–Gans scattering in a waveguiding MNF, Wang et al. calculated the possibility of detection single nanoparticles adsorbed on the surface of an MNF, and showed that, by optimizing the wavelength of the probing light and the diameter of the MNF, nanoparticle-induced scattering intensity can reach detectable level with possibilities for single-molecule detection [145]. In 2007, relying on absorption of molecules adsorbed on the surface of a 500-nm-diameter MNF, Warken et al. reported an ultra-sensitive molecular sensor that was possible to detect sub-monolayers of 3,4,9,10-perylene-tetracarboxylic dianhydride (PTCDA) molecules at ambient conditions [148]. In 2011, by integrating a 900-nm-diameter MNF into a 125-μm-wide microfluidic channel, Zhang et al. demonstrated a MNF sensor for chemical and biological sensing [150] (Fig. 2.13a–e). As shown in Fig. 2.13f, using a broadband white light as probing light with power of about 150 nW, the absorbance of bovine serum albumin (BSA) was clearly observed with a detection limit down to 10 fg mL^{-1}. In addition, Fig. 2.13g provides transmission intensity of the MNF response to analytes cycled with 500 pM methylene blue (MB) solution and ultrapure water, showing good absorbance reversibility of the sensor.

Besides above-mentioned silica-MNF-based sensors, sensors made from polymer MNFs show great flexibility and opportunities for highly selective optical detection in complex chemical or biological environments with a single MNF by doping functional dopants (e.g., quantum dots, and noble-metal nanoparticles) [32, 36, 37, 149]. In 2008, Gu et al. doped bromothymol blue (one of the typical chemical indicators) into poly-(methyl methacrylate) (PMMA) MNFs, and operated the MNF as a fast-response (30 ms) optical gas sensor for NH_3 detection with detection limit down to 1 ppm level [32]. In 2011, by directly drawing doped polymer solutions, Meng et al. demonstrated optical-quality CdSe/ZnS quantum dot-doped polystyrene (PS) MNFs [36] with quantum dot concentration up to 10^4 μm^{-3}. By measuring the waveguided photoluminscence intensity in a single quantum dot/PS MNF, they demonstrated a quantum dot/PS nanofiber optical sensor for relative humidity (RH) detection with fast response and extremely low optical power (100 pW level). In 2012, Wang et al. demonstrated an optical RH sensor based on single gold nanorods (GNR)-doped polyacrylamide (PAM) MNFs with fast response (about 110 ms), small footprint, ultra-low optical power (25 nW level) and high photochemical stability, by measuring the spectral shift of the localized surface plasmon resonance (LSPR) with respect to the index change of the polymer due to the adsorption of water molecules [37] (Fig. 2.14).

(2) Phase sensitive detection

The sensitivity of an MNF sensor can be significantly increased by employing interferometric and resonant structures. In 2005, Lou et al. proposed a Mach-Zehnder-based sensor assembled using two single-mode subwavelength-diameter silica MNFs, with one MNF serving as sensitive arm with a certain length of sensitive area exposed to the measurand [146]. By calculating the refractive-index-dependent phase shift of 325-nm-wavelength light guided in a 200-nm-diameter silica MNF, they predicted a refractive-index detection limit of 6×10^{-4} within a sensitive length of 245 μm. In 2007, Shi et al. proposed using optical MNF loop resonators for ambient refractive index sensing, with optimized structural parameters, they predicted a detection limit down to 10^{-5} refractive index unit (RIU) [78]. In 2008, Guo et al. demonstrated a copper-rod-supported MNF loop resonator for refractive-index sensing using a 2.1-μm-diameter MNF wrapping around a 480-μm-diameter copper rod [83]. By measuring the resonance dip wavelength shift induced by the concentration change of the immersing liquid, sensitivity of refractive-index measurement of 1.1×10^{-4} and 1.8×10^{-5} were obtained in a low-concentration ethanol solution and high-concentration glycerol solution, respectively. Around the same time, Xu et al. reported a loop resonator refractometric sensor embedding in polymer and studied the dependence of sensitivity on the MNF diameter and coating thickness [43]. Later, the same group demonstrated a refractometric sensor based on optical microfiber coil resonator and achieved a sensitivity of about 40 nm/RIU by using a 2.5-μm-diameter MNF wrapping around a 1-mm-diameter rod with a total sensing length of about 50 mm [85]. In 2009,

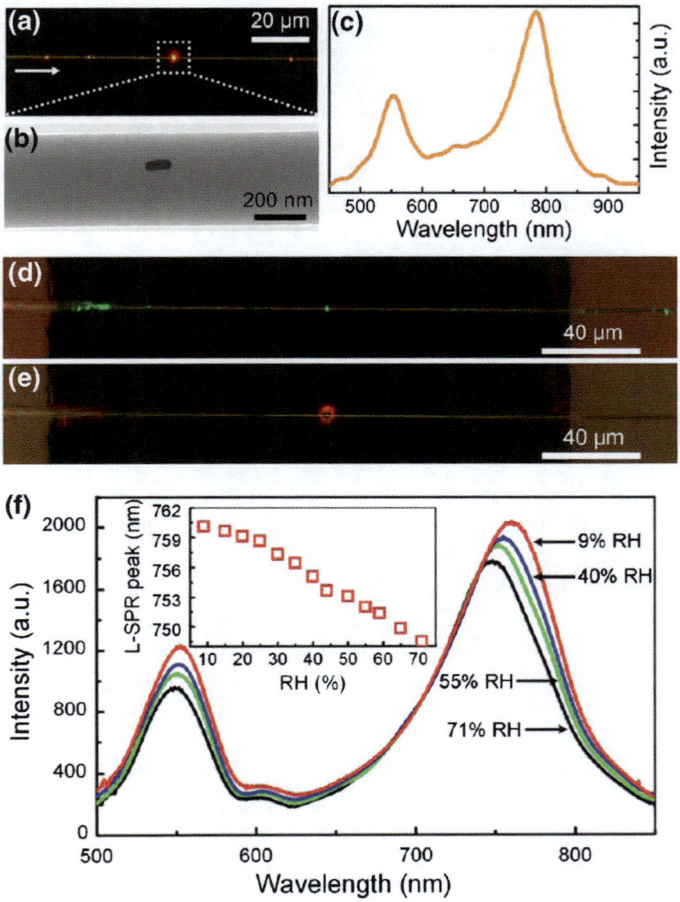

Fig. 2.14 Polymer MNF sensors. **a** Optical microscope image of a single GNR embedded in a waveguiding MNF of 350 nm in diameter and excited by a white light. **b** TEM image of the embedded GNR. **c** Scattering spectrum of the waveguiding excited GNR shown in (**a**). **d**, **e** Optical microscope images of the MNF waveguiding monochromatic lasers with wavelength of 532 and 785 nm. **f** Scattering spectra of the embedded GNR exposed to air of varying RH. *Inset* the dependence of LSPR peak on the RH of ambient air [37]

Scheuer proposed an optical rotation sensor based on a fiber microcoil resonator. Based on the combination of slow-light and conventional propagation effects, an enhancement of the rotation-detection sensitivity by orders of magnitudes (up to 4) was obtained theoretically [197]. In 2011, Wu et al. reported two fiber-optic interferometric humidity sensors based on silica/polymer microfiber knot resonators using silica/polymer MNFs without any humidity-sensitive coating [198]. The silica microfiber knot resonator sensor has a humidity sensitivity of ~12 pm/10 %-RH within a linearity range from 15 to 60 %-RH, while the polymer microfiber knot resonators sensor has a humidity sensitivity of ~88 pm/10 %-RH, with a linearity range from 17 to 95 %-RH. In 2012, Li et al. demonstrated an all-fiber magnetic-field sensor based on a device consisting of a microfiber knot resonator and magnetic fluid [199]. By measuring the change of the resonance wavelength in sensor transmission spectra, they realized a minimal detectable magnetic-field strength of 10 Oe. Recently in 2013, Bo et al. demonstrated an optical microfiber coupler-based refractive index sensor using two 2.5-μm-diamter MNFs to assemble an X-shape coupler with a coupling length of 2 mm [200]. As-assembled MNF coupler can achieve an average sensitivity of 2,723 nm/RIU over the entire refractive index range from 1.3340 to 1.3800 and a highest sensitivity of 4,155 nm/RIU over the range from 1.3340 to 1.3515.

2.5 Outlook

As a valuable technological platform for lab on fiber, the MNF will continue to open up new opportunities for manipulation and utilization of light on micro and nanoscale. Meanwhile, as an ideal waveguide that connects light from atomic to macroscopy scale, the MNF is expected to offer "stronger coupling" between fiber optics, near-field optics, nonlinear optics, plasmonics, quantum/atom optics and optomechanics in the near future, and outgrow new branches in the "tree" shown in Fig. 2.8. Finally, as a fast-growing fiber-optic technology, there are many challenges to be addressed to make the MNF with lower loss, better protection from environmental contamination, greater versatility for active devices and broader material availability, which may offer new possibilities for handling light on wavelength or subwavelength scale, when incorporated with new science, technology and creativity.

References

1. L.M. Tong, R.R. Gattass, J.B. Ashcom, S.L. He, J.Y. Lou, M.Y. Shen, I. Maxwell, E. Mazur, Subwavelength-diameter silica wires for low-loss optical wave guiding. Nature **426**, 816–819 (2003)
2. L.M. Tong, J.Y. Lou, E. Mazur, Single-mode guiding properties of subwavelength-diameter silica and silicon wire waveguides. Opt. Express **12**, 1025–1035 (2004)
3. M. Sumetsky, Y. Dulashko, A. Hale, Fabrication and study of bent and coiled free silica nanowires: Self-coupling microloop optical interferometer. Opt. Express **12**, 3521–3531 (2004)
4. E.C. Magi, H.C. Nguyen, B.J. Eggleton, Air-hole collapse and mode transitions in microstructured fiber photonic wires. Opt. Express **13**, 453–459 (2005)
5. L.M. Tong, J.Y. Lou, R.R. Gattass, S.L. He, X.W. Chen, L. Liu, E. Mazur, Assembly of silica nanowires on silica aerogels for microphotonic devices. Nano Lett. **5**, 259–262 (2005)
6. K.J. Huang, S.Y. Yang, L.M. Tong, Modeling of evanescent coupling between two parallel optical nanowires. Appl. Opt. **46**, 1429–1434 (2007)

7. H.K. Yu, S.S. Wang, J. Fu, M. Qiu, Y.H. Li, F.X. Gu, L.M. Tong, Modeling bending losses of optical nanofibers or nanowires. Appl. Opt. **48**, 4365–4369 (2009)
8. G. Brambilla, F. Xu, P. Horak, Y.M. Jung, F. Koizumi, N.P. Sessions, E. Koukharenko, X. Feng, G.S. Murugan, J.S. Wilkinson, D.J. Richardson, Optical fiber nanowires and microwires: fabrication and applications. Adv. Opt. Photonics **1**, 107 (2009)
9. G. Brambilla, Optical fibre nanowires and microwires: a review. J. Opt. **12**, 043001 (2010)
10. G. Brambilla, Optical fibre nanotaper sensors. Opt. Fiber Technol. **16**, 331–342 (2010)
11. L. Zhang, J.Y. Lou, L.M. Tong, Micro/nanofiber optical sensors. Photonic Sens. **1**, 31–42 (2011)
12. L.M. Tong, F. Zi, X. Guo, J.Y. Lou, Optical microfibers and nanofibers: a tutorial. Opt. Commun. **285**, 4641–4647 (2012)
13. G. Brambilla, V. Finazzi, D. Richardson, Ultra-low-loss optical fiber nanotapers. Opt. Express **12**, 2258–2263 (2004)
14. S. Leon-Saval, T. Birks, W. Wadsworth, P. J Russell, M. Mason, Supercontinuum generation in submicron fibre waveguides. Opt. Express **12**, 2864–2869 (2004)
15. L.M. Tong, J.Y. Lou, Z.Z. Ye, G.T. Svacha, E. Mazur, Self-modulated taper drawing of silica nanowires. Nanotechnology **16**, 1445 (2005)
16. G. Brambilla, F. Xu, X. Feng, Fabrication of optical fibre nanowires and their optical and mechanical characterisation. Electron. Lett. **42**, 517 (2006)
17. X.B. Xing, Y.Q. Wang, B.J. Li, Nanofibers drawing and nanodevices assembly in poly (trimethylene terephthalate). Opt. Express **16**, 10815–10822 (2008)
18. S. Pricking, H. Giessen, Tapering fibers with complex shape. Opt. Express **18**, 3426–3437 (2010)
19. H.F. Xuan, J. Ju, W. Jin, Highly birefringent optical microfibers. Opt. Express **18**, 3828–3839 (2010)
20. L.M. Tong, L.L. Hu, J.J. Zhang, J.R. Qiu, Q. Yang, J.Y. Lou, Y.H. Shen, J.L. He, Z.Z. Ye, Photonic nanowires directly drawn from bulk glasses. Opt. Express **14**, 82–87 (2006)
21. T.E. Dimmick, G. Kakarantzas, T.A. Birks, P.S.J. Russell, Carbon dioxide laser fabrication of fused-fiber couplers and tapers. Appl. Opt. **38**, 6845–6848 (1999)
22. C.F. Bohren, D.R. Huffman, *Absorption and Scattering of Light by Small Particles* (Wiley-Vch, Weinheim, 2008)
23. P. Pal, W.H. Knox, Low loss fusion splicing of micron scale silica fibers. Opt. Express **16**, 11568–11573 (2008)
24. G. Brambilla, F. Koizumi, X. Feng, D.J. Richardson, Compound-glass optical nanowires. Electron. Lett. **41**, 400 (2005)
25. L. Shi, X.F. Chen, H.J. Liu, Y.P. Chen, Z.Q. Ye, W.J. Liao, Y.X. Xia, Fabrication of submicron-diameter silica fibers using electric strip heater. Opt. Express **14**, 5055–5060 (2006)
26. A. Coillet, B. Cluzel, G. Vienne, P. Grelu, F. de Fornel, Near-field characterization of glass microfibers on a low-index substrate. Appl. Phys. B **101**, 291–295 (2010)
27. Y.A. Dzenis, Spinning continuous fibers for nanotechnology. Science **304**, 1917–1919 (2004)
28. D. Li, Y.N. Xia, Electrospinning of nanofibers: reinventing the wheel? Adv. Mater. **16**, 1151–1170 (2004)
29. H.Q. Liu, J.B. Edel, L.M. Bellan, H. Craighead, Electrospun polymer nanofibers as subwavelength optical waveguides incorporating quantum dots. Small **2**, 495–499 (2006)
30. A. Camposeo, F. Di Benedetto, R. Stabile, A.A. Neves, R. Cingolani, D. Pisignano, Laser emission from electrospun polymer nanofibers. Small **5**, 562–566 (2009)
31. S.A. Harfenist, S.D. Cambron, E.W. Nelson, S.M. Berry, A.W. Isham, M.M. Crain, K.M. Walsh, R.S. Keynton, R.W. Cohn, Direct drawing of suspended filamentary micro-and nanostructures from liquid polymers. Nano Lett. **4**, 1931–1937 (2004)
32. F.X. Gu, L. Zhang, X.F. Yin, L.M. Tong, Polymer single-nanowire optical sensors. Nano Lett. **8**, 2757–2761 (2008)

33. X.B. Xing, H. Zhu, Y.Q. Wang, B.J. Li, Ultracompact photonic coupling splitters twisted by PTT nanowires. Nano Lett. **8**, 2839–2843 (2008)
34. Q.H. Song, L.Y. Liu, L. Xu, Lasing action in dye doped polymer nanofiber knot resonator. J. Lightwave Technol. **27**, 4374–4376 (2009)
35. F.X. Gu, H.K. Yu, P. Wang, Z.Y. Yang, L.M. Tong, Light-emitting polymer single nanofibers via waveguiding excitation. ACS Nano **4**, 5332–5338 (2010)
36. C. Meng, Y. Xiao, P. Wang, L. Zhang, Y.X. Liu, L.M. Tong, Quantum-dot-doped polymer nanofibers for optical sensing. Adv. Mater. **23**, 3770–3774 (2011)
37. P. Wang, L. Zhang, Y.N. Xia, L.M. Tong, X. Xu, Y.B. Ying, Polymer nanofibers embedded with aligned gold nanorods: a new platform for plasmonic studies and optical sensing. Nano Lett. **12**, 3145–3150 (2012)
38. P. Wang, Z.Y. Li, L. Zhang, L.M. Tong, Electron-beam-activated light-emitting polymer nanofibers. Opt. Lett. **38**, 1040–1042 (2013)
39. J. Villatoro, D. Monzón-Hernández, Fast detection of hydrogen with nano fiber tapers coated with ultra thin palladium layers. Opt. Express **13**, 5087–5092 (2005)
40. L. Zhang, F.X. Gu, J.Y. Lou, X.F. Yin, L.M. Tong, Fast detection of humidity with a subwavelength-diameter fiber taper coated with gelatin film. Opt. Express **16**, 13349–13353 (2008)
41. G. Vienne, Y.H. Li, L.M. Tong, Effect of host polymer on microfiber resonator. IEEE Photonics Technol. Lett. **19**, 1386–1388 (2007)
42. F. Xu, G. Brambilla, Embedding optical microfiber coil resonators in Teflon. Opt. Lett. **32**, 2164–2166 (2007)
43. F. Xu, V. Pruneri, V. Finazzi, G. Brambilla, An embedded optical nanowire loop resonator refractometric sensor. Opt. Express **16**, 1062–1067 (2008)
44. F. Xu, G. Brambilla, Preservation of micro-optical fibers by embedding. Jpn. J. Appl. Phys. **47**, 6675–6677 (2008)
45. L.M. Xiao, M.D.W. Grogan, S.G. Leon-Saval, R. Williams, R. England, W.J. Wadsworth, T.A. Birks, Tapered fibers embedded in silica aerogel. Opt. Lett. **34**, 2724–2726 (2009)
46. L.M. Xiao, M.D.W. Grogan, W.J. Wadsworth, R. England, T.A. Birks, Stable low-loss optical nanofibres embedded in hydrophobic aerogel. Opt. Express **19**, 764–769 (2011)
47. P. Pal, W.H. Knox, Fabrication and characterization of fused microfiber resonators. IEEE Photonics Technol. Lett. **21**, 766–768 (2009)
48. P. Wang, L. Zhang, Z.Y. Yang, F.X. Gu, S.S. Wang, Q. Yang, L.M. Tong, Fusion spliced microfiber closed-loop resonators. IEEE Photonics Technol. Lett. **22**, 1075–1077 (2010)
49. Y.M. Jung, G. Brambilla, D.J. Richardson, Polarization-maintaining optical microfiber. Opt. Lett. **35**, 2034–2036 (2010)
50. Y. Lizé, E. Mägi, V. Ta'eed, J. Bolger, P. Steinvurzel, B. Eggleton, Microstructured optical fiber photonic wires with subwavelength core diameter. Opt. Express **12**, 3209–3217 (2004)
51. H. Ebendorff-Heidepriem, S.C. Warren-Smith, T.M. Monro, Suspended nanowires: fabrication, design and characterization of fibers with nanoscale cores. Opt. Express **17**, 2646–2657 (2009)
52. S.M. Chuo, L.A. Wang, Propagation loss, degradation and protective coating of long drawn microfibers. Opt. Commun. **284**, 2825–2828 (2011)
53. X. Guo, M. Qiu, J.M. Bao, B.J. Wiley, Q. Yang, X.N. Zhang, Y.G. Ma, H.K. Yu, L.M. Tong, Direct coupling of plasmonic and photonic nanowires for hybrid nanophotonic components and circuits. Nano Lett. **9**, 4515–4519 (2009)
54. X.Y. Li, W. Li, X. Guo, J.Y. Lou, L.M. Tong, All-fiber hybrid photon-plasmon circuits: integrating nanowire plasmonics with fiber optics. Opt. Express **21**, 15698–15705 (2013)
55. Y.H. Li, L.M. Tong, Mach-Zehnder interferometers assembled with optical microfibers or nanofibers. Opt. Lett. **33**, 303–305 (2008)
56. Y.H. Chen, Y. Wu, Y.J. Rao, Q. Deng, Y. Gong, Hybrid Mach-Zehnder interferometer and knot resonator based on silica microfibers. Opt. Commun. **283**, 2953–2956 (2010)
57. B.Y. Li, L. Jiang, S.M. Wang, L.Y. Zhou, H. Xiao, H.L. Tsai, Ultra-abrupt tapered fiber Mach-Zehnder interferometer sensors. Sensors **11**, 5729–5739 (2011)

58. Z.F. Hu, W. Li, Y.G. Ma, L.M. Tong, General approach to splicing optical microfibers via polymer nanowires. Opt. Lett. **37**, 4383–4385 (2012)
59. A. Sulaiman, S.W. Harun, K.S. Lim, F. Ahmad, H. Ahmad, Microfiber Mach-Zehnder interferometer embedded in low index polymer. Opt. Laser Technol. **44**, 1186–1189 (2012)
60. J.H. Wo, G.H. Wang, Y. Cui, Q.Z. Sun, R.B. Liang, P.P. Shum, D.M. Liu, Refractive index sensor using microfiber-based Mach-Zehnder interferometer. Opt. Lett. **37**, 67–69 (2012)
61. M. Sumetsky, Y. Dulashko, J.M. Fini, A. Hale, Optical microfiber loop resonator. Appl. Phys. Lett. **86**, 161108 (2005)
62. X.S. Jiang, L.M. Tong, G. Vienne, X. Guo, A. Tsao, Q. Yang, D.R. Yang, Demonstration of optical microfiber knot resonators. Appl. Phys. Lett. **88**, 223501 (2006)
63. M. Sumetsky, Y. Dulashko, J.M. Fini, A. Hale, D.J. DiGiovanni, The microfiber loop resonator: theory, experiment, and application. J. Lightwave Technol. **24**, 242 (2006)
64. X. Guo, Y.H. Li, X.S. Jiang, L.M. Tong, Demonstration of critical coupling in microfiber loops wrapped around a copper rod. Appl. Phys. Lett. **91**, 073512 (2007)
65. X.S. Jiang, Y. Chen, G. Vienne, L.M. Tong, All-fiber add-drop filters based on microfiber knot resonators. Opt. Lett. **32**, 1710–1712 (2007)
66. G. Vienne, Y.H. Li, L.M. Tong, P. Grelu, Observation of a nonlinear microfiber resonator. Opt. Lett. **33**, 1500–1502 (2008)
67. G. Vienne, A. Coillet, P. Grelu, M. El Amraoui, J.-C. Jules, F. Smektala, L.M. Tong, Demonstration of a reef knot microfiber resonator. Opt. Express **17**, 6224–6229 (2009)
68. Y.M. Jung, G. Brambilla, G.S. Murugan, D.J. Richardson, Optical racetrack ring-resonator based on two U-bent microfibers. Appl. Phys. Lett. **98**, 021109 (2011)
69. L.M. Xiao, T.A. Birks, High finesse microfiber knot resonators made from double-ended tapered fibers. Opt. Lett. **36**, 1098–1100 (2011)
70. P. Wang, Y.P. Wang, L.M. Tong, Functionalized polymer nanofibers: a versatile platform for manipulating light at the nanoscale. Light Sci. Appl. **2**, e102 (2013)
71. C.J. Barrelet, A.B. Greytak, C.M. Lieber, Nanowire photonic circuit elements. Nano Lett. **4**, 1981–1985 (2004)
72. X.S. Jiang, Q. Yang, G. Vienne, Y.H. Li, L.M. Tong, J.J. Zhang, L.L. Hu, Demonstration of microfiber knot laser. Appl. Phys. Lett. **89**, 143513 (2006)
73. Y.H. Li, G. Vienne, X.S. Jiang, X.Y. Pan, X. Liu, P.F. Gu, L.M. Tong, Modeling rare-earth doped microfiber ring lasers. Opt. Express **14**, 7073–7086 (2006)
74. X.S. Jiang, Q.H. Song, L. Xu, J. Fu, L.M. Tong, Microfiber knot dye laser based on the evanescent-wave-coupled gain. Appl. Phys. Lett. **90**, 233501 (2007)
75. W. Li, P. Wang, Z.F. Hu, L.M. Tong, Fusion splicing soft glass microfibers for photonic devices. IEEE Photonics Technol. Lett. **23**, 831–833 (2011)
76. X.Y. He, Z.B. Liu, D.N. Wang, M.W. Yang, C.R. Liao, X. Zhao, Passively mode-locked fiber laser based on reduced graphene oxide on microfiber for ultra-wide-band doublet pulse generation. J. Lightwave Technol. **30**, 984–989 (2012)
77. A. Sulaiman, S.W. Harun, F. Ahmad, S.F. Norizan, H. Ahmad, Tunable laser generation with erbium-doped microfiber knot resonator. Laser Phys. **22**, 588–591 (2012)
78. L. Shi, Y.H. Xu, W. Tan, X.F. Chen, Simulation of optical microfiber loop resonators for ambient refractive index sensing. Sensors **7**, 689–696 (2007)
79. M. Sumetsky, Optimization of optical ring resonator devices for sensing applications. Opt. Lett. **32**, 2577–2579 (2007)
80. M. Sumetsky, Optimization of resonant optical sensors. Opt. Express **15**, 17449–17457 (2007)
81. M. Sumetsky, R.S. Windeler, Y. Dulashko, X. Fan, Optical liquid ring resonator sensor. Opt. Express **15**, 14376–14381 (2007)
82. F. Xu, P. Horak, G. Brambilla, Optical microfiber coil resonator refractometric sensor. Opt. Express **15**, 7888–7893 (2007)
83. X. Guo, L.M. Tong, Supported microfiber loops for optical sensing. Opt. Express **16**, 14429–14434 (2008)

84. G. Vienne, P. Grelu, X.Y. Pan, Y.H. Li, L.M. Tong, Theoretical study of microfiber resonator devices exploiting a phase shift. J. Opt. A: Pure Appl. Opt. **10**, 025303 (2008)
85. F. Xu, G. Brambilla, Demonstration of a refractometric sensor based on optical microfiber coil resonator. Appl. Phys. Lett. **92**, 101126 (2008)
86. A.W. Snyder, J. Love, *Optical waveguide theory*, vol. 190 (Springer, New York, 1983)
87. X.Q. Wu, L.M. Tong, Optical microfibers and nanofibers. Nanophotonics. **2**, 407–428 (2013)
88. M. Sumetsky, Optics of tunneling from adiabatic nanotapers. Opt. Lett. **31**, 3420–3422 (2006)
89. M. Sumetsky, Radiation loss of a nanotaper: singular Gaussian beam model. Opt. Express **15**, 1480–1490 (2007)
90. F.L. Kien, J.Q. Liang, K. Hakuta, V.I. Balykin, Field intensity distributions and polarization orientations in a vacuum-clad subwavelength-diameter optical fiber. Opt. Commun. **242**, 445–455 (2004)
91. A.M. Zheltikov, Birefringence of guided modes in photonic wires: Gaussian-mode analysis. Opt. Commun. **252**, 78–83 (2005)
92. M.A. Foster, K.D. Moll, A.L. Gaeta, Optimal waveguide dimensions for nonlinear interactions. Opt. Express **12**, 2880–2887 (2004)
93. J.Y. Lou, L.M. Tong, Z.Z. Ye, Dispersion shifts in optical nanowires with thin dielectric coatings. Opt. Express **14**, 6993–6998 (2006)
94. C.J. Zhao, Z.X. Tang, Y.X. Ye, D.Y. Fan, L.J. Qian, S.C. Wen, G.H. Chen, Field and dispersion properties of subwavelength-diameter hollow optical fiber. Opt. Express **15**, 6629–6634 (2007)
95. W. Guo, J.L. Kou, F. Xu, Y.Q. Lu, Ultra-flattened and low dispersion in engineered microfibers with highly efficient nonlinearity reduction. Opt. Express **19**, 15229–15235 (2011)
96. A.M. Clohessy, N. Healy, D.F. Murphy, C.D. Hussey, Short low-loss nanowire tapers on singlemode fibres. Electron. Lett. **41**, 954–955 (2005)
97. M. Sumetsky, How thin can a microfiber be and still guide light? Opt. Lett. **31**, 870–872 (2006)
98. M. Sumetsky, Y. Dulashko, P. Domachuk, B.J. Eggleton, Thinnest optical waveguide: experimental test. Opt. Lett. **32**, 754–756 (2007)
99. G.Y. Zhai, L.M. Tong, Roughness-induced radiation losses in optical micro or nanofibers. Opt. Express **15**, 13805–13816 (2007)
100. A.V. Kovalenko, V.N. Kurashov, A.V. Kisil, Radiation losses in optical nanofibers with random rough surface. Opt. Express **16**, 5797–5806 (2008)
101. M. Kolesik, J.V. Moloney, Nonlinear optical pulse propagation simulation: from Maxwell's to unidirectional equations. Phys. Rev. E **70**, 036604 (2004)
102. M. Kolesik, E.M. Wright, J.V. Moloney, Simulation of femtosecond pulse propagation in sub-micron diameter tapered fibers. Appl. Phys. B **79**, 293–300 (2004)
103. M. Foster, A. Gaeta, Q. Cao, R. Trebino, Soliton-effect compression of supercontinuum to few-cycle durations in photonic nanowires. Opt. Express **13**, 6848–6855 (2005)
104. M.A. Foster, J.M. Dudley, B. Kibler, Q. Cao, D. Lee, R. Trebino, A.L. Gaeta, Nonlinear pulse propagation and supercontinuum generation in photonic nanowires: experiment and simulation. Appl. Phys. B **81**, 363–367 (2005)
105. A.M. Zheltikov, Microstructure optical fibers for a new generation of fiber-optic sources and converters of light pulses. Phys. Usp. **50**, 705 (2007)
106. Z. Ma, S.S. Wang, Q. Yang, L.M. Tong, Near-field characterization of optical micro/nanofibres. Chin. Phys. Lett. **24**, 3006 (2007)
107. M.A. Foster, A.L. Gaeta, Ultra-low threshold supercontinuum generation in sub-wavelength waveguides. Opt. Express **12**, 3137–3143 (2004)
108. N.A. Wolchover, F. Luan, A.K. George, J.C. Knight, F.G. Omenetto, High nonlinearity glass photonic crystal nanowires. Opt. Express **15**, 829–833 (2007)

109. M.A. Foster, A.C. Turner, M. Lipson, A.L. Gaeta, Nonlinear optics in photonic nanowires. Opt. Express **16**, 1300–1320 (2008)
110. D.I. Yeom, E.C. Mägi, M.R. Lamont, M.A. Roelens, L.B. Fu, B.J. Eggleton, Low-threshold supercontinuum generation in highly nonlinear chalcogenide nanowires. Opt. Lett. **33**, 660–662 (2008)
111. W.Q. Zhang, V.S. Afshar, H. Ebendorff-Heidepriem, T.M. Monro, Record nonlinearity in optical fibre. Electron. Lett. **44**, 1453–1455 (2008)
112. S.S. Wang, J. Fu, M. Qiu, K.J. Huang, Z. Ma, L.M. Tong, Modeling endface output patterns of optical micro/nanofibers. Opt. Express **16**, 8887–8895 (2008)
113. S.S. Wang, Z.F. Hu, H.K. Yu, W. Fang, M. Qiu, L.M. Tong, Endface reflectivities of optical nanowires. Opt. Express **17**, 10881–10886 (2009)
114. I. Brevik, S.Å. Ellingsen, Transverse radiation force in a tailored optical fiber. Phys. Rev. A **81**, 011806 (2010)
115. F. Le Kien, K. Hakuta, Motion of an atom in a weakly driven fiber-Bragg-grating cavity: force, friction, and diffusion. Phys. Rev. A **81**, 063808 (2010)
116. Y. Chen, Z. Ma, Q. Yang, L.M. Tong, Compact optical short-pass filters based on microfibers. Opt. Lett. **33**, 2565–2567 (2008)
117. Y.M. Jung, G. Brambilla, D.J. Richardson, Broadband single-mode operation of standard optical fibers by using a sub-wavelength optical wire filter. Opt. Express **16**, 14661–14667 (2008)
118. S.S. Wang, Z.F. Hu, Y.H. Li, L.M. Tong, All-fiber Fabry-Perot resonators based on microfiber Sagnac loop mirrors. Opt. Lett. **34**, 253–255 (2009)
119. S.D. Lim, S.G. Lee, K. Lee, S.B. Lee, A tunable-transmission Sagnac interferometer using an optical microfiber. Jpn. J. Appl. Phys. **49**, 2502 (2010)
120. T. Wang, X.H. Li, F.F. Liu, W.H. Long, Z.Y. Zhang, L.M. Tong, Y.K. Su, Enhanced fast light in microfiber ring resonator with a Sagnac loop reflector. Opt. Express **18**, 16156–16161 (2010)
121. M. Sumetsky, Optical fiber microcoil resonators. Opt. Express **12**, 2303–2316 (2004)
122. M. Sumetsky, Uniform coil optical resonator and waveguide: transmission spectrum, eigenmodes, and dispersion relation. Opt. Express **13**, 4331–4340 (2005)
123. M. Sumetsky, Vertically-stacked multi-ring resonator. Opt. Express **13**, 6354–6375 (2005)
124. M. Sumetsky, Y. Dulashko, M. Fishteyn, Demonstration of a multi-turn microfiber coil resonator, in *National Fiber Optic Engineers Conference* (Optical Society of America, 2007)
125. F. Xu, G. Brambilla, Manufacture of 3-D microfiber coil resonators. IEEE Photonics Technol. Lett. **19**, 1481–1483 (2007)
126. F. Xu, P. Horak, G. Brambilla, Optimized design of microcoil resonators. J. Lightwave Technol. **25**, 1561–1567 (2007)
127. F. Xu, P. Horak, G. Brambilla, Conical and biconical ultra-high-Q optical-fiber nanowire microcoil resonator. Appl. Opt. **46**, 570–573 (2007)
128. Y.M. Jung, G.S. Murugan, G. Brambilla, D.J. Richardson, Embedded optical microfiber coil resonator with enhanced high-Q. IEEE Photonics Technol. Lett. **22**, 1638–1640 (2010)
129. M. Ding, P.F. Wang, T. Lee, G. Brambilla, A microfiber cavity with minimal-volume confinement. Appl. Phys. Lett. **99**, 051105 (2011)
130. K.P. Nayak, F. Le Kien, Y. Kawai, K. Hakuta, K. Nakajima, H.T. Miyazaki, Y. Sugimoto, Cavity formation on an optical nanofiber using focused ion beam milling technique. Opt. Express **19**, 14040–14050 (2011)
131. F. Le Kien, K.P. Nayak, K. Hakuta, Nanofibers with Bragg gratings from equidistant holes. J. Mod. Opt. **59**, 274–286 (2012)
132. H.F. Xuan, W. Jin, M. Zhang, CO2 laser induced long period gratings in optical microfibers. Opt. Express **17**, 21882–21890 (2009)
133. H.F. Xuan, W. Jin, S.J. Liu, Long-period gratings in wavelength-scale microfibers. Opt. Lett. **35**, 85–87 (2010)

134. W. Liang, Y.Y. Huang, Y. Xu, R.K. Lee, A. Yariv, Highly sensitive fiber Bragg grating refractive index sensors. Appl. Phys. Lett. **86**, 151122 (2005)
135. X. Fang, C.R. Liao, D.N. Wang, Femtosecond laser fabricated fiber Bragg grating in microfiber for refractive index sensing. Opt. Lett. **35**, 1007–1009 (2010)
136. Y. Zhang, B. Lin, S.C. Tjin, H. Zhang, G.H. Wang, P. Shum, X.L. Zhang, Refractive index sensing based on higher-order mode reflection of a microfiber Bragg grating. Opt. Express **18**, 26345–26350 (2010)
137. R. Ahmad, M. Rochette, C. Baker, Fabrication of Bragg gratings in subwavelength diameter As2Se3 chalcogenide wires. Opt. Lett. **36**, 2886–2888 (2011)
138. J.L. Kou, S.J. Qiu, F. Xu, Y.Q. Lu, Demonstration of a compact temperature sensor based on first-order Bragg grating in a tapered fiber probe. Opt. Express **19**, 18452–18457 (2011)
139. Y.X. Liu, C. Meng, A.P. Zhang, Y. Xiao, H.K. Yu, L.M. Tong, Compact microfiber Bragg gratings with high-index contrast. Opt. Lett. **36**, 3115–3117 (2011)
140. Y. Ran, Y.N. Tan, L.P. Sun, S. Gao, J. Li, L. Jin, B.O. Guan, 193 nm excimer laser inscribed Bragg gratings in microfibers for refractive index sensing. Opt. Express **19**, 18577–18583 (2011)
141. Y. Ran, L. Jin, Y.N. Tan, L.P. Sun, J. Li, B.O. Guan, High-efficiency ultraviolet inscription of Bragg gratings in microfibers. IEEE Photonics J. **4**, 181–186 (2012)
142. F. Xu, G. Brambilla, Y.Q. Lu, A microfluidic refractometric sensor based on gratings in optical fibre microwires. Opt. Express **17**, 20866–20871 (2009)
143. F. Xu, G. Brambilla, J. Feng, Y.Q. Lu, A microfiber Bragg grating based on a microstructured rod: a proposal. IEEE Photonics Technol. Lett. **22**, 218–220 (2010)
144. I.D. Chremmos, N.K. Uzunoglu, Analysis of scattering by a linear chain of spherical inclusions in an optical fiber. JOSA A **23**, 3054–3062 (2006)
145. S.S. Wang, X.Y. Pan, L.M. Tong, Modeling of nanoparticle-induced Rayleigh-Gans scattering for nanofiber optical sensing. Opt. Commun. **276**, 293–297 (2007)
146. J.Y. Lou, L.M. Tong, Z.Z. Ye, Modeling of silica nanowires for optical sensing. Opt. Express **13**, 2135–2140 (2005)
147. P. Polynkin, A. Polynkin, N. Peyghambarian, M. Mansuripur, Evanescent field-based optical fiber sensing device for measuring the refractive index of liquids in microfluidic channels. Opt. Lett. **30**, 1273–1275 (2005)
148. F. Warken, E. Vetsch, D. Meschede, M. Sokolowski, A. Rauschenbeutel, Ultra-sensitive surface absorption spectroscopy using sub-wavelength diameter optical fibers. Opt. Express **15**, 11952–11958 (2007)
149. F.X. Gu, X.F. Yin, H.K. Yu, P. Wang, L.M. Tong, Polyaniline/polystyrene single-nanowire devices for highly selective optical detection of gas mixtures. Opt. Express **17**, 11230–11235 (2009)
150. L. Zhang, P. Wang, Y. Xiao, H.K. Yu, L.M. Tong, Ultra-sensitive microfibre absorption detection in a microfluidic chip. Lab Chip **11**, 3720 (2011)
151. Y. Ding, Q. Yang, X. Guo, S.S. Wang, F.X. Gu, J. Fu, Q. Wan, J.P. Cheng, L.M. Tong, Nanowires/microfiber hybrid structure multicolor laser. Opt. Express **17**, 21813–21818 (2009)
152. Q. Yang, X.S. Jiang, X. Guo, Y. Chen, L.M. Tong, Hybrid structure laser based on semiconductor nanowires and a silica microfiber knot cavity. Appl. Phys. Lett. **94**, 101108 (2009)
153. K. Kieu, M. Mansuripur, Femtosecond laser pulse generation with a fiber taper embedded in carbon nanotube/polymer composite. Opt. Lett. **32**, 2242–2244 (2007)
154. Y.W. Song, K. Morimune, S.Y. Set, S. Yamashita, Polarization insensitive all-fiber mode-lockers functioned by carbon nanotubes deposited onto tapered fibers. Appl. Phys. Lett. **90**, 021101 (2007)
155. K. Kieu, M. Mansuripur, All-fiber bidirectional passively mode-locked ring laser. Opt. Lett. **33**, 64–66 (2008)
156. K. Kashiwagi, S. Yamashita, Deposition of carbon nanotubes around microfiber via evanescent light. Opt. Express **17**, 18364–18370 (2009)

157. K. Kieu, F. Wise, Soliton thulium-doped fiber laser with carbon nanotube saturable absorber. IEEE Photonics Technol. Lett. **21**, 128–130 (2009)
158. M. Fujiwara, K. Toubaru, T. Noda, H.Q. Zhao, S. Takeuchi, Highly efficient coupling of photons from nanoemitters into single-mode optical fibers. Nano Lett. **11**, 4362–4365 (2011)
159. F. Le Kien, K. Hakuta, Deterministic generation of a pair of entangled guided photons from a single atom in a nanofiber cavity. Phys. Rev. A **84**, 053801 (2011)
160. F. Le Kien, K. Hakuta, Triggered generation of single guided photons from a single atom in a nanofiber cavity. Phys. Rev. A **83**, 043801 (2011)
161. V.I. Balykin, K. Hakuta, F. Le Kien, J.Q. Liang, M. Morinaga, Atom trapping and guiding with a subwavelength-diameter optical fiber. Phys. Rev. A **70**, 011401 (2004)
162. F. Le Kien, V.I. Balykin, K. Hakuta, Atom trap and waveguide using a two-color evanescent light field around a subwavelength-diameter optical fiber. Phys. Rev. A **70**, 063403 (2004)
163. G. Sagué, E. Vetsch, W. Alt, D. Meschede, A. Rauschenbeutel, Cold-atom physics using ultrathin optical fibers: light-induced dipole forces and surface interactions. Phys. Rev. Lett. **99**, 163602 (2007)
164. J. Fu, X. Yin, N. Li, L.M. Tong, Atom waveguide and 1D optical lattice using a two-color evanescent light field around an optical micro/nano-fiber. Chin. Opt. Lett. **6**, 112–115 (2008)
165. C. Lacroûte, K.S. Choi, A. Goban, D.J. Alton, D. Ding, N.P. Stern, H.J. Kimble, A state-insensitive, compensated nanofiber trap. New J. Phys. **14**, 023056 (2012)
166. V.G. Minogin, S.N. Chormaic, Manifestation of the van der Waals surface interaction in the spontaneous emission of atoms into an optical nanofiber. Laser Phys. **20**, 32–37 (2010)
167. F. Le Kien, S.D. Gupta, K.P. Nayak, K. Hakuta, Nanofiber-mediated radiative transfer between two distant atoms. Phys. Rev. A **72**, 063815 (2005)
168. E. Vetsch, D. Reitz, G. Sagué, R. Schmidt, S.T. Dawkins, A. Rauschenbeutel, Optical interface created by laser-cooled atoms trapped in the evanescent field surrounding an optical nanofiber. Phys. Rev. Lett. **104**, 203603 (2010)
169. K. Salit, M. Salit, S. Krishnamurthy, Y. Wang, P. Kumar, M.S. Shahriar, Ultra-low power, Zeno effect based optical modulation in a degenerate V-system with a tapered nano fiber in atomic vapor. Opt. Express **19**, 22874–22881 (2011)
170. L. Russell, K. Deasy, M.J. Daly, M.J. Morrissey, S.N. Chormaic, Sub-doppler temperature measurements of laser-cooled atoms using optical nanofibres. Meas. Sci. Technol. **23**, 015201 (2012)
171. T.A. Birks, W.J. Wadsworth, P.J. Russell, Supercontinuum generation in tapered fibers. Opt. Lett. **25**, 1415–1417 (2000)
172. G. Zhou, G. Feng, H. Zhou, G. Deng, Y. Zhang, Z. Ma, Experimental investigation of supercontinuum generated from microfiber loop wound on Al-coated silica rod. Opt. Commun. **284**, 4769–4772 (2011)
173. V. Grubsky, J. Feinberg, Phase-matched third-harmonic UV generation using low-order modes in a glass micro-fiber. Opt. Commun. **274**, 447–450 (2007)
174. E.C. Mägi, L.B. Fu, H.C. Nguyen, M.R.E. Lamont, D.I. Yeom, B.J. Eggleton, Enhanced Kerr nonlinearity in sub-wavelength diameter As2Se3 chalcogenide fiber tapers. Opt. Express **15**, 10324–10329 (2007)
175. S.M. Spillane, G.S. Pati, K. Salit, M. Hall, P. Kumar, R.G. Beausoleil, M.S. Shahriar, Observation of nonlinear optical interactions of ultralow levels of light in a tapered optical nanofiber embedded in a hot rubidium vapor. Phys. Rev. Lett. **100**, 233602 (2008)
176. H. You, S.M. Hendrickson, J.D. Franson, Analysis of enhanced two-photon absorption in tapered optical fibers. Phys. Rev. A **78**, 053803 (2008)
177. A. Coillet, G. Vienne, P. Grelu, Potentialities of glass air-clad micro- and nanofibers for nonlinear optics. J. Opt. Soc. Am. B: Opt. Phys. **27**, 394–401 (2010)
178. S.M. Hendrickson, M.M. Lai, T.B. Pittman, J.D. Franson, Observation of two-photon absorption at low power levels using tapered optical fibers in rubidium vapor. Phys. Rev. Lett. **105**, 173602 (2010)

179. L.B. Fu, M.D. Pelusi, E.C. Mägi, V.G. Ta'Eed, B.J. Eggleton, Broadband all-optical wavelength conversion of 40Gbit/s signals in nonlinearity enhanced tapered chalcogenide fibre. Electron. Lett. **44**, 44–46 (2008)
180. M.D. Pelusi, F. Luan, E. Magi, M.R.E. Lamont, D.J. Moss, B.J. Eggleton, J.S. Sanghera, L.B. Shaw, I.D. Aggarwal, High bit rate all-optical signal processing in a fiber photonic wire. Opt. Express **16**, 11506–11512 (2008)
181. F. Luan, J. Van Erps, M.D. Pelusi, E. Mägi, T. Iredale, H. Thienpont, B.J. Eggleton, High-resolution optical sampling of 640Gbit/s data using dispersion-engineered chalcogenide photonic wire. Electron. Lett. **46**, 223–225 (2010)
182. X.W. Chen, V. Sandoghdar, M. Agio, Highly efficient interfacing of guided plasmons and photons in nanowires. Nano Lett. **9**, 3756–3761 (2009)
183. C.H. Dong, X.F. Ren, R. Yang, J.Y. Duan, J.G. Guan, G.C. Guo, G.P. Guo, Coupling of light from an optical fiber taper into silver nanowires. Appl. Phys. Lett. **95**, 221109 (2009)
184. Y.G. Ma, X.Y. Li, H.K. Yu, L.M. Tong, Y. Gu, Q.H. Gong, Direct measurement of propagation losses in silver nanowires. Opt. Lett. **35**, 1160–1162 (2010)
185. B. Ung, M. Skorobogatiy, Extreme nonlinear optical enhancement in chalcogenide glass fibers with deep-subwavelength metallic nanowires. Opt. Lett. **36**, 2527–2529 (2011)
186. U. Schröter, A. Dereux, Surface plasmon polaritons on metal cylinders with dielectric core. Phys. Rev. B: Condens. Matter **64**, 1254201 (2001)
187. K.E. Roskov, K.A. Kozek, W.C. Wu, R.K. Chhetri, A.L. Oldenburg, R.J. Spontak, J.B. Tracy, Long-range alignment of gold nanorods in electrospun polymer nano/microfibers. Langmuir **27**, 13965–13969 (2011)
188. C.H. Dong, C.L. Zou, X.F. Ren, G.C. Guo, F.W. Sun, In-line high efficient fiber polarizer based on surface plasmon. Appl. Phys. Lett. **100**, 041104 (2012)
189. G. Brambilla, G.S. Murugan, J.S. Wilkinson, D.J. Richardson, Optical manipulation of microspheres along a sub wavelength optical wire. Opt. Lett. **32**, 3041–3043 (2007)
190. L. Zhao, Y. Li, J. Qi, J. Xu, Q. Sun, Quasi 3-dimensional optical trapping by two counter-propagating beams in nano-fiber. Opt. Express **18**, 5724–5729 (2010)
191. F.L. Kien, K. Hakuta, V.I. Balykin, Angular momentum of light in an optical nanofiber. Phys. Rev. A **73**, 053823 (2006)
192. W.L. She, J.H. Yu, R.H. Feng, Observation of a push force on the end face of a nanometer silica filament exerted by outgoing light. Phys. Rev. Lett. **101**, 243601 (2008)
193. M. Mansuripur, A.R. Zakharian, Theoretical analysis of the force on the end face of a nanofilament exerted by an outgoing light pulse. Phys. Rev. A **80**, 023823 (2009)
194. H.K. Yu, W. Fang, F.X. Gu, M. Qiu, Z.Y. Yang, L.M. Tong, Longitudinal Lorentz force on a subwavelength-diameter optical fiber. Phys. Rev. A **83**, 053830 (2011)
195. J.H. Yu, C.Y. Chen, Y.F. Zhai, Z. Chen, J. Zhang, L.J. Wu, F.R. Huang, Y. Xiao, Total longitudinal momentum in a dispersive optical waveguide. Opt. Express **19**, 25263–25278 (2011)
196. J. Yu, R. Feng, W. She, Low-power all-optical switch based on the bend effect of a nm fiber taper driven by outgoing light. Opt. Express **17**, 4640–4645 (2009)
197. J. Scheuer, Fiber microcoil optical gyroscope. Opt. Lett. **34**, 1630–1632 (2009)
198. Y. Wu, T.H. Zhang, Y.J. Rao, Y. Gong, Miniature interferometric humidity sensors based on silica/polymer microfiber knot resonators. Sens. Actuators B **155**, 258–263 (2011)
199. X.L. Li, H. Ding, All-fiber magnetic-field sensor based on microfiber knot resonator and magnetic fluid. Opt. Lett. **37**, 5187–5189 (2012)
200. L. Bo, P.F. Wang, Y. Semenova, G. Farrell, High sensitivity fiber refractometer based on an optical microfiber coupler. IEEE Photonics Technol. Lett. **25**, 228–230 (2013)

Chapter 3
SOI Microring Resonator Sensor Integrated on a Fiber Facet

Cristina Lerma Arce, Katrien De Vos, Tom Claes, Katarzyna Komorowska and Peter Bienstman

Abstract The application of optical fiber technology for sensing has undergone tremendous growth over the last years. Its use for imaging hard-to-reach locations and its property to conduct light to a remote convenient location make of it a suitable tool for in vivo sensing applications, such as endoscopy. Here, we present an optical fiber probe sensor for label-free biosensing based on SOI ring resonators. We describe the operating principle of the device, the technology used to integrate a Silicon-on-insulator (SOI) chip on a fiber facet and discuss some experimental results.

3.1 Introduction

Optical fiber sensors have been established for half a century now during which time they have stimulated a great deal of research and useful practical engineering outcomes. The property of fibers to conduct light to a remote, convenient location makes them ideal for in vivo biosensing applications. Biosensing is a technique

C. Lerma Arce (✉) · K. De Vos · T. Claes · K. Komorowska · P. Bienstman
Photonics Research Group (INTEC), Ghent University, Sint-Pietersnieuwstraat 41, 9000 Ghent, Belgium
e-mail: cristina.lermaarce@intec.ugent.be

K. De Vos
e-mail: Katrien.Devos@intec.ugent.be

T. Claes
e-mail: mail@tomclaes.be

K. Komorowska
e-mail: katarzyna.komorowska@intec.UGent.be

P. Bienstman
e-mail: Peter.Bienstman@UGent.be

A. Cusano et al. (eds.), *Lab-on-Fiber Technology*,
Springer Series in Surface Sciences 56, DOI: 10.1007/978-3-319-06998-2_3,

that provides the ability to study interactions between biomolecules and to detect analytes from body fluids, manufacturing processes or environmental samples. The development of biosensors is challenging, because biomolecules are extremely small and sometimes present within a sample at very low concentrations (pg/ml to fg/ml) that simultaneously contains thousands of other molecules at much higher concentrations. Biological research has historically relied upon attachment of a label to one or more of the molecules, viruses or cells being studied.

Commonly used labels are fluorescent dies, radioactive molecules or nano-particles. These are easily measurable and serve to indicate the presence of the analyte to which they have been attached. Today, commercial optical fiber biosensors require this labeling of one or more biomolecules [1]. In the case of fluorescent dye, once the labeled biomolecule has interacted with its target, the label is excited by the evanescent wave and the resulting fluorescence is then conducted via the fiber to an analytical device that has been equipped with appropriate wavelength filters. This area is the subject of a recent review [2].

While labeled detection methods can be sensitive down to a single molecule, labels can structurally and functionally alter the assay and the labeling process is labor intensive and costly. Quantification is difficult since the label's intensity level is dependent on all working conditions. Moreover a labeled assay can only be performed in an 'end-point' fashion so that no kinetic information on the biomolecular interaction can be obtained. Due to these considerations, there has been a drive to reduce assay cost and complexity while providing more quantitative information with high throughput.

Label-free detection involves a transducer that directly measures some physical property of the biological compound. A so-called 'receptor' or 'ligand' is attached to the surface of the sensor, which responds to the affinity interaction of the receptor with an analyte. The formation of complexes can thus be monitored continuously and many interactions can be followed simultaneously. Emerging applications include point-of-care diagnostics, drug development, environmental control and food monitoring. Point-of-care systems are designed to process clinical samples for a number of different types of biomarkers and to work in a variety of settings.

Optical label-free biosensors make use of the fact that biological molecules reduce the propagation velocity of electromagnetic fields that pass through them. They translate changes in the propagation speed of light into a quantifiable signal proportional to the amount of biological material present on the sensor surface.

Over the past years there has been a vast research effort in the integration of optical label-free biosensors on photonic chips. Photonic integrated circuits (PIC) consist of large variety of optical components that are integrated on one single chip. The integrated approach allows for a significant simplification of the optical system, since the light routing and processing can be accomplished by a waveguide circuit that is fabricated by a collective lithography process. The sensing part is fully decoupled from the optical read-out part and can thus be optimized separately. Extension to multiplexed sensor arrays is simply attained by arraying one sensor next to the other and connecting them through integrated waveguides.

Unlike in electronic integration, where silicon is the dominant material, photonic integrated circuits have been fabricated from a variety of material systems like silica-on-silicon, polymers, 3–5 semiconductors, silicon- on-insulator, etc. Aside from ease and cost of fabrication, the number of components per surface area (the integration density) is a key issue for the choice of material. This is related to the minimal waveguide bend radius that supports guiding without substantial losses and the minimal device separation that avoids mode coupling. Essentially this relates to the degree of confinement in the waveguides. Platforms consisting of semiconductor materials like silicon or III–V semiconductors with refractive index around 3 offer the highest degree of integration, in contrast to glasses or polymers with indices in the range of 1.5–2.

Additional levels of integration with electronic and biologic functions are also possible. The integration of microfluidics and the transducer is of big importance since different sample preparation steps like mixing or centrifugation can be incorporated in the sensor so there is no longer a need for, often expensive, external laboratory equipment to perform these sample preparation tasks. There are many implementations of microfluidics or methods of transporting the analyte to the sensor but in its most standard form, a microfluidic compartment, often referred to as a flow cell, is glued onto the sensor surface [3] and the analyte is flown over the sensor surface, bringing the target molecules in close contact with the receptor molecules.

The ultimate goal is to integrate several laboratory functions in a single chip, a so-called lab-on-a-chip, with a reduced cost, smaller required analyte volumes, compactness and increased ease of use.

However, while a lot of progress has been made to implement such sensors in small and easy-to-use cartridges, many applications still require a sensor probe that can perform measurements at locations that are hard to reach. Over the past years, optical fibers have been also investigated for label-free biosensing. They include fiber bragg gratings, long-period grating biosensors, photonic crystal fiber sensors, hollow fibers in a Fabry-Perot cavity, etc. [4–6]. Most applications however focus on bulk refractive index sensing. For biochemical detection, these sensors cannot compete with other photonic sensors on chip in terms of their detection limit, although their bulk sensitivity can exceed 100 nm/RIU.

The main motivation of this work is to combine the good performance of photonic biosensors on chip with the advantages that optical fibers offer, such as portability, flexibility, low cost, and an excellent light delivery to a remote convenient location, creating a so-called "lab-on-fiber".

In the rest of this chapter, a method to integrate a photonic sensor on the facet of a standard single-mode optical fiber will be described. We will also discuss in detail the performance of the material and biosensor chosen for this approach, as well as the design of the complete photonic circuit, and the experiments performed.

3.2 Silicon-on-Insulator Ring Resonator Biosensors

The working principle of the integrated optical label-free biosensor we studied is illustrated in Fig. 3.1. A ring resonator supports circulating waveguide modes with resonant wavelengths that are related to the effective optical path length:

$$\lambda_{res} = \frac{L \cdot n_{eff}}{m} \tag{3.1}$$

where L is the circumference of the ring, m is an integer and n_{eff} the effective refractive index of the circulating mode. When the refractive index near the ring resonator surface is modified due to the capture of target molecules on the surface, n_{eff} will change which in turn leads to a shift in the resonance wavelength λ_{res}. Taking first order dispersion into account this shift can be calculated as:

$$\Delta\lambda = \frac{\Delta_{env} n_{eff} \cdot \lambda_{res}}{n_g} \tag{3.2}$$

where n_g the group index and $\Delta_{env} n_{eff}$ the effective index shift caused by an environmental change. This wavelength shift is a quantitative measure for the number of binding events near the surface.

We use ring resonators that are fabricated using Silicon-on-insulator (SOI) as a material platform, which offers the advantage of reusing the extensively developed CMOS processing technology. Typical minimum SOI photonic feature sizes are around 100 nm, which fits well within the capacities of today's industrial fabrication tools based on 193 nm optical projection lithography and dry etching.

A SOI wafer consists of a silicon and a buried silicon-oxide layer on a supporting substrate. Silicon is transparent at telecom wavelengths (1.30 and 1.55 μm). SOI waveguides confine light within sub-micron structures because of the high omni-directional index contrast ($n_{Si} = 3.46, n_{SiO_2} = 1.44$) which is achieved by etching fully through the silicon layer. This enables guiding through sharp bends and thus large density sensor integration.

Fabrication of photonic chips with high reliability and with no custom equipment development cost is a major asset for SOI biosensors. It allows for mass production and economics of scale. For point-of-care diagnostics the availability of cheap disposables is a prior requirement. Also in drug discovery where immense numbers of test assays are run, cheap disposables can increase efficiency and keep down the development costs of the drugs.

The high index contrast of SOI thus results in micron- sized ring resonator biosensors, which is important when detecting very small amounts of molecules and beneficial for highly sensitive detection. The resonance wavelength shift does not scale with the size of the cavity, hence the smaller the cavity, the less molecules are needed to cover its entire surface, while the response does not decrease with decreasing surface area.

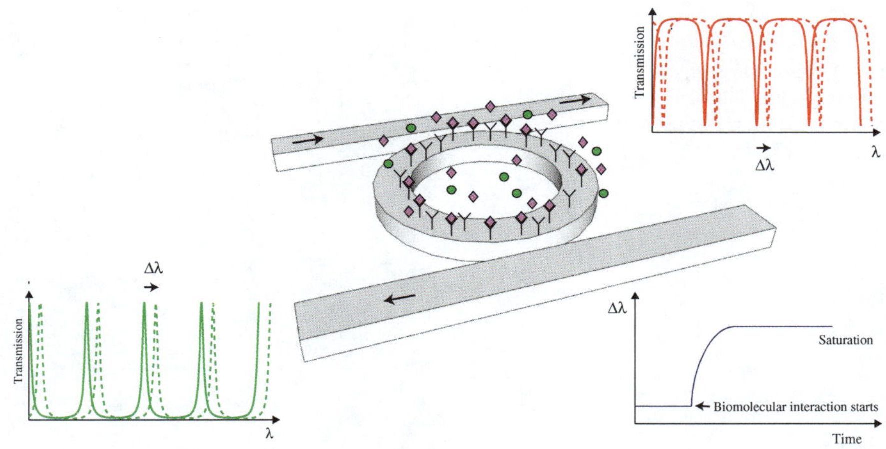

Fig. 3.1 Basic principle of a ring resonator biosensor

Figure 3.2 shows an SEM picture of a fabricated SOI wire with typical dimensions. The silicon layer of our fabricated devices measures 220 nm in height, and mono modal operation is achieved for waveguide widths around 450 nm (depending on the top cladding -often watery solutions in biosensing- and the polarization).

The microring resonator used in this work is a racetrack with 4 μm radius and 4 μm straight section. The racetrack shape is used in order to have more control over the coupling coefficient. An exhaustive study on the optimization of the resonator dimensions was previously performed by De Vos [7].

Selvajara et. al. and Bogaerts et. al. [8–10] give a detailed description on how industrial techniques are adapted for fabrication of high quality SOI photonic integrated circuits. All test devices used in this work are fabricated with the research facilities of imec, Leuven, Belgium.

3.3 Results and Discussion

3.3.1 Design of the Optical Circuit to be Transferred on the Fiber Facet

The structure to be transferred to the fiber tip consists of two parts: the grating coupler and the sensing circuit. We will now discuss each of these components in more detail.

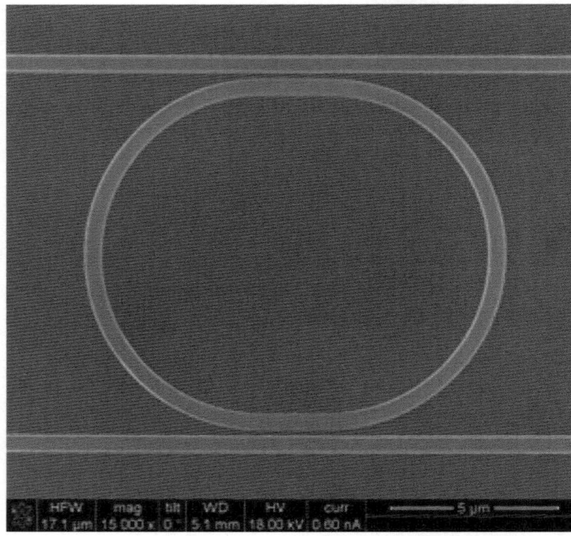

Fig. 3.2 SEM picture of an SOI ring resonator with radius 5 μm and straight section 2 μm. Waveguide width = 450 nm, waveguide height = 220 nm

3.3.1.1 Grating Couplers

Coupling light from an optical fiber with a core diameter of about 10 μm into a typical waveguide dimension (220 × 450 nm) is a non-trivial task. The size mismatch of a few orders of magnitude between the fiber core and a photonic wire, makes of this coupling a very inefficient coupling. Over the years this problem has been investigated thoroughly in literature [11–16] and tackled with different mechanisms. The use of grating couplers is one of the most elegant methods.

Here we briefly overview the working principle of a one-dimensional grating coupler, since it is a substantial part of the optical circuit used in our sensor probe (Fig. 3.3).

Waveguide gratings are structures having a periodic modulation of the refractive index, see Fig. 3.4. According to the Bragg condition, this one-dimensional periodic structure will allow power exchange between particular modes of the structure. For each grating a K-vector along the direction of periodicity can be defined with magnitude:

$$K = \frac{2\pi}{\Lambda} \tag{3.3}$$

in which Λ is the period of the grating. The fundamental waveguide mode with propagation constant β couples to other modes with k_x vectors given by the Bragg condition:

$$k_x = \beta + mK \tag{3.4}$$

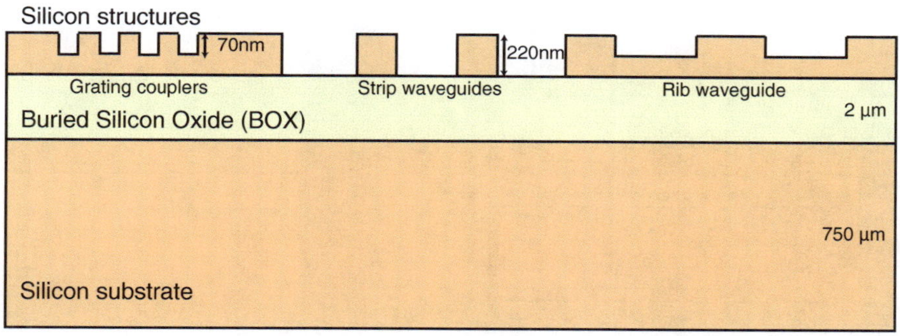

Fig. 3.3 SOI chip. Using Deep UV lithography, a full and shallow etch step are executed on the top silicon layer with respective etch depths of 220 and 70 nm. Most basic photonic structures can be implemented using these two etch steps

with m = …, −2, −1, 0, 1, 2,…. From the vector diagram it is clear that vertical coupling through a grating coupler will go hand in hand with a large second order reflection back into the waveguide. To avoid this, the gratings are designed for slightly off-vertical coupling. In order to optimize the coupling efficiency, the overlap between the field profiles of an optical fiber and the upwards diffracted field profile is maximized. An extensive analysis of the design of grating couplers is given by Taillaert in [11] and by Tamir et al. in [12].

In [13], these gratings show 33 % coupling efficiency at a 1.55 µm wavelength, 40 nm 1 dB bandwidth and 2 µm alignment tolerance for 1 dB excess loss. Typically these gratings measure 10 × 10 µm and are defined in a 10 µm-wide waveguide guaranteeing that most of the light is collected and changed from the vertical to the horizontal direction. However it still has to be converted from a 10 µm-broad to a 450 nm-broad waveguide. This is done by using an adiabatic taper lengthening the structure with at least 150 µm.

This adiabatic transition determines the length of the coupling structure between fiber and chip. Focusing the light would alleviate the need for this long adiabatic transition and would result in a substantial length decrease, and hence a higher degree of integration. Additionally, light could be focused directly on an integrated component, without needing a waveguide.

In [14] Van Laere demonstrates focusing grating coupler structures in SOI. The focusing grating couplers perform equally well to the previously mentioned linear gratings. The fiber-to-fiber loss is 10.5 dB, corresponding with a coupling efficiency of a single grating coupler around 30 %. The most compact structure of this type has a total footprint of 18.5 µm (width of the grating) by 28 µm (grating + focusing section).

We will use these focused grating couplers to couple the light from the fiber to the sensor circuit, since the space available is limited by the dimensions of a mono mode fiber facet.

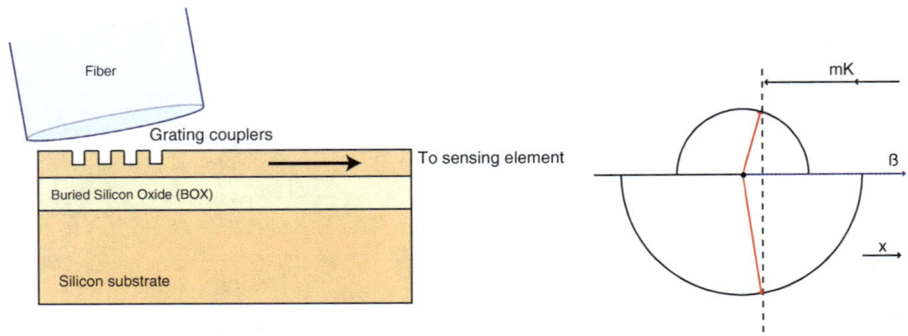

Fig. 3.4 *Left* Light is coupled from the fiber to the photonic circuit by means of the grating couplers from [11]. *Right* Bragg condition for one-dimensional periodic structure

3.3.1.2 Sensing Circuit

Our photonic device is composed of typical 220 nm-high and 450 nm-wide photonic wires with 2.7 dB/cm propagation loss, on top of 2 μm of silicon oxide and 750 μm silicon substrate.

In order to fit on the facet of a standard single-mode optical fiber, the photonic circuit is required to be compact. Therefore it is designed to be 'retro-reflective': using a dedicated light coupling scheme, light will be coupled in and out of the integrated circuit via the same grating coupler and under the same angle, similar to the work presented in [9].

Figure 3.5 shows an schematic of the top view of the optical fiber probe with the photonic integrated circuit aligned to the core of the optical fiber. The cladding of the fiber carries the rest of the integrated circuit comprising a 3 dB multimode interference (MMI) splitter/combiner and a ring resonator biosensor. A focused curved grating coupler focuses the light onto the SOI waveguide, which as already mentioned circumvents the need for space consuming waveguide tapers [14] (Fig. 3.6).

The 3 dB MMI coupler splits the incoupled light into two different waveguides which coincide with both access waveguides of an add-drop ring resonator. Two inverted tapers at the end of this waveguides prevents light not coupled in the ring of being reflected back in the waveguides and the ring.

The microring is a compact racetrack shape with 4 μm bend radius and 4 μm straight sections that are used as directional couplers with a gap of 180 nm between the waveguides. The ring supports modes that resonate at a wavelength λ_{res} (3.1). At this wavelength, both the clockwise and the counterclockwise resonance mode will be excited, causing light to be coupled back in the access waveguides.

The same MMI is used to combine the reflected light, i.e. only wavelengths near the resonance of the ring resonator, towards the grating coupler and the optical fiber. In this way, the circuit is retro-reflective for only the dropped wavelengths.

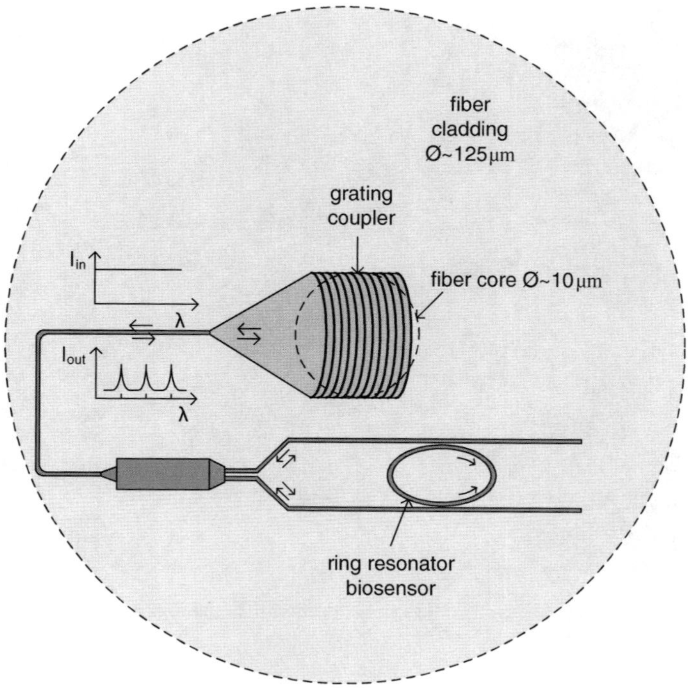

Fig. 3.5 Schematic of the top view of the optical fiber probe with the photonic circuit aligned to the core of the fiber

Light coupled back in the optical fiber will show sharp peaks in the spectrum corresponding to the resonance wavelengths. When biomolecular interaction takes place in the vicinity of the ring sensor, these resonances will shift quantifying the number of binding events on the sensor surface.

3.3.2 Transfer of the Photonic Chip to the Fiber Facet

Our approach consists of detaching the photonic circuit through the back side of the chip, removing every layer under the sensing device. The fabrication process is illustrated in Fig. 3.7 and described here in detail.

First the sample is bonded upside down on a silicon carrier wafer by means of CrystalBond 509 wax from Electron Microscopy Science. This wax releases easily after processing and leaves only little and easy-to-remove residue on the sensor surface.

In a next step the silicon substrate, which is 750 μm thick, must be removed. This is not a straightforward task since many aspects must be taken into account. Three different methods are used for this approach:

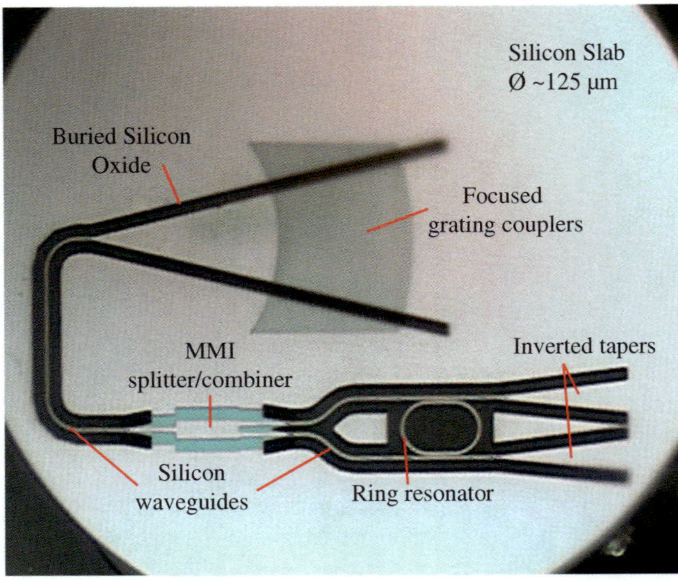

Fig. 3.6 Microscope picture of the fiber probe facet with the photonic circuit aligned to the core of the fiber

First we use mechanical grinding to reduce the substrate to 100 μm. This is done by mounting the wafer on a chuck after which the substrate is pressed down against a rotating pad while a chemical slurry is flown over the pad. The silicon is removed by a combination of abrasive polishing and chemical etching.

The remaining 100 μm of silicon substrate are removed by dry and wet etching. Dry etching of silicon is mainly done using a combination of a fluorine-based gas (like SF_6 or CF_4) and other gases (like O_2 or N_2O). The principle of this etching method is based on generating a plasma of the precursor gasses that comes into contact with silicon. Fluorine ions are responsible for the main mechanism of etching silicon while the oxygen prevents recombination of fluorine ions with fluorosulphur radicals, thus increasing the net amount of fluorine ions and as a consequence, the etch rate [17]. With this method we reduce the thickness of silicon to 50 μm, leaving the rest to be removed by wet etching.

Wet etching of silicon can be done either isotropically (using a mixture of HF, HNO_3, CH_3COOH and water) or anisotropically where the most commonly used etchants are KOH and tetramethylammoniumhydroxide (TMAH).

At first sight it seems logical to opt for KOH and TMAH, which are both anisotropic etchants and show a high selectivity of etching silicon compared to silicon dioxide [18]. The buried oxide layer of our chip acts thus as an etch stop layer guaranteeing the integrity of our sensing device located under this layer. However, this etchants require a temperature of 70–90 °C to achieve a typical etch

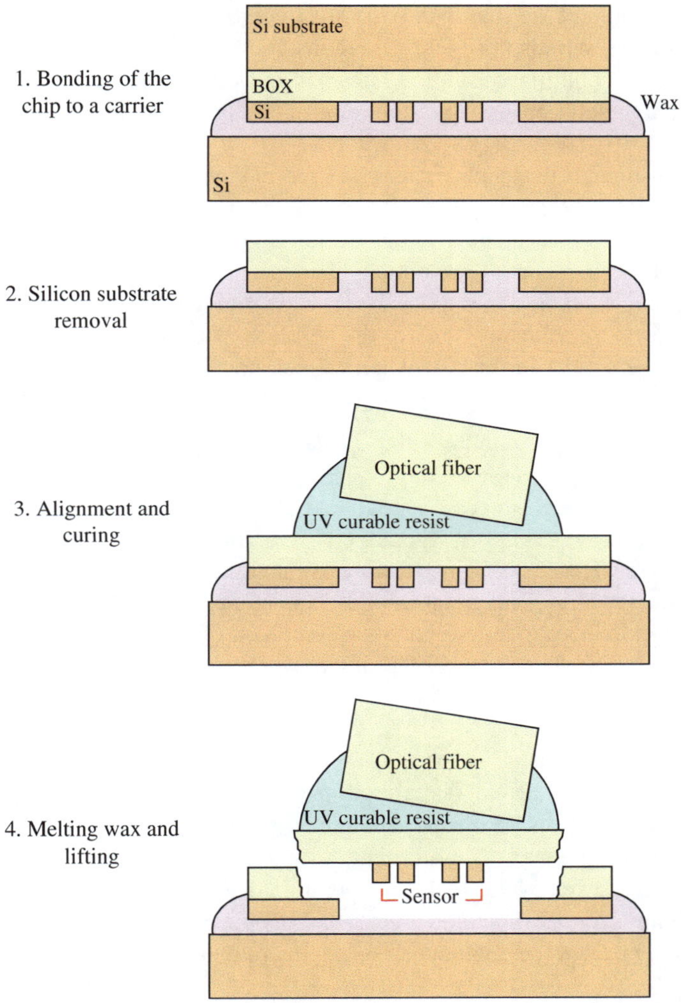

Fig. 3.7 Fabrication steps. (*1*) Upside-down bonding of the SOI chip to a silicon carrier wafer by means of wax. (*2*) Removal of the silicon substrate by subsequent mechanical grinding, dry etching and wet etching. (*3*) Gluing of the photonic circuit to the tip of a fiber after active alignment of the fiber core to the grating coupler. (*4*) The wax that bonds the sensor to the silicon carrier wafer is melted and by lifting the fiber, the thin layer of silicon dioxide around the sensor breaks under the weight of the remaining part of the photonic chip, leaving only the sensor circuit on the tip of the fiber

rate of 0.1–1 μm/min. As a consequence, the wax used to bond the dye to the carrier melts, releasing the 50 μm-thick membrane in the etching solution.

After discarding the previous option, we can only use isotropic etching. This has one main disadvantage: due to the fact that the mixture contains HF, once the substrate has been removed, the etch will not stop at the buried oxide layer. This implies that not only the silicon dioxide layer will be removed but the 220 nm thin top silicon layer will also be etched in an instant. Therefore very precise timing is required and an optimized solution rates had to be investigated.

The etching solution used is a mixture of acetic, nitric and hydrofluoric acid (CH_3COOH:NH_3:HF, 4:5:1) at 20 °C. After a precise removal of the rest of the silicone substrate, a thoroughly rinsing of the dye with DI water is required.

At this point only the silicon circuit and the 2 μm-thick silicon dioxide layer are left on top of the wax, bonded to the carrier wafer.

Once at the optical setup, the optical fiber core is actively aligned to the grating coupler through the silicon dioxide layer. This top cladding layer on the grating coupler will not significantly change the absolute value of the coupling efficiency but will cause a shift in the wavelength spectrum. From the grating period condition $\Lambda = \frac{\lambda_0}{n_{eff} - n_{incident} sin(\theta)}$ where Λ is the period of the grating, λ_0 the wavelength of the incident light in vacuum, n_{eff} the optical mode effective index, θ the incident angle with the normal to the grating surface and $n_{incident}$ the refractive index of the medium of incidence, we can expect that the wavelength spectrum will shift towards shorter wavelengths when increasing the refractive index of the top cladding layer (from 1.0 of air to 1.44 of silicon dioxide).

For the active alignment an ultraviolet (UV) curable resist PAK-01 (Toyo Gosei Co.) with a refractive index very similar to silicon dioxide is applied on the fiber tip. Since the light doesn't travel through an air layer anymore, the index contrast between the incident medium and the silicon dioxide is decreased optimizing then the transmission.

After the alignment is performed, the fiber is glued to the chip by exposing PAK-01 to UV light. Note that the chip and the fiber form an angle of 10°, corresponding to the angle of optimal efficiency and low reflection of the grating.

Subsequently, the wax that holds the sensor to the silicon carrier wafer is melted by heating the carrier wafer to 110 °C with a hot plate. By lifting the fiber carefully, the thin layer of silicon dioxide around the sensor breaks under the weight of the remaining part of the photonic chip, leaving only the sensor circuit on the tip of the fiber.

To facilitate breaking the silicon dioxide layer around the circuit, we found it necessary to selectively thin down that layer from 2 μm to 700 nm by dry etching over a circular area around the circuit. This partial etching is performed prior to the bonding of the sample to the carrier wafer.

Finally, the residues of wax at the sensing surface are removed by immersing the fiber in acetone during 5 min at 75 °C.

3.3.3 Characterization of the Fiber Probe Sensor

To measure the transmission of the sensor integrated on the fiber tip, light from a tunable laser source is coupled into our sensor fiber and this is coupled out and

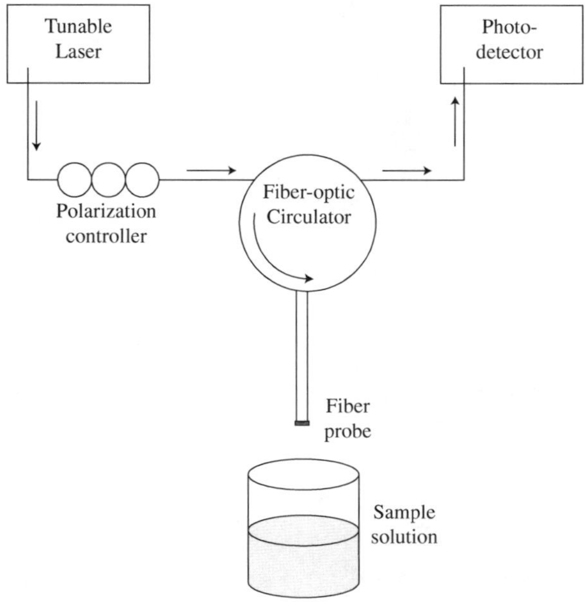

Fig. 3.8 Read-out: The optical fiber probe is connected to a tunable laser source and a photodetector through a fiber-optic circulator. The polarization is optimized using a fiber polarization controller

detected by a photodetector. The bi-directional transmission through the sensor fiber is achieved with a fiber-optic circulator as illustrated in Fig. 3.8. The polarization is optimized for maximal coupling to the quasi-transverse electric (quasi-TE) mode of the SOI circuit using a fiber polarization controller.

When the sensing device is immersed in a sample solution, the change of the refractive index in that solution leads to a shift in the resonance wavelength that can be calculated by (3.2). This shift is monitored by repeatedly scanning the spectrum.

Figure 3.9 displays a measured resonance peak of the resonator integrated on the fiber tip. By fitting a Lorentzian function the quality factor of this resonance was determined to be 2,350, which is the same as before bonding. Note that this is lower than that of optimal SOI ring resonators in water, which could be over 10,000. By optimizing the coupling sections of the resonator and by increasing its bend radius to reduce the bend loss, which can be achieved by making more efficient use of the available space on the fiber tip, we can expect a significant improvement of the quality factor.

As a proof-of-principle to show the capabilities of the fiber probe, we measured the sensitivity for refractive index changes of aqueous solutions and compared it with the one we previously measured for a similar silicon-on-insulator ring resonator sensor that was not integrated on a fiber tip [19]. We used aqueous ethanol solutions with different concentrations, which refractive indices were derived from [20]. No surface chemistry was applied to the sensor surface.

Figure 3.10 illustrates the measured resonance wavelength shift of our sensor as a function of the refractive index change of the solution. A linear fit is depicted

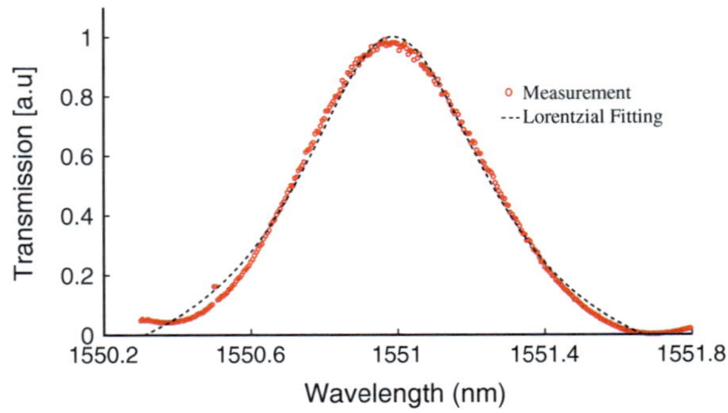

Fig. 3.9 Fiber sensor resonance. In red a measured resonance peak of the resonator integrated on the fiber tip. By fitting a Lorentzian function (*dashed black line*) the quality factor was determined to be 2,350

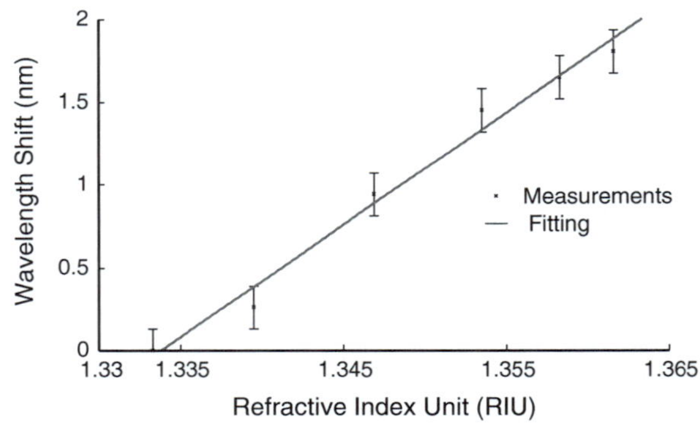

Fig. 3.10 Measured resonance wavelength shift due to the change in refractive index of various aqueous solutions. The mean wavelength shifts over three measurements on the same device are shown with weighted variance error bars. The resonance wavelength shifts linearly with the refractive index of the solution with a sensitivity of 70 nm/RIU

with mean wavelength shift values and weighted variance error bars determined over 3 measurements on the same device.

The sensitivity is determined to be 70 nm/RIU, which corresponds with the value we previously measured in [19]. This shows that the entire fabrication process has no detrimental effect on the performance of the device.

In order to compensate temperature variations in a real environment, a second fiber probe sensor is required, which must be shielded from refractive index changes by a cladding, providing then a reference for the measurement.

After this work, a similar approach has been performed by Shambat et al. in [21] where a semiconductor photonic crystal membrane is transferred to a fiber facet using a different fabrication process, and maintaining as well the performance of the sensor.

3.4 Conclusions

A lot of progress has been made to implement optical label-free biosensors on photonic chips since this allows for a significant simplification of the optical system. The good performance, potential cheap integration and miniaturization that materials like Silicon-on-Insulator offers has led to the ‘lab-on-chip’ concept where the integration of all the different laboratory functions can be implemented in a single chip.

However, many applications still require a sensor probe that can perform measurements at locations that are hard to reach. We have presented the implementation of a fiber probe that combines the good performance of silicon-on-insulator sensors with the high mobility of optical fibers. In this implementation an integrated SOI circuit is transferred to the facet of a single mode optical fiber. The fabrication of the device has been described in detail and its characterization by bulk sensing experiments has shown no degradation when comparing to the same sensor on SOI chip, i.e. not integrated on the fiber tip.

Future work involves chemical modification of the sensor surface for specific sensing which can lead to a robust and portable device for in-vivo label-free biosensing applications such as endoscopy.

References

1. J.P. Golden, C.R. Taitt, L.C. Shriver-Lake, Y.S. Shubin, F.S. Ligler, A portable automated multianalyte biosensor. Talanta **65**(5), 1078–1085 (2005)
2. A. Leung, P.M. Shankar, R. Mutharasan, A review of fiber-optic biosensors. Sens. Actuators B **125**(2), 688–703 (2007)
3. Elveflow, Microfluidic innovation center http://www.elveflow.com/microfluidic-reviews-and-tutorials/microfluidics-and-microfluidic-devices-a-review
4. A.D. Kersey, M.A. Davis, H.J. Patrick, M. LeBlanc, K.P. Koo, C.G. Askins, M.A. Putnam, E.J. Friebele, Fiber grating sensors. J. Lightwave Technol. **15**(8), 1442–1463 (1997)
5. X.D. Fan, I.M. White, S.I. Shopoua, H.Y. Zhu, J.D. Suter, Y.Z. Sun, Sensitive optical biosensors for unlabeled targets: a review. Anal. Chim. Acta **620**(1–2), 826 (2008)
6. A. Cusano, A. Cutola, A. Jaques, *Fiber Bragg Grating Sensors: Research Advancements, Industrial Applications and Market Explotation* (Bentham Science Publishers Ltd, 2011)
7. K. De Vos, Label-free silicon photonics biosensor platform with microring resonators, Ph.D. Dissertation, (Ghent University, 2010)
8. S.K. Selvaraja, P. Jaenen, W. Bogaerts, D. Van Thourhout, P. Dumon, R. Baets, Fabrication of photonic wire and crystal circuits in silicon-on-insulator using 193-nm optical lithography. J. Lightwave Technol. **27**(18), 4076–4083 (2009)
9. W. Bogaerts, R. Baets, P. Dumon, V. Wiaux, S. Beckx, D. Taillaert, B. Luyssaert, J. Van Campenhout, P. Bienstman, D. Van Thourhout, Nanophotonic waveguides in Silicon-on-insulator fabricated with CMOS technology. J. Lightwave Technol. **23**(1), 401–412 (2005)
10. S.K. Selvaraja, W. Bogaerts, P. Dumon, D. Van Thourhout, R.G. Baets, Subnanometer linewidth uniformity in silicon nanophotonic waveguide devices using CMOS fabrication technology. J. Sel. Top. Quantum Electron. **16**(1), 316–324 (2010)

11. D. Taillaert, Grating couplers as Interface between Optical Fibres and Nanophotonic Waveguides, Ph.D. Dissertation, (Ghent University, 2004)
12. T. Tamir, S.T. Peng, Analysis and design of grating couplers. Appl. Phys. **14**(3), 235–254 (1977)
13. D. Taillaert, F. Van Laere, M. Ayre, W. Bogaerts, D. Van Thourhout, P. Bienstman, R. Baets, Grating couplers for coupling between optical fibers and nanophotonic waveguides. J. Appl. Phys. **45**(1), 6071–6077 (2006)
14. F. Van Laere, T. Claes, J. Schrauwen, S. Scheerlinck, W. Bogaerts, D. Taillaert, L. O'Faolain, D. Van Thourhout, R. Baets, Compact focusing grating couplers for silicon-on-insulator integrated circuits. IEEE Photonics Technol. Lett. **19**(23), 1919–1921 (2007)
15. F. Van Laere, W. Bogaerts, P. Dumon, G. Roelkens, D. Van Thourhout, R. Baets, Focusing polarization diversity grating couplers in silicon-on- insulator. J. Lightwave Technol. **27**(5–8), 612–618 (2009)
16. D. Vermeulen, S. Selvaraja, P. Verheyen, G. Lepage, W. Bogaerts, P. Absil, D. Van Thourhout, G. Roelkens, High-efficiency fiber-to-chip grating couplers realized using an advanced CMOS-compatible silicon-on-Insulator platform. Opt. Express **18**(17), 18278–18283 (2010)
17. Y. Tzeng, H. Lin, Dry etching of silicon materials in SF6-based plasmas. J. Electrochem. Soc. **134**(9), 2304–2309 (1987)
18. H. Seidel, L. Csepregi, A. Heuberger, H. Baumgrtel, Anisotropic etching of crystalline silicon in alkaline solutions—orientation dependence and behavior of passivation layers. J. Electrochem. Soc. **137**(11), 3612–3626 (1990)
19. K. De Vos, I. Bartolozzi, E. Schacht, P. Bienstman, R. Baets, Silicon-on- Insulator microring resonator for sensitive and label-free biosensing. Opt. Express **15**(12), 7610–7615 (2007)
20. R. Belda, J.V. Herraez, O. Diez, A study of the refractive index and surface tension synergy of the binary water/ethanol: influence of concentration. Phys. Chem. Liq. **43**(1), 91–101 (2005)
21. G. Shambat, J. Provine, K. Rivoire, T. Sarmiento, J. Harris, J. Vuckovic, Optical fiber tips functionalized with semiconductor photonic crystal cavities. Appl. Phys. Lett. **99**(19), 191102 (2011)

Chapter 4
Monolithic Silicon Photonic Crystal Fiber Tip Sensors

Bryan Park and Olav Solgaard

Abstract Photonic Crystals (PC) enable sensitive and robust sensors for a large number of measurands, including temperature, refractive index, displacement, pressure, acceleration, and rotation. The small volumes of two-dimensional PC sensors also make them ideal for integration onto the facet of optical fibers. In this chapter, we describe the operation, design, and fabrication of Si PC sensors for refractive index and temperature, and the process technology used to integrate the sensors onto the facets of standard single-mode fibers (SMFs). The PC sensors are fabricated on standard Si wafers, using a single photolithography mask and a combination of isotropic and anisotropic etching. Once the sensor fabrication is completed, the miniaturized sensors (50 μm × 50 μm × 0.5 μm) are assembled onto SMF facets. The results are highly-sensitive, robust sensors with submicron sensing layers that are well suited for deployment in harsh environments, including temperatures up to 700 °C.

4.1 Introduction

Fiber optic sensors have been intensively developed and deployed in a wide variety of applications due to their unique advantages: small size, high sensitivity, insusceptibility to electromagnetic interference, chemical and mechanical robustness, wavelength multiplexing, and distributed sensing capabilities. They have been used to measure various physical quantities such as refractive index, temperature, strain,

B. Park · O. Solgaard (✉)
E. L. Ginzton Laboratory, Department of Electrical Engineering,
Stanford University, Stanford, CA 94305, USA
e-mail: solgaard@stanford.edu

B. Park
e-mail: insun@stanford.edu

A. Cusano et al. (eds.), *Lab-on-Fiber Technology*,
Springer Series in Surface Sciences 56, DOI: 10.1007/978-3-319-06998-2_4,

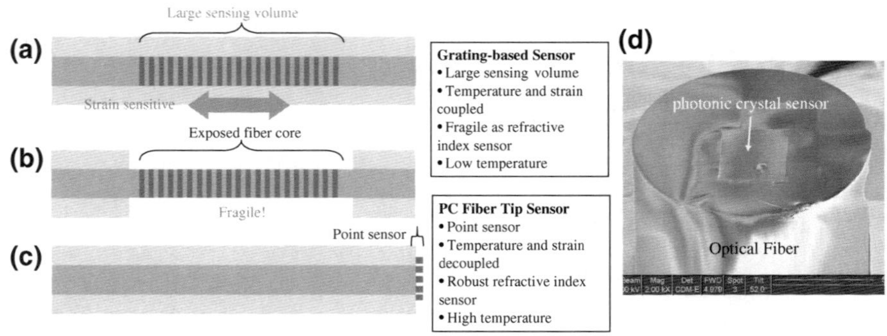

Fig. 4.1 **a** Grating-based fiber sensors require large sensing volumes and are subject to undesired strain sensitivity. **b** core-exposed FBG refractive index sensors for higher sensitivity become fragile. **c** compact and robust PC fiber tip sensors are insensitive to external strain. **d** SEM image of a PC fiber tip sensor. A silicon PC sensor covers the core of the fiber and interacts with the surrounding environment

pressure, and acceleration with different detection methods [1, 2]. They are especially promising for use under harsh conditions, e.g. high temperatures, corrosive, noisy, and high-voltage environments, because their structural materials, typically glasses, can withstand high temperatures, above 500 °C, while conventional electronics fail to operate at such temperatures due to the intrinsic properties of semiconductors and metal contacts/interconnects. Most commonly used fiber sensors are grating-based fiber sensors (Fig. 4.1a), which include fiber Bragg grating (FBG) and long-period fiber gratings (LPFG). In FBG sensors, the silica fiber core is photo-inscribed with a periodic refractive index contrast along the light propagation direction through ultraviolet exposure [3, 4], and the grating structure functions as spectral filter whose characteristics alter according to changes in the surrounding environment. LPFG sensors have long-period gratings ($\sim$100 μm) and couple the fundamental mode into forward-propagating cladding modes that interact with materials outside the cladding and therefore detect changes in ambient conditions [5]. Although grating-based sensors have many advantages compared to other sensors, they also suffer from significant drawbacks. Many periods are needed for grating functionality, and, accordingly, these sensors have large sensing volumes and are vulnerable to unwanted coupling to external strains. In order for FBGs to have better refractive index sensitivity, their fiber cores are directly exposed to the environment by removing their claddings, but this process compromises their structural strength and durability (Fig. 4.1b) [6]. They also have low sensitivity to temperature ($\sim$0.01 nm/°C), and require isolators to prevent back reflections into the laser, which poses significant difficulties in practical system implementation. The LPFG have higher temperature sensitivity ($\sim$0.1 nm/°C) and low back reflections, but their sensing lengths are even longer ($\sim$a few centimeters) due to their long periodicities [7]. In addition, both types have a fundamental problem in high temperature applications; at temperatures above 200 °C, UV-written gratings are erased [8]. FBG sensors with direct etching of gratings in their fiber cores have been demonstrated [9], but this approach complicates the fabrication process and inevitably reduces the mechanical strength of the sensors.

In this chapter, we introduce a hybrid structure where a very compact two-dimensional silicon photonic crystal (PC) is attached on a tip of a single mode optical fiber (Fig. 4.1c, d) in order to overcome the problems of fiber-only sensors (i.e. limited sensitivity, large sensing length, fragile structure) [10, 11]. The small PC membrane (50 μm × 50 μm × 0.5 μm) is completely confined to the fiber facet and it has high sensitivity to changes in refractive index and temperature. The resulting device is a small and robust reflection-based sensor that interacts with the measurand in a small volume, approaching a point sensor at the tip of the fiber. Other hybrid fiber sensors based on plasmonic-resonances of gold nanoparticles also have been demonstrated [12], but the use of the low-temperature metal limits its range of applications because the metal may melt or degrade in corrosive, high temperature, high optical power environments. In contrast, our sensor is made of chemically-robust high temperature dielectric materials, silicon and glass, and is therefore well-suited for operation in harsh environments. The PC-on-fiber configuration is flexible and can be employed in a variety of applications, including inertial sensors and hydrophones [13].

In Sect. 4.2, we introduce the working principle of PC sensors based on the guided resonance phenomenon. Then we describe the fabrication of the PC and the assembly of the PC on the fiber tip in the Sect. 4.3. Section 4.4 includes the experimental characterization of the PC fiber tip sensor, showing high sensitivity to refractive index and temperature. High temperature measurement up to 900 °C is also presented in this section. Lastly, we conclude the chapter in Sect. 4.5.

4.2 Photonic Crystals Fundamentals

PCs are optical nanostructures with subwavelength periodic modulation of the dielectric constants. They provide means to control and manipulate photons in similar ways to how semiconductors control electrons. PCs have photonic band structures determined by their constituent materials and crystal organization, and the presence or absence of modes at a range of wave vectors and wavelengths lead to many interesting and useful phenomena, such as photonic bandgaps, superprism effects, negative refraction, and guided resonances [14–17], allowing them to be used in a variety of applications including waveguides, high-Q cavities, sensors, and mirrors [18–21].

PCs can be fabricated in one, two, or three-dimensional structures. The most well-known one-dimensional (1-D) PCs are Bragg reflectors, in which periodic layers of alternating high and low refractive index are cascaded to form high-quality mirrors (Fig. 4.2a). The Bragg reflectors are particularly useful at optical frequencies where metal mirrors suffer from significant absorption loss, finding their applications in vertical-external-cavity surface-emitting-lasers (VECSELs)

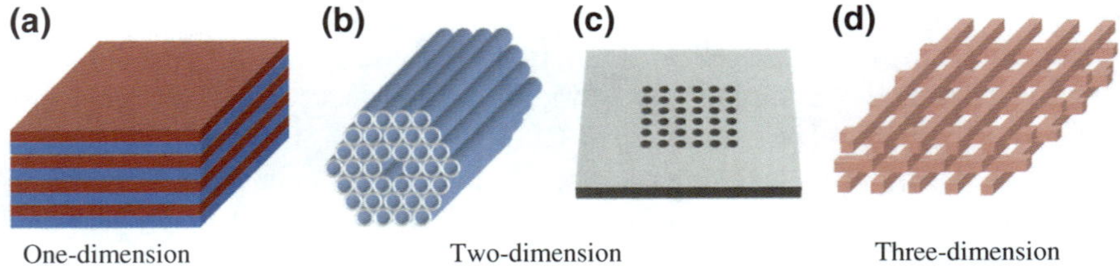

Fig. 4.2 Photonic crystals. **a** One-dimensional Bragg reflector. **b** Two-dimensional PC holey fiber. **c** Two-dimensional PC slab. **d** Three-dimensional woodpile structure

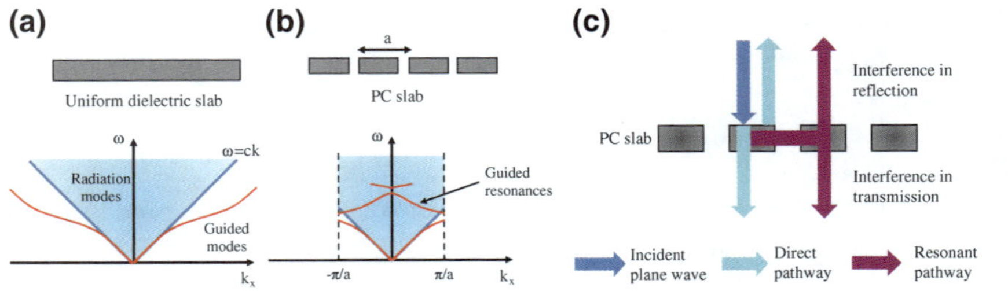

Fig. 4.3 Physical origin of guided resonances in 2-D PC slabs: **a** Band structure for a uniform dielectric slab. **b** Band structure for a 1-D PC slab. In the PC slab, the originally guided modes can couple to radiation modes due to phase matching provided by the periodic holes, resulting in guided resonances with finite lifetimes [28]. **c** Transmission and reflection of the 2-D PC are determined by interference between the direct pathway and resonant pathways arising from coupling to guided resonances

and optical MEMS systems. Two-dimensional (2-D) PCs come in two different versions; fibers and patterned thin films as shown in Fig. 4.2b and c. PC fibers are patterned so that photons at target wavelengths are confined to the core by an in-plane bandgap. For example, holey fibers support one or more guided modes at their center void because surrounding hexagonally periodic structures provide a photonic bandgap (Fig. 4.2b) [22]. The defining example of a thin-film PC is a 2-D PC slab where periodic patterns, typically square or hexagonal lattices, are etched into a thin film of high refractive index surrounded by low refractive index materials (Fig. 4.2c). Such 2-D PC slabs have been employed in many photonic devices because their optical properties, e.g. transmission and reflection, can be designed to support a wide range of applications from highly sensitive sensors to broadband reflectors. Furthermore, their fabrication is relatively straightforward, and their compact size and planar geometry facilitate integration with ICs and MEMS [23]. Three-dimensional (3-D) PCs (e.g. the woodpile structure in Fig. 4.2d) with a complete bandgap offer the ultimate control of photons [24, 25], but their fabrication has proven challenging up to this point although novel methods, such as direct laser writing and self-assembly of nanospheres, have been devised [26, 27].

In this chapter, we focus on a 2-D PC slab with a square lattice of cylindrical air holes in a thin silicon membrane. Our 2-D PC slab-integrated fiber sensors rely on guided resonances of the PCs [17]. In order to understand the physics of the guided resonances, we consider a 1-D PC slab with periodic air holes only in one lateral direction for simplicity. In a uniform dielectric slab without holes (Fig. 4.3a), the band diagram is divided into two regions by the light line, which is defined as $\omega = ck$ where ω is the angular frequency, c is the speed of light in vacuum, and k is the wave vector. A continuum of radiation modes, which can propagate in the surrounding medium, exists above the light line. Modes below the light line are "guided modes" whose electromagnetic field decays exponentially away from the dielectric slab. Therefore, their electromagnetic energies are completely confined to the slab and external radiation cannot couple to these guided modes. However, in the PC slab, scattering from the periodic hole arrays causes the dispersion relation of the guided modes to fold at the first Brillouin zone boundaries, and accordingly, some of the guided modes lie above the light line (Fig. 4.3b). This allows them to have radiative characteristics with finite lifetime. Theses modes, called guided resonances, are still confined within the slab, but can be excited from or couple to plane waves outside the PC slab.

The presence of the guided resonances brings about significant changes in the transmission and reflection of the PC slabs compared to the uniform dielectric slab. Assume a plane wave is normally incident on the PC. As in the uniform dielectric slab, some of the light are reflected back from the PC surfaces or directly propagates through the membrane. This forms the direct pathway through the PC. On the other hand, the incident light can excite guided resonances, and then the coupled light is radiated back to outside the PC with a characteristic lifetime. These resonant pathways interfere with the direct pathway as illustrated in Fig. 4.3c and determine the PC reflection and transmission spectra [29–32]. For example, if the interference is destructive on the transmission side and constructive on the reflection side over a wavelength band of interest, the PC slab can achieve high reflectivity near unity. Such spectral characteristics of the PC are closely related to the physical parameters of the PC: the lattice constant, the slab thickness, and the hole radius. The lattice constant decides the target operation range of the PC. If the material constants do not vary with wavelength, then the physical dimensions scale linearly in wavelength, but material dispersion is significant, so linear scaling is only approximately correct. For instance, lattice constants of 400 nm and 820 nm have been reported for PCs designed for visible and 1550 nm wavelengths respectively [21, 30]. The slab thickness and the hole diameter mainly determine the properties of the guided resonances. In general, larger holes and thinner slabs increase scattering and thereby decrease the lifetimes of the guided resonances. This effect results in broader resonances in the PC spectrum, which is desired in broadband, high-reflectivity mirrors. In contrast, smaller holes and thicker slabs have the opposite effect, leading to longer lifetimes, which yields sharper resonance features. The resonant peaks or dips shift in response to changes in surrounding environment and these shifts are particularly straightforward to detect if the spectra contain sharp features, so long life time guided resonances are well suited to sensing applications.

Due to their compactness and mechanical robustness, 2-D PC mirrors have been incorporated into different platforms for beam steering [21, 23], polarization control [33], optical filtering [34], and sensing. PC sensors can be divided into two groups; Fabry-Perot-based sensors and spectral-shift sensors. In Fabry-Perot based PC sensors one of the mirrors of a Fabry-Perot resonator is a 2-D PC slab reflector. In these devices, broad band response of the PCs is preferable to make the sensors operable over wide range of laser wavelengths. Such sensors have been developed for measurements of acoustic waves [13, 35] and quasi-static pressure [36]. PC Fabry-Perot sensors can also be extended to incorporate near-field coupling, i.e. photon tunneling, effects [37] for added functionality.

Single-layer 2-D PC sensors based on spectral shifts of the PC reflection or transmission spectra also have received a lot of attention due to their potential as compact optical devices that measure a large variety of physical quantities, e.g. refractive index, and bio-molecule associations [38–40]. Their sensitivity to the surrounding media stems from the fact that the guided resonances in the PCs have evanescently decaying fields outside the PC slabs, and these evanescent fields interact with the media and these interactions determine the sensors, characteristics. The shifts or changes of the resonant features in response to changes in the media surrounding the PCs are monitored and analyzed for quantitative description on the measurand. We will explain the fabrication of the PC fiber tip sensors and their performance as sensors for refractive index and temperature in the Sect. 4.3.

4.3 Photonic Crystal Fabrication and Fiber Sensor Assembly

We fabricate the PC fiber tips sensors in two stages: fabrication of 2-D PCs on standard wafers in a silicon foundry and transfer of the PCs onto the facets of single-mode optical fibers.

4.3.1 Photonic Crystal Fabrication

We fabricate 2-D PC membranes on silicon wafers using the Generation of Photonic Elements by RIE (GOPHER) process, which employs a combination of anisotropic and isotropic etch steps together with thermal oxidation [41].

The process starts with thermal oxidation of silicon wafers to grow an oxide hard mask. Next we use a 5x reduction ASM-L i-line stepper to photolithographically inscribe the periodic arrays of cylindrical holes in photoresists spun on the wafers (Fig. 4.4a). A square lattice of circular holes is used as the PC pattern.

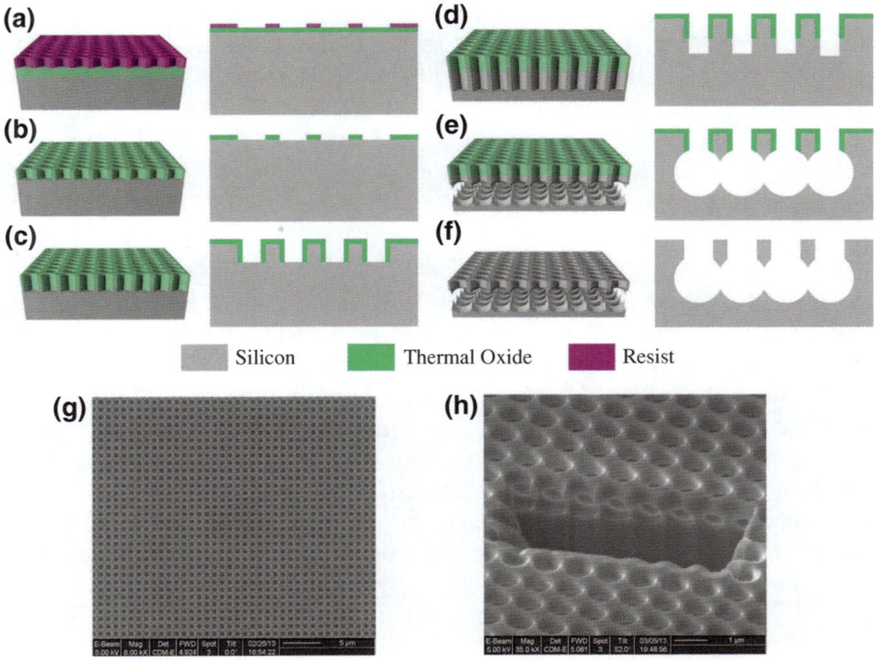

Fig. 4.4 2-D PC fabrication process flow: **a** Thermal oxidation and photolithography of the PC pattern. **b** Oxide layer etch by RIE to transfer the PC pattern from the resist into the oxide hard mask. **c** Si RIE, followed by conformal oxidation, and removal of oxide from the bottoms of the PC holes by RIE. **d** Hole extension into Si by RIE. **e** Isotropic Si etch for PC membrane release. **f** Oxide removal by HF. **g** *Top SEM view* of the fabricated PC. **h** *Crosssectional SEM* image of the PC with a section removed by a focused ion beam to reveal the structure of the PC

The optical stepper enables large-area patterns (e.g. 500 μm by 500 μm) with submicron-features to be patterned on entire wafers very quickly and also enables tuning of the PC optical characteristics by varying the hole size with different exposure doses given a single photomask [21]. Next, we transfer the PC pattern to the oxide hard mask (Fig. 4.4b) and subsequently to the silicon substrate by Reactive Ion Etching (RIE). The etched depth in the silicon layer determines the PC membrane thickness. Then, the exposed Si on the sidewalls and bottoms of the holes are covered with thin, conformal thermal silicon oxide. We perform a series of RIEs to remove the bottom oxide inside the holes without removing the sidewall oxide (Fig. 4.4c) and extend the holes into the silicon substrate (Fig. 4.4d). The side wall protection together with the hole extension beyond the sidewall protection determines the PC thickness after the upcoming isotropic silicon etching. The PC layer is then partially or completely released in a timed isotropic undercut etch. For fiber sensor applications, in which the PC is transferred onto an optical fiber facet, the undercut etch is performed until the PC membrane is completely released from the substrate (Fig. 4.4e). Finally, the oxide layer is removed by vapor HF to minimize stress during release and avoid membrane stiction problems, which can arise from wet oxide etch (Fig. 4.4f). SEM pictures of a GOPHER-PC (periodicity = 1 μm, hole diameter = 700 nm, thickness = 650 nm) are shown as examples in Fig. 4.4g and h.

The GOPHER process has unique advantages compared to other PC fabrication technologies based on silicon-on-insulator wafers or deposited thin films [30, 31]. The fabrication is based on standard Si processing technology and performed on standard single crystalline silicon wafers, yielding a monolithic structure with excellent PC material quality and mechanical properties that far exceeds those of PCs made in deposited films (e.g. LPCVD silicon) or the device layer of the silicon-on-insulator wafers. The absence of material stress mismatch in the GOPHER PCs is particularly important, because it allows for fabrication of large-area devices that can be transferred to any flat substrate. The GOPHER process is also compatible with standard MEMS and CMOS processing, so photonic elements such as waveguides and filters can be integrated with ICs and other photonic devices [42]. Furthermore, the process facilitates fabrication of multi-layered structures with single photomask because the top layer can be used as a self-aligned mask for bottom layers [43]. Multi-layered structures have interesting and unique optical properties posing high potential in various applications including displacement sensing [37], temperature sensing [44], and tunable filters [34].

4.3.2 Fiber Tip Sensor Assembly

After the silicon PCs are fabricated on silicon wafers, we transfer and affix them to the facets of single mode optical fibers using different fixation techniques: point welding with a Focused Ion Beam (FIB) [10], template-assisted gluing with epoxy [44], and transfer-printing [23]. In the FIB method we utilize a FIB tool equipped with an Omniprobe micro-manipulator. First, we partially cut out the PC membrane with the ion-beam in a size large enough to completely cover the fiber core (Fig. 4.5a). Then we bring the Omniprobe needle tip in contact with the PC and weld the tip to the PC by selectively depositing a Platinum (Pt) patch onto the tip (Fig. 4.5b). The welding is done on the corner of the PC in order to prevent the Pt lump from affecting the PC optical property. Next, we lift the probe needle up to pull the PC apart by breaking the remaining connecting sections (Fig. 4.5c). The membrane is then transferred to a cleaved fiber facet and aligned to the fiber core such that light from the core interacts only with the silicon PC (Fig. 4.5d). Pt is deposited at the corners of the PC to bond it onto the fiber tip (Fig. 4.5e), and the probe tip is retracted after being disconnected from the PC by ion-milling. The assembled sensor is shown in Fig. 4.5f. The sensor has the transferred PC membrane size of ~30 μm × 30 μm that covers the mode field size of the single mode input from the fiber (Corning SMF-28e). Larger membranes may be used as long as they fit within the 125 μm-diameter fiber facet.

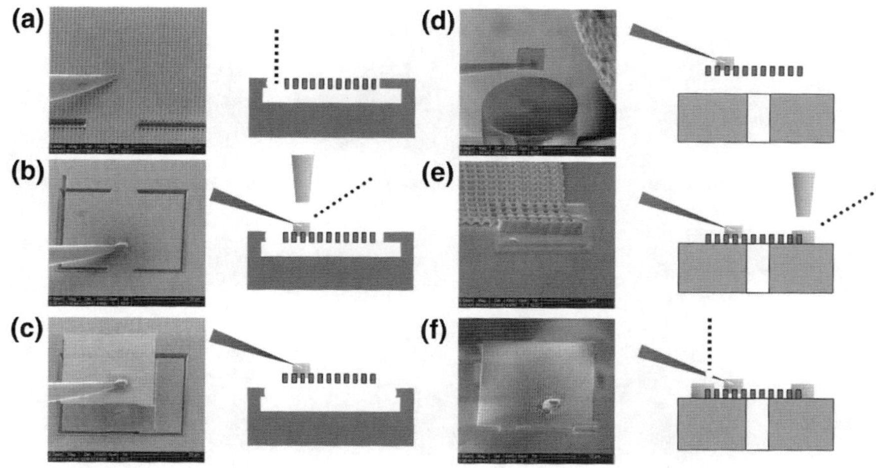

Fig. 4.5 Fiber sensor assembly using the FIB method: **a** a PC membrane is partially cut by ion-milling and the Omniprobe tip is brought in touch with the PC. **b** The probe tip is welded to the PC with Pt deposition. **c** The attached PC is pulled up and released from the substrate. **d** The PC is transferred onto to a fiber tip. **e** The PC is affixed to the fiber facet with Pt welding at its corners. **f** The probe tip is cut off by ion-milling. The *dotted line* represents the ion-beam

Even though the FIB method enables precise assembly of the PC membrane on the fiber tip, it is a serial process, which may limit the fast production of a large number of the sensors. The template-assisted epoxy bonding method is devised to accelerate the sensor fabrication with batch-production capability. We fabricate a silicon wafer template with a penetrating cylindrical hole whose diameter smaller than the fiber diameter (Fig. 4.6a). We apply an epoxy on the template (Fig. 4.6b) and apply suction through the hole, leaving the epoxy only on the boundary of the hole (Fig. 4.6c). The fiber facet is aligned to the hole and brought in contact with the template (Fig. 4.6d) so that the epoxy pattern is transferred onto the facet while keeping the fiber core clean from the epoxy (Fig. 4.6e). Next, the fiber is pressed on the PC membrane (Fig. 4.6f) and lifted up with the PC fixed to the facet after the epoxy hardens. The perforation in the PC membrane facilitates the separation and transfer process. This assembly method allows for economical batch production of the sensors with easy extension to harsh environment applications by replacing the epoxy with ceramic adhesives or glass glues [45]. However, this method may require more sophisticated modeling and analysis in the sensor characterization because a gap between the fiber tip and the PC due to the epoxy thickness and a tilt angle of the PC membrane with respect to the facet may influence the sensor responses. The transfer-printing technique can solve these issues by directly affixing two flat surfaces of separate devices [46]. A PC diaphragm fabricated on a silicon wafer can be picked up using a polymer microstamp, and then transferred and printed on an optical platform, e.g. a fiber facet. A high temperature annealing process is performed to form strong covalent

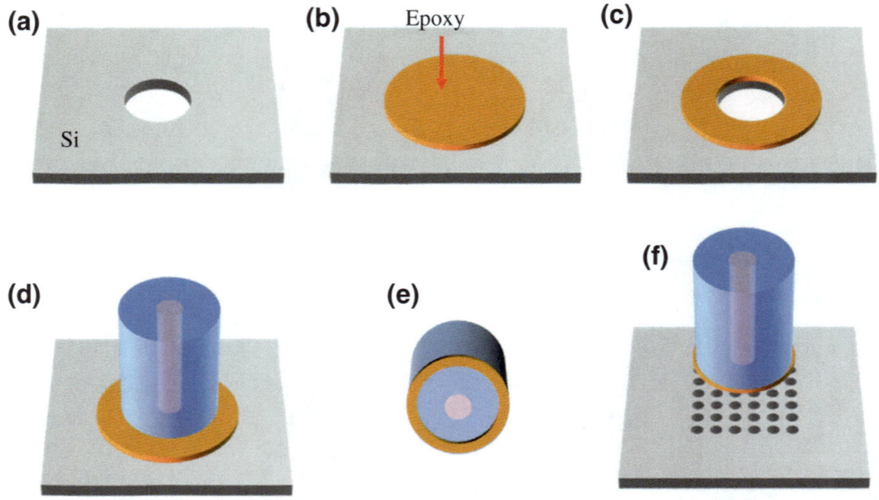

Fig. 4.6 Fiber sensor assembly using the template-assisted epoxy bonding: **a** a through-wafer hole is etched on a silicon wafer using Deep RIE. The hole diameter is smaller than the fiber diameter. **b** A small amount of an epoxy is applied around the hole. **c** Suction through the hole removes the epoxy on the hole and form a *ring pattern* on the hole edge. **d** A cleaved fiber is pressed on the template with fiber core aligned to the hole center. **e** The lifted fiber has the epoxy only on its facet boundary, which allows the guided light in the core to interact purely with the PC. **f** The fiber is pressed on the PC membrane and pulled away with the PC on it after the epoxy hardens

bonds at the interface, which allows for more robust sensor structure than the point-welding and the epoxy-gluing. However, the strict requirement of the flat surfaces on both devices may limit or complicate the design, geometry, and fabrication of the devices.

4.4 Fiber Tip Sensor Characterization

2-D PCs are very sensitive transducers whose reflection/transmission spectra are modulated in response to changes in their surrounding environment, and therefore our PC-mounted fiber sensors enable compact sensing on robust and reliable optical platforms. The incident optical field guided along the fiber is reflected back from the fiber-PC sensor, which is exposed to different measurands, and the modulation of the reflected signal, typically in a form of shifts in the PC reflection spectrum, is used to analyze changes in the internal states and surrounding medium of the PC sensor. In this section, we demonstrate refractive index and temperature sensing with our sensor, showing high sensitivity to both parameters. We also present high temperature performance that illustrates how its robust dielectric material composition makes it promising for applications in harsh environments.

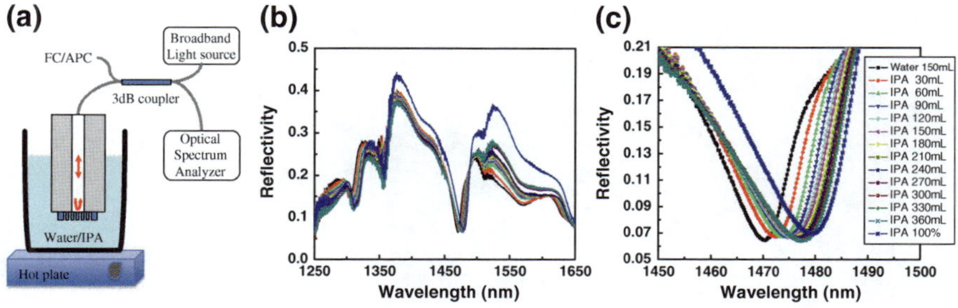

Fig. 4.7 **a** Experimental setup. **b** Measured reflection spectrum of a fiber sensor exposed to different DI water/IPA solution concentrations. **c** Zoomed-in view of the reflection minimum at about around 1475 nm wavelength. The reflection minimum is *red*-shifted as the IPA concentration increases. The graphs show that different parts of the spectra experiences different amounts of shift in response to changes in concentration

4.4.1 Sensitivity to Refractive Index

In order to examine the sensor response to refractive index variations, we immerse the sensor in a de-ionized(DI) water/isopropanol(IPA) solution with different concentrations at room temperature (Fig. 4.7a). The distal end of the fiber, onto which the sensor is integrated, is spliced to a 3 dB-couplers. The input to the 3-dB coupler is provided by a broadband light source (B&W TEK BWC-SLD for 1,250 $\sim$ 1,650 nm). One output of the 3 dB coupler is connected to the sensor fiber, while the other output port is connected to an angled fiber connector to eliminate back reflections from that port. The back reflected optical field from the sensor is detected and spectrally resolved in an optical spectrum analyzer (Yokogawa AQ6370).

The reflection spectrum is measured at each concentration of the DI water/IPA solution. Figure 4.7b and c shows that the reflection spectra have a sharp and prominent minimum at about 1475 nm wavelength, and that the minimum shifts to higher wavelengths as the IPA concentration increases. The center wavelength shift of the reflectance minimum is plotted as a function of the effective refractive index of the solution in Fig. 4.8a. The measured sensitivity is $\sim$213 nm/RIU (Refractive Index Unit), which is about 3 times larger than that of core-exposed FBG sensors ($\sim$71 nm/RIU) [6]. The measured spectra show that different features of the spectra experience different amounts of shift in response to changes in concentration. This is because the guided resonances of the PC sensor, which determines the spectra, have different dependence on the external refractive index variations. Therefore, each guided resonance excited at its resonant wavelength results in the dissimilar spectral shift across the wavelength range. The chosen spectral minimum represent close to the maximum sensitivity to refractive index of this specific sensor. In combination with a spectroscopic measurement system that has picometer resolution, the observed sensitivity leads to a refractive index resolution of $\Delta n \sim 5 \times 10^{-6}$.

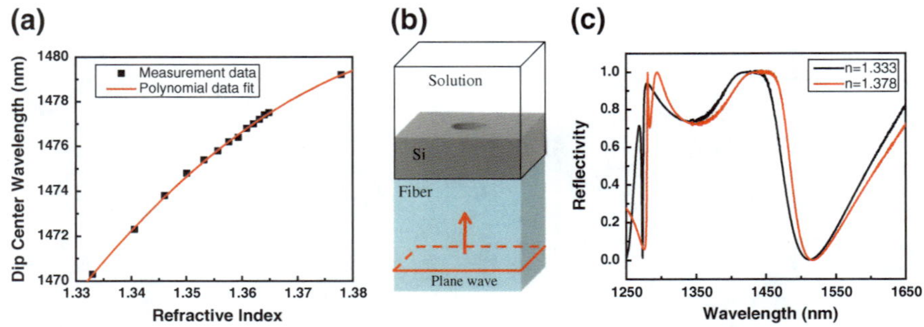

Fig. 4.8 **a** Plot of *center* wavelength shift of the reflection minimum around 1475 nm as a function of the effective refractive index of the solution being measured. **b** Unit cell model used in FDTD simulations of the sensor. The unit cell consists of a uniformly thick silicon slab with single cylindrical hole at its center and a glass substrate. **c** FDTD simulation results showing that the spectrum *red*-shifts as the refractive index increases. As in the measurements, the shift is not uniform over the wavelength range of the simulations

To understand and quantitatively describe the response of the sensor, we compare the experimental result to Finite Difference Time Domain (FDTD) simulations. The simplified simulation model consists of a top silicon layer with a cylindrical hole at its center and a bottom glass layer as shown in Fig. 4.8b. The hole and the region above the silicon layer are filled with a material corresponding to the effective index of the solution to be measured. The glass represents the fiber core, and a normally incident plane wave is used to model the incident optical field on the optical fiber. The sidewall boundaries are treated as periodic while the top and bottom of the simulation cell are terminated with perfectly matched layers. Although the model is a simplified version of the sensor, it predicts a shift of ~222 nm for a refractive index change from 1.333 to 1.378 (Fig. 4.8c), closely matching the observed results.

4.4.2 Sensitivity to Temperature

For temperature measurements, the experimental setup in Fig. 4.7a is slightly modified. We immerse the same fiber sensor in DI water and control the water temperature using a hot plate. The reflection spectra (Fig. 4.9a) are measured while the water is cooled down from 80 °C. They exhibit the identical sharp reflection dip around 1472 nm wavelength shifting to lower wavelengths as the temperature decreases (Fig. 4.9b). The spectral shift of the minimum as a function of the temperature (Fig. 4.9c) shows that the temperature sensitivity of the sensor is ~0.063 nm/°C, which is significantly better than that of FBG sensors [9].

We also compare the temperature results to FDTD simulations. In these simulations, we only consider the thermo-optic effect of the silicon photonic crystal and water, i.e. the temperature dependent refractive index change [47, 48], and we

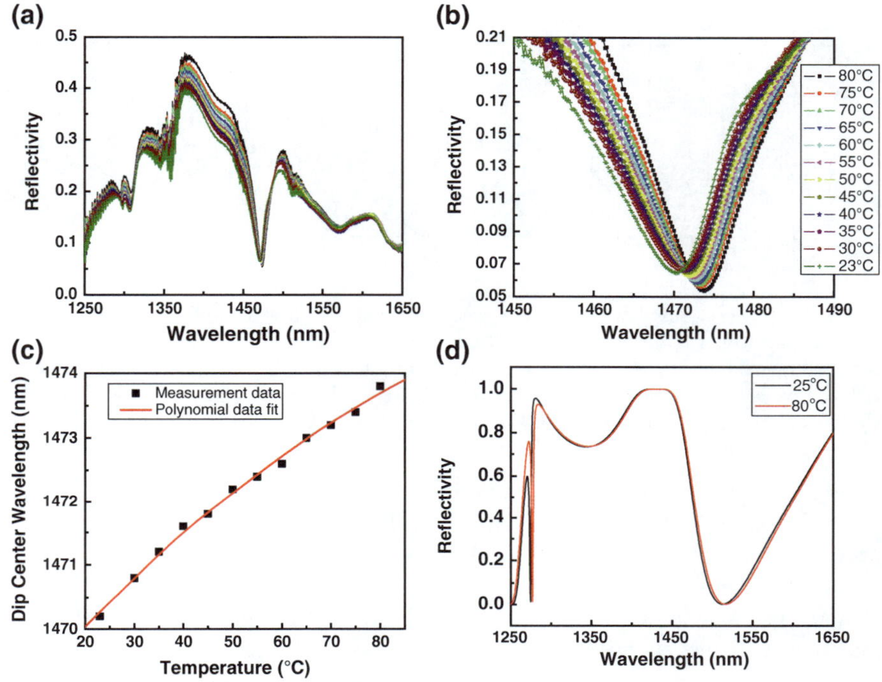

Fig. 4.9 **a** Measured reflection spectra of the fiber sensor (the same sensor as in Fig. 4.7) at different temperatures. **b** Zoomed view of the reflection minimum at about 1472 nm wavelength. **c** Plot of the reflection minimum center wavelength as a function of temperature. **d** FDTD simulation result showing a *red*-shift of 3.6 nm for temperature increase from 25 to 80 °C

ignore dimensional changes. Using this simplified model, we predict a shift of ~3.6 nm from 80 °C to room temperature (Fig. 4.9d), which agrees well with the measured shift of ~3.4 nm. Our model reveals that the refractive indices of the PC and a surrounding measurand are the dominant factors determining the thermal response of the sensor, while other effects such as thermal expansion of the PC and refractive index changes of the fiber are less significant. The fiber is composed of glass whose thermo-optic coefficient is an-order of magnitude smaller than that of silicon [49], which explains its relative insignificance.

4.4.3 *Simultaneous Detection of Refractive Index and Temperature*

The high sensitivity of our sensor to both refractive index and temperature of the environment means that its spectral response is determined by the combined effects of the two. Therefore, it is very important to unambiguously distinguish between them when we measure the sensor output. A sensor with single characteristic

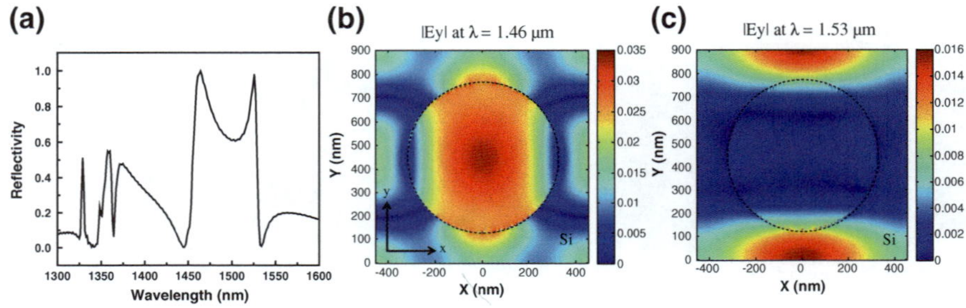

Fig. 4.10 **a** FDTD simulated reflection spectrum of a 2-D PC (periodicity = 900 nm, hole diameter = 640 nm, and thickness = 550 nm) on a fiber using the model in Fig. 4.8b. The surrounding medium is IPA at 25 °C. **b** Electric field distribution($|E_y|$) in the PC symmetry plane at λ = 1.39 μm. **c** Electric field distribution($|E_y|$) in the PC symmetry plane at λ = 1.55 μm. The electric field is calculated with the FDTD simulation. The hole boundary is shown as *dotted line*

resonance does not support this capability because both quantities affect its output. However, 2-D PCs have multiple guided resonances at different wavelengths, and each resonance has a different field configuration inside and outside the PC. Accordingly they are expected to respond differently to internal and external changes in PC; external refractive index variation in the vicinity of the PC body interacts with the outside field whereas the inside field is sensitive to internal refractive index variations, which are mainly due to temperature change.

To clarify this effect, a 2-D PC (periodicity = 900 nm, hole diameter = 640 nm, and thickness = 550 nm) on a fiber immersed in an IPA is simulated using the FDTD simulation model presented in Fig. 4.8b. The simulated reflection spectrum is shown in Fig. 4.10a. For a resonant peak at λ = 1.46 μm the PC has a strong electric field inside the hole and weak field inside the silicon (Fig. 4.10b), whereas for a resonant dip at λ = 1.53 μm the field is very weak inside the hole and strong in the silicon (Fig. 4.10c). As a result, the mode at λ = 1.53 μm is selectively sensitive to temperature changes which affect the refractive index of silicon PC body. On the other hand, the mode at λ = 1.46 μm is sensitive to changes in the refractive index of a surrounding measurand. By examining the independent behaviors of such two modes with selective sensitivity to either temperature or refractive index, we can simultaneously determine two parameters of a medium to be measured. An experimental demonstration of this capability is described in [10]. In comparison, the previous fiber sensor shows maximum refractive index and temperature sensitivity at the same reflection dip wavelength because the corresponding mode is sensitive to both parameters. Therefore, this mode is not suitable for discrimination of two quantities.

4.4.4 High Temperature Measurement

The combination of high sensitivity to refractive index and temperature with its compact construction using robust, high temperature dielectric materials (silicon

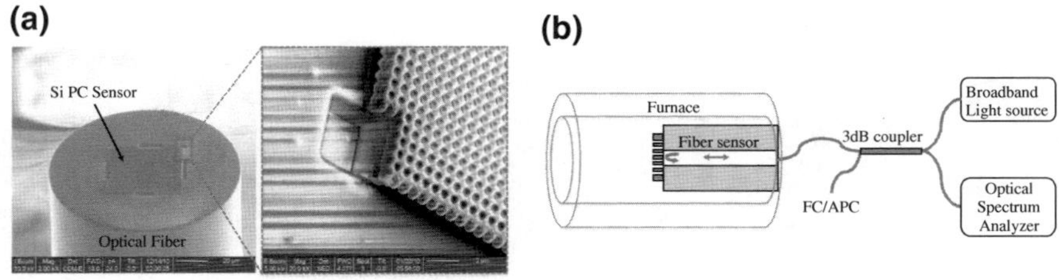

Fig. 4.11 **a** Fiber sensor for high temperature measurement and *zoomed view* of Si-welding bonding at the PC-fiber interface. **b** Experimental setup for high temperature measurement

for the PCs and silicon oxide for the fibers), gives our sensors the potential for operation in harsh environments. In this section, we explore its thermal response and stability in high temperatures up to 900 °C. For this purpose, we integrate our PC sensors onto optical fiber using Si-welding bonding as shown in Fig. 4.11a. This method avoids FIB deposited Pt, which contains organic material from its metal-organic precursor and may decompose at high temperatures. Instead, we mill the corners of the PC membrane with an ion beam, which leads to sputtering and re-deposition of Si and oxide at the PC-fiber interface, yielding strong bonds, which further enhance the sensor's stability at high temperature by avoiding thermal stress mismatch with Pt.

4.4.4.1 Experimental Setup

The high temperature experimental setup is similar to the previous setup except that the sensor is positioned inside a quartz furnace as shown in Fig. 4.11b. We cycle the temperature over 100–700 °C five times and measure the sensor reflection spectra at every 100 °C change. Each cycle consists of heating and cooling over the temperature range. These repeated measurements are designed to check the stability of our sensors at high temperature. We also increase the temperature above 700 °C and perform heating and cooling cycles in the 700–900 °C temperature range with spectral measurements at 50 °C intervals. This higher temperature range is expected to exceed the sensor's stable operation regime. The anomalous behavior of our sensors in this temperature range will be analyzed in detail in the Sect. 4.4.4.2. The upper limit of 900 °C is chosen because the reliability of optical fibers greatly decreases beyond this temperature due to thermal diffusion of the fiber core dopants and glass devitrification [2].

4.4.4.2 Measurement Analysis and Simulation

For measurements in the 100–700 °C range, we observe that a narrow reflection dip shifts from 1500 to 1580 nm wavelength with the same trend as in previous temperature measurements; the spectra red-shifts when the temperature increases

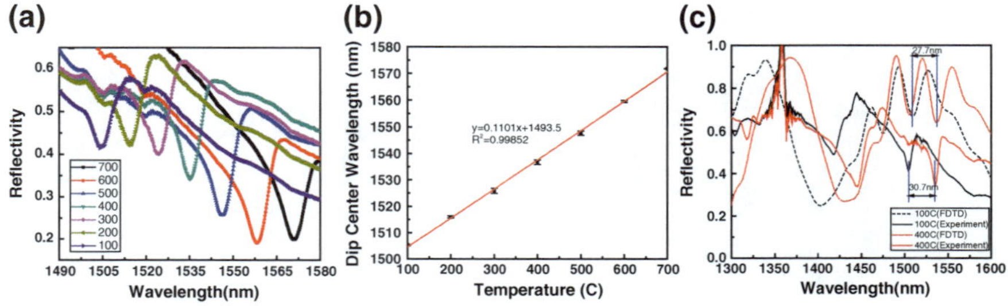

Fig. 4.12 **a** Reflection spectra measured during the fifth cooling cycle (700 to 100 °C). **b** Dip center wavelength with respect to temperature during 100–700 °C temperature cycles. The slope or sensitivity of the sensor is 0.11 nm/°C. **c** Comparison of FDTD simulation and experimental data: a simulated 27.7 nm maximum shift agrees well with an experimental 30.7 nm shift for a temperature change from 100 to 400 °C

while it blue shifts when the temperature decreases. Figure 4.12a clearly shows the blue-shift during the fifth cooling cycle. The center wavelengths of the reflection minimum measured during five cycles are plotted as a function of temperature in Fig. 4.12b. They coincide with each other with a linear sensitivity slope of 0.11 nm/°C. This demonstrates not only stable and repeatable sensor operation in this temperature range, but also the very high temperature sensitivity of the sensor compared to that of high-temperature FBG sensors ($\sim$0.01 nm/°C) [9].

We also run FDTD simulations to clarify the physics of the sensors at high temperature. We extend the simulation model in Fig. 4.8b by incorporating the characteristics of the optical fields on single-mode fibers into the model. In practical single mode fibers, the guided mode profile can be approximated as a Gaussian shape propagating within the core. Therefore, we build a large simulation cell of finite silicon PC hole arrays sitting on a glass cube with a cylindrical core of slightly higher refractive index at the center. Perfectly matched layers are applied to all boundaries, and the input is a Gaussian beam with the mode size matching the mode field diameter of a standard single mode fiber (SMF) at 1550 nm wavelength. The 2-D PC has the following dimensions: periodicity = 850 nm, hole diameter = 700 nm, and thickness = 520 nm. GOPHER-generated 120 nm-high scallops are added to the bottom of the PC.

Again we are only considering temperature-induced changes to the materials of the sensor due to the thermo-optic effect of silicon and glass [47, 49], while we ignore dimensional changes. The simulation results for 100 and 400 °C are presented together with experimental data in Fig. 4.12c. Although the simulated curves do not match exactly, the simulated shift of 27.7 nm closely predicts the experimental shift of 30.7 nm. We expect that the discrepancy is due to the simplifications we make in modeling of the PC-fiber system (e.g. spherical GOPHER undercut, fixed PC dimension, and wavelength-independent input modal size). The relatively close match of experimental and modeling results indicates that the thermo-optic effect of the sensor materials is the main factor determining the sensor's thermal response. Since silicon ($\sim 10^{-4}$ K^{-1}) has an order of magnitude higher thermo-optic coefficient than glass ($\sim 10^{-5}$ K^{-1}) [47, 49], the silicon PC has the largest impact on the temperature sensitivity. The observation that our sensor has 10x higher sensitivity than FBGs can be explained from this material point of view; FBG relies on the thermo-optic effect of glass, the core material of the optical fiber. If higher sensitivity is required for an application, other semiconductor materials with larger thermo-optic coefficient, such as GaAs ($\sim 1.25 \times 10^{-4}$ K^{-1}) and Ge ($\sim 3 \times 10^{-4}$ K^{-1}) can be used for PC sensor fabrication [50, 51]. However, their lower melting points may limit the maximum operation temperature of the sensor.

We test the sensor's operation limit by raising the temperature above 700 °C and cycling heating and cooling over 700–900 °C three times. The dip center wavelengths at every 50 °C interval are plotted as a function of temperature in Fig. 4.13a. Two distinct phenomena, which are different from the previous 700–900 °C results, are observed. First, as the temperature approaches 900 °C in one heating cycle, the dip shift gradually deviates from the previous linear trend towards a decreased slope and consequently lower sensitivity. Second, a permanent spectral drift is observed after each cycle and the amount of the drift per cycle decreases as the cycles repeat.

We first discuss how the sensitivity is reduced above 700 °C. The refractive index of silicon is reported to increase linearly up to 750 °C [52], and this thermo-optic effect of silicon plays a significant role in the sensor's spectral behavior as described previously. Therefore, the lower sensitivity indicates that the effective thermo-optic coefficient of the silicon PC decreases as the temperature is heated to 900 °C. At such high temperatures, silicon thermal oxidation starts and the in situ grown oxide film exerts stress on the silicon PC surface, which compensates for the increase in its refractive index [53]. The net result is smaller thermo-optic effect in the PC and hence smaller dip shifts. We compare the measured dip shift to a simulated value to confirm the effect of oxidation-induced stress on the sensor. The sensor stays above 700 °C for more than 40 min per cycle in our experiment, and we measure the thermal oxide film thickness on a PC sample under one temperature cycle to be 18.2 nm using an ellipsometer (J.A. Woollam M2000 Spectroscopic Ellipsometer) as shown in Fig. 4.13b. The corresponding average oxide film stress at 846 °C is about 300 MPa [54], and we analytically calculate the silicon surface stress to be $\sim$1GPa under the assumption of the ideal bending stress distribution over the cross-section of a composite beam. This leads to stress-induced silicon refractive index decrease of 0.01 [53]. When this effect is considered in FDTD simulations, a 7.34 nm blue-shift is predicted (Fig. 4.13c). This value closely matches the measured offset of 4.5–10 nm compared to the values extrapolated from the 100–700 °C data. In addition to the decreased thermo-optic effect, the PC dimensions slightly changes as thermal oxidation consumes silicon; the PC becomes thinner and its hole diameter becomes larger. Also, the grown oxide layer conformally covers the PC surface. When all these effects (dimension change, oxide coating, stress-induced refractive index change) are taken into account in the FDTD simulations, a 30.6-nm dip shift is predicted as shown in Fig. 4.13d. This is also consistent with the total measured drift after three temperature cycles. The good agreement between the oxidation model and the

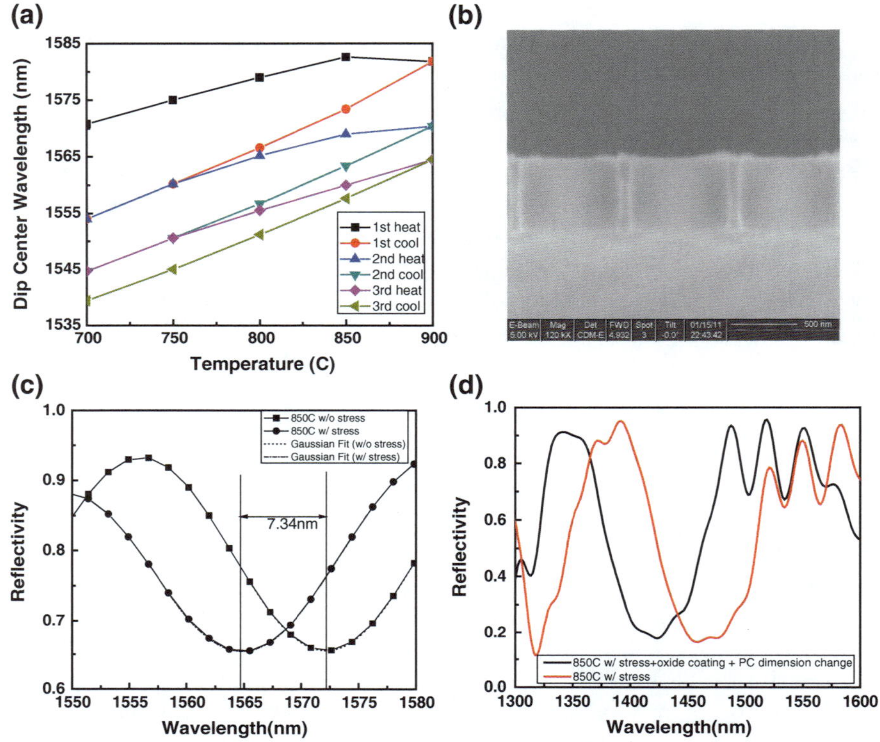

Fig. 4.13 **a** Maximum spectral wavelength shift as a function of temperature during three cycles over 700 ~ 900 °C. The slope of the shift decreases at the *upper end* of the heating cycle and a permanent shift occurs after each cycle **b** SEM image of the PC membrane crosssection where a 18.2 nm-thick oxide layer is grown during one temperature cycle (700 ~ 900 °C) **c** FDTD simulation with and without the thermal oxidation-induced refractive index decrease at 850 °C. The oxidation model predicts the dip shift of 7.34 nm, matching the experimental result above 700 °C. **d** FDTD simulation with and without growth of conformal oxide and the corresponding PC dimensional change. The shift of 30.6 nm closely matches the measurement after three cycles

experimental results is a strong indication that in situ oxide stress is a big factor for the sensitivity reduction above 700 °C.

The second observed anomalous behavior is the permanent drift of the resonance dip after each cycle. During the cooling process, the dip center wavelength shift is linear with a slightly higher sensitivity of 0.12 ~ 0.13 nm/°C. This is likely caused by the fact that the refractive index modulation due to existing surface stress constructively adds to the original thermo-optic effect. The difference in the slope during heating and cooling stages results in the observed drift. Moreover, plastic deformation of the silicon structure may also be a factor because the silicon yield strength decreases at high temperature and is lower than the calculated stress at 700 °C or above [55]. The thermal oxidation and the plastic deformation are irreversible, and therefore lead to a permanent spectral shift

However, the amount of shift decreases after each cycle because of the slow-down of the oxidation process with thicker oxide coverage and relaxation of the oxide film and the silicon.

In summary, our PC fiber tip sensor can reliably operate up to 700 °C with a high sensitivity of 0.11 nm/°C, but thermal oxidation of the silicon PC brings about lower sensitivity and permanent shifts beyond 700 °C. The stable temperature range of the sensor can be extended above 700 °C by applying a conformal oxidation barrier film on the silicon PC or using more robust materials, e.g. silicon carbide (SiC) for PCs [56].

4.5 Conclusion

Photonic Crystal fiber facet sensors are highly sensitive and well suited to operation in challenging environments. The monolithic silicon PCs are fabricated using the GOPHER process, which is based on standard Si processing and standard Si wafers. The batch fabricated PCs are integrated onto fiber facets using one of several micro-assembly techniques optimized for different applications; FIB welding yields robust interfaces, while template-assisted adhesive bonding or transfer-printing are more cost effective. The fiber-facet design enables compact fiber sensors that are mechanically, thermally, and chemically robust in spite of the fact that they are comprised of different materials (glass and Si). The ability to combine different materials without compromising robustness enables the design of high sensitive refractive index and temperature sensors. These sensors are especially promising for harsh environment, such as oil wells, gas turbines, automotive engines, geothermal instruments, and aircraft propulsion systems, where other sensors fail due to corrosive chemicals and high temperature/pressure. Moreover, 2-D PC slab on fiber facet sensor technology facilitates optimization and extension of the sensor performance through several types of modifications of the PCs: the use of different materials (e.g. higher thermo-optic coefficient, higher melting point), surface functionalization or passivation, and multi-layered PC structures.

References

1. E. Udd, Overview of fiber optic sensors. Rev. Sci. Instrum. **66**, 4015–4030 (1995)
2. A. Wang, Y. Zhu, G. Pickrell, Optical fiber high-temperature sensors. Opt. Photon. News **20**, 26–31 (2009)
3. A.D. Kersey et al., Fiber grating sensors. JLT **15**(8), 1442–1463 (1997)
4. G. Meltz, W.W. Morey, W.H. Glenn, Formation of Bragg gratings in optical fibers by transverse holographic method. Opt. Lett. **14**, 823–825 (1989)
5. V. Bhatia, A.M. Vengsarkar, Optical fiber long-period grating sensors. Opt. Letters **21**(9), 692–694 (1996)
6. W. Liang, Y. Huang, Y. Xu, R.K. Lee, A. Yariv, Highly sensitive fiber Bragg grating refractive index sensors. Appl. Phys. Lett. **86**, 151122-1–151122-3 (2005)

7. S. James, R. Tatam, Optical fibre long-period grating sensors: characteristics and application. Meas. Sci. Technol. **14**, R49 (2003)
8. T. Ergodan, V. Mizrahi, P.J. Lemaire, D. Monroe, Decay of ultraviolet-induced fiber Bragg gratings. J. Appl. Phys. **76**, 73–80 (1994)
9. T.L. Lowder et al., High-temperature sensing using surface relief fiber Bragg gratings. IEEE Photonics Technol. Lett. **17**(9), 1926–1928 (2005)
10. I. Jung, B. Park, J. Provine, R. Howe, O. Solgaard, Highly sensitive monolithic silicon photonic crystal fiber tip sensor for simultaneous measurement of refractive index and temperature. J. Lightwave Technol. **29**, 1367–1374 (2011)
11. B. Park, J. Provine, I. Jung, R. Howe, O. Solgaard, Photonic crystal fiber tip sensor for high-temperature measurement. IEEE Sens. J. **11**(11), 2643–2648 (2011)
12. S.-K. Eah, H.M. Jaeger, N.F. Scherer, G.P. Wiederrecht, X.-M. Lin, Plasmon scattering from a single gold nanoparticle collected through an optical fiber. Appl. Phys. Lett. **86**, 031902-1–031902-3 (2005)
13. O.C. Akkaya, M. Digonnet, G.S. Kino, O. Solgaard, Modeling and demonstration of thermally stable high-sensitivity reproducible acoustic sensors. J. Microelectromech. Syst. **21**(6), 1347–1356 (2012)
14. E. Yablonovitch, Photonic band-gap structures. J. Opt. Soc. Am. B **10**(2), 283–295 (1993)
15. H. Kosaka, T. Kawashima, A. Tomita, M. Notomi, T. Tamamura, T. Sato, S. Kawakami, Superprism phenomena in photonic crystals. Phys. Rev. B **58**(16), R10096–R10099 (1998)
16. C. Luo, S.G. Johnson, J.D. Joannopoulos, J.B. Pendry, All-angle negative refraction without negative effective index. Phys. Rev. B **65**, 201104(R) (2002)
17. S. Fan, J.D. Joannopoulos, Analysis of guided resonances in photonic crystal slabs. Phys. Rev. B **65**, 235112 (2002)
18. S. Lin, E. Chow, V. Hietala, P.R. Villeneuve, J.D. Joannopoulos, Experimental demonstration of guiding and bending of electromagnetic waves in a photonic crystal. Science **282**, 274–276 (1998)
19. B. Ellis, M.A. Mayer, G. Shambat, T. Sarmiento, J. Harris, E. Haller, J. Vučković, Ultralow-threshold electrically pumped quantum-dot photonic-crystal nanocavity laser. Nat. Photonics **5**, 297–300 (2011)
20. M. Lee, P.M. Fauchet, Two-dimensional silicon photonic crystal based biosensing platform for protein detection. Opt. Express **15**(8), 4530–4535 (2007)
21. I.W. Jung, S.B. Mallick, O. Solgaard, A large-area high-reflectivity broadband monolithic single-crystal silicon photonic crystal mirror MEMS scanner with low dependence on incident angle and polarization. J. Sel. Top. Quan. Elect. **15**, 1447–1454 (2009)
22. J.C. Knight, T.A. Birks, P.St.J. Russell, D.M. Atkin, Pure silica single mode fiber with hexagonal photonic crystal cladding. Presented at the conference on optical fiber communication (OFC), San Jose, CA, Mar 1996, Postdeadline paper PD3
23. J. Jeong, B. Park, H. Keum, S. Kim, J. Rogers, O. Solgaard, Two-axis MEMS scanner with transfer-printed high-reflectivity, broadband monolithic silicon photonic crystal mirrors. Opt. Express **21**, 13800–13809 (2013)
24. T.G. Euser, A.J. Molenaar, J.G. Fleming, B. Gralak, A. Polman, W.L. Vos, All-optical octave-broad ultrafast switching of Si woodpile photonic band gap crystals. Phys. Rev. B **77**, 115214-1-6 (2008)
25. K.M. Ho, C.T. Chan, C.M. Soukoulis, Existence of a photonic gap in periodic dielectric structures. Phys. Rev. Lett. **65**, 3152 (1990)
26. M. Deubel, M. Wegener, S. Linden, G. von Freymann, S. John, 3D-2D-3D photonic crystal heterostructures fabricated by direct laser writing. Opt. Lett. **31**, 805–807 (2006)
27. R.C. Schroden, M. Al-Daous, C.F. Blanford, A. Stein, Optical properties of inverse opal photonic crystals. Chem. Mater. **14**(8), 3305–3315 (2002)
28. W. Suh, Photonic crystal slabs: theory and applications. Ph.D. dissertation, Stanford Univ., Stanford, CA (2006)

29. W. Suh, M. F. Yanik, O. Solgaard, S.-H. Fan, Displacement-sensitive photonic crystal structures based on guided resonance in photonic crystal slabs. Appl. Phys. Lett. **82**(13), 1999–2001 (2003)
30. K.B. Crozier, V. Lousse, O. Kilic, S. Kim,W. Suh, S. Fan, O. Solgaard, Air-bridged photonic crystal slabs at visible and near-infrared wave lengths. Phys. Rev. B (Condens. Matter Mater. Phys.) **73**(11), 115126-1-14 (2006)
31. V. Lousse, W. Suh, O. Kilic, S. Kim, O. Solgaard, S. Fan, Angular and polarization properties of a photonic crystal slab mirror. Opt. Express **12**, 1575–1582 (2004)
32. O. Kilic, S. Kim, W. Suh, Y.-A. Peter, A.S. Sudbø, M.F. Yanik, S. Fan, O. Solgaard, Photonic crystal slabs demonstrating strong broadband suppression of transmission in the presence of disorders. Opt. Lett. **29**, 2782–2784 (2004)
33. O. Kilic, S. Fan, O. Solgaard, Analysis of guided-resonance based polarization beam splitting in photonic crystal slabs. J. Opt. Soc. Am. A **25**(11), 2680–2692 (2008)
34. W. Suh, S. Fan, Mechanically switchable photonic crystal filter with either all pass transmission or flat-top reflection characteristics. Opt. Lett. **28**, 1763–1765 (2003)
35. O. Kilic, M. Digonnet, G. Kino, O. Solgaard, External fibre Fabry-Perot acoustic sensor based on a photonic-crystal mirror. IOP Publishing Meas. Sci. Technol. **18**, 3049–3054 (2007)
36. X. Wu, O. Solgaard, Short-cavity multimode fiber-tip Fabry-Perot sensors. Opt. Express **21**, 14487–14499 (2013)
37. W. Suh, O. Solgaard, S. Fan, Displacement sensing using evanescent tunneling between guided resonances in photonic crystal slabs. J. Appl. Phys. **98**, article 033102 (2005)
38. J.O. Grepstad, P. Kaspar, O. Solgaard, I.-R. Johansen, A.S. Sudbø, Photonic-crystal membranes for optical detection of single nano-particles, designed for biosensor application. Opt. Express **20**, 7954–7965 (2012)
39. M. El Beheiry, V. Liu, S. Fan, O. Levi, Sensitivity enhancement in photonic crystal slab biosensors. Opt. Express **18**, 22702–22714 (2010)
40. D. Threm, Y. Nazirizadeh, M. Gerken, Photonic crystal biosensors towards on-chip integration. J. Biophotonics **5**(8–9), 601–616 (2012)
41. S. Hadzialic, S. Kim, A.S. Sudbo, O. Solgaard, Two-dimensional photonic crystals fabricated in monolithic single-crystal silicon. PTL **22**(2), 67–69 (2010)
42. C.-M. Chang, O. Solgaard, Monolithic silicon waveguides in standard silicon. IEEE Micro **33**(1), 32–40 (2013)
43. S.B. Mallick, I.-W. Jung, A.M. Meisner, J. Provine, R.T. Howe, O. Solgaard, Multilayered monolithic silicon photonic crystals. IEEE Photonics Technol. Lett. **23**(11), 730–732 (2011)
44. B. Park, I. Jung, J Provine, R.T. Howe, O. Solgaard, Double-layer silicon photonic crystal fiber tip temperature sensor, in *2012 IEEE Photonics Conference (IPC)*, pp. 550, 551, 23–27 Sept 2012
45. S. Sinha, K. Urbanek, A. Krzywicki, R. Byer, Investigation of the suitability of silicate bonding for facet termination in active fiber devices. Opt. Express **15**, 13003–13022 (2007)
46. S. Kim, J. Wu, A. Carlson, S.H. Jin, A. Kovalsky, P. Glass, Z. Liu, N. Ahmed, S.L. Elgan, W. Chen, P.M. Ferreira, M. Sitti, Y. Huang, J.A. Rogers, Microstructured elastomeric surfaces with reversible adhesion and examples of their use in deterministic assembly by transfer printing. Proc. Nat. Acad. Sci. **107**, 17095 (2010)
47. H.H. Li, Refractive index of silicon and germanium and its wavelength and temperature derivatives. J. Phys. Chem. Ref. Data **9**, 561 (1980)
48. A.N. Bashkatov, E.A. Genina, Water refractive index in dependence on temperature and wavelength: a simple approximation. Proc. SPIE **5068**, 393–395 (2003)
49. C.Z. Tan, J. Arndt, Temperature dependence of refractive index of glassy SiO_2 in the infrared wavelength range. J. Phys. Chem. Solids **61**, 1315–1320 (2000)
50. B.J. Frey, D.B. Leviton, T.J. Madison, Temperature-dependent refractive index of silicon and germanium. Proc. SPIE **6273**, 62732J (2006)

51. F.G. Della Corte, G. Cocorullo, M. Iodice, I. Rendina, Temperature dependence of the thermo-optic coefficient of InP, GaAs, and SiC from room temperature to 600 K at the wavelength of 1.5 μm. Appl. Phys. Lett. **77**(11), 1614–1616 (2000)
52. G.E. Jellison Jr, H.H. Burke, The temperature dependence of the refractive index of silicon at elevated temperatures at severed laser wavelengths. J. Appl. Phys. **60**, 841 (1986)
53. S.M. Weiss, M. Molinari, P.M. Fauchet, Temperature stability for silicon-based photonic band-gap structures. Appl. Phys. Lett. **83**, 1980 (2003)
54. C. Yu, P.A. Flinn, J.C. Bravman, In situ stress measurements during dry oxidation of silicon. Mat. Res. Soc. Symp. Proc. **473**, 323–328 (1997)
55. J. Rabier, J.L. Demenet, Low temperature, high stress plastic deformation of semiconductors: the silicon case. Pysica Status Solidi (B) **222**(1), 63–74 (2000)
56. M. Mehregany, C.A. Zorman, N. Rajan, C. Wu, Silicon carbide MEMS for harsh environments. Proc. IEEE **86**(8), 1594–1609 (1998)

Chapter 5
Hybrid Nanoimprint-Soft Lithography for Highly Curved Surface with Sub-15 nm Resolution

Haixiong Ge, Wei Wu and Wen-Di Li

Abstract Nanoimprint lithography is a high-resolution, high-throughput and low-cost technology to pattern nanostructure, but it only works well on planar surface. To solve this issue, a hybrid nanoimprint-soft Lithography (HNSL) was developed to pattern nanostructures on highly curved surfaces (e.g. the sidewall of an optical fiber). Moreover, double transfer UV-curing nanoimprint lithography, an improved version of HNSL, was introduced to enable high fidelity pattern transfer. Optical fibers can be patterned using this technology, and that opened the door to numerous applications.

5.1 Introduction

There are ever increasing demand of having nano-electronic, nano-photonics and chemical sensing devices on non-conventional substrates for applications in healthcare, security and displays. However, the fabrication of nanostructures onto non-planar topography and irregular substrate still remains a huge challenge

H. Ge
Department of Materials Science and Engineering, College of Engineering and Applied Sciences, National Laboratory of Solid State Microstructures, Nanjing University, 210093 Nanjing, People's Republic of China
e-mail: haixiong@nju.edu.cn

W. Wu (✉)
Department of Electrical Engineering, University of Southern California, Los Angeles, CA 90089, USA
e-mail: wu.w@usc.edu

W.-D. Li
Department of Mechanical Engineering, The University of Hong Kong, Pokfulam, Hong Kong, People's Republic of China
e-mail: liwd@hku.hk

A. Cusano et al. (eds.), *Lab-on-Fiber Technology*,
Springer Series in Surface Sciences 56, DOI: 10.1007/978-3-319-06998-2_5,

because the most conventional lithographic techniques are limited to planar surfaces. Optical interference and phase mask lithography have been used to pattern Bragg grating on optical fiber [1–3]. One of the key advantages of those approaches is the infinitely large focus depth, however this approach can only be used to fabricate periodical structures and the resolution is also limited by the exposure wavelength. Electron-beam lithography and focused ion beam have been used to pattern the sidewall of optical fiber [4, 5]. While they have better resolution than optical lithographies, they have relatively low throughput. Nanoimprint lithography (NIL) [6] is a high-throughput and cost-effective nano-patterning technology, but it is based on mechanical contact between the mold and the substrate, which makes it intrinsically difficult to pattern highly non-planar surface, such as the sidewall of an optical fiber. We developed a hybrid nanoimprint-soft lithography (HNSL) [7] technique and demonstrated its capability of patterning highly curved surface (e.g. the surface of a single-mode fiber) with sub-15 nm resolution. Moreover, in order to create uniform nanostructures on highly curved surfaces, such as optical fibers, we also introduced a double transfer UV-curing nanoimprint technique to transfer patterned nanoimprint resist uniformly on highly curved or irregular topography substrates. In this chapter we will cover the HNSL technique, the flexible high-resolution mold, the double transfer UV-curing nanoimprint technique and examples of applications.

5.2 Hybrid Nanoimprint-Soft Lithography

NIL is based on the mechanical deformation of the resist rather than local chemical reaction by radiation, like other lithographies. The information content in the mask, or mold as conventionally called in NIL, is presented in the form of topography, instead of light transmission function as in photolithography. During a NIL process, patterns in the mold are duplicated into the resist by pressing the mold onto the resist (Fig. 5.1) [6]. It has thus three favorable characteristics:

- High resolution: As a mechanical process, NIL has no diffraction limit.
- High throughput: Unlike serial process of e-beam direct write, NIL is a parallel process.
- Low cost: NIL does not require sophisticated optical lithography systems.

NIL has been demonstrated as a high-throughput and low cost lithography technology with sub-10 nm resolution [8]. It has significant applications for various fields, such as nano-electronics, [9–11] nano-photonics, [12, 13] data storage, [14, 15] chemical sensing [16, 17] and biological applications [18]. The NIL mold (also called as "template" or "stamp") is the crucial component for NIL, and was typically fabricated on a rigid material such as silicon, silicon dioxide, quartz or metal. The high-resolution master molds are normally patterned using other lithographical technologies, such as electron beam lithography, Helium ion beam lithography, photolithography and optical interference lithography, followed by

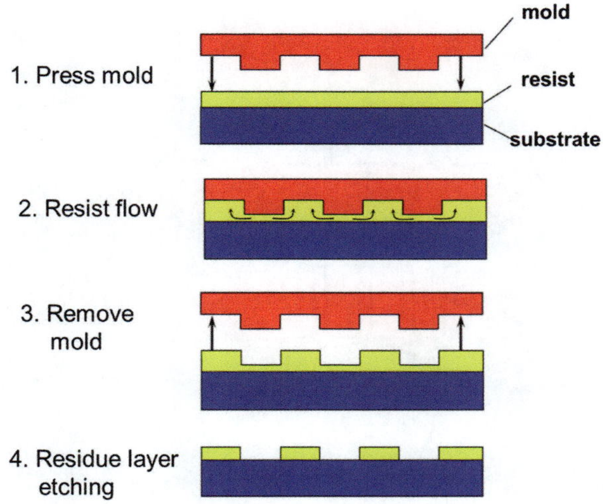

Fig. 5.1 Schematic of Nanoimprint lithography. Modified by [6]

pattern transfer using reactive ion etching (RIE) [19]. Hard materials are used in imprint molds to minimize the deformation of the features in the mold during the imprint process. That is especially important for high-resolution patterning, since the smaller the feature size the more prone to deformation under stress. For the same reason, nanoimprint lithography with soft mold and soft lithography had primarily been used for fabricating structures on the size scale of 500 nm and larger [20].

Soft lithography [21] is a widely used micro-fabrication technique in lab. It, especially distinct from NIL, employs a patterned elastic polymer as the stamp to generate micro-patterns and microstructures. Soft lithography includes several derived methods: micro-contact printing (μCP), replica molding (REM), micro-transfer molding (μTM), micro-molding in capillaries (MIMIC), solvent-assisted micro-molding (SAMIM), cast molding, embossing, and injection molding. The most commonly used stamp material in soft lithography is the flexible polymer, poly(dimethylsiloxane)(PDMS), which enables an intimate physical contact with substrates without applying external pressure and provides an feasible route to pattern non-planar substrate. The elastic stamp is prepared by a simple cast molding method: prepolymers of the elastomer PDMS are mixed and poured over a master with relief structures, then cured and peeled off. However, the extension of conventional soft lithography to sub-100 nm scale realm has been limited by the low elastic modulus of the PDMS stamp. As shown in Fig. 5.2, the structures on the stamp tend to deform, merge or collapse during embossing, inking and printing.

Several approaches have been developed to extend conventional soft lithography's resolution, which has been limited by the low elastic modulus of the commercial PDMS (e.g. Sylgard 184 PDMS) used for fabricating stamps, to produce sub-100 nm scale features [22]. For example, a higher modulus (ca. 8 MPa) "hard-PDMS"(h-PDMS) was developed by Schmid et al. [23] to

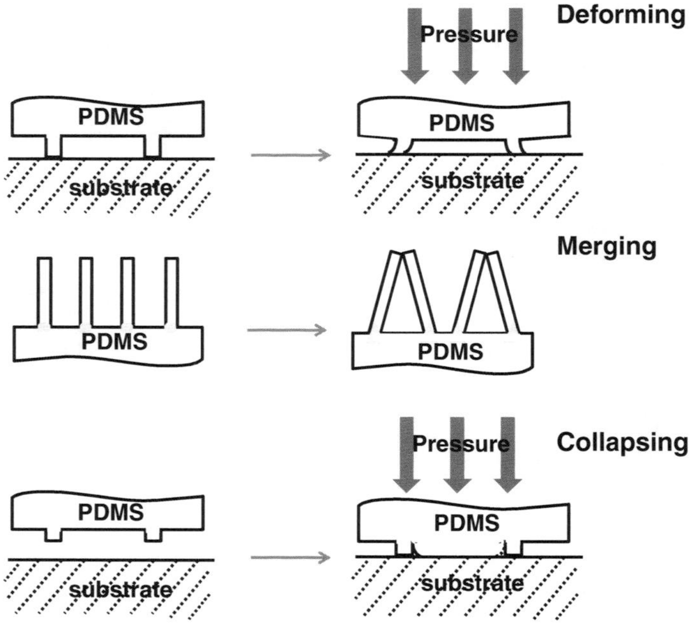

Fig. 5.2 Schematic of technical issues of soft lithography for nano-structure patterning caused by the low elastic modulus of PDMS mold

achieve soft lithography with resolution into the sub-100-nm regime. Although the h-PDMS stamp worked well on flat substrate, but this approach still needs to compromise between high resolution, which requires hard mold, and the ability to pattern curve the surface, which requires soft mold. To overcome these problems, Odom et al. [22] developed a composite stamp composed of two layers: a thin h-PDMS layer supported by a thick flexible PDMS layer, which combined the advantages of a rigid layer to achieve high resolution pattern transfer and an elastic support to enable conformal contact. However, as discussed in [21], some drawbacks still remained in using h-PDMS as a patterning layer for the double layer composite stamp, such as cracking during the stamp fabrication step and degraded conformal contact with the substrate compared with the PDMS material normally used in soft lithography [22].

In order to pattern high-resolution nanostructure on non-planar substrates, we developed a hybrid nanoimprint-soft lithography (HNSL). The key component of this technology is the mold, which consisted of an ultrathin (100 $\sim$ 200 nm) rigid cross-linked patterning layer on an elastic PDMS (thickness up to 2 mm) support. This mold combined the advantages of both a rigid nanoimprint mold to achieve a high resolution pattern transfer and a flexible soft lithography stamp to enable patterning on non-planar surface, such as optical fiber. The top pattern layer is made of a photo-curable crosslinked silicon-containing polymer which was developed for NIL molds, [24] and the thin photo-cured feature layer and the thick elastic PDMS support were fused via an interpenetrating polymer network. The schematic of the mold fabrication and the hybrid nanoimprint-soft lithography are shown in Fig. 5.3 [7]. First, an elastic PDMS support is prepared by casting and thermal curing of a PDMS prepolymer (Sylgard184) between two silicon wafers (Fig. 5.3a), followed by a uniform absorption treatment that involved placing the elastic support onto a film of a silicon-containing multifunctional photo-curable resist spin-coated onto a silicon wafer (Fig. 5.1b). The same silicon-containing photo-curable nanoimprint resist is spin-coated onto a Si master mold with patterns, which are normally generated using nano-lithographies, such as electron-beam lithography. The absorbed elastic support is then placed against the resist-covered master mold, and exposed to UV radiation in a nitrogen ambient (Fig. 5.3c). After separation, the crosslinked photo-curable resist on the master mold is transferred to the elastic support because of the difference in the surface adhesion properties. With a brief exposure to an O_2 plasma, the surface of the duplicated mold is oxidized to form inorganic silica and is then coated with a mold release layer of trichloro(1H, 1H, 2H, 2H-perfluorooctyl)silane (Fig. 5.3d). The resulting mold is ready to be used for hybrid nanoimprint-soft lithography (Fig. 5.3e and f) [7].

The silicon wafers used to fabricate the PDMS support are treated with a self-assembled monolayers (SAM) of trichloro(1H, 1H, 2H, 2H-perfluorooctyl)silane first to lowered the surface energy of the silicon wafers, hence prevent the cured PDMS from stick to the Si wafers [25]. The area and thickness of the elastic support are determined by the size of silicon wafers and the amount of prepolymer respectively. Thicknesses up to 2 mm have been used. The top rigid polymer pattern layer is fabricated using a photo-curable reverse nanoimprint process instead of a regular nanoimprint process. In this way, the thin uniform liquid film of the photo-curable resist was spin coated onto the master mold rather than on the PDMS support to avoid swelling of the PDMS in the resist solution. The photo-curable resist is the same resist that we use for regular photo-curable NIL process. The resist is composed of a low viscosity UV-curable acrylatedpoly(dimethylsiloxane) material, a multifunctional acrylate cross-linker, and a free radical initiator [7]. The film thickness ranges from 10 to 100 s of nm depending on the feature height of the mold [24]. The top patterned layer and PDMS support are bonded together by taking advantage of the intrinsic high permeability of PDMS for gases and organic molecules. The low viscous photo-curable resist is absorb into the top surface of elastic substrate by placing the PDMS substrate onto a thin uniform resist film that was spin-coated onto a Si wafer. The photo-curable molecules are diffused into the crosslinked network of thermal cured PDMS through the interaction of the resist film with the PDMS surface. After an absorption about 10 min, the elastic support is separated from the Si wafer. The PDMS support absorbed with photo-curable resist is then placed onto the master mold spin-coated with a thin resist film. The flexibility of the PDMS enables it to make a conformal contact with the master mold over a large area without external pressure. The resist is cured by UV light in a nitrogen atmosphere because oxygen diffusing through PDMS can inhibit the free-radical polymerization of the photo-curable resist. During the curing process, the resist spin-coated onto the master mold not only self-polymerized but also reacted with the resist molecules absorbed into the

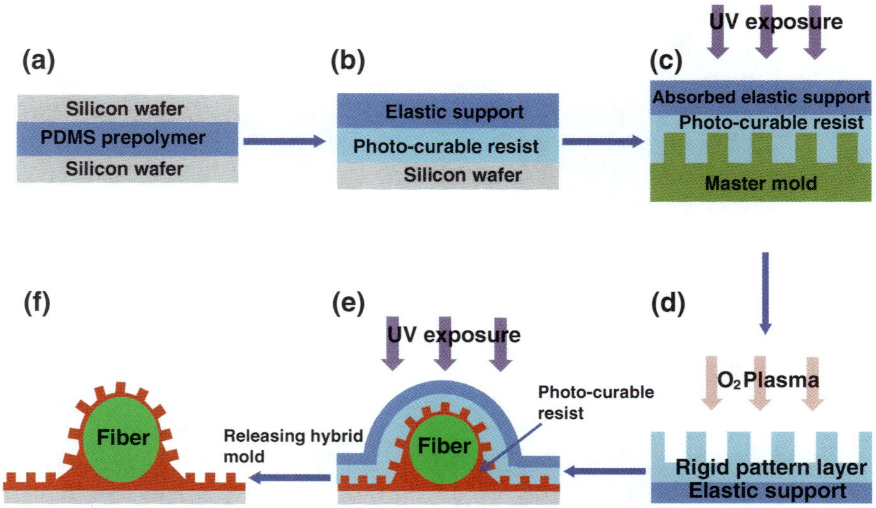

Fig. 5.3 Schematic of the fabrication process of the hybrid mold and the hybrid nanoimprint-soft lithography. **a** preparing an elastic support; **b** Absorption of the photo-curable resist into the PDMS elastic support; **c** fabrication of hybrid mold by spin-coating nanoimprint resist on the master mold, making a conformal contact and curing the UV-curable resist; **d** mold surface treatment by O_2 plasmaoxidizing of the surface and coating with a mold release layer; **e** UV-imprint on curved substrate by hybrid mold and **f** imprinted patterns on curved substrate. Modified by [7]

PDMS to form an integrated interpenetrating polymer network near the surface region of PDMS support, which provided excellent adhesion of the two distinct layers [7]. This strategy prevented the rigid features from peeling off the elastic support during the de-molding process. As shown in Fig. 5.4, the flexible mold is transparent to broadband UV-visible light, so it can be used in UV-curable NIL process.

Similar to conventional NIL mold, the flexible mold is coated with a SAM mold releasing layer trichloro-fluoroalkylsilane to lower the surface energy, hence prevent resist adhesion to the mold during nanoimprint. In order to treat with the trichloro-fluoroalkylsilane, the surface of the rigid polysiloxane layer was oxidized into inorganic silica by short O_2 plasma etching (30 W, 25 mTorr, 5 Sccm, 5 s). The low power and short time plasma treatment only oxidized the near-surface region of the mold without destroying nanoscale features on the mold. In this way silanol groups are formed on the mold surface. The chlorosilane group of the trichloro-fluoroalkylsilane release agent reacts with silanol groups of mold surface to form –O–Si–O– covalent bonds under catalysis of trace amounts of environmental water [26]. The application of the mold release agent was carried out by putting the hybrid mold in the vapor phase of the chlorosilane under reduced pressure [25]. The water contact-angle (113°) shows that the surface energy of the treated flexible mold is indent to that of a standard quartz mold [7, 24].

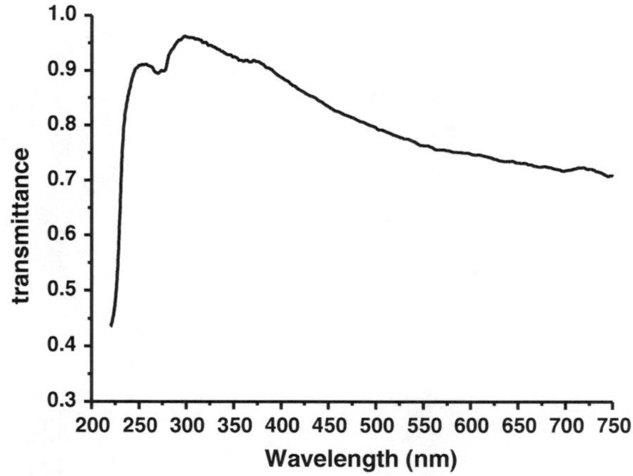

Fig. 5.4 Ultraviolet-visible (UV-vis) transmission spectrum of the polymer mold

The complementary mechanical properties of the rigid cross-linked patterning layer and elastic support of the hybrid mold are crucial factors for ideal high-resolution imprinting on non-planar surface. The high mechanical strength of the rigid patterning layer is the key to achieve high lithographic resolution. On the other hand, a soft and elastic support with high flexibility allows a conformal contact between mold and the non-planar surface. The mechanical properties of the two types of mold layers, rigid and flexible, were estimated by dynamical mechanical analysis (DMA) for the tensile modulus and elongation at breaking with the samples prepared by molding and curing of pure liquid precursors. The tensile modulus can be calculated from the slope of the stress-strain curve in the linear region, and the elongation at breaking can be measured from the value of the strain at which fracture occurs, which corresponds to the abrupt decrease in the stress [27]. The tensile modulus and elongation at breaking for the rigid patterning layer is 216 N/mm^2 and 1.3 %, respectively. Compared with the hard PDMS modulus of $\sim$9 N/mm^2, the much higher modulus of this patterning layer provided sufficient mechanical strength to sustain the high capillary forces and the nanoimprint pressure acting at the nanometer level. The modulus and the elongation at breaking of the flexible support (2.4 N/mm^2 and 236 %) are similar to these reported for sylgard 184 PDMS in other references, [23] which enabled conformal contact with a substrate without external pressure. The mold with 200 nm pitch grating was bent along and across the grating lines at angles as large as shown in Fig. 5.5a. There were no apparent bending effects observed from the SEM images of the grating patterns (Fig. 5.5b), which were replicated from the bent area of the hybrid mold.

5.3 Patterning with HNSL

HNSL are used to pattern nanostructures on both planar and non-planar substrates. For example, sub-15 nm resolution is shown in Fig. 5.6 [7]. Figures 5.4 and 5.6a are scanning electron microscopy (SEM) images of a master Si mold with a dot array at

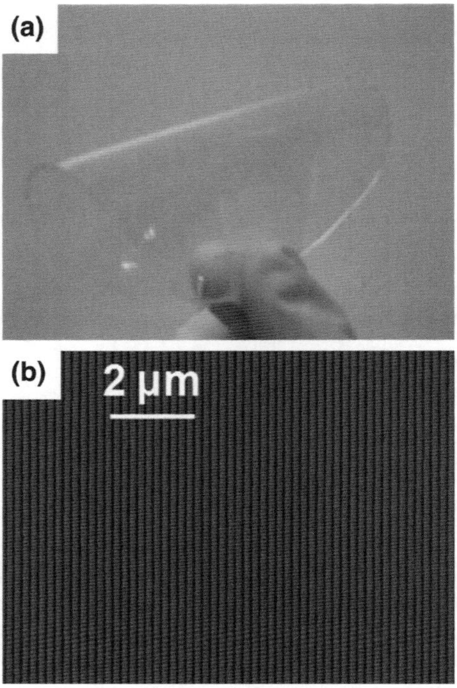

Fig. 5.5 **a** Photograph of the hybrid mold with the 200 nm pitch grating bent over on itself. **b** SEM image of the 200 nm pitch grating replicated from the bent area of the hybrid mold. Modified by [7]

100 nm pitch and 15 nm in diameter, and its flexible mold replica respectively. Figure 5.6c shows that the 15 nm dot array were accurately replicated by HNSL. The excellent fidelity indicates that is should be possible to imprint features much smaller than 15 nm. The resolution in our experiments was limited entirely by the feature size resolution of the master mold. Uniform patterning field up to 4-inch wafer size has been successfully demonstrate [7]. As a demonstration of nano-patterning on a highly non-planar substrate, a segment of a cylindrical single-mode optical fiber (125 μm in diameter) was nanoimprinted using HNSL. The flexible mold was duplicated from master Si mold of 200 nm pitch gratings. The 200 nm pitch gratings with the grating line orientated both perpendicular and parallel to the fiber axis were successfully patterned onto the NIL resist on the fiber surface. The inserted SEM image of the cross-section of a patterned fiber (Fig. 5.7a) [7] shows that the grating structures were uniformly distributed on the cylindrical surface parallel to the fiber. Defects around edge of the cleaved facet were due to the fiber cutting for SEM imaging, not the nanoimprint process. Top views of the grating features parallel with and perpendicular to the fiber show that the quality and uniformity are excellent, as the lines extend for several tens of microns without cracks or observable variation in period (Fig. 5.7a and b). The contrast of the dark and bright area of the top view images indicated that the grating patterns were not located on the same plane. The patterned area extended over 60° around the fiber. The cylindrical surface with 125 μm in diameter is highly curved, but there is no sign showing that it is the ultimate limit of substrate curvature to the hybrid mold [7].

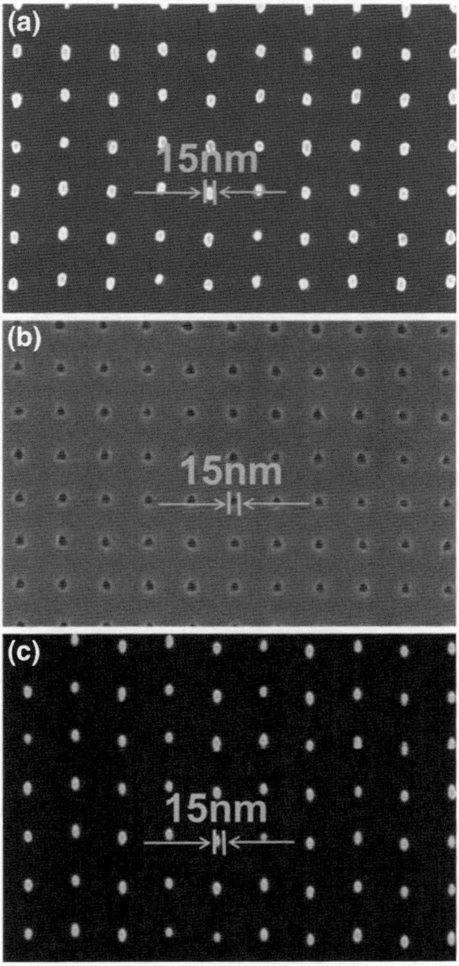

Fig. 5.6 SEM images of the master mold, the replica and the imprint result of the replica **a** silicon *dot* array mold at 100 nm pitch and 15 nm diameter features; **b** the hybrid *hole* array mold; **c** imprint result of the hybrid *hole* array mold. Modified by [7]

5.4 Double Transfer UV-Curing Nanoimprint Lithography

While Figs. 5.5 and 5.6 show that HNSL is capable of generating high-resolution patterns in the resist layer on non-planar surface, such as optical fiber, nano-patterning doesn't stop at patterning the resist; patterns in the resist layer have to be transferred to make real applications. To achieve this goal, a thin, uniform resist film is desired in the lithography process as required by subsequent high-fidelity pattern transfer processes. Conventional resist coating methods are extremely challenging to form such a uniform resist layer on irregular and/or non-planar substrates, therefore the double transfer UV-curing nanoimprint (DTUCN) [28] lithography, an improved version of HNSL, was introduced.

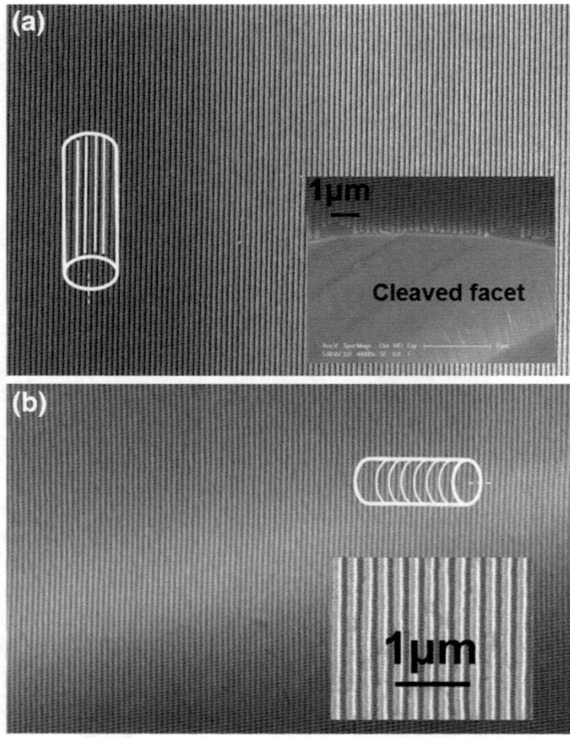

Fig. 5.7 SEM images of 200 nm pitch grating on the surface of 125 micron diameter optical fiber with the direction **a** parallel with and **b** perpendicular to the fiber axis. In (**a**), the *cross-sectional view* of the patterned fiber shows that the grating structures are distributed on a cylindrical surface parallel to the fiber. Modified by [7]

So far a number of resist coating methods have been used to form a resist film on a substrate. Spin coating is the most widely adopted method particularly used for flat substrates in microelectronics and nanotechnologies. It has the advantages of the highest film thickness uniformity and a short coating time. However, spin coating on non-flat and irregular substrates will suffer from the physical topology that could interrupt the resist solution flow driven by the centrifugal force, resulting in an incomplete coverage and significant film thickness variation [29]. Another commonly used method for resist coating is spray coating [30–33], including its variations such as dispensing [34, 35]. They have been demonstrated as promising techniques to coat resist on arbitrarily sized and shaped substrates, even three-dimensional objects, with less resist consumption [36, 37]. However, spray coating typically has a droplet size of micrometers, making it difficult to form a uniform resist layer down to sub-micron thickness. This technique also requires expensive equipment and has special requirements on resist solutions. These disadvantages limit the application of spray coating. There are other less commonly used resist coating methods for non-conventional substrates. For example, dip coating [38, 39] is suitable for arbitrarily shaped large objects but suffers from poor film uniformity; electro-deposition [40, 41] is an attractive method for three-dimensional stacks of devices and interconnects only if a conductive layer is pre-deposited on the substrate [42]. In general, there are few existing methods that can coat a sufficiently thin and uniform resist film on non-planar and irregular substrates to meet the requirement of pattern transfer of nano-structure.

Patterning of sub-100 nm structures in resist on highly-curved surfaces has been demonstrated using HNSL [7]. The HNSL technique was further improved by introducing the DTUCN process featuring an improved resist coating method to form and pattern a thin, uniform resist layer on arbitrary substrates with irregular surface topography. Detailed processing of this DTUCN method is illustrated in Fig. 5.8 and explained in the following. This DTUCN process starts with spin coating liquid UV-curable resist on a carrying silicon wafer. Then, the HNSL mold is placed on top of and then separated from the carrying wafer. UV-curable resist, originally spin coated on the carrying wafer, will be adhered off by the HNSL mode. Subsequently the HNSL mold is placed in contact with a planar or non-planar substrate under UV light exposure, curing and transferring the UV-curable liquid resist to the substrate with relief nano-patterns duplicated from the HNSL mold. Varying the initial thickness of the spin-coated resist on the carrying wafer provides a precise control on the thickness of the transferred resist layer on the substrate. This double transfer process has achieved a uniform sub-50 nm residual layer thickness on the nanoimprinted resist layer, enabling high-fidelity transfer of the imprinted patterns into underlying substrate through reactive ion etching (RIE).

The resist transfer and film formation in the above fabrication process depend on the surface properties of the carrying wafer, the HNSL mold and the substrate. Monitoring the resist morphology adhered on HNSL mold helps understand the mechanism of the above process. This was done in the following process. Once the HNSL mold was separate from the resist-coated carrying silicon wafer, UV light was used to expose and cure the resist adhered on the HNSL mold. Frozen UV resist was then examined under a scanning electron microscope. The morphology of frozen UV resist film can therefore be studied to reveal the surface properties' effects on the resist film formation in the DTUCN process.

Morphologies of adhered resist on HNSL molds with different surface properties are compared in Fig. 5.9 [43]. Figure 5.9a shows the cured resist on a blank HNSL mold coated with a self-assembled monolayer of fluoroalkyltrichlorosilane. UV-curable resist used in this chapter has a contact angle of 59° on the treated mold surface. After the blank HNSL mold is separated from the carrying wafer, UV resist is adhered on its surface and ball up into droplets due to minimization of surface energy. Typical sizes of resist droplets found on the blank surface are a few micrometers while the largest droplets have diameters of about 10 μm, as extracted from the SEM pictures by an image processing software. Statistical analysis of the size distribution of these droplets (Fig. 5.9a) is shown in Fig. 5.10 [43], with a mean diameter of 2.45 μm. The poly dispersity, the ratio of weighted average diameter to number average diameter, is calculated to be 1.13. The droplets adhered on the HNSL mold are planarized during the imprint process to merge into an integrated film sandwiched by the mold and the substrate. The thickness of final planarized resist film can therefore be calculated by the total volume of all droplets as transferred from the carrying wafer.

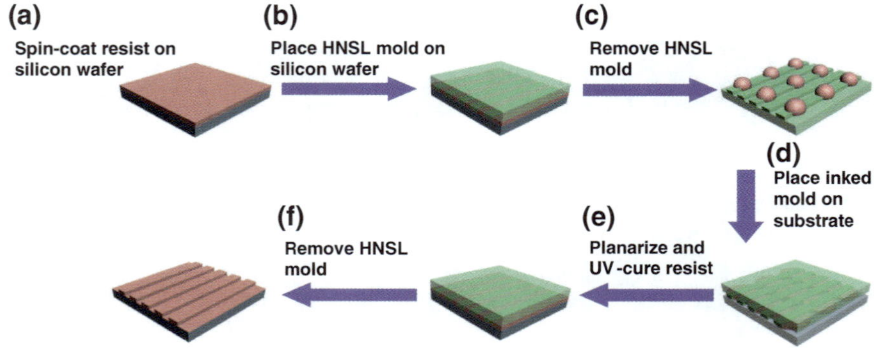

Fig. 5.8 Schematic of the DTUCN fabrication process

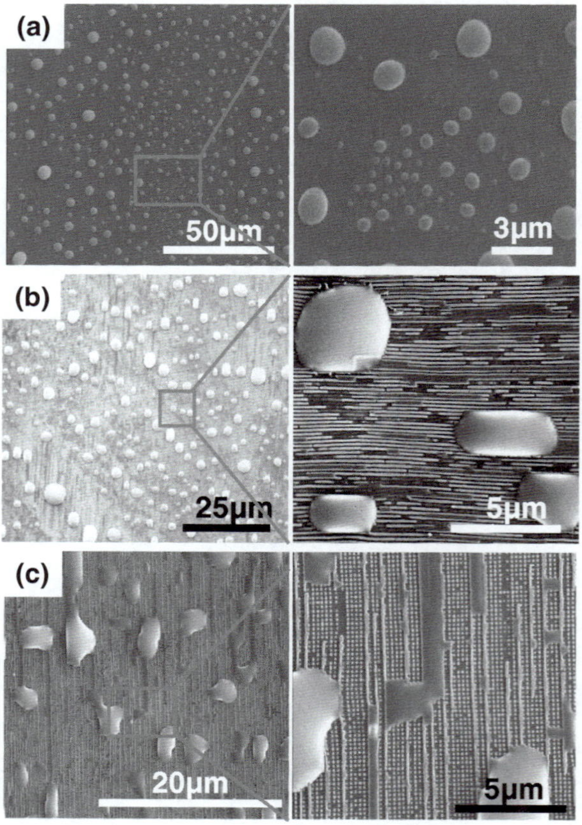

Fig. 5.9 SEM images of morphologies of the adhered resist on HNSL molds with different surface structures after UV-curing. **a** The UV-cured resist on a blank HNSL mold with a smooth surface. **b** The UV-cured resist on a HNSL mold with 200-nm-pitch gratings. **c** The UV-cured resist on a HNSL mold with 200-nm-pitch hole arrays. Modified by [43]

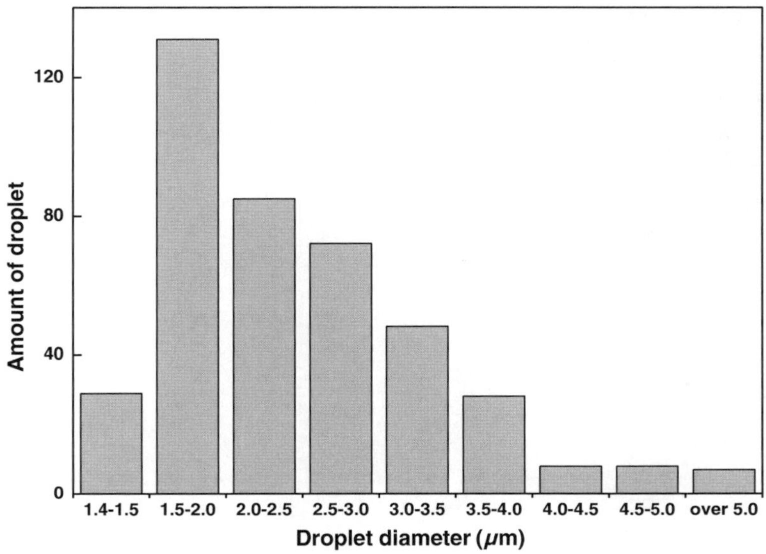

Fig. 5.10 Droplet size distribution of the transferred liquid resist on a blank HNSL mold with a smooth surface. Modified by [43]

Figure 5.9b and c show the morphologies of adhered droplets on two fluoroalkyltrichlorosilane-coated HNSL molds with different geometries, a 200 nm pitch grating and a 200 nm pitch hole array. Significantly different morphology of transferred resist can be observed on the relief nanostructures. The resist on nano-structured HNSL molds features over spilling droplets and extended filaments. Because the mold surface is partially wetting, droplets are found to spill over groove edges and extended filaments are confined to the grooves of the nanostructures due to capillary effects. Anisotropic geometry of the underlying nanostructures strongly affects the shape of the over spilling droplets. Interaction of capillary forces and nanoscale energy barriers induced by the periodic gratings or hole arrays causes the liquid resist to form elongated droplets and spread directionally. A more extended and wetting morphology is seen on the droplets on the nanostructures than that on the smooth surface, implying that nanostructures on HNSL mold may benefit the spreading and merging of droplets during nanoimprint process.

In the DTUCN process, the amount of transferred resist is determined by several factors such as the total resist volume as spin-coated on the carrying silicon wafer, the surface properties and surface relief nanostructures on the HNSL mold. Measurement on the resist film thickness on the carrying silicon wafer before and after resist transfer provides information about the amount of resist transferred to the HNSL mold. Figure 5.11 [43] plots the film thickness of remaining resist on the carrying substrate after resist transfer to three different HNSL mold, a blank one treated with O_2 plasma, a blank one treated with fluoroalkyltrichlorosilane and a fluoroalkyltrichlorosilane coated grating mold with a pitch of 200 nm and a

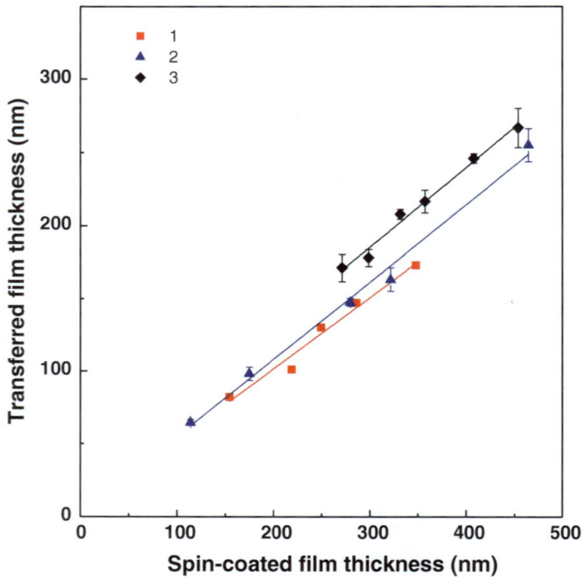

Fig. 5.11 Relation between the *spin-coated resist film thickness* and *the transferred film thickness* under different mold morphologies and surface properties. *Line 1* was obtained by a blank HNSL mold treated with O_2 plasma. *Line 2* was obtained by a blank HNSL mold treated with O_2 plasma and then coated with fluoroalkyltrichlorosilane. *Line 3* was obtained by a nano-grating mold with a pitch of 200 nm and a depth of 100 nm and coated with fluoroalkyltrichlorosilane. Modified by [43]

depth of 100 nm. As shown in Fig. 5.11, the transferred amount of resist is linearly proportional to the initial resist film thickness spin-coated on the carrier wafer, i.e., the partition ratio of the resist between HNSL mold and the carrier wafer is independent of the initial resist thickness. Surface relief patterns can increase the transferred resist amount because the total surface area on the mold is increased. Surface treatment of the HNSL has a less effect on the partition ratio as almost half volume of the liquid resist spin-coated on the carrier wafer can be transferred by the blank HNSL molds with both hydrophobic and hydrophilic surfaces.

5.5 Applications

Figure 5.12a shows an example of a grating at 200 nm period imprinted through the DTUCN process to demonstrate the uniformity of residual layer thickness of the imprinted resist. Cross-sectional SEM pictures taken at 9 points of the imprinted grating were used to inspect the residual thickness (Fig. 5.12b). The residual layer had a thickness of less than 40 nm, and the thickness variation (maximum minus minimum) is less than 10 nm. With such residual layer thickness and variation, faithful pattern transfer to underlying substrates can be realized.

DTUCN was also demonstrated using different nanoscale features to pattern substrates with different morphologies. Figure 5.12c shows a 200-nm-pitch pillar array imprinted on a flat silicon wafer through DTUCN. Figure 5.12d shows a SEM picture of 200-nm-pitch grating patterns imprinted on a 20-μm-diameter glass bead. DTUCN can even be used to pattern nanostructures on a periodic relief

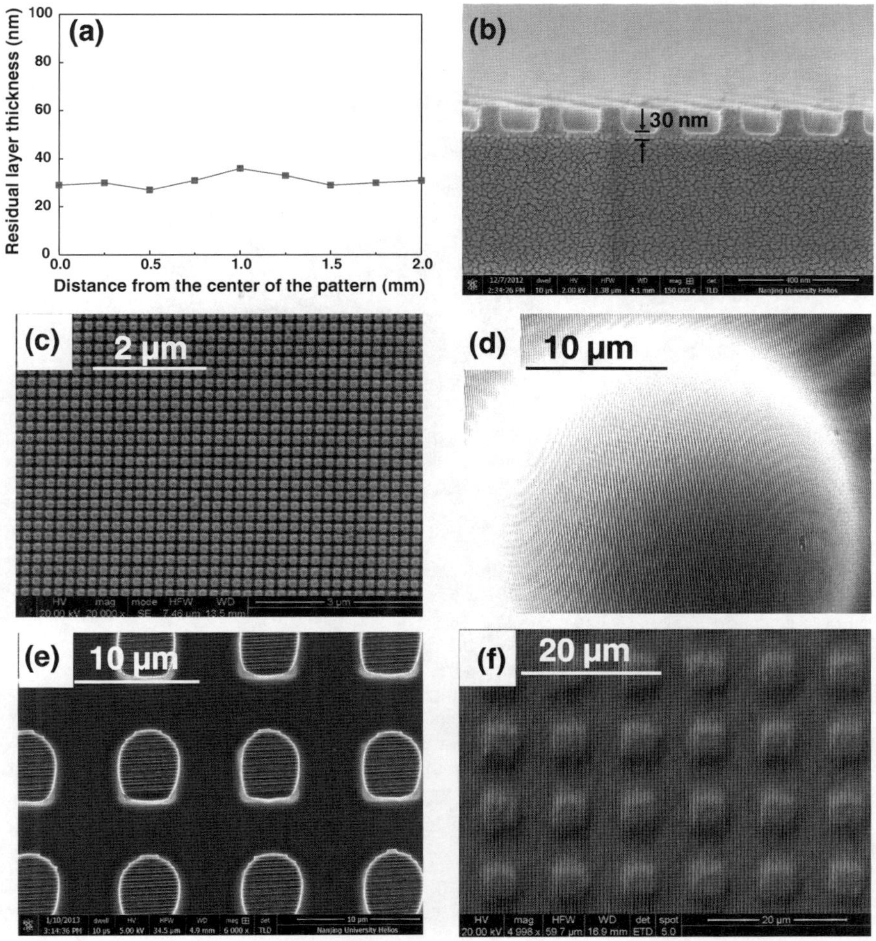

Fig. 5.12 **a** and **b** Measurements of the residual layer thickness of a 200 nm pitch grating in the imprinted resist film. **c** SEM image of a 200 nm pitch post array imprinted on a flat silicon wafer; **d** SEM image of a 200 nm pitch grating on a beadwith a 20 μm diameter; **e** SEM image of a 550 nm pitch grating on top facets of a micro-post array; **d** SEM image of 550 nm pitch gratings imprinted on both the top facets of the micro-posts and the intervening flat portions between the micro-posts

patter with micro-size [28]. For example, as shown in Fig. 5.12e and f, 550 nm period gratings were patterned on a post array with a diameter of 5 μm, a pitch of 10 μm and a height of 500 nm. Without applying external pressure, only the top surfaces of the micro-post array were patterned by the nano-gratings (Fig. 5.12e). With a 3000 Pa pressure applied, both the top surfaces and the flat area between posts could be patterned with the same grating structures, as shown in Fig. 5.12f.

The capability of DTUCN was further demonstrated on the curved surface of a fiber sidewall, with the key steps and key results illustrated in Fig. 5.13. Before

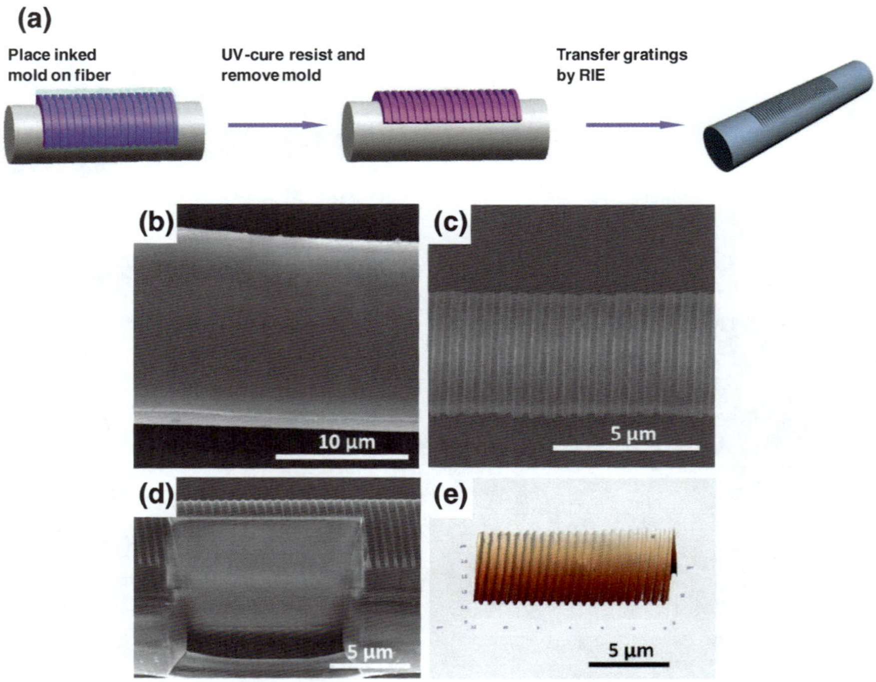

Fig. 5.13 Fabrication of surface relief gratings into an optical fiber by DTUCN. **a** Schematic illustration of the DTUCN-based fabrication process. **b** Top-view SEM image of a blank cured resist film on an optical fiber. **c** Top-view SEM image of a 550 nm pitch grating imprinted onto an optical fiber after DTUCN patterning. **d** SEM image of a transversal cross-section of the surface relief grating. **e** AFM 3D image of the 550 nm pitch grating etched into the optical fiber. Modified by [43]

DTUCN fabrication, a standard optical fiber with a 125 μm diameter was first immersed in a bath of buffered hydrofluoric acid (BHF) and then mounted on a silicon wafer for the fabrication process. In a subsequent DTUCN process, a HNSL mold with grating patterns and inked with UV-curable resist was placed on the etched fiber. The resist was patterned with a 550 nm pitch grating patterns after UV-curing and mold separation. Since a uniform resist thickness was achieved in the DTUCN fabrication, the grating patterns was then successfully transferred into the SiO_2 matrix of the fiber through a multi-step reactive ion etching (RIE) process. In this example, the grating direction was perpendicular to the fiber axis, therefore the gratings in the imprinted resist had vertical sidewalls over the whole imprinted area, making it possible to use RIE to etch the patterns into the cylindrical surface.

Figure 5.13b shows a top view of a blank cured resist layer on the fiber, demonstrating a thin uniform resist layer was formed on the cylindrical fiber surface. Figure 5.13c shows the top view of the 550 nm pitch grating pattern imprinted onto the fiber after the RIE process. Cross-sectional view of the transferred grating pattern by RIE is shown in Fig. 5.13d. Atomic force microscopy (AFM) imaging is also used on the patterned fiber surface to confirm that the grating structures were uniformly distributed perpendicular to the fiber axis on the highly curved cylindrical surface. Successful patterning and transferring of gratings into a fiber demonstrated the capability of DTUCN to form a thin, uniform resist layer on highly curved surfaces. Such nanostructures transferred into a fiber are particularly useful for fiber based bio/chemical sensing devices because analyst molecules can access to the strong electrical field confined in the center of a fiber through deeply etched nanostructures on the fiber sidewall. Physically etched gratings on a fiber can also perform like normal fiber Bragg grating devices with higher stability and longer durability. More novel applications can be developed based on this unprecedented fabrication capability.

5.6 Conclusion

HNSL is capable of patterning of highly curved surfaces (e.g. the sidewall of an optical fiber) at high-resolution and low-cost. This technology enables a wide spectrum of applications in nano-electronic, nano-photonics and chemical sensing.

References

1. K.O. Hill, Y. Fujii, D.C. Johnson, B.S. Kawasaki, Photosensitivity in optical fiber waveguides: Application to reflection filter fabrication. Appl. Phys. Lett. **32**(10), 647–649 (1978)
2. K.O. Hill, B. Malo, F. Bilodeau, D.C. Johnson, J. Albert, Bragg gratings fabricated in monomode photosensitive optical fiber by UV exposure through a phase mask. Appl. Phys. Lett. **62**(10), 1035–1037 (1993)
3. J. Martin, F. Ouellette, Novel writing technique of long and highly reflective in-fibre gratings. Electron. Lett. **30**(10), 811–812 (1994)
4. J. Albert, S. Theriault, F. Bilodeau, D.C. Johnson, K.O. Hill, P. Sixt, M.J. Rooks, Minimization of phase errors in long fiber Bragg grating phase masks made using electron beam lithography. IEEE Photonics Technol. Lett. **8**(10), 1334–1336 (1996)
5. M.L. von Bibra, A. Roberts, J. Canning, Fabrication of long-period fiber gratings by use of focused ion-beam irradiation. Opt. Lett. **26**(11), 765–767 (2001)
6. S.Y. Chou, P.R. Krauss, P.J. Renstrom, Nanoimprint lithography. J. Vac. Sci. Technol. B **14**(6), 4129–4133 (1996)
7. Z.W. Li, Y.N. Gu, L. Wang, H.X. Ge, W. Wu, Q.F. Xia, C.S. Yuan, Y. Chen, B. Cui, R.S. Williams, Hybrid nanoimprint-soft lithography with sub-15 nm resolution. Nano Lett. **9**(6), 2306–2310 (2009)
8. M.D. Austin, H.X. Ge, W. Wu, M.T. Li, Z.N. Yu, D. Wasserman, S.A. Lyon, S.Y. Chou, Fabrication of 5 nm linewidth and 14 nm pitch features by nanoimprint lithography. Appl. Phys. Lett. **84**(26), 5299–5301 (2004)
9. L.J. Guo, Recent progress in nanoimprint technology and its applications. J. Phys. D-Appl. Phys. **37**(11), R123–R141 (2004)

10. E. Menard, M.A. Meitl, Y. Sun, J.-U. Park, D.J.-L. Shir, Y.-S. Nam, S. Jeon, J.A. Rogers, Micro- and nanopatterning techniques for organic electronic and optoelectronic systems. Chem. Rev. **107**(4), 1117–1160 (2007)
11. Q.F. Xia, W. Robinett, M.W. Cumbie, N. Banerjee, T.J. Cardinali, J.J. Yang, W. Wu, X.M. Li, W.M. Tong, D.B. Strukov, G.S. Snider, G. Medeiros-Ribeiro, R.S. Williams, Memristor-CMOS hybrid integrated circuits for reconfigurable logic. Nano Lett. **9**(10), 3640–3645 (2009)
12. W. Wu, E. Kim, E. Ponizovskaya, Y. Liu, Z. Yu, N. Fang, Y.R. Shen, A.M. Bratkovsky, W. Tong, C. Sun, X. Zhang, S.Y. Wang, R.S. Williams, Optical metamaterials at near and mid-IR range fabricated by nanoimprint lithography. Appl. Phys. A Mater. Sci. Process. **87**(2), 143–150 (2007)
13. D.J. Cho, W. Wu, E. Ponizovskaya, P. Chaturvedi, A.M. Bratkovsky, S.Y. Wang, X. Zhang, F. Wang, Y.R. Shen, Ultrafast modulation of optical metamaterials. Opt. Express **17**(20), 17652–17657 (2009)
14. W. Wu, B. Cui, X.Y. Sun, W. Zhang, L. Zhuang, L.S. Kong, S.Y. Chou, Large area high density quantized magnetic disks fabricated using nanoimprint lithography. J. Vac. Sci. Technol. B **16**(6), 3825–3829 (1998)
15. H. Oshima, H. Kikuchi, H. Nakao, K.-I. Itoh, T. Kamimura, T. Morikawa, K. Matsumoto, T. Umada, H. Tamura, K. Nishio, H. Masuda, Detecting dynamic signals of ideally ordered nanohole patterned disk media fabricated using nanoimprint lithography. Appl. Phys. Lett. **91**(2), 22508 (2007)
16. W. Wu, M. Hu, F.S. Ou, Z.Y. Li, R.S. Williams, Cones fabricated by 3D nanoimprint lithography for highly sensitive surface enhanced Raman spectroscopy. Nanotechnology **21**(25), 255502 (2010)
17. M. Hu, F.S. Ou, W. Wu, I. Naumov, X.M. Li, A.M. Bratkovsky, R.S. Williams, Z.Y. Li, Gold nanofingers for molecule trapping and detection. J. Am. Chem. Soc. **132**(37), 12820–12822 (2010)
18. V.N. Truskett, M.P.C. Watts, Trends in imprint lithography for biological applications. Trends Biotechnol. **24**(7), 312–317 (2006)
19. P.F. Murphy, K.J. Morton, Z. Fu, S.Y. Chou, Nanoimprint mold fabrication and replication by room-temperature conformal chemical vapor deposition. Appl. Phys. Lett. **90**(20) (2007)
20. Y. Xia, G.M. Whitesides, Soft lithography. Angew. Chem. Int. Ed. **37**(5), 550–575 (1998)
21. Y.N. Xia, G.M. Whitesides, Soft lithography. Annu. Rev. Mater. Sci. **28**, 153–184 (1998)
22. T.W. Odom, J.C. Love, D.B. Wolfe, K.E. Paul, G.M. Whitesides, Improved pattern transfer in soft lithography using composite stamps. Langmuir **18**(13), 5314–5320 (2002)
23. H. Schmid, B. Michel, Siloxane polymers for high-resolution, high-accuracy soft lithography. Macromolecules **33**(8), 3042–3049 (2000)
24. H.X. Ge, W. Wu, Z.Y. Li, G.Y. Jung, D. Olynick, Y.F. Chen, J.A. Liddle, S.Y. Wang, R.S. Williams, Cross-linked polymer replica of a nanoimprint mold at 30 nm half-pitch. Nano Lett. **5**(1), 179–182 (2005)
25. G.Y. Jung, Z.Y. Li, W. Wu, Y. Chen, D.L. Olynick, S.Y. Wang, W.M. Tong, R.S. Williams, Vapor-phase self-assembled monolayer for improved mold release in nanoimprint lithography. Langmuir **21**(4), 1158–1161 (2005)
26. M.K. Chaudhury, G.M. Whitesides, Direct measurement of interfacial interactions between semispherical lenses and flat sheets of poly(dimethylsiloxane) and their chemical derivatives. Langmuir **7**(5), 1013–1025 (1991)
27. P.J. Yoo, S.J. Choi, J.H. Kim, D. Suh, S.J. Baek, T.W. Kim, H.H. Lee, Unconventional patterning with a modulus-tunable mold: from imprinting to microcontact printing. Chem. Mater. **16**(24), 5000–5005 (2004)
28. Y. Shen, L. Yao, Z. Li, J. Kou, Y. Cui, J. Bian, C. Yuan, H. Ge, W.-D. Li, W. Wu, Y. Chen, Double transfer UV-curing nanoimprint lithography. Nanotechnology **24**(46), 465304 (2013)
29. R. Martinez-Duarte, G.T. Teixidor, P.P. Mukherjee, Q. Kang, M.J. Madou, in *Perspectives of micro and nanofabrication of carbon for electrochemical and microfluidic applications, in microfluidics and microfabrication*, ed. by S. Chakraborty (Springer, US, 2010), pp. 181–263

30. S. Kumagai, H. Tajima, M. Sasaki, Flow analysis of photoresist spray coating towards improving coverage on three-dimensional structures. Jpn. J. Appl. Phys. **50**(10) (2011)
31. N.P. Pham, E. Boellaard, J.N. Burghartz, P.M. Sarro, Photoresist coating methods for the integration of novel 3-D RF microstructures. J. Microelectromech. Syst. **13**(3), 491–499 (2004)
32. V.K. Singh, M. Sasaki, K. Hane, M. Esashi, Flow condition in resist spray coating and patterning performance for three-dimensional photolithography over deep structures. Jpn. J. Appl. Phys. Part 1 (Regular Papers Short Notes and Review Papers) **43**(4B), 2387–2391 (2004)
33. V.K. Singh, M. Sasaki, J.H. Song, K. Hane, Technique for preparing defect-free spray coated resist film on three-dimensional micro-electromechanical systems. Jpn. J. Appl. Phys. Part 1 (Regular Papers Brief Communications and Review Papers) **44**(4A), 2016–2020 (2005)
34. X. Liang, H. Tan, Z. Fu, S.Y. Chou, Air bubble formation and dissolution in dispensing nanoimprint lithography. Nanotechnology **18**(2) (2007)
35. K. Usuki, S. Wakamatsu, T. Oomatsu, K. Kodama, K. Kodama, *in Approaches to rapid resist spreading on dispensing based UV-NIL, Alternative Lithographic Technologies III*, ed. by D.J.C. Herr (2011)
36. J. Linden, C. Thanner, B. Schaaf, S. Wolff, B. Laegel, E. Oesterschulze, Spray coating of PMMA for pattern transfer via electron beam lithography on surfaces with high topography. Microelectron. Eng. **88**(8), 2030–2032 (2011)
37. M. Toepper, M. Wilke, J. Roeder, T. Fischer, C. Lopper, in *IEEE, Coating Techniques for 3D-Packaging Applications*, 2012 IEEE 62nd Electronic Components and Technology Conference, (2012) pp. 1673–1676
38. C.J. Brinker, G.C. Frye, A.J. Hurd, C.S. Ashley, Fundamentals of sol-gel dip coating. Thin Solid Films **201**(1), 97–108 (1991)
39. P. Yimsiri, M.R. Mackley, Spin and dip coating of light-emitting polymer solutions: matching experiment with modelling. Chem. Eng. Sci. **61**(11), 3496–3505 (2006)
40. Y.B. Kim, H.K. Kim, J.W. Hong, Epoxy-acrylic microgels in electrodeposition coating films. Surf. Coat. Technol. **153**(2–3), 284–289 (2002)
41. R. Liu, F. An, S. Zhang, J. Luo, X. Liu, Photosensitive acrylate copolymer for electrodeposition photoresist. Polym. Sci. Ser. A **55**(4), 225–232 (2013)
42. N.P. Pham, D.S. Tezcan, W. Ruythooren, P. De Moor, B. Majeed, K. Baert, B. Swinnen, Photoresist coating and patterning for through-silicon via technology. J. Micromech. Microeng. **18**(12) (2008)
43. Y. Shen, L. Yao, Z. Li, J. Kou, Y. Cui, J. Bian, C. Yuan, H. Ge, W.-D. Li, W. Wu, Y. Chen, Double transfer UV-curing nanoimprint lithography. Nanotechnology **24**(46), 465304 (2013)

Chapter 6
Functional Metamaterials for Lab-on-Fiber

Peter Reader-Harris and Andrea Di Falco

Abstract In this chapter we discuss a versatile method to functionalize the tip of optical fibers for lab-on-fiber applications. At variance with traditional direct writing methods, the photonic functionality is implemented on a flexible, ultrathin support hosting metallic nano-features that is applied on the tip of a fiber. Here we present and discuss several fabrication strategies that combine different substrates with alternative approaches to realize the metallic photonic layer. As a specific application example we focus on a method to filter spectrally the broadband light transmitted through a standard multimode fiber. The filter consists of a metallic nanowire grating on a polymeric substrate, which acts as a guided mode resonance filter with an additional Fano resonance. The advantages and limitations of the presented technique are discussed as well.

6.1 Introduction

Applications and uses of optical fibers have long surpassed the mere delivery of information from a source to a detector. For many applications the most appealing feature of optical fibers is the possibility to take advantage of long propagation distances, granting extended interaction between light and matter. This means that a weak perturbation can be harnessed efficiently at the end of the fiber. For example it is possible to create very efficient distributed Bragg gratings with very small index contrast or obtain strong non-linear response from the fibers, using reduced optical power. Direct fiber writing and photonic crystal fibers are

P. Reader-Harris · A. Di Falco (✉)
SUPA, School of Physics and Astronomy, University of St Andrews, St Andrews, UK
e-mail: adf10@st-andrews.ac.uk

P. Reader-Harris
e-mail: pjrh@st-andrews.ac.uk

A. Cusano et al. (eds.), *Lab-on-Fiber Technology*,
Springer Series in Surface Sciences 56, DOI: 10.1007/978-3-319-06998-2_6,

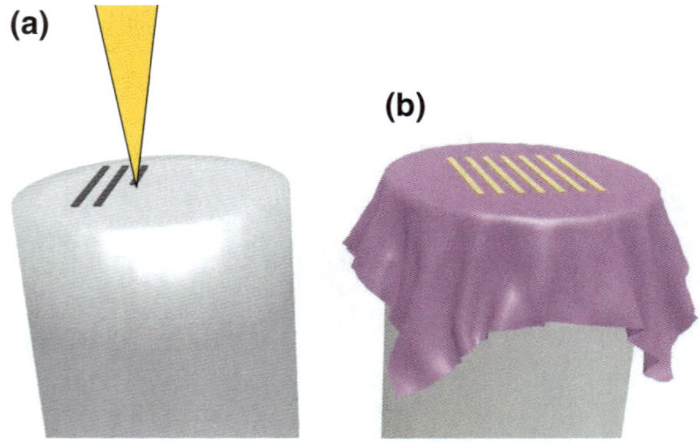

Fig. 6.1 **a** Direct writing approach. **b** Membrane functionalization

extremely successful embodiments of this approach, and supercontinuum sources are a widely available product of this success.

However, for the most advanced photonic functions, like signal routing, optical filtering and sensing, the standard solution has been that of resorting to external, more or less integrated, optical circuits. This approach requires the ability to control reliably and accurately the coupling between signal-delivering fibers and more *complex* circuitry.

Recent technological progress has provided a means to overcome this intrinsic limitation, by adding the desired functionality onto the fiber tip itself.

In this chapter we discuss a promising approach to the lab-on-fiber concept. There exist different methods in the literature to add a photonic function on the tip of a fiber. Successful and versatile examples include using a focused ion beam (FIB) to directly etch the tip of the fiber, with or without metallo-dielectric coating [1, 2]. Another popular method uses electron beam lithography (EBL) to pattern the resist spun on the tip of a fiber, which is then transferred onto the fiber itself with dry or wet etching [3] (see Fig. 6.1a).

At variance with these *direct writing* methods we realize the photonic layer on a separate substrate, which is then applied to the facet of the fiber (see Fig. 6.1b). This allows the fabrication to be achieved using only standard fabrication techniques on a planar substrate. It also means that one fiber can be coated multiple times with different terminating layers, so that a variety of different experiments can be performed with the same fiber.

While it is possible to use rigid substrates, for example semiconductor based photonic crystals [4], we adopt flexible membrane substrates, which offer a series of practical advantages, as discussed in the following sections.

Our membranes belong to the wide class of metamaterials (MMs). MMs are materials with engineered properties. They can be designed to work at optical wavelengths, where they exhibit special properties, like unusual refraction. Their optical behaviour is determined by the properties of their smallest structure (meta-atom), which is artificially created, e.g. via lithographic means or self-organization, rather than by the mere materials of which they are made. MMs are effective index materials, so the properties depend on the ensemble response to light. Typically they consist of periodic distributions of metal resonators with a typical size which is a fraction of the operating wavelength. Ordinary MMs are bulky and rigid, since traditional approaches to realize nano features are inherently planar. However this legacy has recently been surpassed, and the field of MMs has undergone a paradigm shift through the introduction of flexibility in a traditionally flat and static world. Flexibility has already been covered in other fields, such as electronics or organic LEDs, electronic foldable paper [5], flexible displays [6], and smart fabrics [7, 8]. In optics, this approach has recently been gaining growing attention [9–11], because of the possibility to mechanically tune the optical properties of the system [12–14], while retaining the degree of freedom of design typical of metamaterials. Crucially, a supple substrate can be easily wrapped over complex topologies, including fibers [15, 16], and their tips [17]. Additionally, the substrates can be chosen between a vast range of materials, like graphene or dye doped polymers, which would grant additional advanced optical and electrical properties.

Single surface MMs, which are typically called metasurfaces [18–23] need to satisfy the criteria of homogenisation and emergent properties typical of MMs. However, for practical purposes in this chapter we will abandon, as required, this stringent nomenclature requirement [24]. In this sense we will refer to flexible nanoplasmonics and frequency selective surfaces [25], rather than strictly to metasurfaces.

Flexible metasurfaces facilitate a broad range of potential applications for lab-on-fiber technology. Simply applying a patterned membrane to the end of the fiber provides an easy route for its functionalization, without permanently modifying the fiber itself. This allows for reusability as well as decoupling the problem of the fiber hardware and the fabrication of the functionalized termination. This is a versatile and robust method because even very thin membranes can withstand a negligibly small bending radius and are independent of the fiber's properties, whether it is GRIN, step index or even a photonic crystal fiber. The ability to introduce any metasurface device onto fibers will certainly contribute to the development of lab-on-fiber applications, including filtering, sensing and coupling as sketched in an artist impression in Fig. 6.2.

In the following sections we first present complementary viable fabrication routes of flexible metasurfaces and focus on the application example of tailored filters based on guided mode resonances. We conclude the chapter with a discussion of the advantages and the limits of our approach.

6.2 Flexible Metasurfaces for Lab-on-Fiber

Depending on the specific application at hand, the photonic membranes have a series of opto-mechanical requirements. A non-exhaustive list includes:

- **Nanoplasmonic Features**: one of the most versatile approaches to control light behaviour at a fraction of visible wavelength is nanoplasmonics. At the

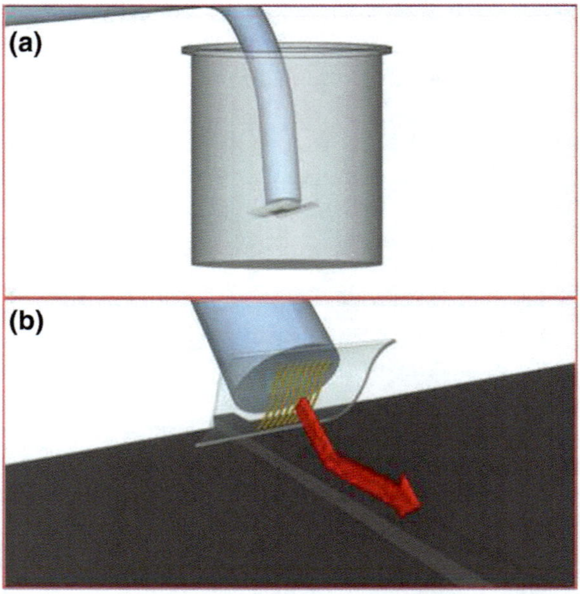

Fig. 6.2 Artistic impression of lab-on-fiber applications. **a** Sensing. **b** Waveguide coupler

nanometer scale, coupling between the free oscillating electrons in metals and incoming photons can be used e.g. to obtain extremely high intensity local fields, strong dependence from the environment and optical phase control. To fabricate such features both top down (e.g. lithography) and bottom-up (involving chemical procedures) approaches can be efficiently used. The application of these techniques can be more or less complicated, given that the substrate is a typically dielectric membrane.

- **Flexibility and Conformability**: The membrane substrate material has to be chosen in order to grant the correct flexibility and conformability to the tip of the fiber. An excessively stiff membrane might not make good contact with the fiber facet, which in turn might not be perfectly flat. On the other hand if the membrane is too supple, it might crease and deteriorate the optical function.
- **Elasticity**: A certain degree of elasticity is highly desirable, as it simplifies the mounting of the membrane onto the fiber. This could solve the adhesion issue and straighten potential irregularities of the membrane's surface.
- **Free standing**: In some cases, like in the application example of the next section, the membranes must be free standing, i.e. the designed photonic response might require a certain distance between the fiber's facet and the membrane itself.

In this section we provide an overview of the technical steps required to fabricate and apply a metasurface on the tip of a fiber. We first focus on the membrane and then on the mounting technique.

6.2.1 Fabrication Techniques

We have classified the fabrication approaches in four categories, according to the way in which the photonic layer is realized.

6.2.1.1 Top-Down Directly on Substrate

This fabrication technique involves depositing the membrane substrate onto a rigid material, realizing the desired metallic nanofeatures and then releasing the membrane from the host material. A sacrificial layer could be used to mediate this final step.

Substrate

There exist several solutions to obtain flexible substrates. For the sake of brevity we focus only on polymer-based and self-assembly monolayer materials. The first are extremely versatile; the latter permits extremely reduced thickness, down to one molecule thick.

Polymer-Based Substrates

Solution processing makes many polymers very easy to handle, and spin coating provides an easy route to the realization of highly uniform polymer layers of known thickness on a rigid, planar wafer. The polymers can also be deposited with other methods, like spray and blade coating or drop casting. If required the same technique can be used to deposit a sacrificial layer first, with a subsequent membrane layer on top. After this the membrane solution can be induced to cure or polymerize, forming a solid layer, typically via optional exposure to ultraviolet (UV) light and heat. A variety of polymers have been used, such as polydimethylsiloxane (PDMS), polyimide, and Microchem SU8. It is important to select polymers which can withstand the required steps for the rest of the fabrication procedure, which can be limiting. For example, often lithographic processes require baking the photoresist at high temperatures. If the temperature is too high, the polymer can be damaged before lifting the membrane from the rigid substrate. It is also important to take into account the wetting and anchoring properties of the metal (for the meta-atoms) on the specific polymer of choice. Failing to do so may result in a discontinuous metallic layer, prone to flake off.

Self-Assembly Monolayers

An alternative method to spin coating polymers onto wafers is to fabricate a monolayer of suitable molecules, e.g. 1,1-biphenyl-4-thiol [26, 27]. This approach gives thinner substrates down to less than 1 nm. The membranes are typically realised by a self-assembly method, as traditional deposition techniques do not have the fine control to create uniform layers thinner than a few tens of nanometers. Using a specially chosen molecule which has a thiol group that readily absorbs onto a surface, a single layer of these molecules can form on the substrate.

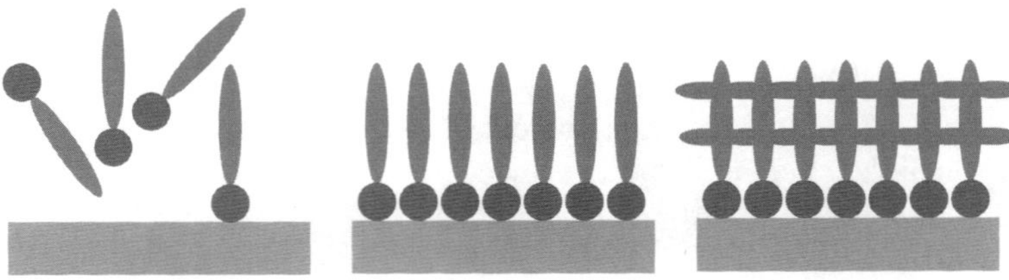

Fig. 6.3 Fabrication of a self-assembled monolayer. The molecules absorb onto the surface of the substrate because of the carefully selected head group, while the tail groups can then be cross-linked to form a monolayer [26]

Fabrication of a nanosheet can then proceed by electron beam exposure which cross-links adjacent molecules together. A schematic of this technique is shown in Fig. 6.3. The membrane can also host metallic patterns on top, before being transferred to a suitable support frame (like a TEM grid). It has been shown to be possible to suspend gold structures supported by a 1 nm thick membrane across distances of tens of microns [28].

Metallic Features

Once a suitable polymer has been selected and spun onto a rigid substrate there are a number of ways to fabricate subwavelength features on top. Here we review etch back and lift off, which are based on electron beam lithography and thus inherit the fast prototyping inherent in this approach. It should be noted that nanoimprinting techniques can be used to scale up the fabrication volumes at a fraction of the cost, but always require a lithographically produced master from which to work. These methods are preferred to focused ion beam (FIB) milling because it can damage the membrane during fabrication.

Etch Back

Further layers of material can be deposited by both spin coating and evaporation, before a patterned layer is created on top. This acts as a mask which selectively protects parts of the lower layers from being etched from the top. The full fabrication steps (including the sacrificial layer) are shown in Fig. 6.4. A rigid substrate is used to form the base onto which the membrane can be built up, before being released. Spin coating is used to create highly uniform layers of polymers of known thickness. The first layer which is spun is the sacrificial release layer, on top of which is spun a second layer of the membrane polymer. The metal is evaporated on top of this, before a final photoresist layer is spun on the top. This can then be patterned via ultraviolet or electron-beam lithography. This selectively polymerises the photoresist layer, so that upon sample development, areas of the photoresist which have not been exposed will be washed away. The sample is then etched in an ion etcher, which removes the unprotected parts of the metal layer. In this way a metal pattern can be transferred onto the top of the membrane. This technique works with thin membranes because the etching can be finely controlled, but care must be taken that the etch step does not affect the quality of the membrane.

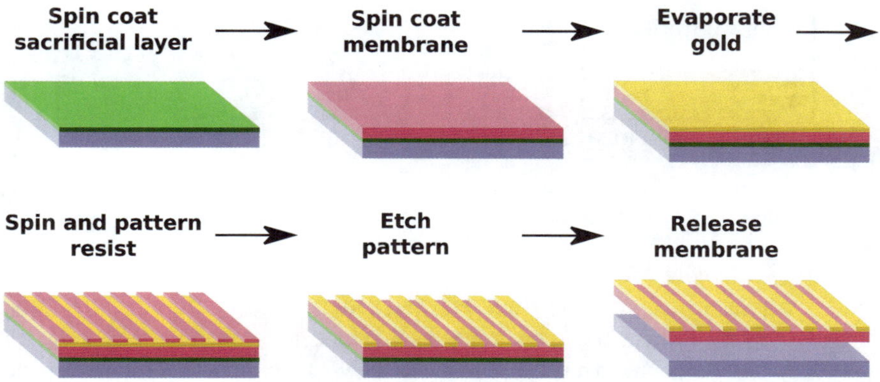

Fig. 6.4 The full steps to fabricate a membrane using the etch back and release layer dissolution methods. A series of layers are stacked by spin coating and metal evaporation, before lithographically patterning the top resist layer to create a mask. The features are transferred in the layer beneath via ion plasma etching

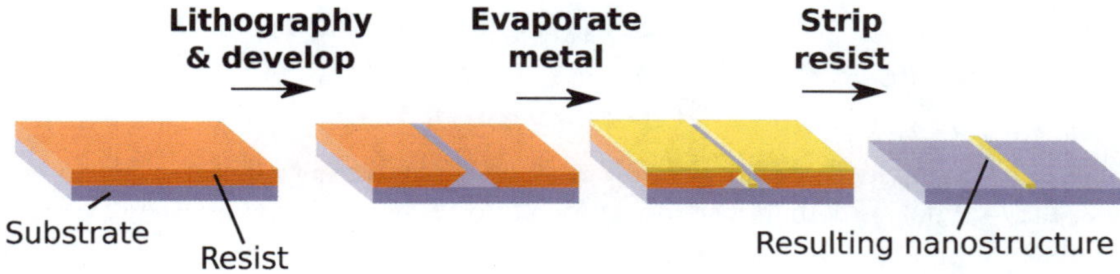

Fig. 6.5 A typical lift off process, which can be coupled with one of the substrate fabrication techniques above

Lift Off

Where directly etching the sample is not possible because of the potential damage caused, lift off can be used to create features similar to etch back. The approach is shown in Fig. 6.5. The resist is spun onto the wafer first, and patterned directly by the electron beam before development. For a positive resist (e.g. PMMA), after development this leaves gaps in the resist where the resist has been exposed. Typically either a thick resist, or more than one layer of resist, is used such that an undercut is formed, i.e. the size of the feature at the bottom of the hole is bigger than the top. This means that when the metal is evaporated onto the sample, it can deposit straight onto the wafer through the gaps, whereas the rest of the sample is protected by the resist. The resist is then stripped with another solvent, leaving the substrate with just the desired pattern left on it. The undercut ensures that the metal does not form a continuous film, which would not allow the solvent reaching and dissolving the resist. Since the electron beam step requires a conducting pathway to prevent charging, this method is unsuitable for insulating polymeric membranes (unless an additional conductive layer is deposited on the resist before exposure), but is ideal for self-assembly monolayers [28].

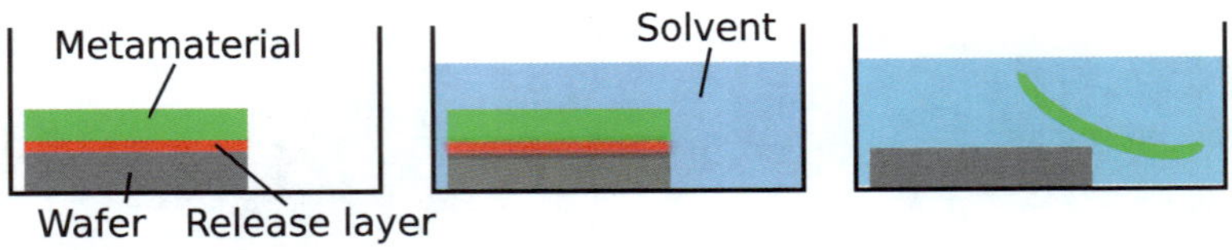

Fig. 6.6 The release of a membrane from a wafer using a sacrificial layer. The membrane is left in the solution ready to be transferred either to another solution, or mounted on a frame

Membrane Release Methods

Once a polymer membrane has been spun and patterned on a rigid substrate, then it is necessary to release the membrane so that it can be manipulated as a free standing and flexible metasurface. The method by which this can be done depends on the thickness and robustness of the membrane.

Peel Off

For thick substrates it has been shown that it is possible to simply peel the structure from the wafer. Here *thick* means of the order of a few microns. For instance, this has been demonstrated with a 8 μm thick terahertz metamaterial absorber based on a polyimide membrane [29]. It is unclear how scalable this method would be, as it does require considerable care to remove the membrane from the substrate.

Chemical Release

Although peeling methods are possible and have the advantage that they are solvent free, they do not provide a route to fabricate thin membranes which are typically fragile and prone to tearing. However, thin membranes are required if the structure thickness is to be smaller than the wavelength of visible light, or if very small bending radii are desired. In this case a sacrificial layer release method is required [9, 15]. In this scheme before the membrane is fabricated on a rigid wafer, an initial sacrificial layer is first deposited. Anything that is subsequently fabricated on top of this layer will be removed from the wafer when the sacrificial layer is dissolved in a solvent, and left suspended in the solution, as shown in Fig. 6.6. Once the membrane is suspended in the solution it is then possible to transfer the membrane by scooping it onto another object, such as a mount, as detailed at the end of this section. This forms a robust protocol which can realize membranes down to less than 200 nm thickness.

Critical Point Drying

Ultrathin substrates, such as the self-assemble monolayers above, can be very fragile and easily torn. It has been found that the yield is increased by using a critical point dryer (CPD). In this process, the delicate membrane sample is attached for mechanical robustness to a polymer (e.g. polymethyl methacrylate, PMMA,) which is soluble in a solvent that does not damage the membrane itself (e.g. acetone). The membrane/polymer sheet is then transferred to the CPD chamber where it is submerged in the solvent, to dissolve the supporting polymer. The solvent is then replaced by CO_2 which is brought around its critical point by adjusting temperature and pressure. Gaseous CO_2 is finally slowly purged away form the chamber, without surface tension forces on the membrane [28].

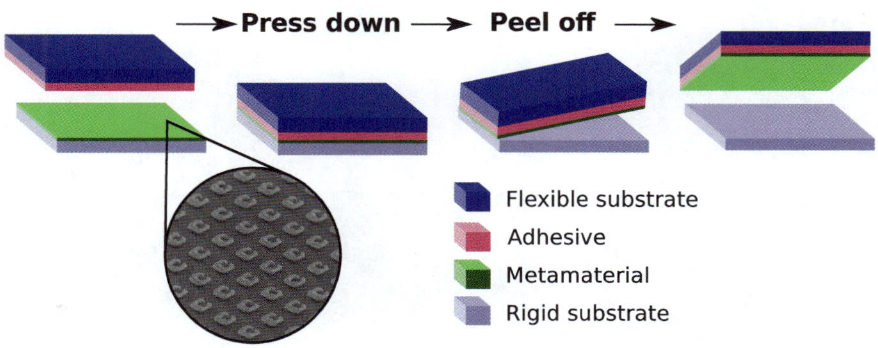

Fig. 6.7 Flip chip transfer. The metamaterial is fabricated on a wafer, before being transferred to a flexible substrate, by promoting greater adhesion to the new substrate rather than the old, so it can be peeled off [11]

6.2.1.2 Peel Off Transfer

There are some occasions when it is not possible to write directly onto a polymeric substrate, due to fragility or incompatibility of the desired materials' processing requirements. In this case it can be useful to create the nanostructures on one substrate and then transfer them later, either onto another substrate, or directly onto an object of interest. A method to achieve this is called peel off, or flip chip, transfer [11].

As shown in Fig. 6.7, this is a general method to transfer any pattern from one surface to another. Any fabrication technique can be used to realize the metamaterial structures. Sometimes an extra layer is placed between the rigid substrate and the features. However this is to reduce, rather than promote, adhesion so that that features do not remain bonded to the substrate. This technique has been demonstrated for lab-on-fiber technologies by transferring metallic nanofeatures onto the end of a fiber [30].

Another way peel off transfer can be used is by growing the features using a template. The typical procedure is sketched in Fig. 6.8a, extracted from [31].

A host substrate is coated with SAM that can be cross linked locally, via electron beam exposure. In this example we used biphenyl-based molecules on mica samples coated with gold. For these molecules at an electron beam irradiation dose between 40 and 1000 mC/cm^2 the SAM crosslinks. The sample is then immersed in a electrochemical deposition chamber, and the desired metal (copper is this case) is allowed to grow where the membrane is not crosslinked. With opportune deposition parameters, the metal can be grown to form continuous features that can be defined down to a scale of a few tens of nanometers [31]. The metallic patterns can then be transferred onto an epoxy stamp via simple peeling off, as sketched in Fig. 6.8b. The main advantage of this approach is that the transfer does not destroy the SAM. The metallic deposition can then be repeated multiple times, without having to repeat the slow and expensive writing step.

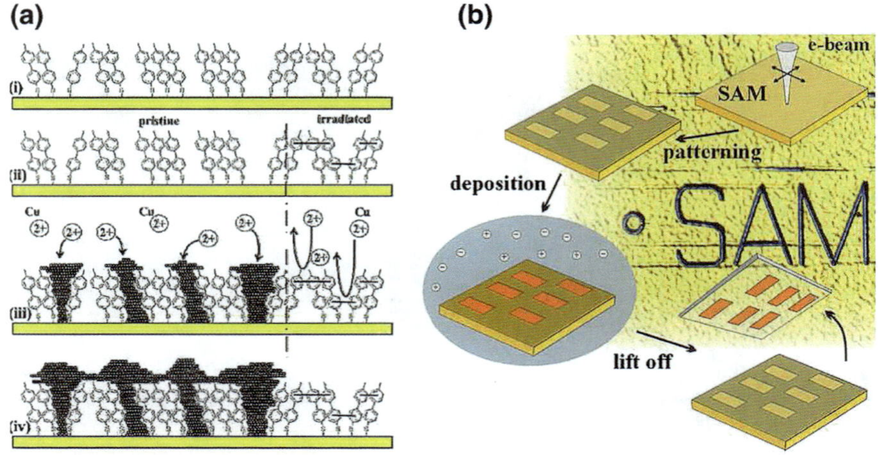

Fig. 6.8 **a** Fabrication procedure: a SAM layer is deposited on a suitable substrate *(i)*, cross linked locally with electron beam exposure *(ii)* and metal is deposited with electrochemical deposition *(iii)* to form continuous metallic features *(iv)*. **b** Sketch of peel off transfer scheme. The background image is a false colour AFM image of a typical sample after metal deposition with minimum features size down to 20 nm [31]

6.2.1.3 Nanoskiving

Although the other methods in this section assume the fabrication of plasmonic elements on either sides of a flexible membrane, nanoskiving is an alternative technique which allows the fabrication of the nanostructures to be embedded into the substrate [32, 33]. It is also possible to then remove the substrate by exposure to an oxygen plasma, leaving behind only the nanostructures. This has been demonstrated for use for lab-on-fiber applications [34].

The most important step in the fabrication procedure is being able to slice, or skive, an epoxy based block of polymer into very thin layers (down to 30 nm in thickness). This is achieved using a suitably sharp knife, usually a diamond edged knife, and careful selection of materials which can withstand such sectioning [35]. The pattern results from a section of a block built up by any soft lithography method, which can include the evaporation of a metal film, an example of which can be seen in Fig. 6.9. The structures are then embedded in epoxy before being sectioned with an ultramicrotome, and applied to where the structures need to be, before subsequently removing the epoxy with an oxygen plasma.

6.2.1.4 Printing with Metallic Ink

The final method that we discuss combines the resolution control of a top-down lithographic approach and the flexibility of a bottom-up technique. A polymeric stamp is obtained via drop casting on a previously patterned master. In the

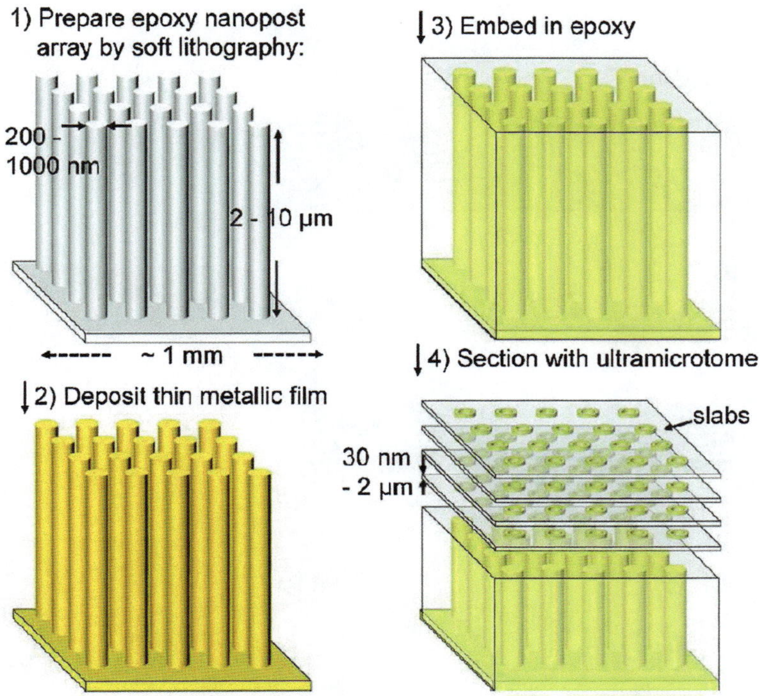

Fig. 6.9 Epoxy based structures can be fabricated by any soft lithography method, before evaporation of gold and subsequent slicing into layers which can be deposited on the fiber facet. The epoxy can then be removed by an oxygen plasma leaving the gold features behind on the fiber facet. Reprinted with permission from [34]. Copyright 2011 American Chemical Society

example of Fig. 6.10 a silicon master was used and a PDMS polymer for the stamp [36]. A SAM of metallic particles is then created on the surface of a DI water pool and used as metallic ink for the stamp. The features are then transferred on the desired target. Panel f of Fig. 6.10 shows a set of metallic nano gratings on a piece of commercial cling film. This method allows replication of a metallic nano pattern virtually on any target surface, including the tips of fibres. Additionally, the flexibility of the PDMS grants the conformability of the pattern to potential complex shaped targets, which would be very difficult to pattern otherwise.

6.2.2 Fiber Mounting

After fabrication the membrane must be transferred to the target object. The released membrane is randomly oriented in the release solvent, and can possibly curl or twist. In order to be able to transfer the membrane in a controlled manner, it is possible to suspend the membrane on the surface of a liquid either by buoyancy

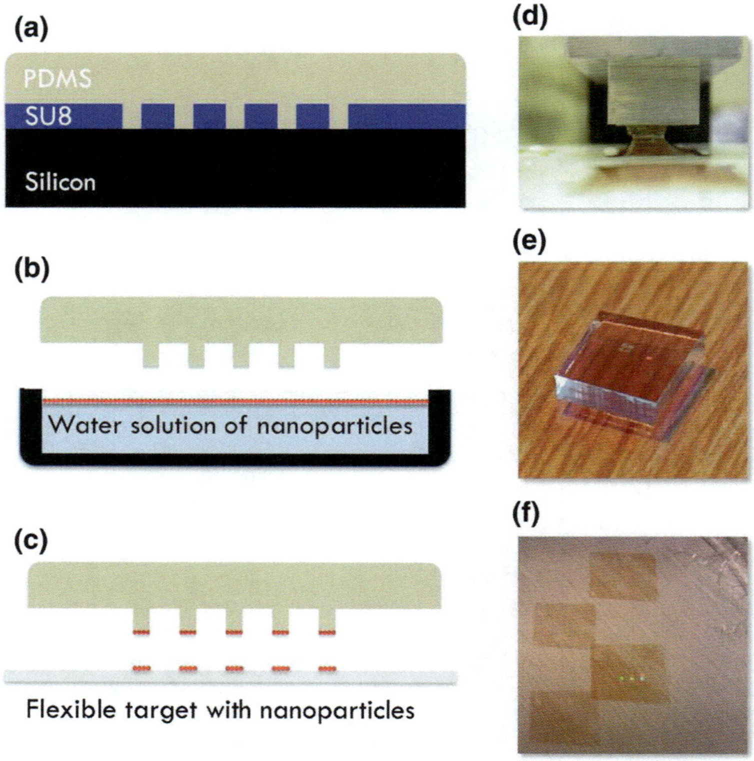

Fig. 6.10 Sketch of the fabrication technique steps (**a**–**c**) and corresponding photographs (**d**–**f**)

or by the hydrophobic effect. Then the membrane can be transferred onto a flat surface [16], or mounted free standing onto a frame, or other object.

Suspension over Fiber
A free standing membrane which has been released and mounted can be placed over the fiber easily by wrapping or otherwise suspending it over the end of the fiber. One example of this is found in Fig. 6.14, where by mounting the membrane on an aluminium foil frame, the membrane could be suspended over the end of a collimator package terminating a fiber. It is envisaged that if the fiber size was known in advance, then a rubber hollow cylinder could act as an endcap containing the membrane, enabling very easy on and off use of the membrane. This could also enable a variety of different functionalized ends to be used with one fiber, merely by changing the endcap.

Direct Application to Fiber
The alternative to this approach is to permanently deposit the functionalized end onto the fiber. This can be done simply by cutting the membrane to size and placing on the end of the fiber or, as demonstrated by Limpomi et al. [34], by

Table 6.1 Three examples of different membrane thickness, with the required SU8 dilution and spin speed

Desired thickness	Mixture	Spin speed (rpm)
4 μm	SU8 2000.5:SU8 2050 1:1	5000
1 μm	SU8 2000.5:SU8 2050 3:1	5000
190 nm	SU8 2000.5:cyclopentanone 1:2	1000

floating the membrane on top of a liquid and pushing the fiber down, submerging the fiber and membrane. Allowing this to dry attaches the membrane to the end of the fiber, probably by Van der Waals forces.

6.3 Demonstration of Use on Fibers

In order to demonstrate the use of the techniques above in a concrete situation, we give the example of a filter mounted on the end of a fiber. This application could be especially useful e.g. in a pump-probe type experiment, if a weak signal is expected amongst a strong background of a different wavelength. Filtering light before it enters the fiber prevents noise scattering in the fiber.

6.3.1 Guided Mode Resonance Filter

A proof of concept of a fiber mounted metasurface is the guided mode resonance filter (GMR). These filters are a class of filter initially developed in the 1990s [37, 38], which can have very narrow linewidths, while still maintaining very good visibility. Typically the substrate of a metasurface can undesirably affect the properties of the plasmonic materials, but for the GMR the substrate is a necessary part of the construction of the device.

6.3.1.1 GMR Fabrication

The GMR can be fabricated using the etch back protocol above [17], shown in Fig. 6.4. The sacrificial layer, Microchem Omnicoat, was firstly spun at 1000 rpm onto a silicon substrate before baking at 230 °C for 1 min. Microchem SU8 is used for the membrane layer because of its compatibility with the rest of the process, and its biocompatibility. It is commercially available in a number of different molecular weights which can be spun to different thickness, or diluted as desired. Table 6.1 shows three example membranes of thickness between 4 μm and 190 nm with the SU8 mixture and spin speed required. The SU8 is then soft baked for 5 min, using a ramp of temperature from 60 to 100 °C, to prevent bubbling. The membrane is then polymerized with UV light for 2 min, followed by a further

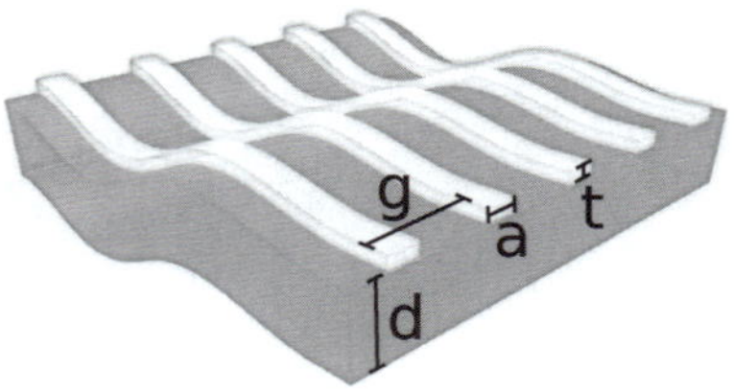

Fig. 6.11 Guided mode resonance filter, geometrical parameters. Reprinted from [17]

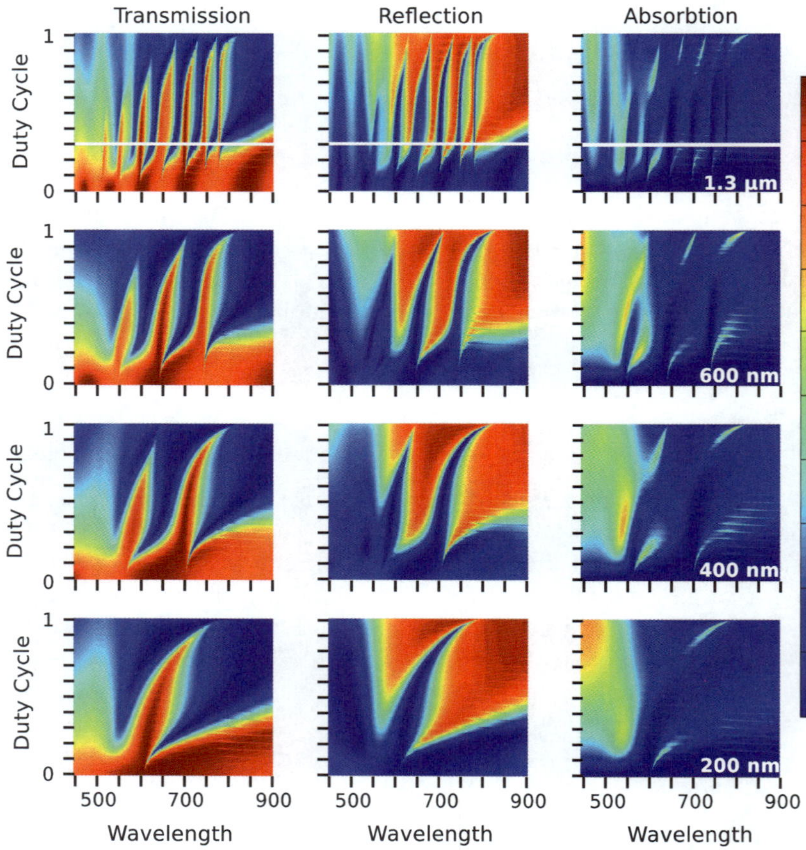

Fig. 6.12 How the geometrical properties of the GMR affect the optical properties: each graph shows how the spectrum (*transmission, reflection and absorption*) changes with the duty cycle, and from *top* to the *bottom* different thicknesses of substrate are shown. The thickness of gold was held at 30 nm and the period at 500 nm for all these simulations, in TM polarization

bake at 100 °C for 2 min. 30 nm of gold is deposited onto the surface of the membrane by electron beam evaporation. The photoresist for the lithography step is then spun. For low duty cycle gratings SU8 itself is a high resolution negative tone photoresist, so that what is written by the electron beam is kept. This is spun to about 90 nm by the same mixture shown in the bottom of Table 6.1 but with a spin speed of 5000 rpm. This thickness is chosen as a trade off between the resolution provided by thin films and the etch resistance required to mask 30 nm of gold. Instead of UV induced polymerization it is patterned by the a 30 kV electron beam with a dose which depends on the structure being written. The sample is then baked for 2 min at 100 °C to increase the contrast between areas which are exposed and unexposed, before being developed in EC solvent for 45 s. This leaves only the SU8 etch mask on top of the gold coated membrane. The samples is then etched with Argon ions, at a power of 20 W, pressure 5×10^{-2} Pa and flow rate of 10 sccm, for 8 min. The final step is to cleave the edges of the whole sample, including the silicon substrate, so that the solvent can reach the sacrificial layer effectively, before placing it in *N*-methyl-2-pyrrolidone (NMP) for about an hour. Once the membrane has detached from the silicon it can be transferred to a water bath where it suspends hydrophobically on the surface, and can be manipulated or transferred onto a frame or other object.

Alternatively, for high duty cycle gratings a positive tone resist is desirable. ZEP520A, from Zeon Chemicals, is used to as a high resolution positive tone resist, so that what is written is later removed. The recipe is the same as that given above, except instead of the SU8 layer on top of the gold, undiluted ZEP520A is spun at 5000 rpm, before baking at 140 °C for 20 min. The pattern is written, there is no post exposure bake, and development is performed with xylene at 23.5 °C for 45 s. The same etch recipe is used, although some ZEP does remain, and has to be removed by rinsing several times with tricholroethylene and exposure to UV light.

6.3.1.2 Filter Operation

The filter works by coupling light into waveguide modes within the flexible substrate. A subwavelength grating is required so that only the zeroth order can be reflected or transmitted. This prevents any light being lost by propagating in a direction which isn't collinear with the incident light, while still allowing the other evanescent modes to couple into the waveguide modes. The light which is coupled into the waveguide slab can then be coupled back out into the same propagation direction as the incident light.

The geometric parameters of the device, shown in Fig. 6.11, define the operation of the filter. Changing these allows a range of different filters to be fabricated. This is illustrated by the simulations shown in Fig. 6.12, where each graph shows how the spectrum (transmission, reflection and absorption) changes with the duty cycle. From top to the bottom different thicknesses of substrate are shown. In Fig. 6.12 the period (g) was fixed at 500 nm so that the resonances fell in the middle of the visible regime, and a gold thickness of 30 nm was chosen. Changing the thickness of the gold layer does not affect the operation of the filter drastically, unless much thicker gold is used when the absorption becomes much higher. The simulations were based on the rigorous coupled wave analysis (RCWA) method

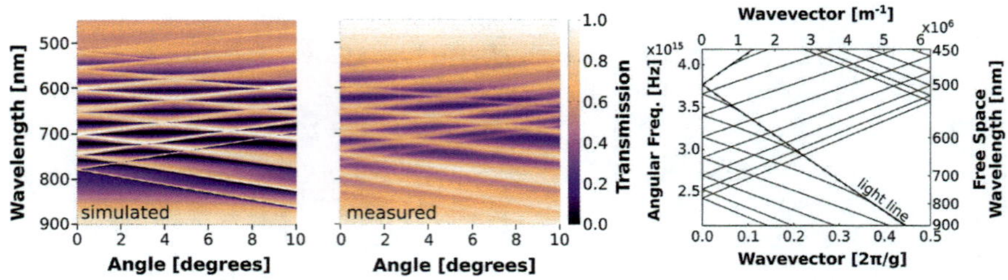

Fig. 6.13 Angular dependence and bandstructure of the GMR membrane. Reprinted from [17]

[39], assuming a plane wave incoming with electric field perpendicular to the grating wire length (TM polarization) and the Johnson and Christy values were used for the refractive index of gold [40]. A single filter has a spectrum which is a horizontal cut at constant duty cycle of one of these graphs.

This is a versatile way to fabricate filters. It can be seen that the duty cycle (see Fig. 6.11) controls whether the filter is primarily transmissive, or reflective. At high duty cycle a bandpass filter is produced in transmission, with a notch filter in reflection, and at low duty cycle these are swapped. The resonances do shift when going from low duty cycle to high, which shows that the position of the waveguide is spectrally changing. At low duty cycle the location of the resonances can be calculated with the Helmholtz equation and the empty lattice approximation, which states that the bandstructure simply folds back at the Brillouin zone edge [41]. However, at high duty cycle the empty lattice approximation no longer holds and the presence of the metal alters the bandstructure. This effect is stronger as the membrane thickness decreases, as the guided mode has a stronger overlap with the metal. It can also be seen that as the thickness of the membrane decreases fewer modes are available in the waveguide into which the incident light can couple. This controls the free spectral range (which for thin membranes can be very large), which can be seen by the reduction in the number of vertical lines in the transmission and reflection plots.

Figure 6.13 shows the angular resolved spectrum of a single filter, in correspondence of the white line from the topmost graph in Fig. 6.12. Both simulated and measured spectra are shown, alongside the bandstructure (as calculated only using the Helmholtz equation and the empty lattice approximation). As the incident angle is varied, this probes different k-vectors parallel to the plane of the dielectric slab, and so the shape of the bandstru cture is traced out, confirming that the dielectric slab guided modes are causing the filtering effect observed.

6.3.1.3 Fiber Mounted Operation

In order to show an initial demonstration on a fiber, the filter was transferred onto a aluminium foil and wrapped around a collimator terminated fiber. This was achieved by puncturing a hole in the aluminium foil and then mounting the flexible

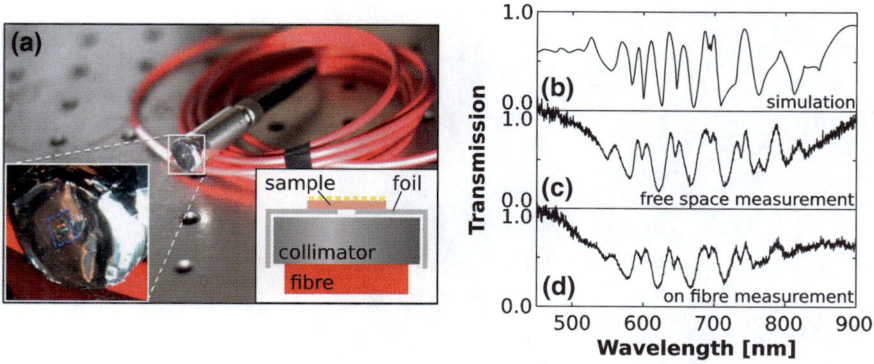

Fig. 6.14 An example of a use of a fiber terminated with a filter. Reprinted from [17]

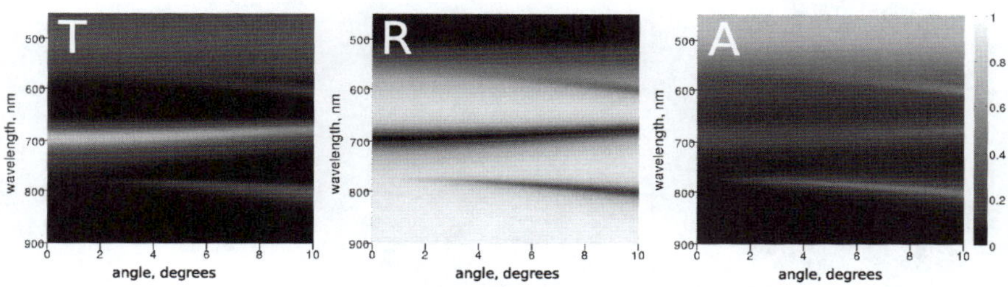

Fig. 6.15 RCWA simulation of a thin membrane showing angular robustness. The membrane is 190 nm thick, with a periodicity of 500 nm, duty cycle of 80 % gold and a gold thickness of 40 nm

filter over the hole, by suspending it on the surface of a water basin by the hydrophobic effect and scooping it up. As shown in Fig. 6.13 the filter does not have a very high angular robustness. To facilitate the application of this filter onto the end of a standard optical fiber, first the fiber was terminated with a collimator package from Thorlabs (F260SMA-B) with a polarizer (LPVISE2X2) cut to size and placed in the collimator. The results of this can be seen in Fig. 6.14, with the accompanying spectra comparing the simulated filtering action compared to the free space and fiber mounted configurations, showing a favourable agreement.

6.3.2 Angular Robustness

One of the key problems with the guided more resonance filter is that typically the device is highly angularly dependant, as shown in Fig. 6.13, preventing direct application to the termination of a fiber.

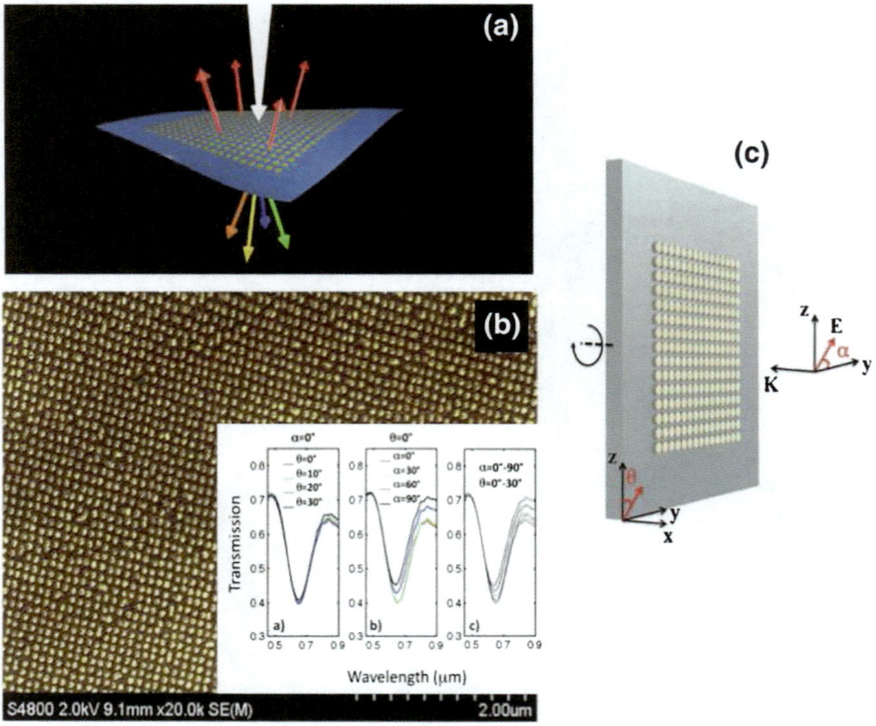

Fig. 6.16 **a** Sketch, **b** SEM and inset, **c** Sketch setup. Reprinted with permission from [44]. Copyright 2011, AIP Publishing LLC

Manipulating the Bandstructure
Methods to engineer the bandstructure for increased angular tolerance have been suggested, for instance using doubly periodic gratings [42, 43]. It is, however, also possible to engineer the bandstructure with only one grating, which is convenient because doubly periodic gratings are harder to fabricate. Evidence for this is that at high duty cycles the location of the guided modes change due to the presence of the metal, as in the previous section. Thin, high duty cycle membranes exhibit this effect the greatest, and by selecting the parameters carefully an anti-crossing at the Γ-point of the Brillouin zone (corresponding to normal incidence) is achieved. Figure 6.15 shows such a membrane: 190 nm thick, with a periodicity of 500 nm, duty cycle of 80 % gold and a gold thickness of 30 nm. Two bands can be seen, the top of which crosses normal incidence at 700 nm. The bottom band cannot couple to normally incident light due to symmetry, but can clearly be seen to be anti-crossing with the top. This occurs because the electric field in the top band is mainly in air, but for the lower it is mainly in the dielectric, forming a small bandgap. This small bandgap is enough to provide angular robustness up to 10°, enough to apply directly onto the end of a fiber with a numerical aperture of 0.17, which covers a broad range of fibers. This allows a filter to be fabricated which

does not require to be applied to a collimator, rather is can be directly to the end of the fiber itself.

All Angle Reflector

An alternative approach to control the scattering properties of a photonic membrane that does not depend on the thickness of the substrate resorts to periodic metallic patterns with scale much smaller than the wavelength. In this case it is possible to design metasurfaces that reflect one specific frequency for all angles and polarisations, as sketched in Fig. 6.16a. In this example the surface of a 4 μm thick SU8 membrane is patterned with gold disks of 80 nm diameters at 100 nm from each other [44]. With this parameters' choice, the resonance condition is only determined by the shape of the gold features. The scattering of the membrane has a large tolerance with respect to the angle of incidence and the lattice arrangement grants independence from the polarisation. The inset of Fig. 6.16b shows the experimental results for different experimental conditions, following the angle definitions of Fig. 6.16c.

6.4 Conclusions and Outlook

Functionalizing the termination of fibers with flexible metasurfaces is an excellent way to bring all the properties of metamaterials, and the unique optical properties afforded by these structures, into the realm of lab-on-fiber technologies. This chapter has shown the versatility of such an approach, which is not limited to a particular geometry, or even a particular fabrication route. This avoids fabricating the required structures on the end of a fiber facet in situ, which is very difficult, usually requiring very specialized equipment.

This method is not without challenges. One consideration is the fragility of the membranes themselves. While nanometrically thin membranes do present fabrication difficulties, in general once mounted they are robust enough to require only the care with which standard optical components are handled. Different groups have found that by suspending the membranes on the surface of water the membrane can be manipulated without damage occurring [16, 17, 34].

Most of the techniques in this chapter use electron beam lithography at some point in the process, which is a quick and accurate way to generate many samples for fast prototyping and device development in the laboratory, but it is expensive. Other techniques, such as nano-imprint lithography could easily be used to fabricate such devices. This is a method of making many identical copies of a nanostructure very quickly, as long as one master stamp can be fabricated, which is then impressed into subsequent polymers, patterning them. Once the master stamp is fabricated then the process becomes very cheap to replicate many devices, so it is hoped that the fabrication protocols mentioned here could all be adapted to this platform.

The main advantage of the scheme presented here is that it removes the constraint of having to fabricate the system directly on the end of the fiber. This allows standard fabrication techniques to be used on a planar substrate, rather than custom made equipment which is used only for fibers. After the fabrication is complete, then the material can be transferred to the end of the fiber. This paradigm should, we hope, enable a cheap, scalable and efficient fabrication route for lab-on-fiber applications.

A further benefit of this method is that one fiber can be coated multiple times with different terminating metasurface membranes. Thus, one set of expensive equipment is used for the light source and data acquisition, but a variety of different experiments can be performed in succession using one fiber. It is hoped that this method will simplify life not only in the fabrication but also for the end user, who can simply change the functionalized ending of the fiber without having to alter the rest of the experiment.

We believe that bringing together the light delivery of optical fibers and the light manipulation capabilities of metamaterials promises to facilitate and further development of lab-on-fiber technologies.

References

1. A. Dhawan, M. Gerhold, J. Muth, Plasmonic structures based on subwavelength apertures for chemical and biological sensing applications. IEEE Sens. J. **8**(6), 942–950 (2008)
2. W. Yuan, F. Wang, A. Savenko, D.H. Petersen, O. Bang, Note: optical fiber milled by focused ion beam and its application for Fabry–Pérot refractive index sensor. Rev. Sci. Instrum. **82**(7), 076103 (2011)
3. M. Consales, A. Ricciardi, A. Crescitelli, E. Esposito, A. Cutolo, A. Cusano, Lab-on-fiber technology: toward multifunctional optical nanoprobes. ACS Nano **6**(4), 3163–3170 (2012)
4. G. Shambat, J. Provine, K. Rivoire, T. Sarmiento, J. Harris, J. Vuckovic, Optical fiber tips functionalized with semiconductor photonic crystal cavities. Appl. Phys. Lett. **99**(19), 191102 (2011)
5. A.C. Siegel, S.T. Phillips, B.J. Wiley, G.M. Whitesides, Thin, lightweight, foldable thermochromic displays on paper. Lab Chip **9**(19), 2775–2781 (2009)
6. D. Lochun, M. Kilitziraki, D. Harrison, I. Samuel, Manufacturing flexible light-emitting polymer displays with conductive lithographic film technology. Smart Mater. Struct. **10**(4), 650–656 (2001)
7. D. Marculescu, R. Marculescu, Electronic textiles: a platform for pervasive computing. Proc. IEEE **91**(12), 1995–2018 (2003)
8. J.A. Rogers, T. Someya, Y. Huang, Materials and mechanics for stretchable electronics. Science **327**(5973), 1603–1607 (2010)
9. A. Di Falco, M. Ploschner, T.F. Krauss, Flexible metamaterials at visible wavelengths. New J. Phys. **12**(11), 113006 (2010)
10. D. Chanda, K. Shigeta, S. Gupta, T. Cain, A. Carlson, A. Mihi, A.J. Baca, G.R. Bogart, P. Braun, J.A. Rogers, Large-area flexible 3D optical negative index metamaterial formed by nanotransfer printing. Nat. Nanotechnol. **6**(7), 402–407 (2011)
11. G.X. Li, S.M. Chen, W.H. Wong, E.Y.B. Pun, K.W. Cheah, Highly flexible near-infrared metamaterials. Opt. Express **20**(1), 397 (2011)
12. K.J. Kim, J.W. Kim, M.C. Oh, Y.O. Noh, H.J. Lee, Flexible polymer waveguide tunable lasers. Opt. Express **18**(8), 8392–8399 (2010)
13. I.M. Pryce, K. Aydin, Y.A. Kelaita, R.M. Briggs, H.A. Atwater, Highly strained compliant optical metamaterials with large frequency tunability. Nano Lett. **10**(10), 4222–4227 (2010)

14. M.G. Millyard, F. Min Huang, R. White, E. Spigone, J. Kivioja, J.J. Baumberg, Stretch-induced plasmonic anisotropy of self-assembled gold nanoparticle mats. Appl. Phys. Lett. **100**(7), 073101 (2012)
15. N. Gibbons, J.J. Baumberg, C.L. Bower, M. Kolle, U. Steiner, Scalable cylindrical metallodielectric metamaterials. Adv. Mater. **21**, 3933–3936 (2009)
16. M. Kolle, B. Zheng, N. Gibbons, J.J. Baumberg, U. Steiner, Stretch-tuneable dielectric mirrors and optical microcavities. Opt. Express **18**(5), 4356–4364 (2010)
17. P. Reader-Harris, A. Ricciardi, T. Krauss, A. Di Falco, Optical guided mode resonance filter on a flexible substrate. Opt. Express **21**(1), 1002–1007 (2013)
18. N. Engheta, Thin absorbing screens using metamaterial surfaces, in *Proceedings of the IEEE Antennas and Propagation Society International Symposium*, pp. 392–395 (2002)
19. C.L. Holloway, A. Dienstfrey, E.F. Kuester, J.F. O'Hara, A.K. Azad, A.J. Taylor, A discussion on the interpretation and characterization of metafilms/metasurfaces: the two-dimensional equivalent of metamaterials. Metamaterials **3**(2), 100–112 (2009)
20. A. Alù, Mantle cloak: invisibility induced by a surface. Phys. Rev. B **80**(24), 1–5 (2009)
21. A. Alù, N. Engheta, in *Optical wave interaction with two-dimensional arrays of plasmonic nanoparticles*, ed. by A.A. Maradudin. Structured Surfaces as Opt. Metamaterials, Chap. 3. (Cambridge University Press, Cambridge, 2011), pp. 58–93
22. Y. Zhao, A. Alù, Manipulating light polarization with ultrathin plasmonic metasurfaces. Phys. Rev. B **84**(20), 205428 (2011)
23. F. Aieta, P. Genevet, M.A. Kats, N. Yu, R. Blanchard, Z. Gaburro, F. Capasso, Aberration-free ultrathin flat lenses and axicons at telecom wavelengths based on plasmonic metasurfaces. Nano Lett. **12**(1702), 4932–4936 (2012)
24. A. Sihvola, Metamaterials: a personal view. Radioengineering **18**(2), 90–94 (2009)
25. B.A. Munk, *Frequncy Selective Surfaces: Theory and Design* (Wiley, New York, 2000)
26. A. Beyer, A. Godt, I. Amin, C.T. Nottbohm, C. Schmidt, J. Zhao, A. Gölzhäuser, Fully cross-linked and chemically patterned self-assembled monolayers. Phys. Chem. Chem. Phys. **10**(48), 7233–7238 (2008)
27. C.T. Nottbohm, A. Turchanin, A. Beyer, R. Stosch, A. Gölzhäuser, Mechanically stacked 1-nm-thick carbon nanosheets: ultrathin layered materials with tunable optical, chemical, and electrical properties. Small **7**(7), 874–883 (2011)
28. A. Beyer, A. Turchanin, C.T. Nottbohm, N. Mellech, M. Schnietz, A. Gölzhäuser, Fabrication of metal patterns on freestanding graphenoid nanomembranes. J. Vac. Sci. Technol. B Microelectron. Nanom. Struct. **28**(6), C6D5–C6D10 (2010)
29. H. Tao, C. Bingham, A. Strikwerda, D. Pilon, D. Shrekenhamer, N. Landy, K. Fan, X. Zhang, W. Padilla, R.D. Averitt, Highly flexible wide angle of incidence terahertz metamaterial absorber: design, fabrication, and characterization. Phys. Rev. B **78**(24), 241103 (2008)
30. E.J. Smythe, M.D. Dickey, G.M. Whitesides, F. Capasso, A technique to transfer metallic nanoscale patterns to small and non-planar surfaces. ACS Nano **3**(1), 59–65 (2009)
31. Z. She, A. Di Falco, G. Hähner, M. Buck, Electron-beam patterned self-assembled monolayers as templates for Cu electrodeposition and lift-off. Beilstein J. Nanotechnol. **3**, 101–113 (2012)
32. Q. Xu, B.D. Gates, G.M. Whitesides, Fabrication of metal structures with nanometer-scale lateral dimensions by sectioning using a microtome. J. Am. Chem. Soc. **126**(5), 1332–1333 (2004)
33. Q. Xu, R.M. Rioux, M.D. Dickey, G.M. Whitesides, Nanoskiving: a new method to produce arrays of nanostructures. Acc. Chem. Res. **41**(12), 1566–1577 (2008)
34. D.J. Lipomi, R.V. Martinez, M.A. Kats, S.H. Kang, P. Kim, J. Aizenberg, F. Capasso, G.M. Whitesides, Patterning the tips of optical fibers with metallic nanostructures using nanoskiving. Nano Lett. **11**(2), 632–636 (2011)
35. D.J. Lipomi, R.V. Martinez, R.M. Rioux, L. Cademartiri, W.F. Reus, G.M. Whitesides, Survey of materials for nanoskiving and influence of the cutting process on the nanostructures produced. ACS Appl. Mater. Interfaces **2**, 2503–2514 (2010)

36. C. Blackley, L. Drummond, A. Fikouras, P. André, A. Di Falco, Hybrid Bottom-up Approach for Flexible Plasmonics, in *PECSX 10th International Symposium on Photonic Electromagnetic Crystal Structure*, vol. 1 (2012)
37. R. Magnusson, S.S. Wang, New principle for optical filters. Appl. Phys. Lett. **61**(9), 1022–1024 (1992)
38. S. Tibuleac, R. Magnusson, Reflection and transmission guided-mode resonance filters. J. Opt. Soc. Am. A **14**(7), 1617–1626 (1997)
39. M.G. Moharam, E.B. Grann, D.A. Pommet, T.K. Gaylord, Formulation for stable and efficient implementation of the rigorous coupled-wave analysis of binary gratings. J. Opt. Soc. Am. A **12**(5), 1068–1076 (1995)
40. P.B. Johnson, R.W. Christy, Optical constants of the noble metals. Phys. Rev. B **6**(12), 4370–4379 (1972)
41. G. Kichin, T. Weiss, H. Gao, J. Henzie, T. Odom, S. Tikhodeev, H. Giessen, Metaldielectric photonic crystal superlattice: 1D and 2D models and empty lattice approximation. Phys. B Condens. Matter **407**(20), 4037–4042 (2012)
42. A.L. Fehrembach, A. Talneau, O. Boyko, F. Lemarchand, A. Sentenac, Experimental demonstration of a narrowband, angular tolerant, polarization independent, doubly periodic resonant grating filter. Opt. Lett. **32**(15), 2269–2271 (2007)
43. E. Sakat, S. Héron, P. Bouchon, G. Vincent, Metaldielectric bi-atomic structure for angular-tolerant spectral filtering. Opt. Lett. **38**(4), 425–427 (2013)
44. A. Di Falco, Y. Zhao, A. Alù, Optical metasurfaces with robust angular response on flexible substrates. Appl. Phys. Lett. **99**(16), 163110 (2011)

Chapter 7
Multifunctional Fiber Optic Plasmonic Nanoprobes

A. Crescitelli, M. Consales, E. Esposito, G. Quero, A. Ricciardi and A. Cusano

Abstract The ‘‘Lab-on-Fiber’’ technology essentially envisages a novel class of advanced, multifunctional photonic devices and components arising from the integration onto optical fibers of different materials at micro and nano-scale. This emerging technology is opening the way for the developing of novel miniaturized “all-in-fiber” devices exploitable for both communication and sensing applications, with unique and unprecedented performances with respect to traditional devices. Although the benefits provided by such a technology can be easily understood, many research efforts are still required to translate the vision in a consolidated technological reality. In this context, we present a reliable fabrication process involving conventional deposition and nano-patterning techniques (typically used for planar devices) adapted to directly operate on optical fiber substrates. Following this approach, we demonstrate a first technological platform based on the integration onto the optical fiber tip of two-dimensional hybrid metallo-dielectric nanostructures supporting localized surface plasmon resonances (LSPRs). By means of

A. Crescitelli · M. Consales · G. Quero · A. Ricciardi · A. Cusano (✉)
Optoelectronic Division, Department of Engineering,
University of Sannio, 82100 BN, Italy
e-mail: a.cusano@unisannio.it

A. Crescitelli
e-mail: alcresci@unisannio.it

M. Consales
e-mail: consales@unisannio.it

G. Quero
e-mail: giuquero@unisannio.it

A. Ricciardi
e-mail: aricciardi@unisannio.it

A. Crescitelli · E. Esposito
Institute for Microelectronics and Microsystems (IMM),
National Council of Research, 80131 NA, Italy
e-mail: emanuela.esposito@cnr.it

A. Cusano et al. (eds.), *Lab-on-Fiber Technology*,
Springer Series in Surface Sciences 56, DOI: 10.1007/978-3-319-06998-2_7,

experimental measurements and full-wave numerical simulations, we characterize the resonant phenomenon and investigate the underlying physics. Furthermore, in order to demonstrate the versatility of the proposed technology, starting from the basic structure, we show how it is possible to exploit all the degrees of freedom it offers, achieving a complete control on the number and the field distribution of the resonant modes (and the consequent spectral features) that determine the device performances for the specific application. In particular in this chapter we present some experimental results demonstrating how the proposed device can work effectively as an optical probe for label free chemical and biological sensing, as a microphone for acoustic wave detection and as an in-fiber polarization-sensitive device. Overall, our study highlights the enormous potentialities and achievements of the Lab-on-Fiber technology that are leading it to be one of the hottest topics in photonics community.

7.1 Introduction

Optical fiber technology has experienced a tremendous growth and advancement over the past several decades, not only in transmission systems for communications but also in the sensing field [1]. Although fiber optic sensors have in many cases completely replaced and outperformed their counterparts based on more conventional technologies, there is still a continuous and ever increasing demand for advanced performance and additional functionalities. These needs could be efficiently fulfilled through new fabrication strategies allowing the integration of optical fibers with all those materials, devices and components that provide a full control of the light at nanoscale. Indeed, new features and functionalities may be conferred to the final device through the integration onto optical fibers of photonics crystals, plasmonic structures, and metamaterials [2, 3] that offer unique properties for sub-wavelength field confinement [4] and resonant field enhancement [5].

It is thus clear that solving the technical issues related to the integration of this kind of structures with optical fibers will definitely open the way to novel multifunctional "all-in-fiber" nanoprobes useful in many strategic sectors such as optical processing, environment, life science, safety and security.

These considerations are pushing many prestigious research groups in the photonic community to fuse together the world of nanotechnologies with optical fibers, leading to envision a novel and intriguing technology known as "Lab on Fiber" [6].

Many fabrication strategies and processes have been recently proposed to adapt modern nanotechnologies facilities to properly work in case of unconventional substrates such as the case of optical fibers [7–16]. These approaches can basically be divided into two main categories: nano-transferring and direct writing. The former essentially consists on the previous fabrication of dielectric or metallic nanostructures onto planar substrates (by means of standard and well-assessed nanofabrication techniques), and their successive transfer to the optical fiber substrate [7–13]. Alternatively, simpler and less time consuming approaches based on the direct-write patterning have been demonstrated [14–16]. This strategy relies on the use of deposition and nanofabrication techniques (typically used for planar devices) suitably adapted to directly operate on the fiber tip. On this line of argument, in this work we report on a reliable fabrication process which enables the realization of both dielectric and metallic nanostructures directly on the fiber tip, involving conventional nanotechnology techniques actually specialized to correctly operate on the tip of optical fibers. The validation of the proposed fabrication process has been carried out through the realization of a first Lab-on-Fiber technological platform based on the integration onto the fiber facet of a 2D hybrid metallo-dielectric nanostructure supporting LSPRs.

Then we focus the attention on the versatility and potentiality of the proposed platform. In particular we demonstrate how it is possible to tailor the field distribution of the plasmonic mode enabling the control on the refractive index sensitivity. With a view towards biological sensing, we experimentally observe that the proposed platform is able to detect the formation of nano-sized overlays over very limited active areas. Moreover, our device shows good potentialities for acoustic wave detection. Finally, the last part of the chapter is devoted to the plasmonic resonance engineering; we discuss about three different approaches to obtain the control on the number and the field distribution of the excited LSPR. Specifically this can be achieved by (i) acting on the dielectric slab thickness, (ii) integrating a high RI overlay, and (iii) breaking the circular symmetry of the crystal nanostructure at both unit cell or entire lattice levels, in order to create novel polarization sensitive wavelength-selective optical fiber devices.

7.2 Fabrication Technique

Our fabrication process basically consists of three main technological steps, as schematically represented in Fig. 7.1a: (i) spin coating deposition of dielectric overlay (electron-beam resist, ZEP 520A, in this case) with accurate thickness control and flat surface over the fiber core region; (ii) nanopatterning via Electron Beam Lithography (EBL) process; and (iii) superstrate deposition of different functional materials (metallic or non metallic) by using sputtering method. It is to be noticed that the spin-coating process typically produces edge beads around the perimeter of the substrates, and this effect is particularly significant when very small substrates are considered (as the optical fiber tip), where the edge bead can be as large as the substrate itself, thus preventing the use of standard lithographic techniques [7, 9].

One of the peculiarities and main innovations of our approach relies on the capability to deposit dielectric layers on the cleaved end of standard optical fibers, with controllable thickness and flat surface area closest to the fiber core. In order to adapt well-assessed nano-scale deposition and patterning techniques to directly

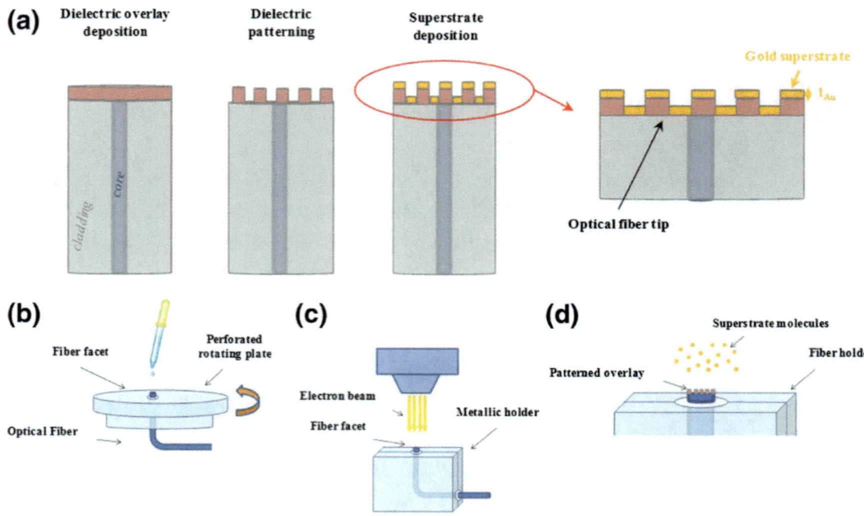

Fig. 7.1 **a** Schematic of the technological steps involved in the fabrication process and **b–d** of the customized chuck and holders for the spin coating, overlay patterning and superstrate deposition

operate on the fiber tip, we realized a customized chuck (see Fig. 7.1b) to host the fiber for the spin coating process and a set of metallic holders (see Fig. 7.1c–d) enabling a perpendicular fixing of the fiber facet for the overlay patterning and superstrate deposition [17].

The resist spinning process allows the repeated deposition of ZEP overlays with different thicknesses depending on the rotating speed (in the range 2,000–6,000 rpm). Once the dielectric overlay is deposited, the consequent step relies on its sub-wavelength patterning by EBL. The optical fiber is lodged on a customized metallic holder (see Fig. 7.1c) allowing a correct exposure process. Moreover, to dissipate the charging effect due to the electron interactions (inevitable when operating on a small glassy substrate), we deposit a metal layer with thickness of about 10 nm on the top of our device, by means of a DC magnetron sputtering in a vacuum system at a base pressure of 8×10^{-5} Torr.

The pattern is written by the EBL system (Raith 150) on the deposited resist. The acceleration voltage is typically 20 kV and the dose ranges from 50 to 60 $\mu C/cm^2$. After the gold removing and ZEP developing, the realized structure is coated by a new gold layer via DC magnetron sputtering to obtain the final hybrid metallo-dielectric device.

The integration of additional functional (either dielectric or metallic) overlays over the patterned areas can also be carried out by using standard coating techniques (such as sputtering, thermal evaporation, etc.). Overall, the proposed fabrication process allows rapid prototyping with a 90 % yield, thanks to the reliable spin coating process and moreover, our nanostructures show good adhesive strength also resulting in reusable devices.

7.3 Structure Design and Analysis

In order to test the capability of the proposed fabrication process, we focused the attention on a first technological platform based on a hybrid metallo-dielectric nanostructure supporting LSPRs. The two-dimensional (2D) structure considered in this study is schematically represented in Fig. 7.2a. It essentially consists of a dielectric (ZEP) 200 nm thick layer patterned with a square lattice of holes and covered with a 40 nm thick gold film deposited on both the ridges and the grooves. The lattice period was $a = 900$ nm and the holes radius was $r = 225$ nm, corresponding to a filling factor (radius to period ratio) $r/a = 0.25$. When such a structure is illuminated in out-of-plane configuration, as in the case of single mode fiber illumination in the paraxial propagation regime, plasmonic and photonic resonances are expected to be excited due to the phase matching condition between the scattered waves and the modes supported by the hybrid structure, depending on the geometric features of the structure itself [18]. For the numerical analysis we used the commercial software COMSOL Multiphysics (RF module) based on the finite element method [19]. Following the same approach of [20], by exploiting the crystal translational and mirror symmetries, the computational domain can be reduced to one quarter of unit cell, terminated with perfectly-electric-conducting (PEC) and vertical perfectly-magnetic-conducting (PMC) boundary conditions, placed two by two in the opposite walls. The resulting structure supports a transverse-electromagnetic (TEM) wave emulating the normally-incident plane-wave; we used this kind of excitation instead of a proper fiber mode with Gaussian profile to simplify our simulations. The refractive index data used for modeling both gold and silica in the IR region were taken from [21]. The ZEP refractive index was taken directly from its data sheet. In Fig. 7.2b we plot the theoretical reflectance spectrum (solid black line) of the hybrid metallo-dielectric fiber tip device; the high reflectivity baseline is interrupted by a resonance dip centered at 1,369 nm with a Q-factor of $\sim$47. Figure 7.2c shows the near field map in correspondence of the resonant wavelength. It is evident that the field is highly localized both around the gold pillar (on the bottom) and the gold slab holes (on the top), meaning that is present a sort of interaction between two plasmonic modes. To assess the origin of the resonant phenomenon we decided to split up the whole structure into its single parts and separately analyze their contribution to the reflectance spectrum. To this aim, we first calculated the reflectance spectrum of the structure where only the gold nanopillars are present on the fiber tip, which is what we called the 'only *gold*' structure. The result is shown in Fig. 7.2b as a dashed blue curve; a very broad reflectance peak (Q $\sim$ 8) appears due to the excitation of LSPR, as in the case of [15]. Then we calculated the reflectance of the '*gold + ZEP*' structure, where just the ZEP slab is grown around the gold nanopillars; in this case, the presence of the dielectric medium causes only a reflectance peak red shift of about 158 nm (see the dotted red curve in Fig. 7.2b) according to the *modification* of the phase matching conditions. Finally, when we consider the '*complete*' structure (black curve), a significant increase is observed in the reflectance baseline with the creation of a well defined spectral

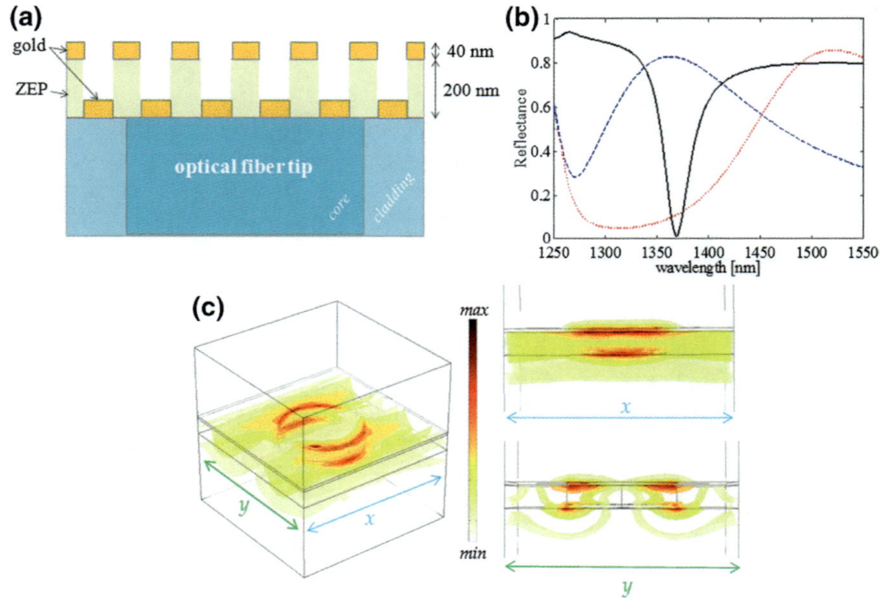

Fig. 7.2 **a** Schematic (*cross section view*) of the hybrid metallo-dielectric PC structure integrated on the optical fiber tip; **b** Theoretical reflectance spectra of the 'complete' (*solid black*), 'only gold' (*dashed blue*) and 'gold + ZEP' (*dotted red*) structures; **c** Electric field intensity distributions of the resonant modes evaluated at the reflectance dip wavelength pertaining to the spectrum plotted with solid black curve in (**b**). Reprinted with permission from [17]. Copyright 2012 American Chemical Society

feature with a Q factor significantly higher than those obtained using simply gold pillars. Hence from our analysis it results that the resonant dip in the reflectance spectrum is not merely due to a simple interaction among the single elements constituting the entire structure, thus meaning that the whole is better than the sum of the parts. It is important to underline that in this specific configuration only localized plasmonic modes have been excited in the analyzed spectral range because of the low refractive index value of ZEP. A richer spectrum in terms of resonant dips is expected to be achieved by using a higher refractive index material such as silicon or a thicker dielectric layer able to give rise also to the excitation of photonic *quasi-guided* modes (guided resonances).

7.4 Structure Fabrication and Characterization

The 2D hybrid metallo-dielectric structure was realized by using the previously described fabrication process. In Fig. 7.3a we show a scanning electron microscope (SEM) image (top view) of the fabricated fiber tip device and a magnified

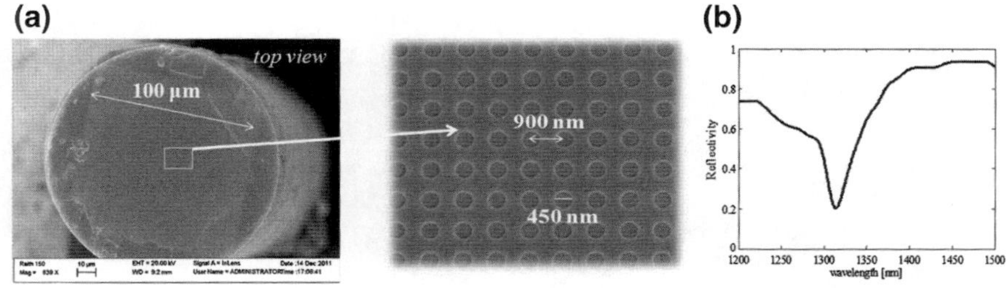

Fig. 7.3 **a** SEM image of the 100 × 100 µm^2 matrix of circular holes patterned by the EBL on the ZEP layer, (*inset*) zoomed detail of **a**; **b** Experimental reflectivity spectrum of the hybrid metallo-dielectric structure integrated on the fiber tip. Reprinted with permission from [17]. Copyright 2012 American Chemical Society

image of the structure, respectively. The lattice periodicity and the hole radius were estimated to be 900 and 450 nm respectively.

The measurement setup essentially consists of a 3 dB directional coupler placed between a broadband light source and the optical fiber nanoprobe, with one return end sent to the optical spectrum analyzer (OSA). In Fig. 7.3b we show the experimental reflectance spectrum (solid black line) of our device; a resonance dip centered at 1,311 nm can be clearly seen, characterized by a Q-factor of about 23. From a comparison with the numerical results a blue shift of 58 nm and a reduction of both the visibility and the Q-factor are found. These discrepancies can be attributed to the fabrication tolerances and to the approximation considered in the numerical analysis where we assumed a plane wave excitation on an *infinite* structure in the periodicity plane [22]. We point out that although no specific parameter optimization was performed, the Q-factor exhibited by our hybrid metallo-dielectric structures is larger than most of those observed in typical plasmonic crystal configurations.

Of course, the excitation wavelengths of LSPRs are very sensitive to environment modifications [23], thus any change in the surrounding refractive index (SRI) around the fiber tip, either local or bulk, causes a wavelength shift of the resonant dip due to a change of the phase matching condition. This makes our lab-on-fiber platform particularly suited for biological and chemical sensing.

7.5 Biological and Chemical Sensing

This section deals with the analysis of the sensitivity to the surrounding refractive index changes; in particular, we experimentally study the dependence of the bulk refractive index sensitivity on physical and geometrical parameters including metal and dielectric thickness as well as the lattice tiling (including the aperiodic case). Successively, by having in mind practical scenarios of label-free biological sensing, we experimentally evaluate the surface sensitivity of the device (i.e. its response when nano-sized overlays are deposited on the top of its structure) and prove that the sensitive area can be reduced down to $20 \times 20\ \mu m^2$ without affecting the device performances.

7.5.1 Bulk Refractive Index Sensitivity

With reference to the schematic representation in Fig. 7.2a, it can be easily noticed how the gold layer deposited on the top of the fiber tip device shields the plasmonic mode excited within the hybrid crystal from the external environment, thus limiting the device sensitivity. Thence, is reasonable to expect that a reduction of the gold overlay thickness would promote an enhancement of the light-matter interaction with the surroundings, improving the SRI sensing performances.

To experimentally verify this assumption, we tested three fiber optic probes with different gold layer thickness, $t_{Au} = 20$, 30 and 40 nm, keeping constant the other geometrical parameters ($a = 900$ nm, $r = 225$ nm and $t_{ZEP} = 200$ nm). In order to avoid small discrepancies in the hole radii and in the ZEP layer thickness, that may occur from sample to sample due to fabrication tolerances, the same fiber tip probe was used, initially coated by a 20 nm-thick gold layer, on which two further gold deposition of $\sim$10 nm each, have been successively carried out. Successively we analyzed the SRI sensitivity by immersing the probes in liquid solutions with different RI, i.e. water ($n = 1.333$), ethanol ($n = 1.362$) and isopropyl alcohol ($n = 1.378$), and measured the resonant wavelength shift ($\Delta\lambda_{min}$) occurring upon immersion.

The characteristic curves ($\Delta\lambda_{min}$ vs. SRI) of the three samples having $t_{AU} = 20$, 30 and 40 nm respectively, have been reported in Fig. 7.4a. Obtained results, confirm that SRI sensitivity increases as a consequence of a reduction of t_{AU} and are in line with our numerical predictions [17]. Indeed, as reported in Fig. 7.4a, the sensitivity increases from 36 to 98 nm/RIU by passing from a gold layer thickness of 40 to 30 nm and from 98 to 125 nm/RIU by passing from $t_{AU} = 30$ nm to $t_{AU} = 20$ nm. In the linear fitting approximation (red line in Fig. 7.4b), a sensitivity increase of $\sim$4.5 nm/RIU per nanometer of gold layer has been estimated. As a consequence, an additional improvement could be hypothetically obtained up to $\sim$170 nm/RIU with $t_{AU} = 10$ nm. However, the reduction of the gold layer thickness has already been demonstrated to produce a red-shift of the resonant dip, combined to its bandwidth widening and visibility reduction [17]. Although with a gold layer thickness of 20 nm, the reflectance dip in the spectrum is still present and visible, a further reduction of the gold thickness would produce an additional decrease in the resonant visibility, thus compromising the capability of detecting the resonant wavelength shift occurring upon SRI changes.

Nevertheless, at a first comparison with respect to the state of the art of LSPR-based fiber tip-devices, our sensitivity results almost comparable to that demonstrated in literature by a fiber optic sensor based on ordered array of gold nanodots

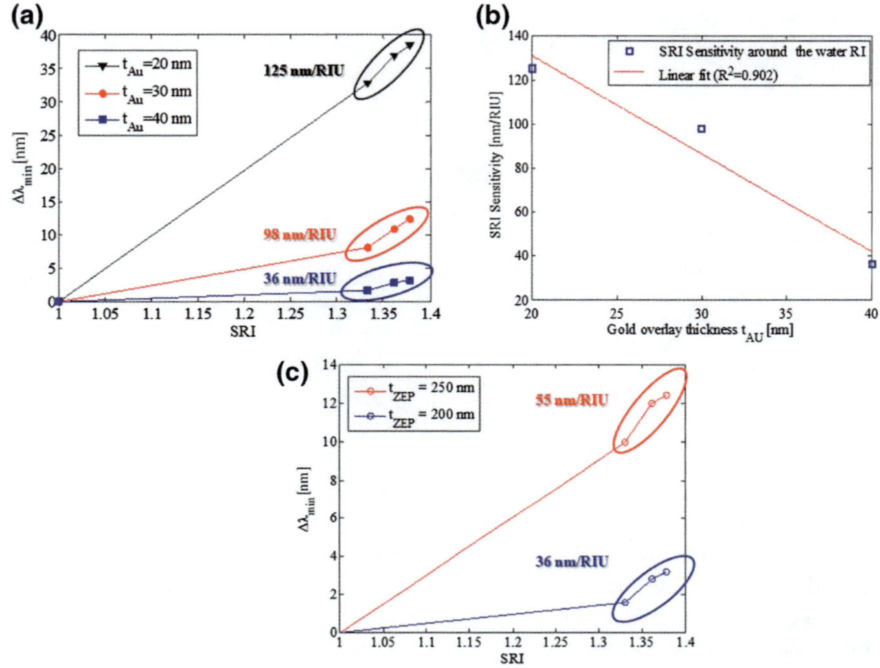

Fig. 7.4 a $\Delta\lambda_{min}$ as a function of the SRI for three fiber optic probe characterized by $t_{Au} = 40$ nm (*blue squares*), $t_{Au} = 30$ nm (*red circles*) and $t_{Au} = 20$ nm (*black triangles*); **b** Variation of the SRI Sensitivity as a function of t_{Au}, around the water RI; **c** $\Delta\lambda_{min}$ as a function of the SRI for two fiber optic probe characterized by $t_{Au} = 40$ nm, a = 900 nm, $r/a = 0.25$, and having two different values of the ZEP overlay thickness, 200 nm (*blue triangles*) and 250 nm (*red circles*). Modified by [24]

(195 nm/RIU) [15] and fourfold lower to that exhibited by a fiber tip-sensor based on arrays of gold sub wavelengths apertures (500 nm/RIU) [14].

In our case, the resonance shows a much higher Q factor (of about 90 %) and has a significant smaller active area, since we use single-instead of multi-mode optical fibers. In light of these considerations, it is important to remark, as it will be better discussed in Sect. 7.5.2.1, that when dealing with sensing platforms for label free bio-sensing, also the active area (in addition to the sensitivity and the Q-factor for the resolution) plays a fundamental role since for a given local RI change, a larger amount of bound biomolecules is required for larger active area platforms. Moreover, differently from Dhawan et al. [14] the resonance visibility remains unaltered when the local refractive index increases [17]. Finally, thanks to the versatility of our technological platform, we can act also on the large set of available parameters (e.g. ZEP layer thickness, suitable dielectric patterns, size of the active area, etc.) to further improve the final device characteristics.

For example, as shown in Fig. 7.4c, a bulk SRI sensitivity of 55 nm/RIU was found for a structure having a slightly thicker ZEP overlay ($t_{ZEP} = 250$ nm) with the same period ($a = 900$ nm), filling factor ($r/a = 0.25$) and gold thickness ($t_{AU} = 40$ nm). Hence an improvement of ~40 % has been obtained with an increase of ~50 nm in the ZEP layer thickness.

7.5.1.1 Sensitivity Enhancement with Quasi-Crystals

In order to improve the SRI sensitivity of the proposed device and on the basis of some recent results inherent the capability of hybrid metallo-dielectric quasi-crystals (QCs) to outperform their periodic counterparts in chemical sensing applications [25, 26], we also investigated the use of quasi-periodic tiling. In particular, we show some results of a metallo-dielectric QC directly realized on the fiber tip according to the Ammann-Beenker (octagonal) tiling [27].

The top view SEM image of the fabricated structure (with an equivalent period $a = 900$ nm, $t_{ZEP} = 250$ nm and $t_{Au} = 40$ nm) is shown in Fig. 7.5a. In Fig. 7.5b we have reported the relative wavelength shifts of the LSPR dip of such a structure as a function of the SRI and also the wavelength shift obtained with its periodic counterpart, characterized by same lattice period, ZEP and gold overlay thickness as well as same filling factor. It turned out that for a given filling factor, QCs based devices are able to outperform (SRI sensitivities of 102 nm/RIU) their periodic counterpart (SRI sensitivities of 55 nm/RIU), with a sensitivity improvement of about a factor 2.

Reported results, lay the foundations for the development of a fiber optic nanoprobes based on aperiodic geometries, opening up different perspectives to obtain a full resonance engineering and control [20]. As a matter of fact, by using aperiodically-ordered QCs lattices it could be possible to achieve a richer resonant phenomena with respect to the periodic case, enabling the selection of number and spectral localization of plasmonic resonances, which in turn can be designed to exhibit high quality factors, high sensitivities and tuning efficiencies [20].

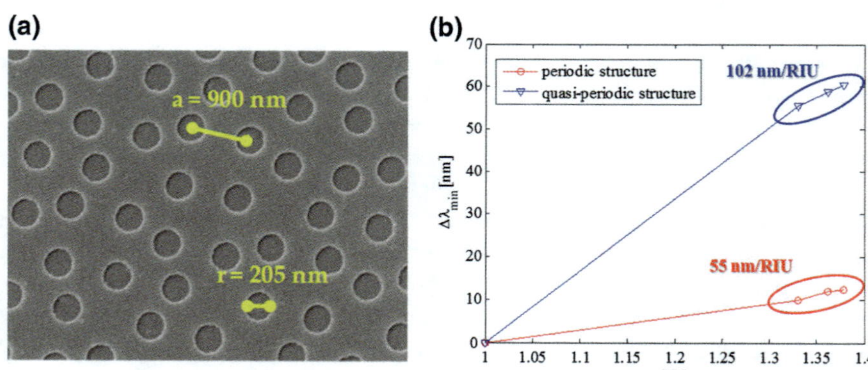

Fig. 7.5 **a** SEM image of the hybrid metallo-dielectric QC structure realized on the fiber tip; **b** $\Delta\lambda$ as a function of the SRI obtained with a quasi-periodic and periodic structures having same filling factor ($r/a = 0.25$), $a = 900$ nm, $t_{ZEP} = 250$ nm and $t_{Au} = 40$ nm. Modified by [24]

Moreover QCs may provide a better control on resonant coupling/uncoupling mechanisms by selectively breaking the lattice mirror symmetries via a judicious introduction of point-defects [28].

7.5.2 Surface Sensitivity

In practical scenarios related to label-free chemical and biological sensing applications, the top metal surface is functionalized with bio-receptors, chosen for their specific affinity toward a given biomolecule to provide the bio-recognition feature. As a consequence, because of the binding of an analyte molecules layer, the RI changes are restricted to local modifications occurring at the sensor surface. Sensors based on LSPR, thanks to the highly localized character of its field distribution at the resonant wavelength, are well suited for detecting these local environment changes. Accordingly, in this section, we experimentally evaluate the surface sensitivity of our optical fiber nano-probe by depositing on its surface nano-sized overlays mimicking local RI changes. To this aim, by using the RF sputtering deposition technique, we deposited on the top of the structure a ~ 100 nm thick layer of SiO_2 (at a pressure of 3 mTorr with a rate of 3.5 Å/sec.), whose refractive index ($n \sim 1.45$) well resembles that of most biological molecules [29, 30]. The sample, in this case, is characterized by $t_{ZEP} = 70$ nm and $t_{Au} = 20$ nm. The reflectance spectra of SiO_2-coated and uncoated fiber tip device are shown in Fig. 7.6, where the experimental data have been also compared with those obtained by a numerical analysis. A significant red-shift ($\Delta\lambda \sim 35$ nm) of the resonant wavelength occurs as a consequence of the surface layer deposition on both the ridges and the grooves of the structure. This result reveals the promising capability of the proposed platform for detecting molecular monolayer at nanoscale; the resonance shift per nanometer of deposited overlay turns out (assuming a

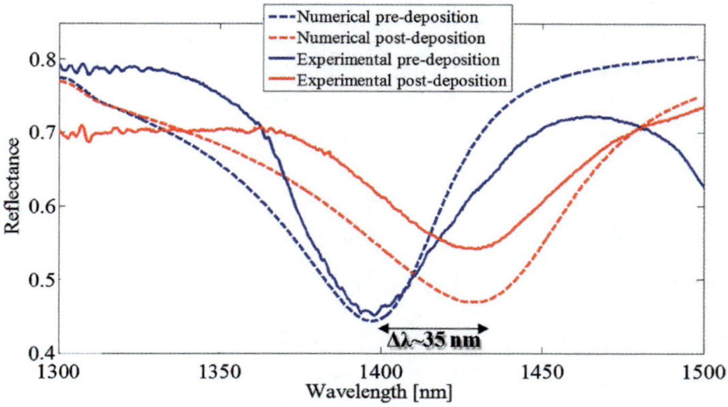

Fig. 7.6 Numerical and experimental spectra of a hybrid nanostructure with $t_{ZEP} = 70$ nm and $t_{Au} = 20$ nm, before and after the SiO_2 coating deposition. Modified by [24]

linear relationship between the wavelength shift and the overlay thickness) as high as 0.35 nm per nm of deposited SiO_2. This surface sensitivity is comparable to that achieved with nanostructured metallo-dielectric crystals realized on planar substrates, for which values up to $\sim$0.4 nm per nm of deposited SiO_2 have been found [25]. Results here provided once more demonstrate the great potentialities of the proposed Lab on Fiber platform, which allows to move devices from planar substrates to fiber tip (with all the related advantages) without affecting their performances. Considering the typical sizes of biological molecules (3.8–5.2 nm), it can be inferred that the binding of a single biological monolayer to the sample surface is able to generate a LSPR shift of approximately 1.3–1.8 nm, which may be easily detected via low-cost commercial spectrophotometer. Although a concrete demonstration of the capability of our probe to work as a biosensor would require the demonstration of its ability to detect the formation of an ultra-thin biomolecular layer upon its functionalized surface (via a detectable resonant wavelength shift), nevertheless the achieved surface sensitivity is the first important step proving the potentialities of our platform for bio-sensing applications.

7.5.2.1 Sensing Area

As previously mentioned, when dealing with sensing platforms for label free biosensing, the active sensitive area is a very important parameter to be carefully considered; indeed, for a given local refractive index change, a larger amount of bound target bio-molecules is required for devices with larger active areas.

Moreover, one might wonder how many periods of the photonic/plasmonic structure are necessary to achieve a strong coupling effect or equivalently a well-defined spectral feature in the reflectance spectrum.

For these considerations, in the following, we discuss the effects induced by the confinement of the ZEP overlay on the spectral resonant feature. We used an UV-laser micromachining to gradually reduce, in successive steps, the patterned area of the device (characterized by $a = 850$ nm, $t_{ZEP} = 200$ nm and $t_{Au} = 40$ nm) to 50 × 50 μm^2, 20 × 20 μm^2, and 10 × 10 μm^2 square region. The reflectance spectra relative to each confinement step, are shown in Fig. 7.7. We can observe how, in case of 50 × 50 μm^2 and 20 × 20 μm^2 confinements, the reflectance spectra (red and green curves) do not undergo any substantial modification with respect to the original one (blue curve); so the patterned area reduction does not modify the field distribution of the excited LSPR, which remains confined in the horizontal plane in that area. Instead, when the patterned surface over the fiber core (diameter $\sim$8 μm) is reduced down to 10 × 10 μm^2, it is possible to observe a significantly variation of the corresponding reflectance spectra (black curve). From the above analysis, it can be comprehended that patterned area on the optical fiber tip can be reduced down to 20 × 20 μm^2 without any significant variations in the reflectance spectrum and consequently on the SRI sensitivity. Compared to other LSPR-based fiber tip-devices, our sensing area is almost comparable (as also the bulk SRI sensitivity is) to that of the fiber

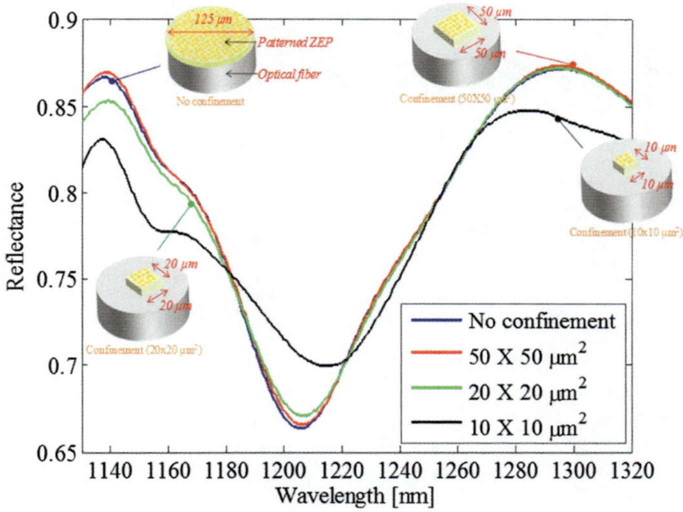

Fig. 7.7 Reflectance spectra of the hybrid nanostructure, having $a = 850$ nm, $t_{ZEP} = 200$ nm and $t_{Au} = 40$ nm, before and after the patterned area confinement. Modified by [24]

optic sensor based on ordered array of gold nanodots [15] which uses single-mode optical fibers, whereas it is much smaller (400 μm^2 vs at least 2,500 μm^2) than that exhibited by the fiber tip-sensor based on arrays of gold sub wavelengths apertures [14]. In this case, even though our device is characterized by a four time lower SRI sensitivity, however it has a significant smaller sensing area which allows to improve the biomolecules limit of detection.

7.6 Acoustic Detection

As a further sensing application, we also show the capability of our nanoprobe to detect acoustic waves. Indeed, by taking advantage from the typical low Young's modulus of the patterned ZEP, significant variations in the geometrical characteristics of the patterned dielectric slab are expected in response to an applied acoustic pressure wave, hence promoting a consequent shift of the resonant wavelength. In this application, the role played by the dielectric structure is fundamental to determine the acoustic performance of the final device. The acoustic experiments have been carried out by testing the sample characterized by a period a = 900 nm (whose reflectance spectrum is shown in Fig. 7.3b), using an experimental setup described in detail in an our previous work [17]. In particular a reference microphone was placed in close proximity to the fiber sensor in order to obtain information about the actual incident acoustic pressure. In Fig. 7.8a (upper curve) is reported the typical time response of the fiber nano-device to a 4 kHz

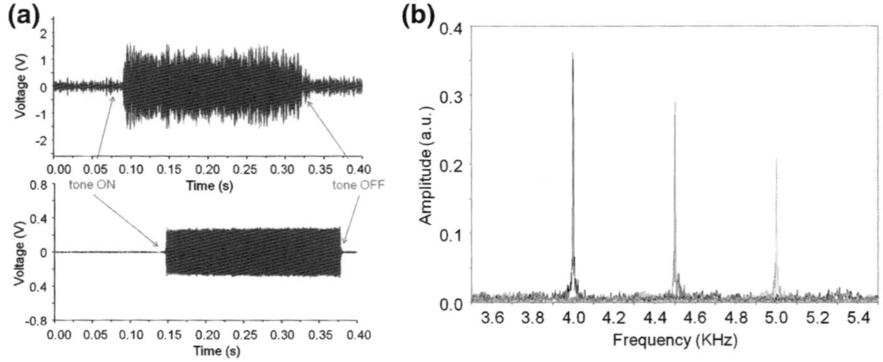

Fig. 7.8 **a** Typical time responses of the hybrid metallo-dielectric fiber-tip device (*top*) and reference microphone (*bottom*) to a 4 kHz acoustic pressure pulse with a duration of 250 ms; **b** FFT spectra obtained by the optical fiber sensor in response to three acoustic tones with increasing frequency (4, 4.5, and 5 kHz). Reprinted with permission from [17]. Copyright 2012 American Chemical Society

acoustic tone with a duration of about 250 ms. Moreover, in the same figure (bottom curve), is reported, for comparison, the response of the reference microphone. The shown data, reveal the capability of the in-fiber device to detect acoustic waves in good agreement with the reference microphone. It is to be noticed that the electrical signal is delayed with respect to the optical counterpart (see Fig. 7.8a); this is due to the slightly greater distance at which the reference microphone is located from the acoustic source as well as to the different electronic processing systems employed. It is also worth noting that, although a relatively high noise level is visible in Fig. 7.8a (the standard deviation of the sensor signal, σ noise, in the absence of the acoustic wave is nearly 0.1 V), attributable to the instability of the utilized tunable laser, the output voltage of the fiber device was found to be more than an order of magnitude higher than the noise level [17]. Finally, we have tested the capability of the optical device to discriminate among different acoustic tones. The FFT spectra obtained by the same optical fiber microphone in response to three acoustic tones with different frequencies (4, 4.5, and 5 kHz) are shown in Fig. 7.8b [17]. The different amplitude of the sample response to the three tones is only due to the different amplification factor applied.

7.7 Resonance Engineering

The analysis carried out so far is based on a single resonant dip occurring in the reflection spectra and associated to the excitation of a single LSPR. If small perturbations of the geometrical parameters in the metallo dielectric structure are introduced, then a slight change of the phase matching condition for the single LSPR occurs. Nevertheless, if larger perturbations are taken into account, multiple

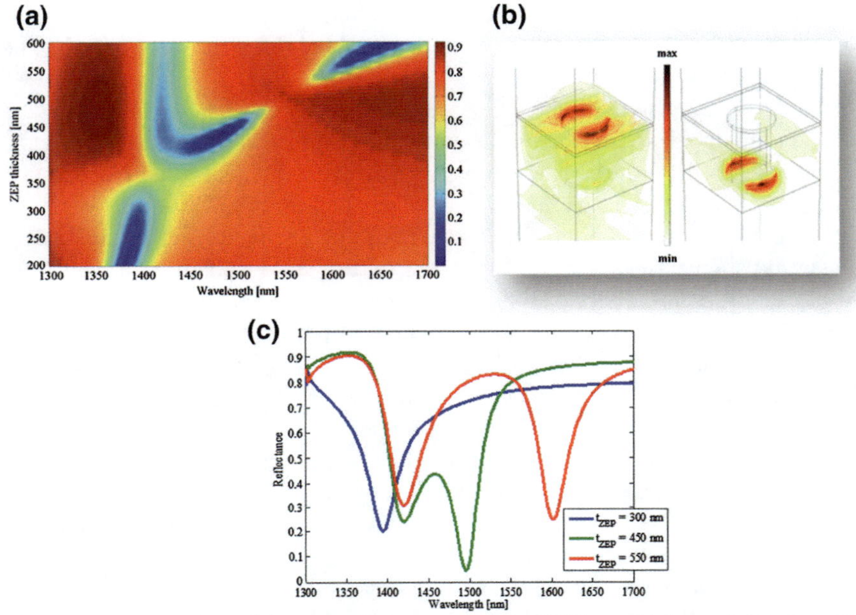

Fig. 7.9 **a** Contour plots of fiber tip reflectance versus the ZEP layer thickness; **b** 3D view of unit cell and electric field distributions evaluated at the reflectance dip wavelengths for $t_{ZEP} = 550$ nm; **b** reflectance spectra extracted from the contour plot for $t_{ZEP} = 300$, 450 and 550 nm. Modified by [24]

LSPRs can be excited at different wavelengths, each one characterized by a specific field distribution and thus by a specific light matter interaction level, leading to the possibility of tailoring and eventually improving the device performances for the specific application. Therefore in the following sub-sections, we show three different approaches to obtain a resonance engineering, acting on the geometrical parameters of our device and also through the integration of high RI overlay onto our nanoprobe.

7.7.1 Slab Thickness Impact

Here, a resonance engineering approach able to influence both the number and modal distribution of the resonances (excited in a given wavelength range), is presented, by considering ZEP overlay thicknesses higher than those previously considered. In Fig. 7.9, we report the results of a numerical analysis aimed at investigating the evolution of the spectral response of our fiber tip device with ZEP overlay thickness ranging between 300 and 600 nm (the other parameters are $a = 900$ nm, $t_{Au} = 30$ nm, $r/a = 0.25$). In particular, we show the contour plot of our device reflectance versus the ZEP layer thickness (see Fig. 7.9a) combined with the 3D views of unit cell and electric field distributions (see Fig. 7.9b) evaluated at the reflectance dip wavelengths for three different values of t_{ZEP}, i.e. 300, 450 and 550 nm.

As it can be noticed, until the ZEP layer thicknesses is below 400 nm, only one reflection dip appears within the reflectance spectrum, corresponding to only one resonance excited by the hybrid nanostructure (characterized by an electric field distribution localized both on the grating ridges and grooves). Instead, when t_{ZEP} values are higher than 400 nm, richer spectra can be obtained, with two distinctive resonant dips appearing in the reflectance spectrum, corresponding to the excitation of two different LSPRs (see Fig. 7.9c): the first one, at smaller wavelengths, exhibiting (as evident from its field distribution reported in Fig. 7.9b) an electric field localized at the top plasmonic crystal (gold voids-like structure), which is independent from t_{ZEP} (the resonance is centered at $\lambda = 1{,}420$ nm for $t_{ZEP} = 450$ and 550 nm as shown Fig. 7.9c); the second one, at longer wavelengths, is characterized by a field distribution localized at the bottom (gold pillars-like) structure (see Fig. 7.9b); the resonant wavelength λ_{min} in this case moves towards higher wavelengths as the ZEP layer increases (specifically from 1,495 to 1,600 nm when t_{ZEP} changes from 450 to 550 nm).

Our results demonstrate that by choosing thicker enough dielectric overlays, we are able to split the single LSPR into two resonances characterized by different electric field distributions and, consequently, different sensitivities to the SRI variations. This phenomenon could have a practical implication also for biological sensing; since immobilization/binding of surface bio-receptors/target molecules can occur in different forms to specific sites on the nano-grating surface (i.e. over the grating rides and grooves), the excitation of different resonant modes with a distinctive near field distributions could be very useful also for gathering information on the geometrical arrangement of bioreceptors/target biomolecules layer. In fact a significant wavelength shift is expected only for that resonances whose near field distribution strongly interact with the immobilized/binded bio-layer. Overall, the possibility to control the electric field distribution of the resonant mode with the ZEP thickness, can be used for achieving different degrees of sensitivity to the external medium variation: for example the insensitive resonance could be used as reference channel or as a compensator in sensing applications.

7.7.2 *Effect of the High RI Overlay*

A different approach, to achieve the resonance engineering, involves the integration of different materials with specific optical properties. In this case, the ZEP patterned slab act as a template for the next deposition of either metallic or dielectric (or a combination of the two) materials able to enlarge new functionalities to our device. Indeed, the main effect of depositing a high RI overlay has the effect of exciting a pure photonic resonance (the so called guided resonance) in place of a plasmonic one [31]. This resonance arises when the orders diffracted by

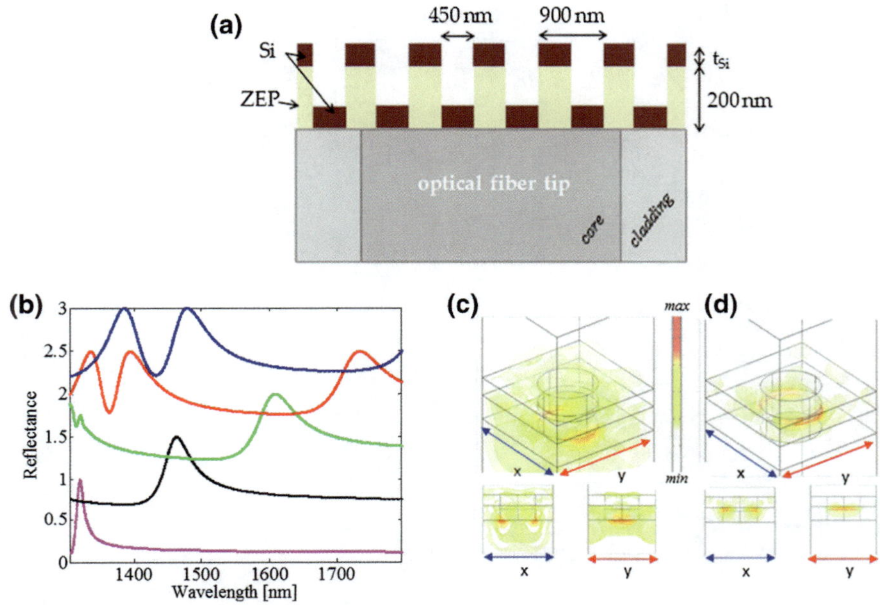

Fig. 7.10 **a** Cross section schematic view of the considered Si-coated nanostructure; **b** evolution of the reflectance spectra for increasing values of t_{Si} from 30 to 150 nm with a step of 30 nm (for clarity, the spectra are offset on the y-axis of 0.5); **c–d** electric field distribution of the two resonances observed for t_{Si} = 150 nm, evaluated at λ = 1,384 nm and λ = 1,476 nm. Modified by [34]

a normally incident wave are coupled to the leaky waveguide modes (i.e. the guided resonances) supported by the slab. As a consequence the presence of an high RI coating on the top of the ZEP layer allows for an increase of the average dielectric slab RI, guaranteeing the RI contrast needed for trapping the photonic mode inside the dielectric slab. For this reason, by appropriately designing the physical and geometrical parameters, it is possible to set the number of modes involved in the coupling process and thus to engineer the resonant spectral behaviour for the specific application. For example, we have demonstrated a high reflectivity broadband mirrors working in the NIR by creating a PC slab in SOI [32] and the evidence of the guided resonance phenomenon in quasi-crystal gratings [20, 33], where the resonances engineering is relatively simpler for to the presence of different inequivalent defect states [28]. Here, we first analyze the effects of a silicon (instead of gold) overlay deposition of thickness t_{Si} on a 200 nm-thick ZEP layer patterned with a square lattice of holes of period a = 900 nm and filling factors r/a = 0.25 (see Fig. 7.10a). The evolution of the reflectance spectra for increasing values of t_{Si} (the Si is supposed to be deposited on both the grating ridges and grooves) from 30 to 150 nm with a step of 30 nm is shown in Fig. 7.10b. For clarity, the spectra are offset on the y-axis of a constant value of 0.5. In our simulations we have considered a lossless and not dispersive

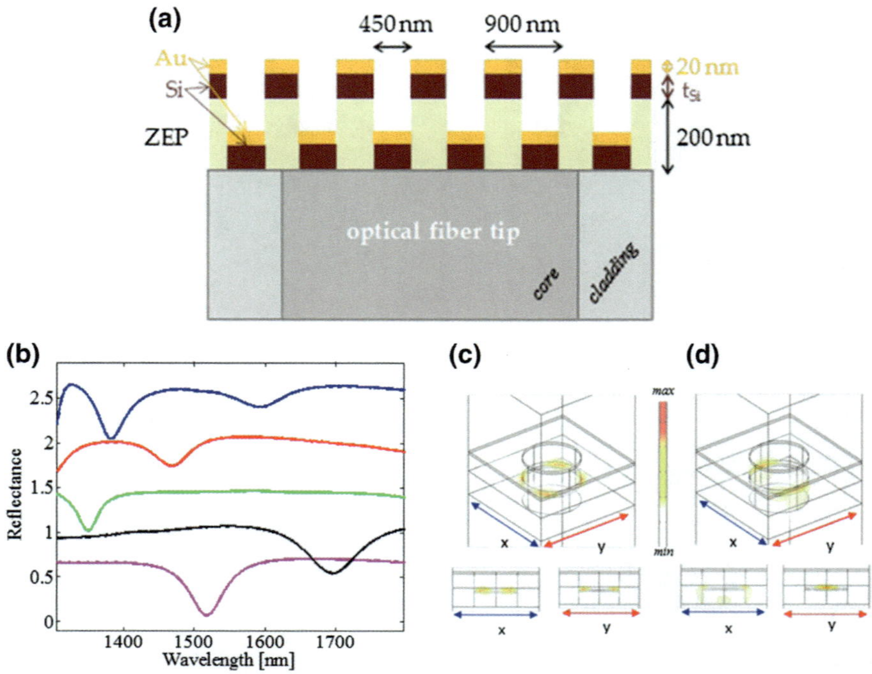

Fig. 7.11 **a** Cross section schematic view of the considered Si–Au-coated nanostructure; **b** evolution of the reflectance spectra for $t_{Au} = 20$ nm and for increasing values of t_{Si} from 30 to 150 nm with a step of 30 nm (for clarity, the spectra are offset on the y-axis of 0.5); **c–d** electric field distribution of the two resonances observed for $t_{Si} = 150$ nm, evaluated at $\lambda = 1{,}382$ nm and $\lambda = 1{,}593$ nm. Modified by [34]

silicon RI model (n = 3.4). It can be noticed that, the absence of the gold layer causes a low reflectivity baseline interrupted by a resonance peak, due to the excitation of the guided resonance. When $t_{Si} = 30$ nm the resonant peak is centered at 1,320 nm. For increasing values of t_{Si}, the peak shifts towards longer wavelengths (it is centered at 1,734 nm when $t_{Si} = 120$ nm) and gradually the Q-factor decrease. When $t_{Si} = 120$ nm, two new resonant peaks appear in the spectrum at 1,336 and 1,394 nm, moving towards longer wavelengths for $t_{Si} = 150$ nm. In Fig. 7.10c, d, we can notice that each excited resonance has a distinctive electric field distribution inside the structure. Therefore it is possible to make each resonance more or less sensitive to local RI changes depending on its modal distribution: indeed if the RI changes in proximity of a strong field enhancement then a significant wavelength shift is expected. Finally, we conclude this section by analyzing a different structure where a double deposition of both dielectric and metallic overlays is made. Such structure can in principle support both photonic and plasmonic (or hybrid) resonances thus offering new possibilities for the resonance engineering. The considered structure is shown in Fig. 7.11a, where, above the ZEP patterned slab, first a silicon layer and then a thin gold overlay with thickness $t_{Au} = 20$ nm, are deposited on the grating ridges and grooves.

In Fig. 7.11b is shown the evolution of the reflectance spectra for increasing values of t_{Si} from 30 to 150 nm with a step of 30 nm. Also in this case, the spectra are vertically offset of 0.5. Differently from the previous structure, the presence of the gold layer implies that the reflectance baseline takes values around 70 % and the resonances appear in the form of a reflectance dip. When $t_{Si} = 30$ nm a resonance is present at 1,517 nm. As t_{Si} increases, the wavelength shift increases accordingly, and a new resonance at 1,350 nm is also excited for $t_{Si} = 90$ nm (the other resonance moves out the analyzed wavelength range). When $t_{Si} = 150$ nm two resonances appears at 1,382 and 1,593 nm, whose electric field distribution are reported in Fig. 7.11c, d. The first one can be considered a plasmonic resonance, being the field localized at the gold/ZEP interface of the bottom pillar. The second resonance looks like an hybrid photonic-plasmonic resonance as the field extends also in the dielectric media. Obviously, enlarging the set of materials to be integrated would open the way to completely control light flow and confinement.

7.7.3 *Polarization Dependent Nanostructures*

The results presented so far pertain to a polarization independent structure; as matter of fact, the spectral response of the device does not depend on the polarization state of the incoming light because of the rotational symmetry of the 2D lattice placed on the optical fiber tip. It is clear that polarization independent operation is desirable in most cases. However, the polarization control is crucial to avoid signal fading in fiber interferometric sensors or to create polarization modulators. Our technological platform lends itself well to the polarization management by breaking the crystal circular rotational symmetry. This can be attained by following two different approaches (eventually combined together): the first one concerns the crystal structure at unit cell level through the definition of elliptical holes in such a way that the two perpendicular polarization states can interact with different crystal radii. The other possibility is to act at the entire lattice level creating a period asymmetry along the two perpendicular direction of the crystal. In the following sub-sections, we discuss about the plasmonic resonance effect relative to these two configurations and show how they differently affect the spectral response of the final device.

7.7.3.1 Holes Ellipticity

We have fabricated and tested a first device (see the top-view SEM image in Fig. 7.12a) consisting of a square lattice of *elliptical* holes with lattice period $a = 900$ nm and radii along the x and y axes respectively of $r_x = 225$ nm and

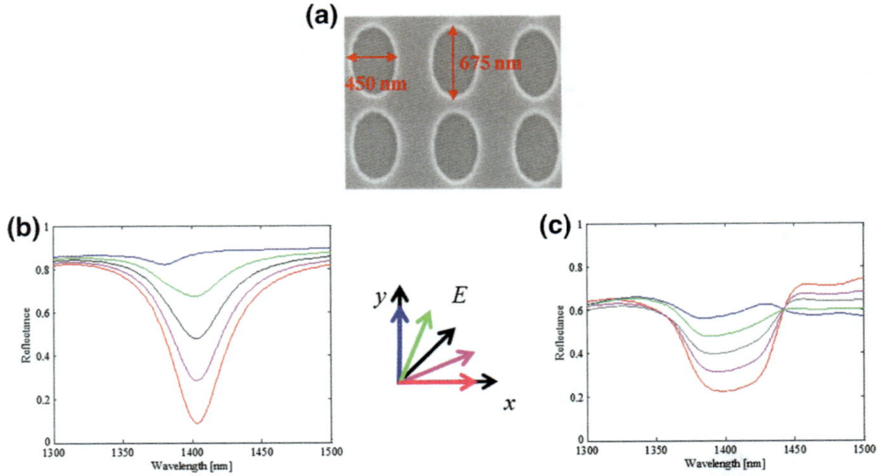

Fig. 7.12 **a** SEM image (*top view, zoomed*) of the probe with elliptical holes ($r_x = 225$ nm, $r_y = 337.5$ nm); **b** numerical and **c** experimental reflectance spectra of the sample for different incident wave polarizations: *blue* and *red lines* refer to POL_x and POL_y. Modified by [25]

$r_y = 337.5$ nm (ellipticity coefficient $\eta = 1.5$). The dielectric and metal thicknesses are $t_{ZEP} = 100$ nm and $t_{Au} = 20$ nm.

In Fig. 7.12b–c we report the numerical and experimental reflectance spectra of the sample for different incident wave polarizations, demonstrating a good agreement. When the electric field is polarized along the *x* axis (POL_x) the reflectance spectrum shows one well defined dip (red line) centered at around 1,400 nm. Vice versa, the spectrum referring to POL_y (blue line) exhibits high reflectivity in the whole analyzed wavelength range without any significant resonant effect. This behavior can be explained by the fact that the two different polarizations experience the same lattice period, but two different values of the holes radius (along the two symmetry axes); specifically in the first case, the *x*-polarized incoming light interact with the 'optimized' structure ($r/a = 0.25$) supporting the LSPR and guaranteeing the maximum resonance visibility. For the *POLy*, since the radius is higher, an almost uniform gold layer is created on the optical fiber tip, thus causing the incident light to be strongly reflected with very small interaction with the external structure. If the incident electric field polarization gradually moves from POL_x to POL_y, the resonant phenomenon becomes weaker with a consequent significant reduction of the resonance visibility, also accompanied by a slight blue wavelength shift.

By changing the polarization input light, the reflected light intensity can be linearly tuned from ~20 to ~60 %. Overall, if hole ellipticity is introduced in the crystal lattice, our device works as a polarization sensitive in-fiber mirror, that strongly reflects only one polarization component and absorbs the other.

7.7.3.2 Period Asymmetry

We have also characterized a second sample based on a lattice of circular holes spaced by two different distances along the two perpendicular axes, namely $a_x = 920$ nm and $a_y = 980$ nm (see SEM top-view image in Fig. 7.13a). The hole diameter is kept constant to 450 nm.

In Fig. 7.13b–c we report the numerical and experimental reflectance spectra for different incident wave polarizations, in good agreement with each other. Differently from the above case, since the two orthogonal polarization states interact with different lattice period, the plasmonic resonance excitation occurs at two different wavelengths. In particular two different LSPRs are excited at ~1,430 and ~1,510 nm for POL_x to POL_y respectively. This is in accordance to the fact that the smaller is the period the smaller is the wavelength at which the resonant phenomenon take place. Clearly, when the incident electric field polarization gradually moves from one polarization to the other, we observe a sort of energy balance effect between the two resonant dips in the reflection spectrum: they basically undergo changes in their shape and visibility, appearing and disappearing at the two resonant wavelengths, as a function of the polarization. In this case, our fiber probe works as polarization-dependent wavelength-selective device, that could be used as a tunable polarization filter as well as a modulator.

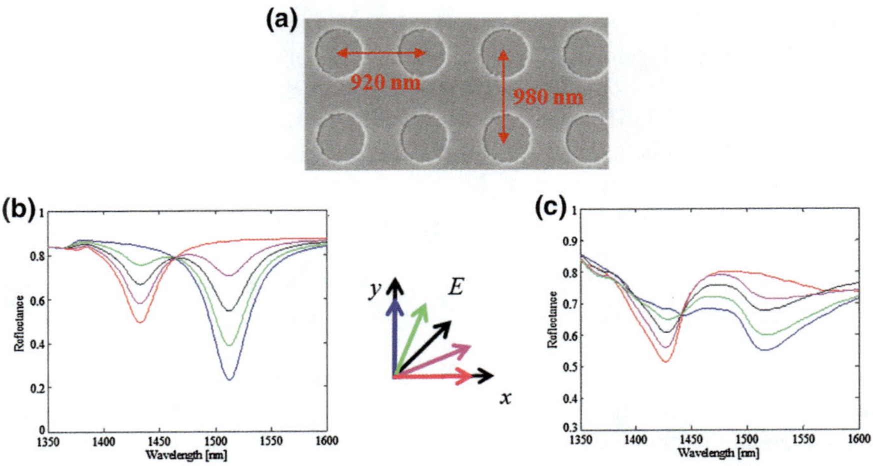

Fig. 7.13 **a** SEM image (*top view, zoomed*) of the probe with period asymmetry ($a_x = 920$ nm, $a_y = 980$ nm); **b** numerical and **c** experimental reflectance spectra of the sample for different incident wave polarizations: *blue* and *red lines* refer to POL_x and POL_y. Modified by [25]

7.8 Conclusions and Outlook

To summarize, we have demonstrated that by means of a proper customization of standard nanoscale deposition and patterning tools, typically exploited for planar devices, we have been able to define a valuable fabrication path enabling the integration of functional dielectric and metallic nanoscale structures directly on the distal end of standard single mode fibers. The actual effectiveness of the proposed technological path has been demonstrated through the realization of a miniaturized fiber tip device based on a two-dimensional hybrid metallo-dielectric nanostructure supporting LSPR. Both experimental and full-wave numerical analyses have been carried out to characterize the resonant phenomena. We have presented some experimental results demonstrating how the proposed device can work effectively as an optical probe for label free chemical and biological sensing, as a microphone for acoustic wave detection and as in-fiber polarization-sensitive device.

First, we have demonstrated how by selecting a specific dielectric and metal layer thickness it is possible to make the fiber tip device more or less sensitive to the external environment refractive index, thus tailoring its performances for the specific applications. Many degrees of freedom can be further exploited to increase the sensitivity: for example, we have preliminarily found that by exploiting a quasi-periodic tiling it is possible to significantly increase (up to a factor 2) the SRI bulk sensitivity with respect to the periodic case.

We have also investigated the surface sensitivity of the proposed devices, which in turn can be considered the key parameter when dealing with chemical and biological applications. Experimental results have revealed the capability to detect nano-sized overlays deposited on the sensor substrate, demonstrating the effectiveness of the proposed platform to be used as high sensitivity label free biosensor. We have analyzed the effect of the metallo-dielectric crystal size on the spectral features exhibited by the integrated platform demonstrating that the sensitive area (i.e. the patterned area on the fiber tip) can be reduced down to only $20 \times 20\ \mu m^2$ without affecting the sensor operability; this foresees improved limits of detection in terms of number of bound biomolecules on the functionalized sensor surface for a given local refractive index change.

Moreover, we have also shown the surprising capability of the proposed platform to detect acoustic waves taking advantage from the low elastic modulus of the patterned polymer.

Concerning the plasmonic resonance engineering, as a result of a numerical analysis, it has been found that by acting on the dielectric layer thickness or integrating high refractive index materials it is possible to increase the number and tailor the near field distributions of different plasmonic modes excited at different wavelengths. We have finally shown that the resonance engineering could be also achieved by acting on the lattice tiling; by breaking the circular symmetry of our device (with creating elliptical holes or different periods in the two perpendicular directions) it is possible to obtain polarization sensitive devices.

It is clear that the results discussed in this work, although they demonstrate the great versatility of the lab on fiber technology, reveal only a little part of what can be effectively achieved using this technology. To give an example, in addition to metallic structures supporting SPR, also metamaterials could be integrated on the optical fiber tip; thus, the creation of periodic distributions of sub-wavelength sized metal resonators could potentially open the way to novel devices exhibiting the exotic optical properties offered by metamaterials such as superlensing and cloaking [35].

One senses that the potentialities offered by this technology, have not yet been entirely exploited. Under many points of view, we are only at the beginning, especially in the fields of medical and biological applications. First, fiber optic nanoprobes could be judiciously integrated with microfluidics components to provide new Lab on chip implementations, taking advantage from the easier connection of the optical chain to complex lighting systems as well as sophisticate interrogation units. Moreover the real point of strength of this technology is certainly the property of allowing in vivo analysis; since optical fibers can be easily inserted inside medical needles, the remote interrogation of optical devices integrated on the fiber tip can provide in situ, real time, monitoring of different biological events such as molecular binding or tissues characterization. Indeed, as demonstrated in [16, 36], the incorporation of an array of gold optical pillars onto the fiber facet can give rise to optical probes for surface enhanced Raman scattering (SERS) detection. Furthermore, by integrating and patterning soft materials (with a low Young modulus), it could be possible to realize very sensitive pressure sensors for elastometric measurements and characterization of human tissues. As already partially demonstrated in [17] the pressure wave impinging on the fiber tip causes cyclical physical deformations of the nanostructured layer, resulting in a resonance shift due to changes in the phase-matching condition. This phenomenon could be in principle exploited also for ultrasound detection, leading to the creation of fiber based ultrasonic receivers for novel in situ ecography systems.

Lab on fiber technology devices could provide significant improvements also in the field of in vivo optical imaging techniques [37]. Nanoscale slit or holes arrays in a metallic film integrated on the optical fiber tip could be used as planar lenses whose focusing effect could further improve the imaging resolution [38]. Anyway, also in this latter case, differently from free space approaches, fiber based devices (as those discussed in this chapter) and interrogation setups strongly facilitate in vivo applications. On the same line of argument, concerning non-invasive optical imaging method, polarization sensitive fiber based devices could be effectively used in polarization sensitive optical coherence tomography (PS-OCT) which extends the concept of OCT providing cross-sectional images of tissue structure in real time. The change in polarization state of back-scattered light (with respect to the polarization state of incident light) provides additional information about the properties of tissues and image contrast enhancement [39]. In order to fully take advantage of the filtering properties, however, it could be necessary to introduce electrical contacts in such a way to realize active device guaranteeing a better polarization control.

To conclude, although further investigations lie ahead and there is still so much to prove, it is fair to predict that the development of the lab on fiber technology (including all the related fabrication methods) is pointing to a novel class of multifunctional nanoprobes, with unique properties in terms of functionality, performances and miniaturization.

References

1. B. Culshaw, A. Kersey, J. Light. Technol. **26**, 1064–1078 (2008)
2. S.A. Maier, *Plasmonics: fundamentals and applications* (Springer, Berlin, 2007)
3. A. Andreone, A. Cusano, A. Cutolo, V. Galdi, *Selected Topics in Photonic Crystals and Metamaterials* (World Scientific Publishing Co., Singapore, 2011)
4. W.L. Barnes, A. Dereux, T.W. Ebbesen, Nature **424**, 824 (2003)
5. D.A. Genov, A.K. Sarychev, V.M. Shalaev, A. Wei, Nano Lett. **4**, 153 (2004)
6. M. Consales, M. Pisco, A. Cusano, Phot. Sens. **2**(4), 289 (2012)
7. E.J. Smythe, M.D. Dickey, G.M. Whitesides, F. Capasso, ACS Nano **3**, 59 (2009)
8. D.J. Lipomi, M.A. Kats, P. Kim, S.H. Kang, J. Aizenberg, V. Capasso, G.M. Whitesides, Nano Lett. **4**, 4017 (2010)
9. D.J. Lipomi, R.V. Martinez, M.A. Kats, S.H. Kang, P. Kim, J. Aizenberg, F. Capasso, G.M. Whitesides, Nano Lett. **11**, 632 (2011)
10. I.W. Jung, B. Park, J. Provine, R.T. Howe, O. Solgaard, J. Light. Tech. **29**, 1367 (2011)
11. S. Scheerlinck, P. Dubruel, P. Bienstman, E. Schacht, D. Van Thourhout, R. Baets, J. Light. Tech. **27**, 1415 (2009)
12. G. Shambat, J. Provine, K. Rivoire, T. Sarmiento, J. Harris, J. Vuckovic, Appl. Phys. Lett., J. Light. Tech. **99**, 191102 (2009)
13. C. Lerma Arce, K. De Vos, T. Claes, K. Komorowska, D. Van Thourhout, P. Bienstman, IEEE Phot.Tech. Lett **23**, 890 (2011)
14. A. Dhawan, M. Gerhold, J. Muth, IEEE Sens. J. **8**, 942 (2008)
15. Y. Lin, Y. Zou, R.G. Lindquist, Biomed. Opt. Exp. **2**, 478–484 (2011)
16. X. Yang, N. Ileri, C. Larson, T. Carlson, J. Britten, A. Chang, C. Gu, T. Bond, Opt. Exp. **20**, 24819 (2012)
17. M. Consales, A. Ricciardi, A. Crescitelli, E. Esposito, A. Cutolo, A. Cusano, ACS Nano **6**, 3163–3170 (2012)
18. D. Rosenblatt, A. Sharon, A.A. Friesem, IEEE J. Quant. Electron. **33**, 2038–2059 (1997)
19. COMSOL Multiphysics, http://www.comsol.com
20. A. Ricciardi, M. Pisco, A. Cutolo, A. Cusano, L. O ' Faolain, T. Krauss, G. Castaldi, V. Galdi, Phys. Rev. B **84**, 085135 (2011)
21. E.D. Palik, *Handbook of Optical Constants of Solids* (Academic Press, Orlando, 1985)
22. D.W. Peters, S.A. Kemme, G.R. Hadley, JOSA A **21**, 981–987 (2004)
23. M.E. Stewart, C.R. Anderton, L.B. Thompson, J. Maria, S.K. Gray, J.A. Rogers, R.G. Nuzzo, Chem. Rev. **108**, 494 (2008)
24. A. Ricciardi, M. Consales, G. Quero, A. Crescitelli, E. Esposito, A. Cusano, ACS Photon. 2013. doi: 10.1021/ph400075r
25. A. Crescitelli, A. Ricciardi, M. Consales, E. Esposito, V. Galdi, A. Cutolo, A. Cusano, Adv. Funct. Mat. **22**, 4389 (2012)
26. M. Pisco, A. Ricciardi, I. Gallina, G. Castaldi, S. Campopiano, A. Cutolo, A. Cusano, V. Galdi, Opt. Exp. **18**, 17280 (2010)
27. M. Senechal, in *Quasicrystals and Geometry*, ed. by M. Senechal (Press Syndicate of Cambridge University, New York, 1995), Ch.5

28. I. Gallina, M. Pisco, A. Ricciardi, S. Campopiano, G. Castaldi, A. Cusano, V. Galdi, Opt. Exp. **17**, 19586 (2009)
29. A. Tsargorodskaya, A.V. Nabok, V. Ray, Nanotechnology **15**, 703 (2004)
30. D. Piscevic, W. Knoll, M.J. Tarlov, Supramol. Sci. **2**, 99 (1995)
31. S.H. Fan, J.D. Joannopoulos, Phys. Rev. B **65**, 2351121–2351128 (2002)
32. A. Ricciardi, S. Campopiano, A. Cusano, T.F. Krauss, L. O'Faolain, IEEE Photon. J. **2**, 696–702 (2010)
33. A. Ricciardi, M. Pisco, I. Gallina, S. Campopiano, V. Galdi, L. O'Faolain, T.F. Krauss, A. Cusano, Opt. Lett. **35**, 3946–3948 (2010)
34. A. Ricciardi, M. Consales, G. Quero, A. Crescitelli, E. Esposito, A. Cusano, Opt. Fib. Technol. **19**(6), 772–784 (2013)
35. R.C. Mc Phedran, I.V. Shadrivov, B.T. Kuhlmey, Y.S. Kivsha, Metamaterials and metaoptics. NPG Asia Mater. **3**, 100–108 (2011)
36. E.J. Smythe, M.D. Dickey, J.M. Bao, G.M. Whitesides, F. Capasso, Nano Lett. **9**, 1132–1138 (2009)
37. V. Ntziachristos, Annu. Rev. Biomed. Eng. **8**, 1–33 (2006)
38. Y. Fu, X. Zhou, Plasmonics **5**, 287–310 (2010)
39. J.F. de Boer, T.E. Milner, M.J.C. van Gemert, J.S. Nelson, Opt. Lett. **22**, 934–936 (1997)

Chapter 8
Miniaturized Optical Tweezers Through Fiber-End Microfabrication

Carlo Liberale, Gheorghe Cojoc, Vijayakumar Rajamanickam, Lorenzo Ferrara, Francesca Bragheri, Paolo Minzioni, Gerardo Perozziello, Patrizio Candeloro, Ilaria Cristiani and Enzo di Fabrizio

Abstract Optical tweezers represent a powerful tool for a variety of applications both in biology and in physics, and their miniaturization and full integration is of great interest so as to reduce size (towards portable systems), and to minimize the

---

C. Liberale (✉) · V. Rajamanickam · L. Ferrara
Nanostructures, Istituto Italiano di Tecnologia, Via Morego 30, 16163 Genoa, Italy
e-mail: carlo.liberale@iit.it

V. Rajamanickam
e-mail: Vijayakumar.Rajamanickam@iit.it

L. Ferrara
e-mail: lorenzo.ferrara@iit.it

G. Cojoc
Max Planck Institute of Molecular Cell Biology and Genetics MPI-CBG,
Pfotenhauer street 108, 01307 Dresden, Germany
e-mail: cojoc@mpi-cbg.de

F. Bragheri · P. Minzioni · I. Cristiani
Dipartimento di Ingegneria Industriale e dell'Informazione, Universita' di Pavia,
Via Ferrata 1, 27100 Pavia, Italy
e-mail: francesca.bragheri@ifn.cnr.it

P. Minzioni
e-mail: paolo.minzioni@unipv.it

I. Cristiani
e-mail: ilaria.cristiani@unipv.it

E. di Fabrizio
PSE and BESE divisions, King Abdullah University of Science
and Technology (KAUST), Jeddah, Saudi Arabia
e-mail: enzo.difabrizio@kaust.edu.sa

G. Perozziello · P. Candeloro · E. di Fabrizio
BioNEM Lab, Dipartimento di Medicina Sperimentale e Clinica,
Universita' Magna Graecia di Catanzaro, 88100 Catanzaro, Italy
e-mail: gerardo.perozziello@unicz.it

P. Candeloro
e-mail: patrizio.candeloro@unicz.it

A. Cusano et al. (eds.), *Lab-on-Fiber Technology*,
Springer Series in Surface Sciences 56, DOI: 10.1007/978-3-319-06998-2_8,

required intervention from the operator. Optical fibers represent a natural solution to achieve this goal, and here we review the realization of single-fiber optical tweezers able to create a purely optical three-dimensional trap.

8.1 Introduction

Optical Tweezers (OT) are based on the effect of the forces exerted by a strongly focused optical beam on micro- or nano-sized particles. Seminal works by Ashkin et al. demonstrated the importance of OT by performing a wide variety of experiments ranging from trapping and cooling of single atoms to contact-less trapping and manipulation of viruses, bacteria and cells [1–4]. Since then, important achievements in biomedical sciences have been obtained through the mechanical manipulation without physical contact of biological specimens and by the combination of OT with optical diagnostic techniques such as two-photon fluorescence, Raman scattering, and coherent anti-Stokes Raman spectroscopy [5–11].

The optical force experienced by a dielectric particle can be described as originating from the momentum transfer due to the refraction and reflection of the incident photons. The optical force can be considered as the sum of two components: a *scattering force*, which pushes the particle along the propagation direction of the incident light, and a *gradient force*, directed as, and proportional to, the intensity spatial gradient, which pulls the particle towards the highest intensity region. Stable three-dimensional (3D) trapping is achieved when exists a point in space where the gradient force counterbalances the scattering force. To satisfy this condition a steep spatial gradient of the beam intensity is needed [12]. Indeed, the "standard" OT scheme consists of a freely propagating laser beam that is tightly focused, by means of a high Numerical Aperture (NA) microscope objective (usually NA > 1), inside the medium where the particles to be trapped are immersed [13, 14].

Despite the relevant results achieved in many fields, the bulky structure of standard OT still limits their utilization in several environments. In addition, the use of standard OT in turbid media or in thick samples presents significant challenges, being difficult to achieve the tight focusing necessary for optical trapping [15]. The realization of OT based on optical fibers would allow a miniaturized, versatile and handy tool to be obtained, suitable for many applications relevant to biology and fundamental physics, such as in vivo biological manipulation or in-vacuum single-particle X-ray spectroscopy. Moreover, these fiber-based OT could be easily integrated into microfluidic circuits.

In this chapter, we describe a new approach to obtain a fiber-Optical Tweezers (fiber-OT) through fiber-end microfabrication. Moreover, we show two different on-fiber fabrication approaches to obtain this configuration, one based on Focused

Ion Beam milling, the second one based on Two-Photon Lithography. Finally, we show optical trapping experiments and a microfluidic chip where our fiber-OT has been integrated.

8.2 Fiber-Optical Tweezers: Working Principle

Various approaches for the realization of fiber-OT have been reported in literature. A typical approach makes use of two separated fibers, aligned so that the laser beams exiting from the fibers are counter-propagating along a common optical axis. In this case, the axial scattering forces are counterbalanced and a full 3D optical trapping is achieved, but the set-up requires critical alignment [16, 17]. Some authors proposed OT based on a single microlensed optical fiber, but the NA that can be obtained using this configuration is usually quite low, so that such a solution allows a two-dimensional only trapping to be achieved [18–20]. Three-dimensional trapping has been demonstrated with a metalized and chemically etched fiber [21], however the optical forces were not able to trap the particles by themselves, but needed to be balanced by electrostatic interactions between the fiber tip and the object to be trapped. A purely optical 3D trapping by means of single-fiber has been achieved through highly tapered fibers [22] but, in this case, the trapping point is very close to the fiber tip, making impossible to trap particles of large size without physical contact.

All of these reported approaches featured some unsatisfying aspect, fostering the research for a different configuration able to overcome the above mentioned issues. Indeed, a new single fiber-OT has been presented more recently by our group [23], and its working principle is sketched in Fig. 8.1. The fiber-OT has been designed by successfully combining two concepts: the use of non-standard fibers, in which light propagates through an annular core structure, and the achievement of light focusing with high NA by using total internal reflection (TIR) (instead of refraction) at the interface between fibers and surrounding medium. By properly shaping the fiber-end at an angle θ, as shown in Fig. 8.1a, the light beam carried in the annular core (the grey annulus in Fig. 8.1b) experiences TIR at the fiber/medium interface. The beam is then deflected through the fiber cladding, and then out of the fiber, converging in a point positioned along the probe axis, thus producing a "focalization effect" corresponding to that obtained using an objective with an equivalent NA (in the ray-optics approximation) given by $NA_{eq} = n_F \sin(2\theta)$, where n_F is the fiber refractive index. In the proposed scheme, the trapping is intrinsically favoured by the beam propagation geometry: in fact, considering a strongly focused gaussian beam in the ray-optics approximation, only the external (and strongly tilted) rays are mainly responsible for the gradient force in the axial direction, whereas the central rays contribute essentially to the scattering force, yielding a negligible contribution to the axial gradient force [12]. In the scheme reported in Fig. 8.1a, the trap is created only by rays inclined with

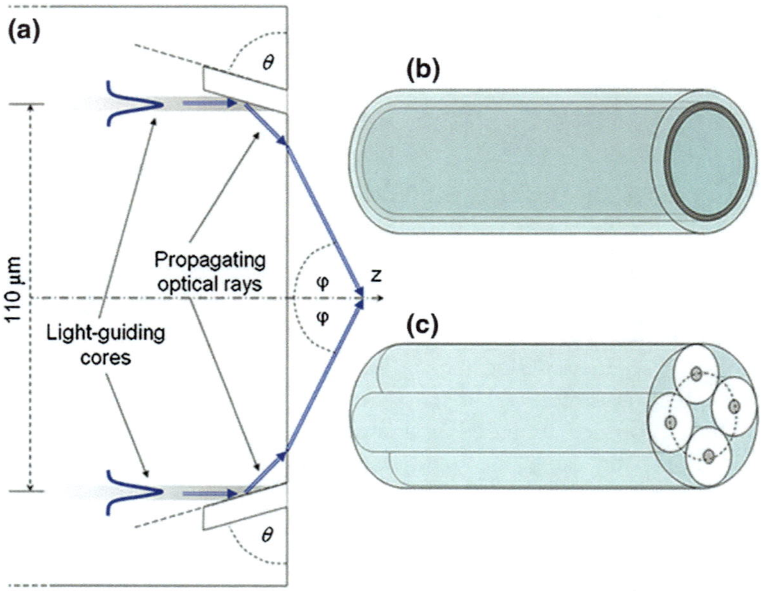

Fig. 8.1 Fiber structures and fiber-OT working principle. **a** Cross-section of the fiber tweezers, in which the light propagates through the annular core. The fiber end-face is properly cut in such a way that the light propagating through the core is reflected at the core/outer-medium interface, propagates inside the fiber, and is finally deflected towards the focal point. **b** Annular-core fiber. The light is guided through the grey annular core. **c** Example of an optical-fiber bundle composed of four fibers. The basic principle depicted in **a** can be applied to both structures (With permission from [23])

respect to the axis, thus weakening the impact of the scattering force and relaxing the constraints on the NA_{eq} required to achieve efficient trapping [24]. This approach provides 3D trapping at a large distance from the fiber-end by using a single fiber. Furthermore, both the NA_{eq} and the trapping distance can be easily tuned by changing the angle θ and the annulus diameter. As an example, taking $n_F = 1.45$ and $n_M = 1.33$ as the refractive index of the surrounding medium (water), and by cutting the fiber surfaces at an angle θ slightly beyond the critical angle for TIR ($\theta_c = 66.5°$), then $NA_{eq} = 1.06$, a value very close to that of the typical objectives used in bulk optical trapping arrangements. Considering the above parameters and a diameter of the core annulus of, as an example, 110 μm, the trapping position is about 35 μm away from the fiber-end, thus allowing a high degree of freedom in sample manipulation, imaging and analysis.

A trapping effect analogous to that obtained through an annular-core fiber can be achieved by using a fiber-bundle structure, such as that shown in Fig. 8.1c. If the size of the structure and that of the fibers composing the bundle are properly chosen, a fairly symmetrical annular distribution of fibers is obtained, as outlined by the dashed line in Fig. 8.1c. The fiber-bundle approach allows obtaining easily a proper fiber cores geometry starting from standard single mode fibers and has some advantages compared to that based on the annular-core fiber, because it allows independent control of the different portions of the optical beam emitted by the structure, and offers the opportunity to specialize each of the used fibers for different purposes (e.g. excitation and collection of spectroscopic signature of the trapped object).

8.3 On-Fiber Fabrication

The previously described geometry for the fiber-OT has been obtained by using two different fabrication approaches, that will be described in this section. In both fabrication techniques we start from a fiber bundle fabricated following the description reported in [23]. Specialty single-mode fibers (produced on request by J-Fiber, Jena) with low index contrast (mode field diameter $\approx 9\ \mu m$ at $\lambda = 1070$ nm) and 80 μm cladding-diameter were assembled, resulting in a miniaturized probe with an external diameter equal to 320 μm.

8.3.1 FIB Fiber-End Fabrication

In a first approach, the fibers' end-faces were microstructured through Focused Ion Beam (FIB) milling, which allows a controlled nanomachining of the surfaces using a beam of accelerated gallium ions, focused on the target region by means of electrostatic lenses.

FIB is one of the most efficient top-down nanofabrication techniques due to the nanometric dimension of the ion beam source that enables both sub-micro/nano imaging (with generated secondary electrons), localized milling and deposition. In combination with a Scanning Electron Microscope (SEM), it can exploit high-spatial resolution imaging, milling, nanomanipulation, deposition and structural analysis.

A "dual beam" system consists of two columns, as shown in the Fig. 8.2, for simultaneous FIB milling and SEM imaging. Gallium ions are accelerated down to the column applying a differential voltage and the ion beam is then focused on the sample surface through electrostatic lenses, with dimension varying from a few to several nm. Magnetic plates provide the deflection of the beam, thus allowing the scanning of the sample for imaging. With pattern generator software it is possible to define a precise scanning path and milling depth.

The beam diameter can be set using different apertures of the ion column. With bigger diameters, the ion current will increase, thus fastening the removal of the material with lesser amount of carbon deposition during the Ga^+ sputtering. However, both the nanofabricated pattern and the redeposition effect will increase. Smaller ion currents will reduce the redeposition and will make finer structures.

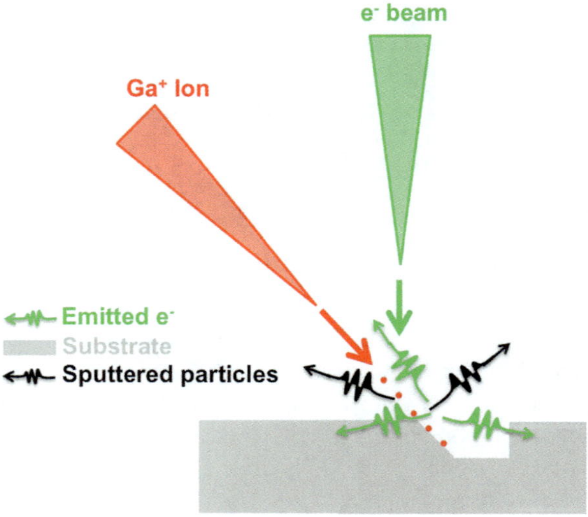

Fig. 8.2 FIB-SEM dual beam system. Sketch of the FIB milling process

The microstructuring of the fiber-ends was realized by a NOVA 600i (FEI Company) FIB-SEM dual beam system to shape the core regions at the fiber surfaces as required by the scheme in Fig. 8.1a, in such a way as to obtain TIR at the fiber-core/water interface. As a first step, the fiber bundles' end-faces were sputtered with a 20-nm-thick gold layer, to avoid electrostatic charge on the sample during FIB operation. The bundle was then mounted in the FIB vacuum chamber on a specifically designed support and, by milling using a 20 nA ion current, apertures were dug on the surfaces corresponding to the fiber cores at the proper angles to the fiber axis (Fig. 8.3). The beam was scanned over a trapezium-like area (height 10 μm, area 150 μm^2) to obtain a large aperture, thus enabling the escape of the sputtered material from the hole. Finally, the remaining gold layer deposited on the top of the probe was removed, using FIB, in correspondence of the exit point for the reflected beams, in order to avoid back-reflections.

The image (Fig. 8.3f) of the final micromachined probe, taken using a SEM, show that the overall structure presents a very good symmetry and that the surfaces are of excellent quality. The θ angles realized in the fabricated structures are about 70°, giving $NA_{eq} = 0.93$.

8.3.2 Two-Photon Lithography Fiber-End Fabrication

8.3.2.1 Two-Photon Lithography

Fabrication of 3D micro- and nano-structures has always represented an intrinsic limit of the most used lithographic techniques, like electron-beam and optical lithography. Generally, in such techniques, a planar two dimensional (2D) pattern

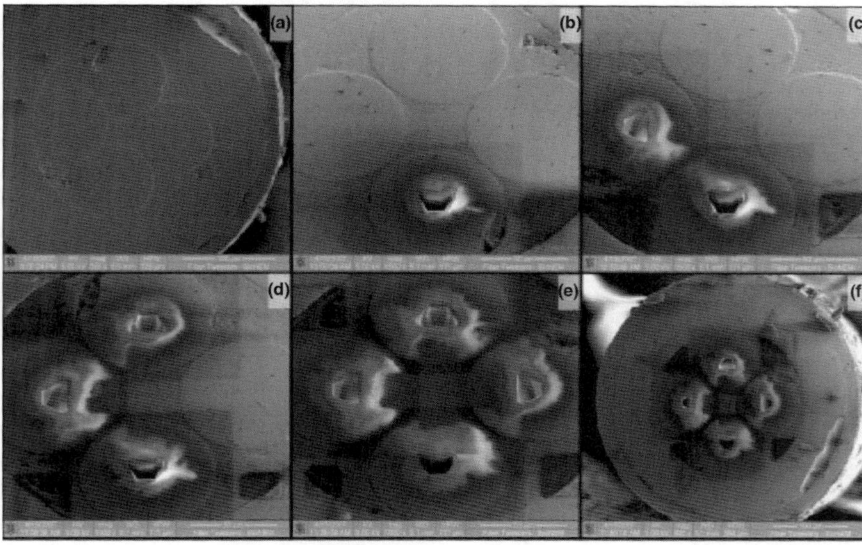

Fig. 8.3 This picture shows the different stages of FIB milling probe micro-machining. The symmetry and surface quality is first verified (**a**). After this check, the probe is rotated, so that one fiber is in the lower possible position, and that fiber is then drilled at the proper angle (**b**). The probe is subsequently rotated by 90° steps and the procedure is repeated on the remaining fibers (**c**–**e**). The images highlight the strong reproducibility of the procedure and the high quality of the surfaces. The final result on the probe is shown in (**f**) (With permission from [23])

is exposed into a resist layer; then the resist mask is transferred to the underlying sample via etching processes or metallic deposition (lift-off process). To obtain 3D structures with these common lithographies, the straightforward way is to reiterate several times the resist exposure step followed by the resist mask transfer. Obviously this approach requires a large experimental effort. Alternatively, direct electron-beam writing can achieve 3D structures carefully modulating the exposure doses and exploiting proximity effects, as reported, for example, by Prasciolu et al. [25]. In this case, however, several 2D patterns with different doses have to be computed for the exposure of a single 3D profile, and undoubtedly this approach represents a demanding and advanced application of electron-beam writing.

Conversely to the above mentioned techniques, two-photon lithography (TPL) is intrinsically a 3D structuring process. In this technique the pulse train generated by a Near-Infrared (NIR) femtosecond laser oscillator is directly focused, with a high NA microscope objective, inside a suitable resin composed by a proper mixture of a photo-initiator and a monomer. The photo-initiator has an absorption spectrum peaked in the UV, therefore it can be excited only by simultaneous absorption of two-photon from the NIR laser. As the rate of two-photon absorption is proportional to the square of the light intensity, the NIR laser beam is strongly absorbed only within the laser focal spot volume, where the resin is then polymerized. Consequently, as the laser focal spot is scanned inside the resin according to the desired structure, arbitrary 3D shaped micro- and nano-structures can be easily fabricated, as widely reported in the last years [26]. The advantages of this technique rely on the straightforward process that allows creating structures which would be very complex, or even impossible, to fabricate with conventional lithographic methods. Moreover, as recently demonstrated [27–29], the spatial resolution attainable using TPL is sub-100 nm using laser wavelengths around 800 nm, that is well below the diffraction limit. Applications range from 3D microdevices fabrication to plasmonic devices, MEMS and micro-photonics structures [30–33].

Micro-optics fabrication appears to be a natural application for TPL due to freedom in geometrical shape design, good optical properties of the photosensitive materials, and rapid fabrication process which consists of very few steps [34–36]. Additional interest is attracted by the fabrication of micro-optics on top of optical fibers, for which the other demonstrated fabrication methods are very different, and specific for each geometrical shape of the desired micro-optics [25, 37–41].

8.3.2.2 Two-Photon Lithography Setup

In Fig. 8.4 is shown the TPL setup we have used for micro-optics fabrication. As the laser excitation source, we have started by using a 100 fs pulsewidth, 80 MHz Ti:Sapphire laser oscillator (Tsunami, Spectra-Physics) that can tune the wavelength over a wide near-infrared (NIR) range, thus being compatible with most photo-initiators used in UV photolithography. More recently, we have chosen to adopt the emerging technology of the less expensive femtosecond fiber lasers. Indeed, our optical setup now is based on a mode-locked Er-doped fiber laser (FemtoFiber pro NIR, Toptica) lasing at 780 nm, with a repetition rate of 80 MHz and a 100 fs pulse width. This laser oscillates at a fixed NIR wavelength for which not all of the available UV photo-initiator absorb efficiently, and then we have also developed suitable materials with proper optical properties and simple processing, that are efficient and optimized for two-photo polymerization with femtosecond fiber lasers.

The output laser power is decreased to a few mW at the back focal plane of the microscope objective by using a variable attenuator made by an achromatic half-waveplate and a Glan–Taylor calcite linear polarizer. The exposure time for each pixel is controlled through a computer-driven mechanical shutter. The beam is expanded and collimated by a telescope in order to obtain overfilling of the focusing microscope objective. Further, the laser beam is reflected by a 45° dichroic mirror directly onto the objective back aperture. The dichroic mirror reflects most of the NIR laser beam and transmits part of the visible spectrum (400×550 nm^2). A semi-apochromatic microscope objective (Olympus, LUC-PlanFLN 60×, N.A. = 0.70) equipped with a spherical aberration correction collar focuses the beam into the resin. A CCD camera and a LED illuminator are used to monitor in real time the fabrication process. In order to avoid the polymerization of

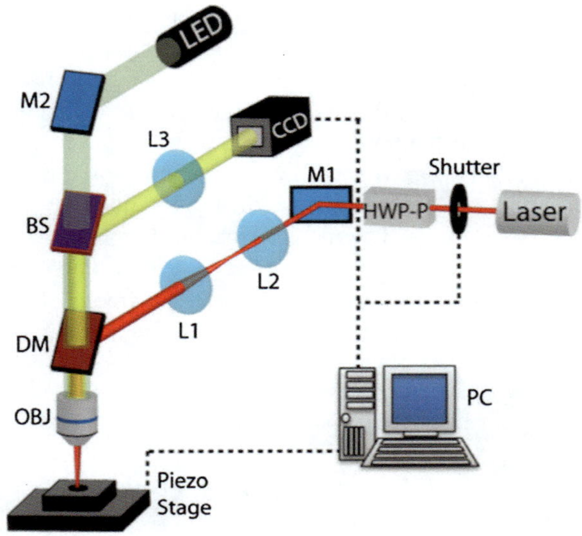

Fig. 8.4 Two-photon lithography setup diagram. *HWP-P* variable attenuator; *L1–L2* beam expander; *L3* imaging tube lens; *BS* beamsplitter; *DM* dichroic mirror; *OBJ* microscope objective; *M1–M2* mirrors (With permission from [42])

resist during the illumination, a longpass filter (FEL0500 from Thorlabs) is inserted after the LED.

A specific fiber holder is mounted on a XYZ piezoelectric driven stage (P-563 PIMars, Physik Instrumente) with a scan range of 300 μm in all of the coordinates and a custom made software translates the structure-points to piezo stage positions and controls the synchronization of the movements with the shutter, in order to achieve the desired local dose. In order to speed up the fabrication process, only the external surfaces of the desired structures are scanned, and then polymerized, by TPL. After removal of the unpolymerized external resin with a suitable solvent, the developed structures are then UV exposed to complete the polymerization of the resin inside the inner volume. The entire system is mounted on a vibration isolation table.

When using the Ti:Sapphire femtosecond laser, we choose to use NOA 63 (Norland Optical Adhesive) as a photopolymer, because of its good adhesion to glass, easy processing, refractive index comparable to that of optical fibers (1.56 for the polymerized resin), and very low cost. For efficient two-photon photopolymerization of this material, the laser wavelength was set to 720 nm.

When we have started to use the femtosecond fiber laser, to overcome the constraint of the fixed 780 nm wavelength, we have found different photo-initiators with an absorption spectrum peaked around 400 nm, thus compatible with our setup. In particular, as the photo-initiators we have selected isopropyl Thioxanthone (ITX, Sigma), 7-diethylamino-3-thenoylcoumarin (DETC, Exciton, USA), and 4,4′ Bis(diethylamino)benzophenone (BDEB, Sigma), since their absorption spectra match well with the laser wavelength. Moreover these photo-initiators grant efficient radical generation, reactivity and high solubility in acrylic monomers. These photo-initiators are mixed with an Acrylic monomer (BPA-EDA, Sigma) with the following characteristics: simple preparation by drop casting, are soluble in common solvents, high cross-linking to resist swelling, low shrinkage, good mechanical properties, high transparency in the visible when polymerized [43]. After the completion of exposure for desired structures, the fiber is retracted from the resin droplet and the not-exposed photo-curing material (that did not experience cross-linking) is removed with a few seconds of rinse in methanol and Isopropanol, leaving the 3D structure attached to the fiber top.

When exposed to the laser beam, the photo-initiator starts the polymerization with a minimum average-power threshold of around 1 mW. If we increase the optical power by a few mW, the polymerized structures achieve a good 3D mechanical stability, even with only the external surface exposed.

8.3.2.3 Micro-optics Fabrication

We have fabricated different kinds of micro-optics structures to demonstrate the feasibility and versatility of the proposed technique [34, 35]. To assess the optical quality of the fabricated structures, we have then performed a characterization of the micro-structures by using a Yb-doped fiber laser (IPG Photonics), coupled to the test fiber, and an imaging setup consisting of a 40× objective, a CCD camera and a translation stage with micrometric resolution to move the fiber end with respect to the objective object plane. Images of the beam at different distances from the fiber end are then acquired and analyzed.

Single mode optical fibers (SM980—Fibercore) with a 6.2 μm Mode Field Diameter at 1064 nm are used for fabrication of their cleaved ends. We have fabricated a spherical micro-lens with various radii of curvature. A 7 μm curvature radius is shown in Fig. 8.5a. Propagation from the fiber end in this case shows a Gaussian beam which is focused to a waist of 1.6 μm (intensity-FWHM) at nearly 15 μm from the fiber, in good agreement with the approximated formula for the focal length $f = R/(n - 1)$ where R is the spherical lens radius and n is the lens refractive index. The intensity FWHM for this beam as a function of propagation distance, compared with a fiber without micro-optics is shown in Fig. 8.5b.

A second kind of structure is a conical lens, we have been able to fabricate at different, arbitrary, apex angles. SEM images of two conical lenses with apex angle of 80° and 120° are shown in Fig. 8.6. Beam propagation from these structures shows, as expected, a Bessel-like profile for a range of distances from the fiber end which is also dependent on the apex angle [34].

To assess the resistance to optical damage of the fabricated micro-structures, an increasing amount of power has been coupled into the fiber, showing no problems even for 100 mW of average power measured in output from the fiber.

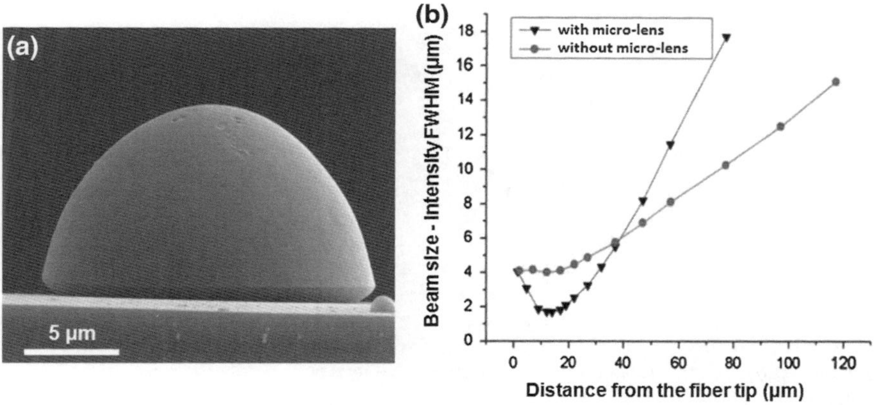

Fig. 8.5 **a** SEM image of a fabricated spherical micro-lens. **b** Intensity FWHM measured for bare (*red line*) and micro-lensed fiber (*black line*)

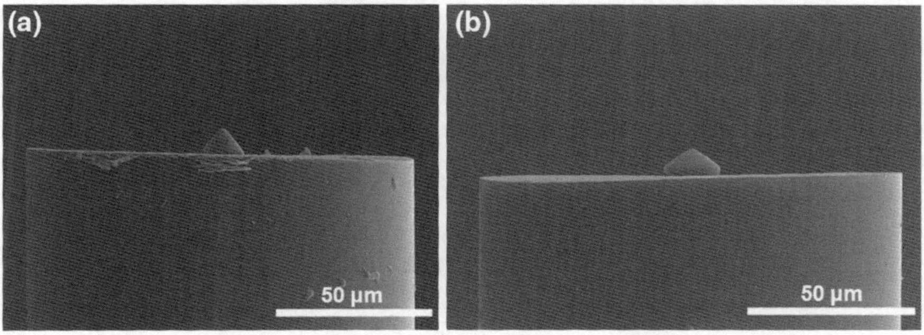

Fig. 8.6 **a** SEM image of a fabricated conical lens with 78° apex angle. **b** SEM image of a fabricated conical lens with 117° apex angle

8.3.2.4 Micro-prisms Fabrication

Owing to the excellent capability to create micro-optics allowed by TPL, we have decided to introduce the beam shaping required to obtain a fiber-OT by fabricating polymeric micro-prism reflectors on the fiber facets (Fig. 8.7). This approach makes the device fabrication significantly easier compared to the FIB technique described in the previous section. Moreover, FIB fiber-OT fabrication does not allow the realization of structures with complex geometries and has high running costs. Finally, we note that the fabrication of reflectors, which protrude with respect to the fiber facet, allows obtaining devices with a longer trapping distance. Notwithstanding these advantages of the TPL fabrication, the FIB fabrication approach still retains interesting and unique features as it can be employed when fiber-OT are required for use at high temperatures (above polymer glass transition

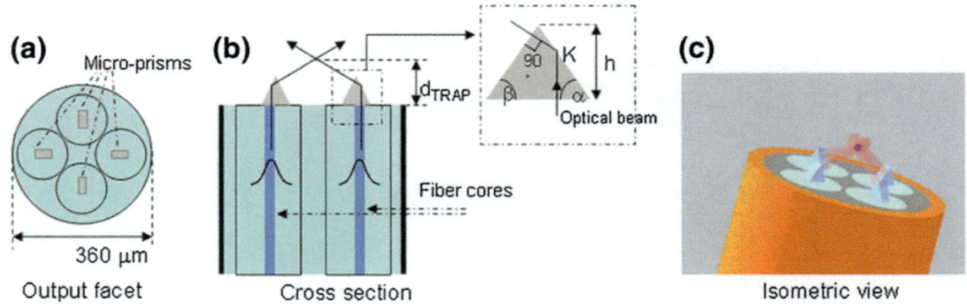

Fig. 8.7 Fiber-OT with micro-prisms. **a** front-view of the output facet, showing the four fibers with micro-prisms. **b** Cross-section of the fiber probe, the beams propagating in two symmetrically positioned fibers are reflected at the interface between micro-prisms and the outer medium; as reported in the inset, the prism angles are chosen so as to produce normal incidence on the second interface. **c** Isometric representation of the optical tweezers trapping a particle (With permission from [42])

temperature) or when it is required to have no parts extending beyond the fiber facet.

As depicted in Fig. 8.7b the output beams are deflected by TIR occurring at the interface with the outer medium of four micro-prisms positioned on the fiber facets: the four deflected beams cross each other in the same spot, thus creating the optical trap.

The micro-prisms realized by TPL were perfectly aligned on the fiber surface and, with reference to the the inset of Fig. 8.7b, were designed with an angle α so as to induce TIR at the first lateral surface of the prism (point K), and with an angle β yielding nearly normal incidence at the second lateral surface of the prism.

With a refractive index of 1.56 for the polymerized resin, and considering water as outer medium, we calculate the minimum angle to have TIR as $\alpha = 58.5°$ and we chose to fabricate micro-prisms with $\alpha = 60°$ as a safety value to compensate for fabrication tolerances. The time needed to expose a single micro-prism was typically around 5 min with only external shell TPL exposure; as a comparison, about half an hour was generally required to drill each hole with FIB milling. In Fig. 8.8 the details of a single prism fabricated with different photo-initiators are shown. The surface exhibits a very good optical quality with a typical roughness below 100 nm.

With this design we calculated an effective NA = 1.15 and a trapping distance from the probe surface $d_{TRAP} = 46$ µm. In the several runs of fabrication we noticed a spread in the actual values of α and β that can vary in a range of about 2° with respect to the nominal value.

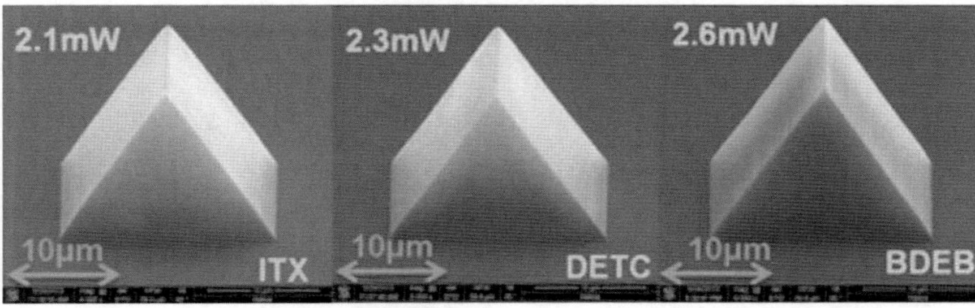

Fig. 8.8 SEM images of micro-prisms fabricated with three different photo-initiators

8.4 Trapping Experiments with Fiber-OT

The integration of all-fiber optical tweezers with the above described fabrication technologies gave birth to a powerful and multipurpose probe that easily adapts to optics and microfluidic applications. In this section we present the trapping experiments performed with our FIB-fabricated and TPL-fabricated fiber-OT with mesoscopic particles, showing the results in term of trap stiffness, multi-particle trapping and controlled microtranslations. Then we will show the implementation of the fiber-OT with a microfluidic devices for lab-on-chip applications like optical trapping, Raman and fluorescence spectroscopy of single cells.

We have assessed the optical quality of the fabricated probes by an all-fiber setup (Fig. 8.9). The radiation emitted by a Ytterbium doped fiber laser at $\lambda = 1070$ nm was coupled to the four single-mode fibers composing the bundle with a 1×4 fiber coupler, and the optical power distribution among the four fibers was finely tuned by means of variable optical attenuators, allowing an extremely compact and stable setup to be obtained. After checking the convergence of the four beams (Fig. 8.10 shows the beam convergence for a TPL-fabricated probe), trapping capabilities of the fiber-OT were verified by mounting the probe, held by a xyz micro-translator, parallel to the xy translation stage of an inverted microscope (Nikon TE2000).

The fiber probe-end was placed on the object plane of a $20\times$ objective, guaranteeing a large field of view, while a CCD camera recorded the trapping action. The optical trapping capability of the fabricated fiber-OT was tested with both polystyrene beads and live in-suspension cells.

The trapping effectiveness was tested by depositing on a cover-slip a water suspension of polystyrene spheres having diameters of 10 μm. The probe, mounted on microtranslators, was able to move the particles that were attracted and trapped by the probe as soon as they entered the range of action of the optical force.

As a first test, we deposited a drop of a water suspension of 10 μm diameter polystyrene beads on a cover-slip and we observed trapping and moving of single beads. In Fig. 8.11 we show a sequence of frames, collected by the CCD camera,

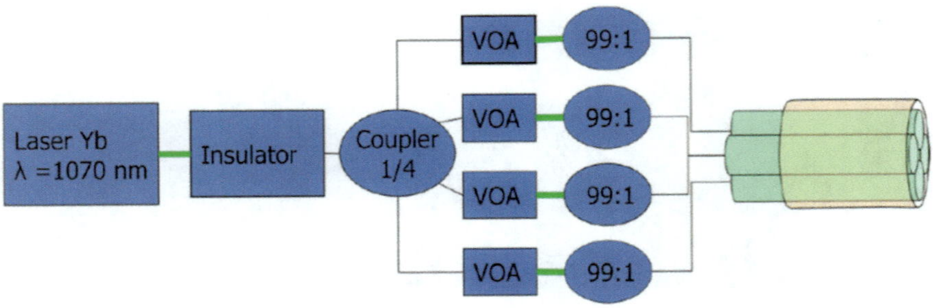

Fig. 8.9 Experimental all-fiber setup. A 1070 nm Ytterbium-doped fiber laser is coupled to an optical insulator and to a 1 × 4 coupler. Each path is then connected to a variable optical attenuator and a coupler 1 × 3 for power monitoring. The four paths are finally coupled to four fibers composing the fiber-OT

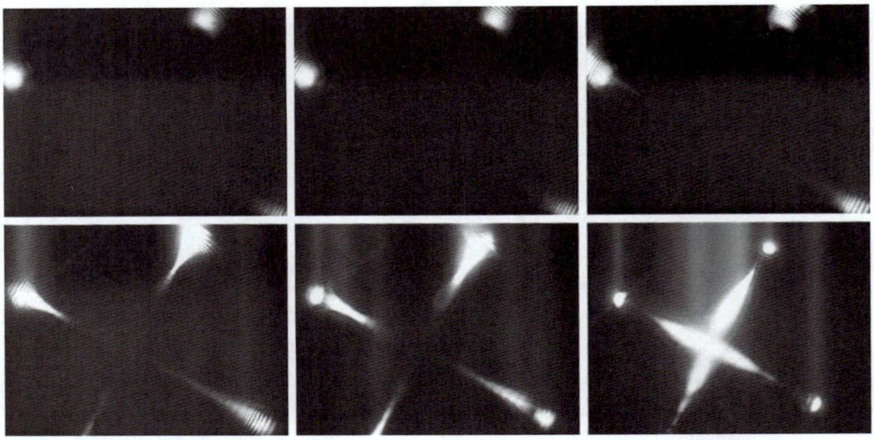

Fig. 8.10 Sequence of images showing the convergence of the beams reflected from the four prisms on top of a TPL-fabricated fiber bundle

in the case of a FIB-fabricated probe that was moved inside the field of view of the microscope, translating two trapped beads.

We checked whether the trap formation could be ascribed only to the optical forces. To this end, we kept the probe at a large distance (few millimetres) from the bottom of the cover slip, and after trapping a particle by using only two opposing beams of the fiber bundle, we verified that the trap vanished as soon as one of the beams was switched off.

We tested the trapping capabilities of fiber-OT also with different biological samples, namely red blood cells (RBC) diluted in hypotonic or isotonic solution and tumor cells (colon cancer cells) diluted in phosphate buffered saline solution. Here we show the cell trapping obtained through our TPL-fabricated fiber-OT. We were able to stably trap single cells by setting the power output at each prism at

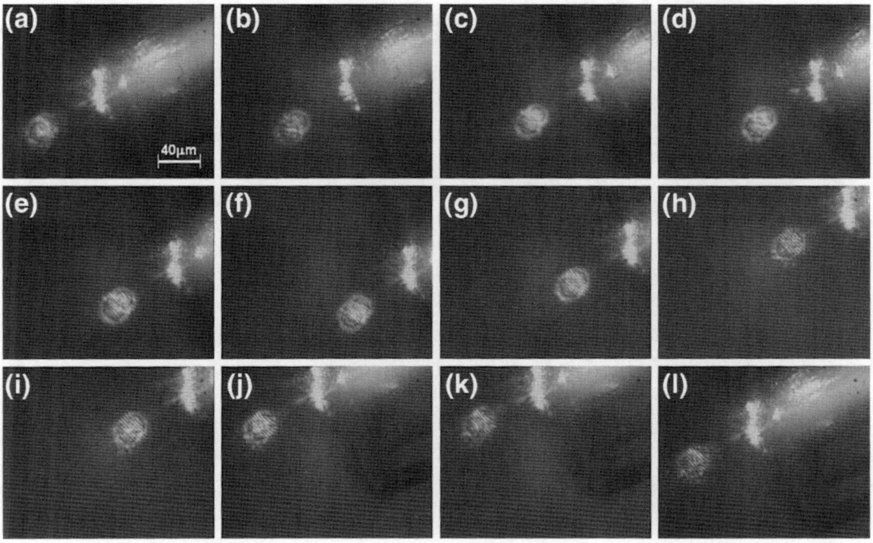

Fig. 8.11 Trapping image sequence: two polystyrene particles are trapped by the FIB-fabricated fiber-OT. The sequence has been recorded by a CCD camera and using a 20× objective. The trapped spheres scatter the light coming from the fiber probe in the infrared range, which is then captured by the CCD. The rings visible around the trapped particles are due to diffraction of the scattered light, whereas the pattern visible in the region where the spheres are overlapping is the result of interference between the light scattered from the particles themselves. The probe is moved inside the suspension by means of microtranslators without losing the trapping: movement *right* (panels **a–d**), *up* (**e–h**), *left* (**i–k**) and *down* (**l**) (With permission from [23])

5 mW. The sequence of frames in Fig. 8.12 shows a trapped RBC, moved inside the solution by translating the probe. It can be noticed that the trapping distance from the probe-end is quite large (>50 μm), thus leaving a great degree of freedom in monitoring the cell, and possibly illuminate it with additional optical beams for further analysis.

We also tested the trapping capabilities of fiber-OT when using only two optical beams emitted by two opposite prisms. We alternatively switched off one couple of opposing beams and, in both cases, we could still observe stable trapping of RBC.

Moreover, by changing the surface angle of one couple of prisms, we created two trapping points, thus we were able to trap two different polystyrene beads at the same time (Fig. 8.13).

Fiber-OT trapping efficiency calibration was carried out by video-tracking the Brownian motion of the trapped particle in order to evaluate the trap stiffness by using polystyrene particles with a diameter of 7, 10 and 15 μm. The beads were held in the trap for a couple of minutes and a video was recorded at a frame rate of 16 fps. The video was then analyzed and the coordinates of the bead's position in each frame were extracted. The optical potential was estimated by using

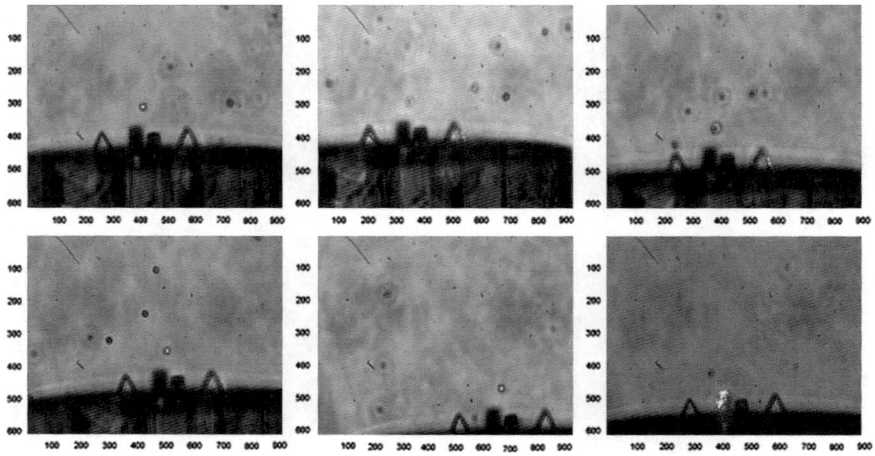

Fig. 8.12 Sequence of trapping and translation of a red blood cell by using a TPL-fabricated fiber-OT

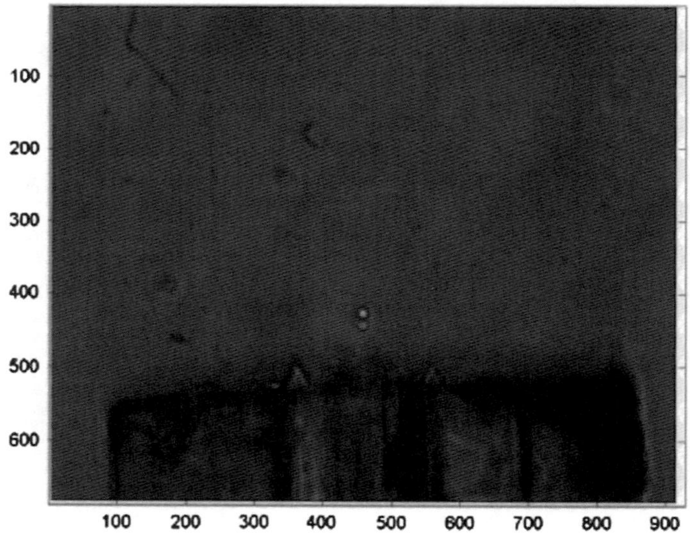

Fig. 8.13 Image of two 10 μm polystyrene beads trapped by an optical tweezers with different couples of micro-prisms

Boltzmann statistics and by fitting the potential well with a parabola, and the trap stiffness was then retrieved. Comparing the obtained values with those reported for highly-focussed beams, we find that the proposed tweezers are ideally suited for particles in the size range from 10 to 20 μm that matches the typical size of biological cells. Moreover, compared to standard OT, the beams emitted by the fibers are not focused nor collimated, hence the intensity of the beams is very low ($<10^3$ W/cm^2), thus minimizing any alteration of biological samples viability due to thermal effects induced by the trapping radiation.

We have also demonstrated the simultaneous particle trapping and optical analysis by using a FIB-fabricated fiber-OT. The probe-end was immersed in a water suspension of 10 μm-diameter fluorescent beads (Fluoresbrite Yellow Green Microspheres, Polysciences). The experimental setup is shown in the inset of Fig. 8.14. By using two dichroic mirrors, both trapping and fluorescence excitation radiations ($\lambda = 1070$ and 408 nm, respectively) were coupled into the fiber probe. The fluorescence signal emitted by the trapped bead was collected by the probe itself, transmitted by the dichroic mirrors, and then detected through a spectrometer with 10-nm spectral resolution (32-PMTs Nikon Spectral Detector). The optical spectra measured with and without a fluorescent bead trapped by the fiber-OT are reported in Fig. 8.14. The optical analysis function is very effective thanks to the short distance between the bead and the probe-end, which acts both as excitation source and signal collector.

8.4.1 Microfluidic Chip Integration

In this section, we show that our fiber-OT can be easily integrated in a microfluidic system in such a way as to enable trapping and analysis of biological specimens flowing inside the channel [42]. The microfluidic device consists of one PMMA layer (2 cm wide, 3 cm long and 2 mm thick) with three inlets enabling hydrodynamic focusing of the flowing particle, one outlet and a calibrated hole for the integration of the fiber-OT. The PMMA layer is bonded to a glass coverslip (0.16 mm tick, 24 mm wide and 40 mm long) laminated with a 50 μm thick dry photoresist (Ordyll SY300 from Elga Europe) in which a microfluidic channel was fabricated by photolithography. The coverslip bottom layer offers full compatibility with standard microscope-based imaging techniques, even with short working distance high-NA microscope objectives (see Fig. 8.15).

TPL-fabricated fiber-OT were then optically aligned in order to level their end to the top surface of the microfluidic channel, so that the optical trap was positioned about 10 μm above the bottom surface of the channel, corresponding to the typical height at which most of the cells flow in stable regime conditions. Once aligned, epoxy glue was used to fix the fiber bundle, seal the channel, and give structural rigidity. The inlets of the microfluidic channel were connected through silicone tubes to reservoirs containing the samples under test, whereas the outlet was connected to a waste reservoir. The tubes and the system were filled with PBS by means of a syringe. The flow was controlled by exploiting the communicating vessel principle: the pressure in the microchannel was tuned by changing the relative position in height between the input and output reservoirs, thus carefully driving the samples towards the trapping position. The samples flowing in the micro-channel were easily and stably trapped by the fiber-OT. Additionally, to test

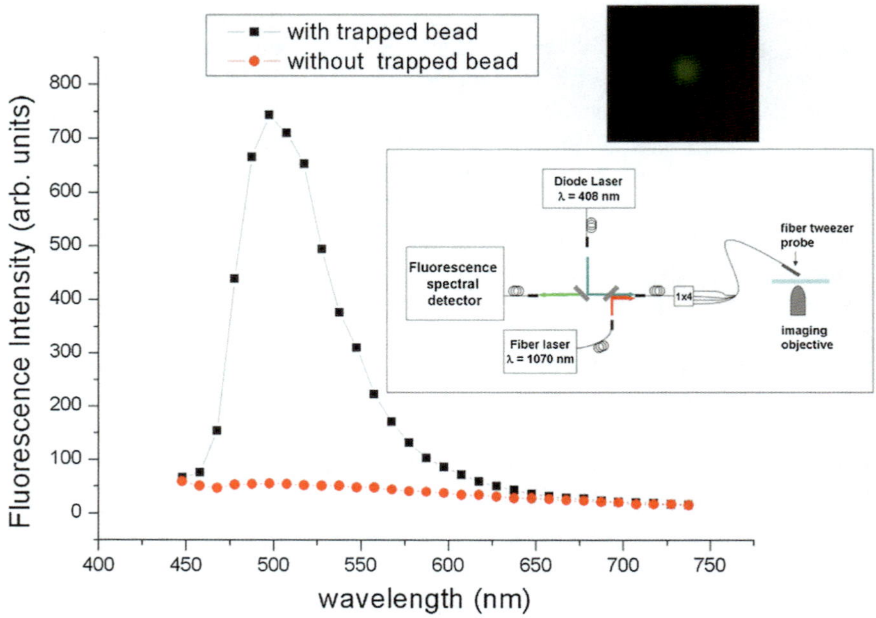

Fig. 8.14 Fluorescence spectra measured by the fiber-OT bundle, with and without the trapped fluorescent bead. *Black squares* fluorescence intensity with a trapped bead; *red dots* without a trapped bead. *Inset* sketch of the experimental set-up. The trapping laser source, emitting at 1070 nm, is an ytterbium-doped fibre laser. The radiation at 408 nm used to excite the fluorescent bead is obtained using a laser diode (With permission from [23])

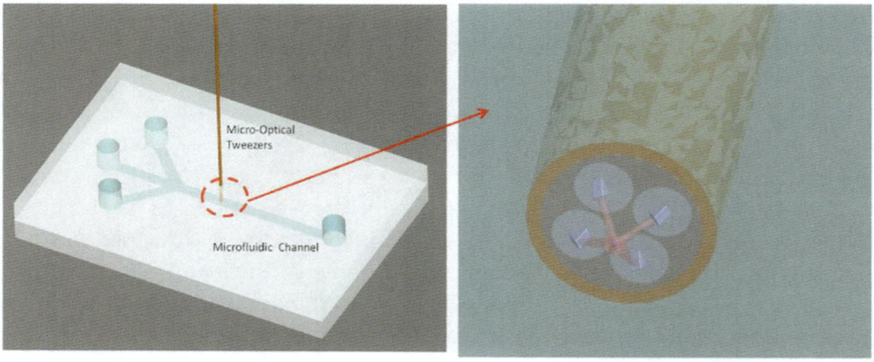

Fig. 8.15 3D design of the microfluidic chip with the fiber-OT inserted in the central inlet

the device capability to perform fluorescence spectroscopy analysis, we mounted the microfluidic system on the object plane of an inverted microscope (TE2000, Nikon), and the experiments were carried out by slightly modifying the all-fiber optical setup. The 1070 nm trapping radiation was coupled to only two of the four

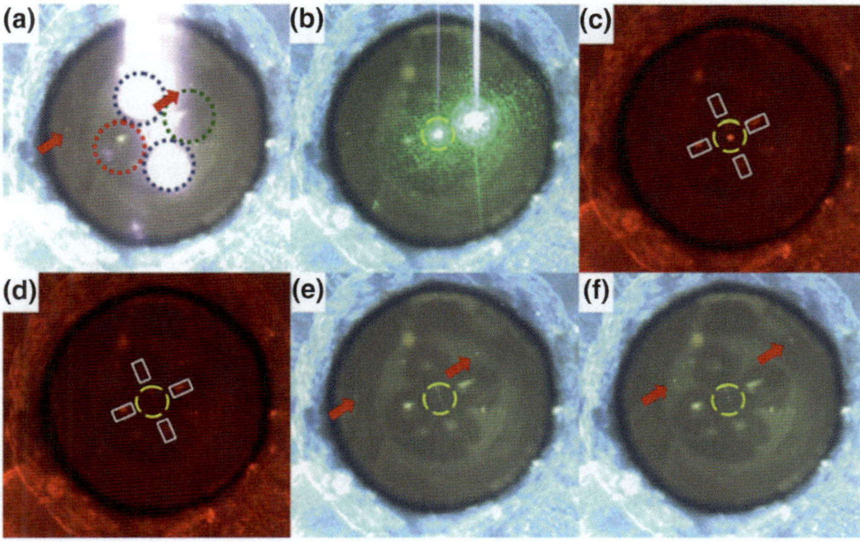

Fig. 8.16 Fluorescence spectroscopy in the microfluidic integrated system. **a** *Front-view* of the probe: the two fibers indicated by the *blue dotted line* emit the infrared radiation used to trap the particle. The fiber surrounded by the *green circle* is connected to the 532-nm laser source, while the one surrounded by the *red circle* is unused. *Red arrows* indicate the position of two non-trapped beads. **b** After inserting an IR-filter and switching on the 532-nm laser the green light scattered by the trapped particle (inside the *yellow*, *dashed circle*) is evident. **c** A long-pass filter is introduced in the light path to remove the 532-nm radiation from the image and the trapped bead (7.5 μm in diameter) shows an evident red fluorescence. **d** When the 532-nm laser is switched off the fluorescence disappear: the red light visible in the picture is given by coaxial white-light illumination used to illuminate the scene. **e** When the long-pass filter is removed, and the IR-filter is used, it is possible to observe the trapped bead in the centre of the *yellow circle*, as well as two additional beads indicated by the *red arrows*. **f** When the flow is restored the trapped bead remains stably trapped, while the other beads flow onwards the upper right corner (With permission from [42])

fibers composing the tweezers, corresponding to two opposing micro-prisms. A 532 nm radiation of 50 mW emitted by a frequency-doubled Nd:YAG laser source was coupled to a third fiber of the bundle. Experiment has been performed first by trapping 7.5 μm fluorescent polystyrene beads (Bangs-Lab, FS06F/10007 532 nm/640 nm) diluted in water, hence by switching on the excitation radiation at 532 nm. After filtering out the green radiation by a notch filter, the red fluorescence of the beads emitted at 640 nm could be easily recorded through the CCD camera connected to the microscope (Fig. 8.16). We note that the device operation is independent of the radiation wavelength and the prisms output is always aligned in such a way as to impinge on the trapped sample. As a final test, we recorded the Raman signature of live cells trapped in the microfluidic channel by exciting the sample at 633 nm through one of the fiber of the bundle, as for fluorescence excitation. The incident power at the sample plane was 3 mW. The Raman

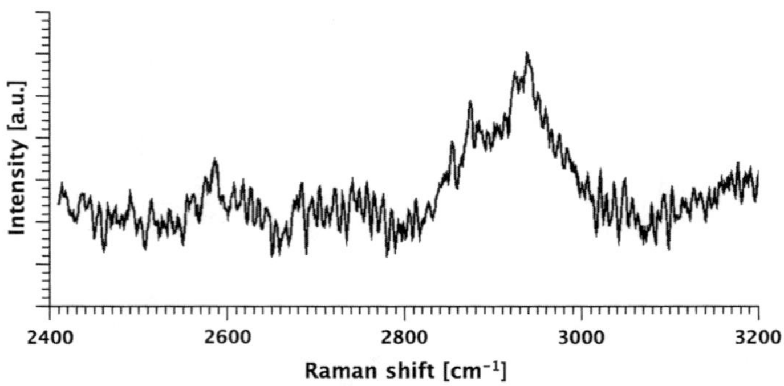

Fig. 8.17 Raman signature in the C–H stretch Raman band of a colon cancer cell optically trapped into the microfluidic channel by using an integrated TPL-fabricated fiber-OT. The Raman scattering has been collected through a microscope objective and measured by a commercial spectrometer (With permission from [42])

scattering from sample was collected by a 60×, 0.95 NA microscope objective and directed to the spectrometer of a commercial (InVia, Renishaw) Raman microscope. As an example, the Raman signature of a colon cancer cell in the C–H stretch Raman band, which is typically dominated by the lipidic and protein content, is shown in Fig. 8.17.

The technology can be further improved, since the microchannel lateral size can be, in principle, reduced to be equal to the size of the fiber optical tweezers probe. In turn, the μOT could be reduced in size by reducing the diameter of the fibers' claddings and the size of the external capillary, being the distance among the cores (around 100 μm) the important parameter to be maintained. The capability to obtain a stable trap with just two prisms offers the chance to design two separate traps in different positions at the fiber-OT output, thus making possible to simultaneously trap multiple cells at a close and controlled distance. Moreover, our microsystem allows the insertion, as a part of the fiber bundle, of specialized optical fibers (e.g. for specific wavelengths, or having large cores for efficient scattered light collection) or even micro-capillaries (e.g. to introduce a proper stimulus on a trapped cell). The here reported device can be a key building block for the realization of microfluidic flow cytometers and cell sorters, when long integration times for the analysis require a stable position for the sample. At the same time, the device can be used as a tool for clinical assays. In fact, the sorted and trapped cells can be manipulated, subjected to stimuli of various nature or to drugs and can be monitored and analyzed by the same OT.

8.5 Conclusions and Outlook

In conclusion, we have reviewed the design, fabrication and experimental demonstration of miniaturized OT, based on optical fibers with fiber-end microfabricated structures, for trapping, manipulation and analysis of microsamples. This device is a result of the combination of a fibre bundle with the TIR mechanism and we have shown two different on-fiber fabrication approaches to obtain this configuration, one based on Focused Ion Beam milling, the second one based on Two-Photon Lithography.

Finally, we have shown the integration of such miniaturized OT into a microfluidic device for on-chip trapping, Raman and fluorescence spectroscopy of single cells.

The realization of OT based on optical fibers provides a miniaturized tool suitable for many applications relevant to biology and fundamental physics, such as in vivo biological manipulation or in-vacuum single-particle X-ray spectroscopy. Additionally, the integration of these fiber-based OT into microfluidic circuits demonstrates their great potential to be a key building block for realization of microfluidic flow cytometers [44, 45] and cell sorters [46, 47].

References

1. S. Chu, J.E. Bjorkholm, A. Ashkin, A. Cable, Phys. Rev. Lett. **57**, 314–317 (1986)
2. A. Ashkin, Phys. Rev. Lett. **24**, 156–158 (1970)
3. A. Ashkin, J.M. Dziezic, J.E. Bjorkholm, S. Chu, Opt. Lett. **11**, 288–290 (1986)
4. A. Ashkin, J.M. Dziezic, Science **235**, 1517–1520 (1987)
5. D.G. Grier, Nature **424**, 810–816 (2003)
6. U. Bockelmann, P. Thomen, B. Essevaz-Roulet, V. Viasnoff, F. Heslot, Biophys. J. **82**, 1537–1553 (2002)
7. G. Whyte, G. Gibson, J. Leach, M. Padgett, D. Robert, M. Miles, Opt. Express **14**, 12497–12502 (2006)
8. D. Cojoc, F. Difato, E. Ferrari, R.B. Shahapure, J. Laishram, M. Righi, E.M. di Fabrizio, V. Torre, PLoS ONE **2**, e1072 (2007). doi:10.1371/journal.pone.0001072
9. C.M. Creely, G. Volpe, G.P. Singh, M. Soler, D.V. Petrov, Opt. Express **12**, 6105–6110 (2005)
10. M. Goksor, J. Enger, D. Hanstorp, Appl. Opt. **43**, 4831–4837 (2004)
11. J.W. Chan, H.K. Winhold, S.M. Lane, T. Huser, IEEE J. Sel. Top. Quant. Electron **11**, 858–863 (2005)
12. A. Ashkin, Biophys. J. **61**, 569–582 (1992)
13. K.C. Neumann, S.M. Block, Rev. Sci. Instrum. **75**, 2787–2809 (2004)
14. D.G. Grier, Y. Roichman, Appl. Opt. **45**, 880–887 (2006)
15. Y. Xin-Cheng, L. Zhao-Lin, G. Hong-Lian, C. Bing-Ying, Z. Dao-Zhong, Chin. Phys. Lett. **18**, 432–434 (2001)
16. A. Constable, Jinha Kim, J. Mervis, F. Zarinetchi, M. Prentiss, Opt. Lett. **18**, 1867–1869 (1993)
17. M.-T. Wei, K.-T. Yang, A. Karmenyan, A. Chiou, Opt. Express **14**, 3056–3064 (2006)
18. K. Taguchi, H. Ueno, T. Hiramatsu, M. Ikeda, Electron. Lett. **33**, 1413–1414 (1997)

19. E.R. Lyons, G.J. Sonek, Appl. Phys. Lett. **66**, 1584–1586 (1995)
20. Z. Hu, J. Wang, J. Liang, Opt. Express **12**, 4123–4128 (2004)
21. R.S. Taylor, C. Hnatovsky, Opt. Express **11**, 2775–2782 (2003)
22. Z. Liu, C. Guo, J. Yang, L. Yuan, Opt. Express **14**, 12511–12516 (2006)
23. C. Liberale, P. Minzioni, F. Bragheri, F. De Angelis, E. Di Fabrizio, I. Cristiani, Nat. Photonics **1**, 723–727 (2007)
24. A.T. ONeil, M.J. Padgett, Opt. Commun. **193**, 45–50 (2001)
25. M. Prasciolu, D. Cojoc, S. Cabrini, L. Businaro, P. Candeloro, M. Tormen, R. Kumar, C. Liberale, V. Degiorgio, A. Gerardino, G. Gigli, D. Pisignano, E. Di Fabrizio, R. Cingolani, Microelectron. Eng. **67**, 169–174 (2003)
26. S. Maruo, J.T. Fourkas, Laser Photon. Rev **2**, 100–111 (2008)
27. K. Takada, H.-B. Sun, S. Kawata, Appl. Phys. Lett. **86**, 071122 (2005)
28. J.-F. Xing, X.-Z. Dong, W.-Q. Chen, X.-M. Duan, N. Takeyasu, T. Tanaka, S. Kawata, Appl. Phys. Lett. **90**, 131103–131106 (2007)
29. D. Tan, Y. Li, F. Qi, H. Yang, Q. Gong, X. Dong, X. Duan, Appl. Phys. Lett. **90**, 071103–071106 (2007)
30. Z. Bayindir, Y. Sun, M.J. Naughton, C.N. LaFratta, T. Baldacchini, J.T. Fourkas, J. Stewart, B.E.A. Saleh, M.C. Teich, Appl. Phys. Lett. **86**, 064105 (2005)
31. R. Kiyan, C. Reinhardt, S. Passinger, A.L. Stepanov, A. Hohenau, J.R. Krenn, B.N. Chichkov, Opt. Express **15**, 4205 (2007)
32. S. Maruo, K. Ikuta, H. Korogi, J. Micromech. Syst **12**, 533–539 (2003)
33. S. Klein, A. Barsella, G. Taupier, V. Stortz, A. Fort, K.D. Dorkenoo, Appl. Surf. Sci. **252**, 4919–4922 (2006)
34. C. Liberale, G. Cojoc, P. Candeloro, G. Das, F. Gentile, F. De Angelis, E. Di Fabrizio, IEEE Photon. Technol. Lett **22**, 474–476 (2010)
35. G. Cojoc, C. Liberale, P. Candeloro, F. Gentile, G. Das, F. De Angelis, E. Di Fabrizio, Microelectron. Eng. **87**, 876–879 (2010)
36. M. Malinauskas, A. Zukauskas, K. Belazaras, K. Tikuisis, V. Purlys, R. Gadonas, A. Piskarskas, Eur. Phys. J. Appl. Phys. **58**, 20501 (2012)
37. F. Schiappelli, R. Kumar, M. Prasciolu, D. Cojoc, S. Cabrini, M. De Vittorio, G. Visimberga, A. Gerardino, V. Degiorgio, E. Di Fabrizio, Microelectron. Eng. **73**, 397–404 (2004)
38. Y.J. Yu, H.N.M.-H. Hong, H.-R. Noh, Y. Arakawa, W. Jhe, Opt. Comm **267**, 264–270 (2006)
39. K.M. Tan, M. Mazilu, T.H. Chow, W.M. Lee, K. Taguchi, B.K. Ng, W. Sibbett, C.S. Herrington, C.T.A. Brown, K. Dholakia, Opt. Express **17**, 2375–2384 (2009)
40. J.K. Kim, Y. Jung, B.H. Lee, K. Oh, C. Chun, D. Kim, Opt. Fiber Technol **13**, 240–245 (2007)
41. K.-R. Kim, S. Chang, K. Oh, IEEE Photon. Technol. Lett **15**, 1100–1102 (2003)
42. C. Liberale, G. Cojoc, F. Bragheri, P. Minzioni, G. Perozziello, R. La Rocca, L. Ferrara, V. Rajamanickam, E. Di Fabrizio, I. Cristiani, Sci. Rep **3**, 1258 (2013)
43. C.N. LaFratta, J.T. Fourkas, T. Baldacchini, R.A. Farrer, Angew. Chem. Int. Ed. **46**, 6238–6258 (2007)
44. F.S. Ligler, J.S. Kim, *The Microflow Cytometer* (Pan Stanford Publishing Pte. Ltd, Singapore, 2010)
45. G. Boer, R. Johann, J. Rohner, F. Merenda, G. Delacrtaz, Ph Renaud, R.-P. Salath, Rev. Sci. Instrum. **78**, 116101 (2007)
46. X. Wang, S. Chen, M. Kong, Z. Wang, K.D. Costa, R.A. Li, D. Sun, Lab Chip **11**, 3656–3662 (2011)
47. F. Bragheri, P. Minzioni, R. Martinez Vazquez, N. Bellini, P. Pai, C. Mondello, R. Ramponi, I. Cristiani, R. Osellame, Lab Chip **12**, 3779–3784 (2012)

Chapter 9
Hydrogen Detection Using a Single Palladium Nano-Aperture on a Fiber Tip

Steven J. McKeown and Lynford L. Goddard

Abstract This chapter discusses the development of fiber optic hydrogen sensors. A motivation for these sensors is given followed by an explanation of the underlying physics of the palladium-hydrogen system. Research results and the strengths and weaknesses of several different fiber optic hydrogen sensor types are discussed. Specifically, the Pd fiber mirror, tapered fiber, Fabry–Pérot interferometer, Fiber Bragg grating, and long period grating sensor architectures are reviewed. Next, a new sensor topology that uses a nano-aperture patterned onto the tip of a Pd coated fiber is presented. The nano-aperture enhances sensitivity because it not only confines light tightly to the Pd surface, but it also creates a Fabry–Pérot resonant structure. Thus, the power shifts in transmission and reflection due to hydrogen induced optical and mechanical changes to the Pd film are amplified. Finally, some conclusions and suggestions for future work are given.

9.1 Introduction

Interest in the development of low cost and accurate trace gas detection platforms has grown recently due to their potential use in a variety of commercial, agricultural, and security industries. Small footprint, unobtrusive, point-based detection of gases or solutes is particularly important since such a device could easily be integrated into existing systems and would allow for high spatial accuracy and modeling for distributed systems. Optical fiber meets these desired criteria since it is cheap and robust, and can have a very small footprint. Optical sensors also offer an inert detection paradigm, have low risk of sparking, negligible electrical

S. J. McKeown · L. L. Goddard (✉)
University of Illinois at Urbana-Champaign, Urbana, IL, USA
e-mail: lgoddard@illinois.edu

S. J. McKeown
e-mail: mckeown3@illinois.edu

A. Cusano et al. (eds.), *Lab-on-Fiber Technology*,
Springer Series in Surface Sciences 56, DOI: 10.1007/978-3-319-06998-2_9,

interference, a robust material platform that can survive in harsh environments such as high temperature and humidity [1], and the ability to be distributed as a point sensor network covering long distances.

Concern over diminishing fossil fuels and the environmental and ecological issues raised by their use has intensified interest in developing and implementing more sustainable sources of energy. Solar, wind, and geothermal have all seen substantial support as technologies to supplant fossil fuels as our primary source of power generation; however, none of these represent a viable alternative to petroleum based fuels in portable applications. Being both energy dense (by weight) and easily transportable, hydrogen gas is one of the most promising candidates for a replacement fuel and as a storage medium for intermittent energy sources. Hydrogen is also used in methanol and ammonia production, among other applications. A major concern with hydrogen gas is its flammability in air, with a lower explosive limit of 4 % [2]. This is compounded by the fact that hydrogen is a small molecule and hard to contain, necessitating early leak detection. Furthermore, whether in a combustion engine, reactor vessel, or some other process, it is often required to know the exact concentration present.

Unlike other gases, hydrogen does not have a strong optical absorption spectra, making absorption spectroscopy difficult. Due to this, one of the primary methods for optical hydrogen detection is the use of a functional layer such as a palladium or platinum film [3]. Pd alloys with Ag, Au, and WO_3 have also been studied [4–6]. These films catalyze the dissociation of the hydrogen molecule through adsorption, at which point it diffuses into the material. This reaction can change many characteristics of the film such as the conductivity, mechanical properties, and optical properties [7]. For optical sensors, the latter two are important, notably the changes in refractive index and size of the film. The refractive index describes how light travels through the medium, affecting the optical phase and loss during propagation. Mechanical changes include deformation of the film, which for nanostructures, can have a significant impact on device operation.

9.2 Palladium-Hydrogen System

Palladium is known for its ability to absorb large amounts of hydrogen, up to 900 times its own volume [2]. This property makes it attractive for hydrogen storage applications. Furthermore, palladium and platinum films also catalyze the dissociation of diatomic hydrogen, leading to a high absorption rate or facilitating diffusion into other materials. For ideal films, this process is limited by surface adsorption described by the Langmuir equation [8], which for diatomic molecules, the fractional coverage of the surface, θ, is given by:

$$\theta(H) = \frac{\theta_{\max}\sqrt{H/H_0}}{1+\sqrt{H/H_0}}, \tag{9.1}$$

where H is the hydrogen gas concentration and θ_{max} and H_0 are constants representing the maximum coverage and the concentration for half maximum coverage, respectively. However, for rough films, the Freundlich equation, which is a power law fit:

$$\theta(H) = kH^n, \tag{9.2}$$

where k and n are constants, is more suitable [9, 10]. Even more importantly, this process is also reversible. Once hydrogen is removed from the external atmosphere, the hydrogen will diffuse out of the palladium and the optical properties of the film can be recovered.

When hydrogen diffuses into the metal lattice, it bonds to interstitial sites creating palladium hydride (PdH). PdH can exist in three states, the α phase which occurs at low concentrations, the β phase which exists at high concentrations, and the α + β phase that exists in the miscibility gap between the two, which is present at normal temperature and pressure. The transition from α to β phase can occur anywhere from 0.1–2 % hydrogen and is dependent on the quality and structure of the film [7, 11].

As hydrogen is incorporated into the lattice, the electrical, optical, and mechanical properties change. For the nano-aperture sensor, the changes in optical constants and mechanical structure are most important. The optical constants change as a result of an increase in the Fermi level. The hydrogen atoms bonded at interstitial sites act as electron donors and lead to orbital hybridization with the 4d band of the palladium atoms. The overall result is a lowering of the density of states at the Fermi level and hence a reduction in the optical transition rate. This causes a decrease in reflection [7, 11, 12].

The change in mechanical structure is a result of hydrogen induced lattice expansion (HILE) in the palladium film. The incorporation of hydrogen into the lattice results in an increase in the lattice constant from the bulk value of 3.890Å to a value of 4.025Å at the onset of the β phase [7]. However, for nanoscale devices, the effects of strain resulting from bonding and deformation mechanics at vertices or edges, such as along the perimeter of the aperture, play a large role. The entire film cannot expand by equally in all directions, as this would amount to several microns of expansion laterally across the bonded film. Since Pd bonds weakly to oxides, this could result in buckling of the film in places and an unknown exact change in shape of any nano-scale features. Even an alteration of a few nanometers could have a large impact on the spectral response of small resonant structures. Some work is being done to investigate the mechanics of HILE for thin films and nano-scale structures [13, 14]. Figure 9.1 shows the time dependent change in height of a 250 nm thick Pd micropillar on glass measured using diffraction phase microscopy during and after exposure to various hydrogen concentrations. Beyond the α-β phase transition, more drastic changes occur in the Pd lattice, which result in permanent damage through plastic deformation. However, even though the structure of the film is irreversibly altered, the hydrogen can still completely

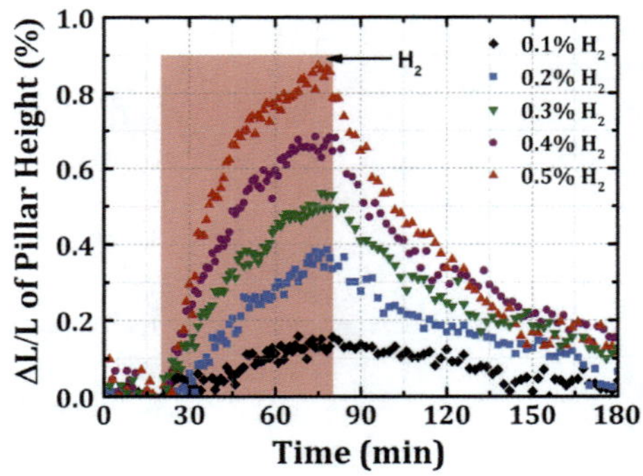

Fig. 9.1 Measured change in the height of a 250 nm thick Pd micropillar on glass using diffraction phase microscopy. Adapted from [13]

diffuse out. Prolonged exposure to high concentrations of hydrogen in metals can also lead to further degradation as the repeated deformations result in embrittlement or delamination of films [11, 15].

9.3 Fiber Optic Hydrogen Sensors

The Pd system has been used in the development of a variety of optical hydrogen sensors ranging from plain film monitoring [16] and functionalized lasers [17–20] to specialized nanostructures such as apertures [21] or antennas [22]. There has also been significant work on developing fiber optic based hydrogen sensors. Most designs use a thin Pd layer either along the fiber length or at the fiber tip to alter the amplitude or phase of the light interacting with the layer. One of the earliest designs, a Mach–Zehnder fiber optic interferometer with a Pd coated fiber as the sensing arm, was realized in 1984 by Butler [23].

The most basic fiber sensor is simply a cleaved facet coated with palladium or some other reactive material. These "fiber mirror" designs are some of the simplest and were demonstrated in 1994 by Butler [24]. Such devices were fabricated with varying film thicknesses and the change in reflected power was measured as the fiber was exposed to a wide range of hydrogen concentrations. Results showed around a 3 % power decrease at 0.5 % hydrogen and up to a 20 % power decrease at 2 % hydrogen in nitrogen for a 10 nm thick film.

More interesting however, was the identification of an additional contribution to the reflected power change, besides the change in the refractive index. Due to film expansion, micro-blistering and micro-cracking of the films were seen. The micro-blistering depended highly on surface adhesion between the Pd film and the substrate. Further, it was reversible, at least for some time. The presence of this effect did not depend on film thickness, but rather was a function of bonding, and hence deposition conditions. The contribution of these two effects to the reflectivity change was very large, with a $\Delta R/R = 80$ % at 10 % hydrogen for some films. By depositing films with random thicknesses, other values for $\Delta R/R$ of 23 and 7 % were found, which most likely correspond to varying states of film clamping.

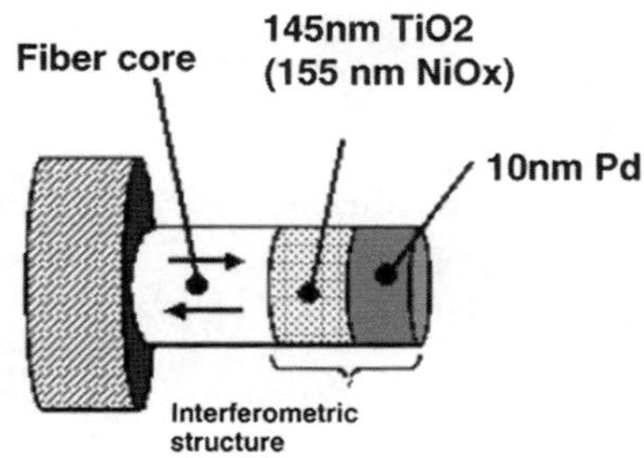

Fig. 9.2 A Fabry–Pérot based fiber sensor using TiO_2 or NiO_x as both the cavity and sensing media. Courtesy of [25]

To test this, Butler also fabricated sensors with a Ni adhesion layer to provide a solid bond between the Pd layer and the substrate. These sensors showed a reduced response in reflectivity, regardless of Ni layer thickness. This suggests that the reduction was due to a bonding effect, rather than some kind of alloying. The sensitivities of these fully bonded sensors also matched very closely to the $\Delta R/R = 7$ % for the plain Pd films, suggesting those were samples with abnormally high adhesion between the SiO_2 and Pd. These simple facet mirror based sensors required very little fabrication and demonstrated a large sensitivity of nearly 10 % per percent of H_2 when measured in reflection. Response times were not discussed, but should be comparable to other Pd hydrogen sensors with similar film thicknesses.

Another type of facet based hydrogen sensor relies on the formation of a Fabry–Pérot cavity on the fiber tip. E. Maciak and Z. Opilski demonstrated a Fabry–Pérot interferometer (FPI) hydrogen sensor that was composed of a dielectric TiO_2 layer on a fiber facet forming the FPI cavity with a Pd layer on top [25]. A schematic of the device can be seen in Fig. 9.2. Here, the fiber/TiO_2 interface is the first mirror, which has a fixed reflectivity, and the Pd layer is the second mirror.

TiO_2 and NiO_x films are gasochromic and serve as the sensing layer in this design. The thin Pd layer acts to catalyze dissociation of the hydrogen and allow diffusion into the sensing layer. As the hydrogen diffuses into the active films, the Ni or Ti metal ions are reduced by the formation of metal bronzes, which result in coloring or bleaching of color from TiO_2 and NiO_x respectively. This results in a change in the optical properties, and therefore a change in the reflectivity of the FPI.

Since the FPI is a resonant structure, the presence of hydrogen not only causes a change in the measured response at a certain wavelength, it also changes the overall shape of reflection spectrum. This can be seen in Fig. 9.3, which shows the measured reflection spectra for a 155 nm NiO_x sensor at 1 and 2 % hydrogen in simulated air normalized to the base reflection spectrum. As a result of shifts in the finesse and free spectral range of the cavity, the reflection increases at some wavelengths, but decreases at others. The maximum measured response was around 60 % for the NiO_x sensor at 2 % hydrogen for a wavelength slightly longer than 600 nm. The wavelength of maximum response and the magnitude of the response could most likely be optimized through simulations involving cavity and

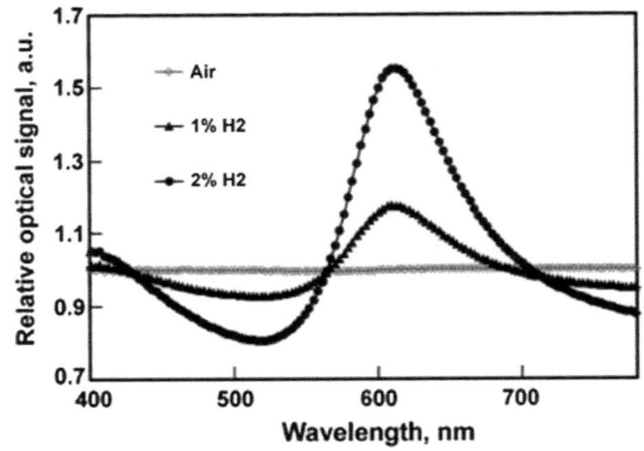

Fig. 9.3 Change in the reflection spectrum of the FPI sensor normalized to 0 % hydrogen for various hydrogen concentrations. Courtesy of [25]

mirror thickness, but to our knowledge, such an analysis was not carried out by the authors. The response times for the sensors were on the order of a couple minutes, with a 4 min response time at 2 % hydrogen and a 1 min recovery time.

The authors also employed a self-referencing normalization method for reducing noise that used the ratio of the responses at two nearby wavelengths, as this would reduce error due to bending loss. Using 545 and 585 nm, a maximum response of 40 % at 3 % hydrogen in air was demonstrated with this method. Another problem, common to many Pd based hydrogen sensors as well, is degradation of the sensor in air or with exposure to water vapor. These devices were stated to only work for several cycles in air. A proposed solution was to use a protective layer such as polymethylpentene to allow diffusion of the relatively small hydrogen molecule while protecting the Pd layer from air and water vapor.

Park et al. also fabricated an integrated FPI hydrogen sensor, using only the Pd mirror as the sensing mechanism, with the cavity replaced by a piece of hollow optical fiber (HOF) spliced between two single mode fiber sections (SMF) [26]. There is reflection resulting from the mode mismatch between the HOF and SMF sections at both splice points, and also from the palladium mirror on the facet of the final SMF section. Assuming low values for coupling, this effectively results in three cavities being formed, the HOF cavity, the SMF cavity, and the HOF + SMF cavity. Since only two of these cavities will be affected by hydrogen, this allows for self-referencing to eliminate fluctuations in intensity not caused by hydrogen. Their devices showed a response amplitude change of up to 10 % at 4 % hydrogen in nitrogen.

FPIs have the unique advantage of simple built-in self-referencing, either by comparing multiple wavelengths, or by simply adding additional cavities. They also have a small footprint since the active sensor area is at the tip of the fiber. Fabrication complexity is minimal. However, neither of these sensor designs was optimized, so it is unknown what level of control of the thickness tolerance is needed to achieve maximum sensitivity. Response times depend on the thickness of the functional layer. Since the layer needs to be reflective, the films need to be

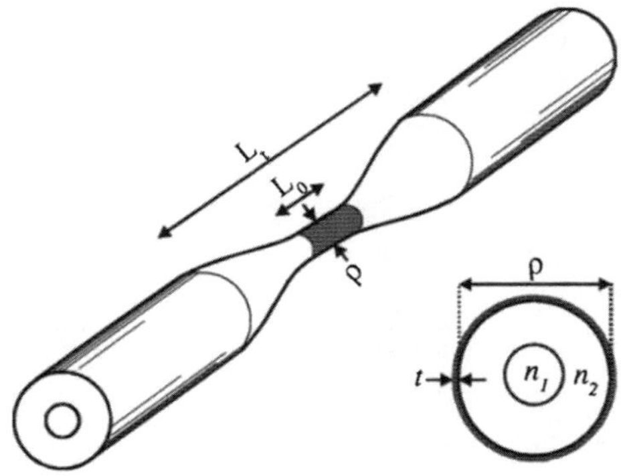

Fig. 9.4 Schematic of a tapered fiber sensor. Courtesy of [27]

thicker than for tapered fiber sensor designs. Although the first FPI sensor used TiO_x and NiO_x as the sensing media, the response times were comparable to Pd films, i.e., on the order of several minutes. The response amplitude of the TiO_x/NiO_x sensor was very high, but none of these devices demonstrated high sensitivity at very low concentrations. The selectivity of these materials was also not given. Overall, the Pd FPI had relatively low response amplitudes, but there may have been room for improvement by optimizing the layer structure.

In addition to facet based sensors, there have also been several iterations that use sidewall coated fiber to improve sensitivity by increasing the interaction length between the light and the functional layer. These typically involve etching away the cladding of the optical fiber and replacing this guiding layer with a palladium film. The loss in this region will then be greatly affected by changes in the refractive index of the film. These are called evanescent or tapered fiber sensors.

Villatoro et al. reported on such a device consisting of a palladium coated multimode fiber operating at 850 nm for various taper sizes and hydrogen concentrations ranging from 0.6 up to 3.5 % in nitrogen [27]. In this case, the fiber taper was fabricated by heating and stretching a fiber. Since the fiber is stretched and not etched, the cladding remains intact, and both the core and cladding shrink in size. This reduces the number of modes in the core, and forces them to propagate in the cladding, where they can be guided by the cladding-Pd interface. A schematic of this sensor can be seen in Fig. 9.4. As hydrogen incorporation changes the optical properties of the Pd film, the loss in these bound cladding modes will change, allowing for detection. Since the taper diameter will affect how much of the light is confined to the cladding, and thus overlaps with the Pd layer, a study on the taper length and diameter was carried out. As expected, it was found that by decreasing the taper diameter, the response of the sensor increased. However, this increase in response was non-linear. In other words, the response increased more for some concentrations than others as the taper diameter was decreased. Figure 9.5a shows experimental data of the time dependent response for various concentrations and Fig. 9.5b shows a summary contrasting various taper diameters. The overall response of the sensor was only around a 0.4 % change at the maximum tested concentration with a response time of around 75 s, but later

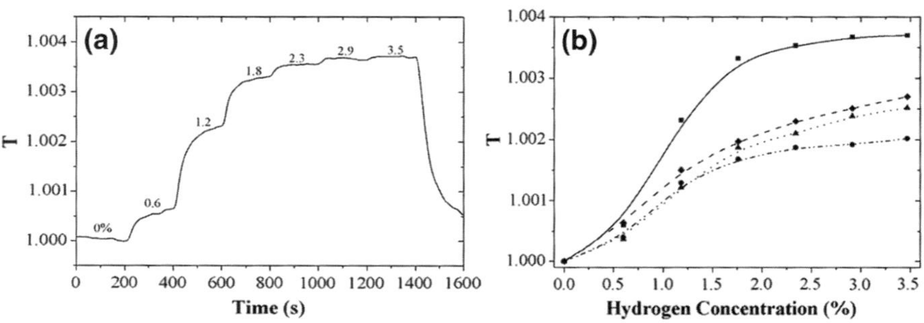

Fig. 9.5 **a** Hydrogen response versus time for a tapered fiber sensor. **b** Response for various taper diameters. Courtesy of [27]

published results showed up to a 60 % change in power at 10 % hydrogen for a slightly different taper design [28].

Further work was published by the same group for nano-tapers 1.3 μm in diameter with 4 nm Pd films, which greatly improved the response time [29]. Several other groups have also published work on tapered fiber designs, with variations in the sensor structure, such as side polishing a fiber [30], adding the sensor to a fiber laser cavity [31], and using fiber heterostructures [32]. Tapered fiber designs offer relatively fast response times from the use of thin films and have a low cost. Fabrication complexity and durability depend highly on the taper size, length, and method of fabrication such as etching versus stretching, as well as the fiber material. However, fabrication complexity is on average fairly low for these devices with the exception of extremely small tapers. For high sensitivities, the tapers are relatively small, around 50 μm, and the protective coatings must be removed to allow hydrogen diffusion. This makes them more susceptible to damage. Combined with bending radius and active region length, the footprint of these sensors is at least several square centimeters. To obtain higher sensitivities, longer interaction lengths are needed, reducing the spatial accuracy of the sensor.

Fiber Bragg gratings (FBGs) are structures that can have a tailored reflection spectrum that is very sensitive to the refractive index and pitch of the grating. A sensor can be created by coating a FBG in palladium, allowing strain from hydrogen induced lattice expansion to change the Bragg wavelength. Such a device was first reported by Sutapun et al. in 1999 [33]. A schematic of the device can be seen in Fig. 9.6a. A spectrometer measured the transmission in the 800–1000 nm range. The data was then fit to calculate the exact Bragg wavelength. In addition, stress in the film was monitored using a cantilever probe system, where film expansion caused bending of the cantilever which could then be measured.

The transmitted spectra of the sensor for hydrogen concentrations ranging from 0.21 to 2.18 % are shown in Fig. 9.6b. A shift in the Bragg wavelength is seen for the minimum concentration of 0.21 % with a corresponding wavelength shift of around 15 pm and up to nearly 50 pm at 1.8 %. Above this concentration, the

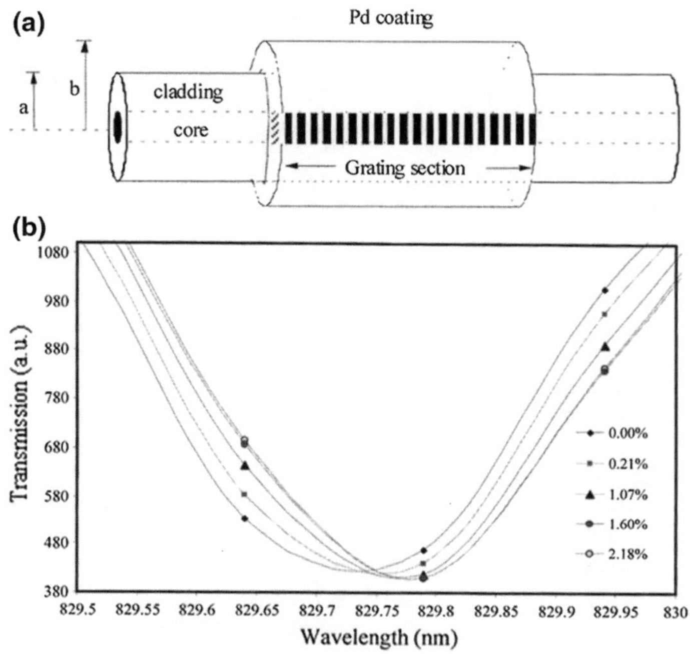

Fig. 9.6 **a** Schematic of a Pd coated FBG sensor. **b** Transmission spectra for various hydrogen concentrations. Courtesy of [33]

response of the sensor began to drop and the sensitivity permanently deteriorated, suggesting the onset of the β phase. The authors imaged the sensors and found that the film began to delaminate at these concentrations. Using the stress measurements, they were also able to calculate a theoretical shift in the Bragg wavelength and make a comparison to experimental values.

Other groups have also reported on FBG based hydrogen sensors with studies being done for various temperatures and light sources, such as using an infrared (IR) source to induce heating and improve response time [34]. Most notable is the work done on long period gratings which greatly increased the measured wavelength shift from the tens of picometers for FBGs to a few nanometers for the long period grating [35–37]. Response times were again comparable to other sensors with similar film thicknesses: under a minute for films 20 nm or less and up to a couple minutes for films in the 20–70 nm range for hydrogen concentrations above 1 %.

Overall, the different types of fiber optic sensors show similar values for sensitivity and response time, but there are unique trade-offs and advantages for each. Fiber Bragg grating, FPI, and other interferometric sensors have the potential to be more sensitive than tapered fiber or mirror designs due to their resonant nature. However, these sensors tend to have a much higher fabrication complexity, especially for FBGs. Some of the interferometric sensors produce a wavelength shift. Thus, they are less affected by misalignment or drift in loss than a sensor that produces a power shift; however, they require a more expensive measurement system such as a spectrometer or a tunable laser. Tapered fiber designs can have much lower response times since a thinner film can be used if the interaction length is increased, but they have larger footprints and are less robust. Alternatively, facet mirror designs have a high durability, low footprint, and show good repeatability, along with having a very low fabrication complexity. Some more thorough reviews and comparisons of hydrogen sensors in general can be found in [3, 38, 39].

9.4 Metallic Nano-Apertures and Extraordinary Transmission

In an aperture with dimensions comparable to, or well above, the wavelength of light in the enclosed media, the majority of the propagating mode's energy resides in the cavity formed by the aperture. However, for extremely sub-wavelength apertures, the transmission drops rapidly as a function of w/λ^4, where w is the size of the confining direction and λ is the wavelength of light in the aperture cavity media [40, 41]. For sub-wavelength metallic apertures, or nano-apertures in the visible-NIR regime, propagating surface plasmon polaritons (SPPs) allow for enhanced throughput, or so-called extraordinary transmission. SPPs are collective oscillations of surface charge on a dielectric-metal interface that result from a coupling of the electromagnetic field to the charges on the metal surface [42]. In certain configurations, such a wave can propagate along a metal interface for several microns or more at optical frequencies, despite the loss introduced by the metal [43]. Furthermore, due to the high permittivity of the metal, the wavelength of a SPP can be extremely small leading to a highly confined field inside the aperture. Specifically, this wave propagates along the interface, placing the majority of the energy along the aperture boundaries, rather than in the aperture cavity [44]. This coupling of the field to the SPP aperture mode can be highly resonant in the appropriate frequency range depending on the metal and aperture dimensions [45].

These metallic nano-apertures and cavities are interesting from a sensing standpoint due to the highly localized fields, potentially strong resonance, and high power throughput [46]. For a sensor based on a functionalized metal coating, this is ideal because the field transmitted through the aperture is localized primarily along the metal interface, rather than in the aperture cavity, which is away from the functional material. Not only does the nano-aperture confine the light to the functional layer, but due to resonant effects resulting from SPP coupling and excitation, changes in the transmission and reflection due to a change in optical properties of the metal or physical dimensions of the aperture are amplified.

The resonance of the aperture is determined by the aperture dimensions and the material properties of the metal [47]. The effects of aperture dimensions can be broken down into three sections. First, there is the scattering and excitation of surface plasmons from the light incident on the aperture. Second, there is the propagation of the light through the aperture, which can be thought of as a small metallic waveguide. Finally, there is back-scattering from the exit port of the aperture and decoupling of the SPPs back into a transmitted field. The modes of the metallic waveguide primarily determine the resonance of the aperture [45]. For rectangular waveguides, these can be broken down into metal-dielectric-metal (MDM) or metallic slot waveguides, and dielectric-metal-dielectric (DMD) or metallic ridge waveguides [48]. MDM waveguides have higher loss, but more localized confinement of the field, which is more useful for a Pd based hydrogen sensor. For sub-wavelength MDM waveguides, the field is localized along the metal interfaces that are perpendicular to the incident polarization. This results from sub-wavelength dimensions being needed to maintain coupling between the SPPs on each metal interface, and from the zero tangential electric field boundary condition on Maxwell's equations. For the fundamental mode, transverse electric (TE_{10}), the cut-off frequency is determined by the length of the longest side of the rectangle. Hence, the transmission through a rectangular nano-aperture can be spectrally tuned by changing the size parallel to the incident polarization. An optimal gap distance can also be found to maximize the transmission for a given wavelength.

Based on this waveguide formulation for nano-aperture transmission, several variations on rectangular apertures were proposed, notably the 'C' and 'H' apertures [45, 49]. The C aperture was shown to have highly resonant transmission and a large increase in power throughput compared to a rectangular aperture with the added benefit of producing a largely circular output beam. For these reasons, the C aperture was chosen for use in designing a single nano-aperture palladium based hydrogen sensor.

Sensor designs using a plain palladium facet mirror offer several advantages over FBGs or tapered fiber designs. They are simple to fabricate and can work at multiple wavelengths for self-referencing or can be measured with a broadband light source without the need for a spectrometer or tunable laser. FBGs can be expensive depending on the desired wavelength and reflection spectrum, and tapered fiber designs can involve wet etching of the cladding which must be done in a controlled manner. Furthermore, tapered fiber designs result in an inherently fragile sensor since the protective layers of the fiber have been etched away. By themselves, facet mirrors do not offer the sensitivity of FBGs or tapered fiber sensors; however, by incorporating nanostructures onto the facet, higher sensitivity can be achieved while maintaining the small form factor of a facet at the cost of some fabrication complexity. The remainder of this chapter will focus on the design and experimental analysis of such a device as presented in [21, 50–53].

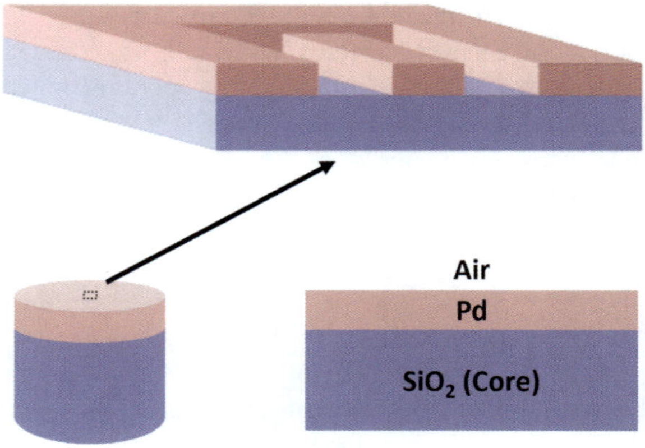

Fig. 9.7 Schematic of the C-shaped nano-aperture optical fiber hydrogen sensor. Adapted from [52]

9.5 Nano-Aperture Sensor Structure and Layout

The on-fiber layout for the nano-aperture sensor is shown in Fig. 9.7. By integrating the sensor onto the fiber facet, as opposed to along its axis, the footprint of the actual sensing region is greatly reduced and spatial accuracy is enhanced. Furthermore, a bundled array of sensors with different functionalized metals can be placed in close proximity for the determination of the various constituent concentrations in a gas mixture. Since the optical properties of sub-wavelength apertures are highly dependent on the material properties, changes in the palladium layer will couple to changes in transmission and reflection by affecting the guided surface plasmons in the nano-aperture. The transmission and reflection spectra of the aperture are simulated to determine the optimal dimensions for a given aperture type or shape in order to maximize the transmitted or reflected power, improve signal-to-noise ratio (SNR), or improve the sensitivity; the latter requiring the optical properties of the film as a function of the analyte to be well known. The C aperture demonstrates extraordinary transmission, providing both high power as well as resonant characteristics. This aperture is also asymmetric, which produces different responses and sensitivities for different polarizations of light and thus has a functional polarization dependent loss (PDL). Such a mechanism allows for self-referencing and the reduction or elimination of common mode noise along the fiber. As discussed in prior chapters, there are several methods for fabricating micro and nano-structures on a fiber tip. For accurate and versatile prototyping, focused ion beam (FIB) milling can be used to etch the aperture. This technique allows for in situ monitoring of the fabrication, as well as the fabrication of sensors with a variety of parameters.

Along with the C aperture, circle apertures and plain-film sensors also show a hydrogen response and can be used as controls to measure the effect of aperture shape and to demonstrate the presence of a polarization dependent sensitivity. The devices can be used in both transmission and reflection by using a detector opposite the facet and an optical circulator, respectively. To experimentally test the apertures, special care needs to be taken so that the gas flow and pressure can be accurately controlled. As a sensing metric, the fractional change in transmitted or reflected power can be measured. In transmission mode, the C aperture demonstrates a sensitivity much larger than that of both the circle and plain film designs, showing the importance of using structures with resonant or extraordinary transmission. The magnitude of the change was about 3 times larger compared to the circle aperture and eight times larger compared to the plain film, improving the SNR and allowing for the detection of lower concentrations of hydrogen. Interestingly, in reflection mode, which offers a much lower footprint, both the C and the circle apertures show an order of magnitude improvement in sensitivity over the plain film. A change in reflected power of almost 20 % at 0.25 % hydrogen was demonstrated for the C aperture, over double the value for the same sensor when measured in transmission.

9.6 Nano-Aperture Sensor Design

To use this information to maximize the performance of a structured optical device, such as a nano-aperture, simulation techniques such as the finite-difference time-domain (FDTD) method and the finite element method (FEM) can be employed to model the spectral response as a function of device parameters. The reflection and transmission properties of C apertures have been studied in general, but the properties of the materials used play an important role. In the case of the hydrogen sensor described in Fig. 9.7, the optical properties of the palladium layer, as well as the three dimensional (3D) nature of the problem must be taken into account. A fundamental problem with simulating this structure in 3D is the widely varying dimensional scales of the device.

To accurately simulate the fiber, a cross section several tens of microns in diameter must be modeled. Coupled with the nanometer scale of the aperture, this represents an extremely large computational domain when meshed, which is unfeasible to simulate on most existing computer systems. Given the large mismatch of modes and the radiative nature of the aperture, mode matching is also too difficult to employ. However, reflection from a plain film is easily understood in terms of the Fresnel equations [54], and the aperture only has a localized effect on reflection and transmission. Therefore, the overall transmission and reflection can be thought of as a coherent superposition of the surrounding plain film region and the aperture region.

To model the aperture, a thin cross section of the facet can be simulated using FDTD and FEM. For FDTD simulations, the dispersion of the palladium film can be taken into account through a Lorentz–Drude oscillator model [55]

$$\varepsilon_r(\omega) = 1 - \frac{f_0\omega_p^2}{\omega(\omega - i\Gamma_0)} + \sum_{j=1}^{k} \frac{f_j\omega_p^2}{(\omega_j^2 - \omega^2) + i\omega\Gamma_j}, \tag{9.6.1}$$

where ε_r is the relative permittivity, ω is the optical frequency, Γ_j is a damping term, ω_p is the plasma frequency, and f_j is the oscillator strength for the jth interband resonance; f_0 and Γ_0 relate to intraband effects. Since the aperture overlaps a small portion of the center of the fundamental mode, the guided mode can be replaced by a plane wave, removing the need to simulate the entire optical fiber. Thus, only a small area around the aperture is simulated. Furthermore, by using a Gaussian impulse excitation and taking Fourier transforms, the transmitted and reflected power can be found across a range of frequencies with a single FDTD simulation.

FEM can also be used to model the transmission and reflection by using a frequency sweep and the plane wave excitation can be modeled by a current sheet. Both techniques can also be used to obtain the electric and magnetic fields of the modes propagating inside the aperture. Electric field profiles along with the corresponding aperture design can be seen in Fig. 9.8. Bulk values for the refractive index of the palladium film are used [55] and the thickness is set to 150 nm. As expected, the simulations show the field strongly confined to the air-palladium interfaces that are perpendicular to the incident polarization. There is also a polarization dependence given the asymmetry of the aperture. These two modes are referred to as quasi-transverse electric (quasi-TE) and quasi-transverse magnetic (quasi-TM), where quasi-TE has a majority E_x field and quasi-TM has a majority E_y field. However, these modes are hybrid modes and E_x and E_y cannot actually be fully decoupled. Excitation of the quasi-TE and quasi-TM depends on the input polarization, with an E_x polarized input field producing a quasi-TE mode, and an E_y input producing a quasi-TM mode. Since these two modes propagate in different parts of the aperture, it is conceivable that they would have different magnitudes of their response to hydrogen-induced changes. This is indeed the case and has been experimentally shown [21].

The smaller aperture in Fig. 9.8b is a typical design for a C aperture. However, for a sensor based on transmission changes, it is useful to maximize both the total transmitted power to improve the SNR as well as the differential response due to hydrogen induced changes. Modeling hydrogen induced changes requires specific knowledge about how such a nanostructured film expands, as well as accurate refractive index data as a function of hydrogen concentration. While general trends are known, these are both areas of continuing research [56, 57]. It is easy, however, to optimize the structure so that there is a transmission peak at or near the target wavelength without hydrogen present. To maintain low cost, it is desirable to operate in the telecom band near 1550 nm. As was mentioned previously, the transmission peak can be altered the most by changing the arm length and waist gap. A second aperture design is the longer aperture shown in Fig. 9.8a where the transmission peak has been shifted to be near 1550 nm for the quasi-TM mode. The quasi-TM reflection and transmission spectra for this new design can be seen in Fig. 9.9 compared to the original C aperture in Fig. 9.8b. This particular aperture was also designed to emphasize the polarization difference and resolve a fabrication problem that arose from having too short of a ridge.

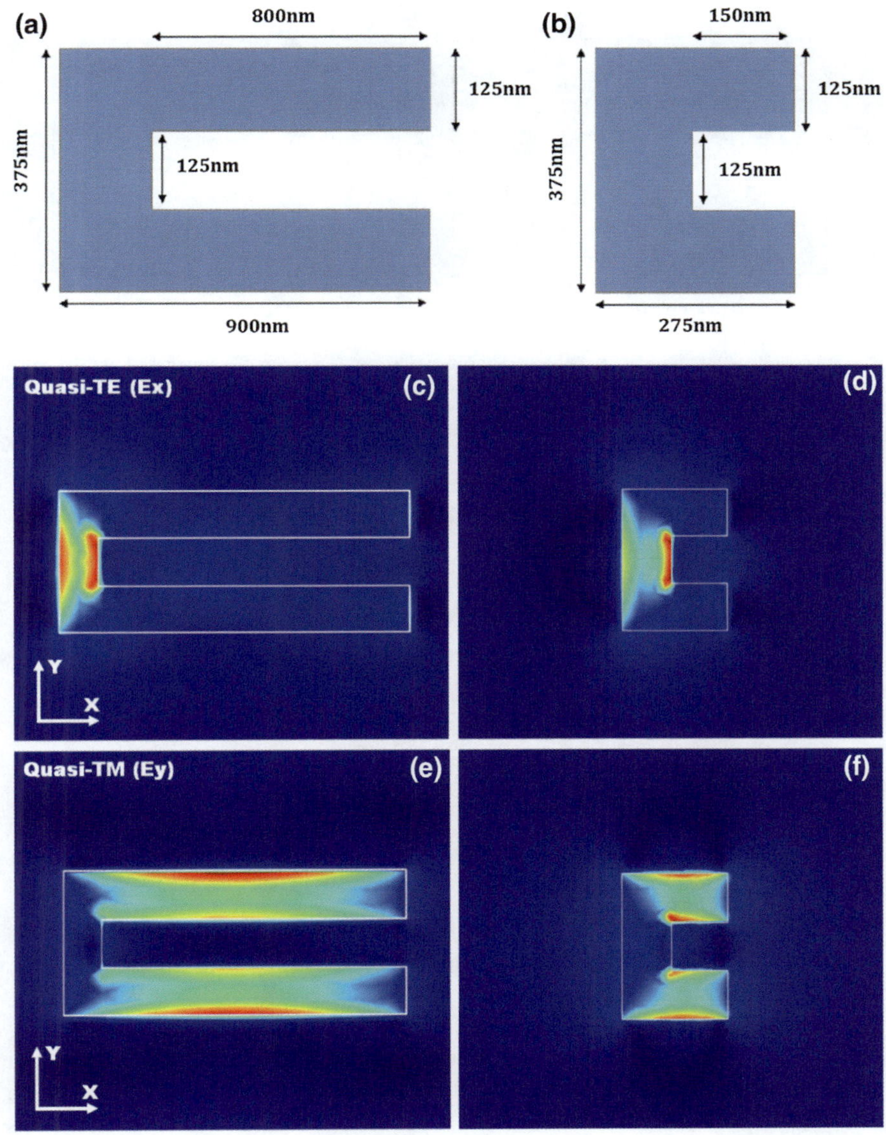

Fig. 9.8 **a** Nano-aperture design used to fabricate devices for experiment. **b** Typical C aperture design. **c**–**f** Electric field profiles for the two aperture designs. Adapted from [21, 52]

While the exact characteristics of hydrogen induced lattice expansion for nanoscale structures are not well established, the film will expand with hydrogen [7]. While changes in the lateral dimensions of the aperture will have a large effect, even vertical expansion will have some impact on the aperture transmission by altering the Fabry–Pérot resonances of the aperture cavity as demonstrated in

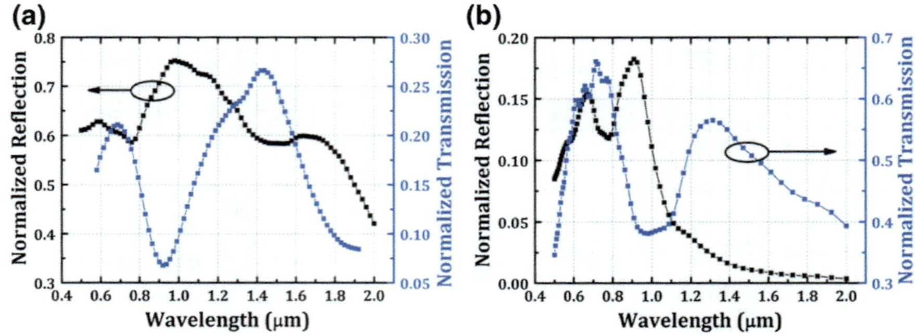

Fig. 9.9 Reflection and transmission of the quasi-TM mode for the **a** exaggerated and **b** typical C aperture. Adapted from [52]

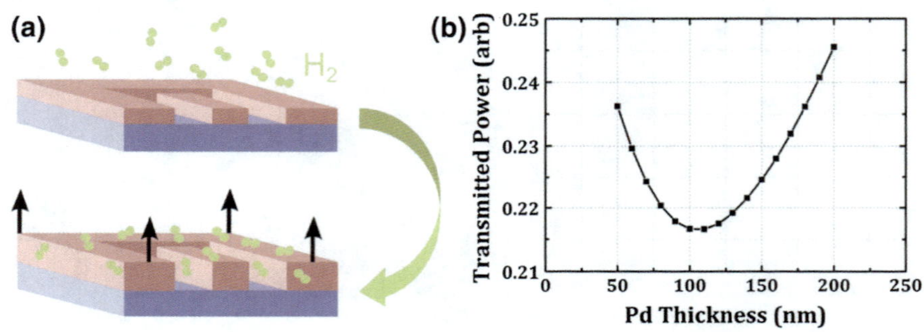

Fig. 9.10 **a** Illustration of hydrogen induced lattice expansion. **b** Transmitted power as a function of Pd layer thickness for a nano-aperture and a small surrounding area. Adapted from [52]

Fig. 9.10. This figure shows simulation results for the exaggerated C aperture shown in Fig. 9.8a as a function of aperture thickness and uses the same truncation of the simulation domain. There is still an exponential decay as thickness increases, but it is modulated by the cavity resonances. This phenomenon only occurs because of the extraordinary transmission resulting from surface plasmon propagation. Typically, as the film thickness increases the loss will also increase, resulting in a decrease in transmission. However, since SPPs in the C aperture have longer propagation lengths than simple plane wave transmission, the Fabry–Pérot effect becomes visible. Note that this simulation does not take into account any light transmitted through the film, which will be present for thin films. To fully model the device, sensitivity plots such as Fig. 9.10 need to be simulated using FEM and/or FDTD for every device dimension as a function of the refractive index. Optimization algorithms can then be employed to determine the optimal device structure.

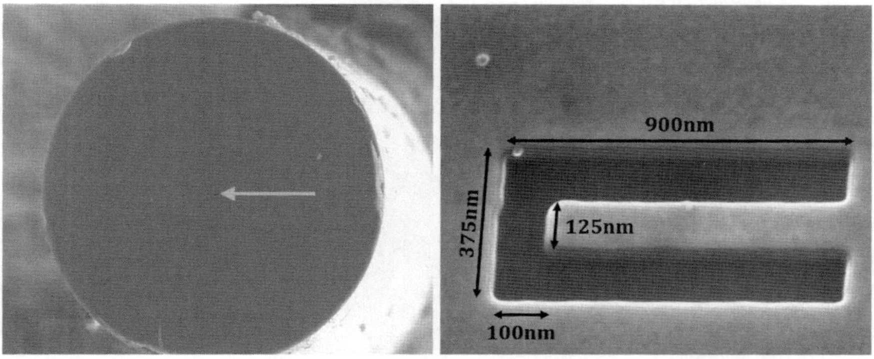

Fig. 9.11 Scanning electron microscope (SEM) images of a fabricated Pd nano-aperture sensor with design dimensions. The film thickness was 150 nm

9.7 Fabrication and Measurement of a Fiber Nano-Aperture Hydrogen Sensor

Based on these concepts, transmission based hydrogen sensors using plain films and facet-etched nano-apertures have been fabricated and experimentally verified. Fabrication was done by cleaving pieces of optical fiber and coating them with palladium using electron beam evaporation. For the aperture sensors, apertures were milled using a FIB. Care must be taken when milling because ion implantation into the fiber substrate from the beam can result in an increase in loss in the SiO_2 layer [58], which would be detrimental to sensor performance. Also, due to the insulating substrate, this charge build-up can result in uneven or over-etching of the aperture, particularly the ridge. Images of a fabricated C aperture with the same dimensions as Fig. 9.8b can be seen in Fig. 9.11.

Since the light in the aperture is going from a highly confined mode to free space, there will be significant diffraction resulting in a diverging beam at the output. The radiation patterns for such an aperture can be seen in Fig. 9.12 for the two modes. In order to capture this light, a high numerical aperture (NA) lens near the sensor head is required. The light is then focused onto a detector, but could also be coupled back to an output fiber for remote measurements.

A typical measurement setup is shown in Fig. 9.13. An optical splitter is used to monitor input power so that the output can be normalized. An optical isolator is used to prevent feedback into the laser cavity and a polarization controller or polarization scrambler is used to respectively set or randomize the input polarization of the light going to the sensor. An optical circulator can then be used to collect reflection data in addition to transmission. After the sensor head is spliced back to fiber, the tail end is connected to the laser input and the head is placed in a flow chamber for testing with a detector to monitor the transmission. Gas flow and data acquisition from the detectors can then be controlled and recorded through the use of LabVIEW. In a standard pulse measurement, the sensor is exposed to a

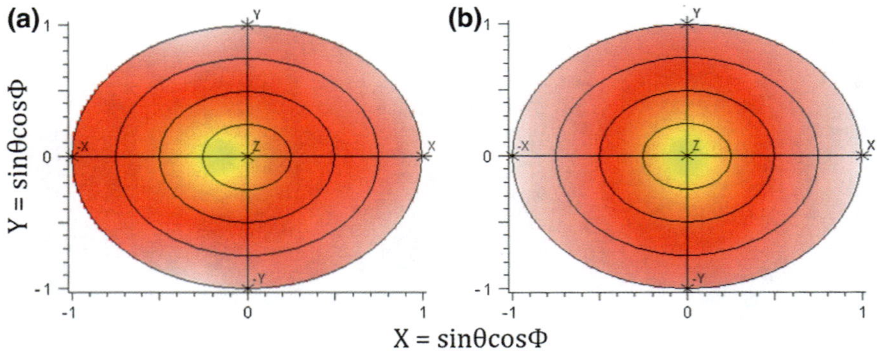

Fig. 9.12 Simulated far-field radiation patterns along the z-axis for the **a** quasi-TE and **b** quasi-TM modes of the fabricated C aperture. Adapted from [52]

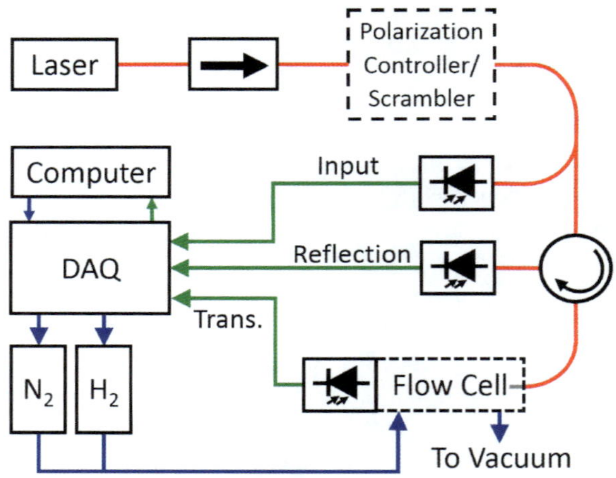

Fig. 9.13 Schematic of the typical experimental setup. Red denotes fiber paths, green denotes data acquisition, and blue denotes gas flow and gas flow control

certain concentration of hydrogen in nitrogen and then purged with nitrogen for a set amount of time. This is then repeated for as many concentrations or pulses as needed.

9.8 α Phase Transmission Measurements

Using this control and measurement setup, the nano-aperture sensor can be investigated in several different ways. To start, we will simply look at the behavior of the sensor at low concentrations in the α phase. As a metric, the change in transmitted power is measured relative to the power in nitrogen ambient after being mounted in the flow chamber. The input polarization is kept fixed using a

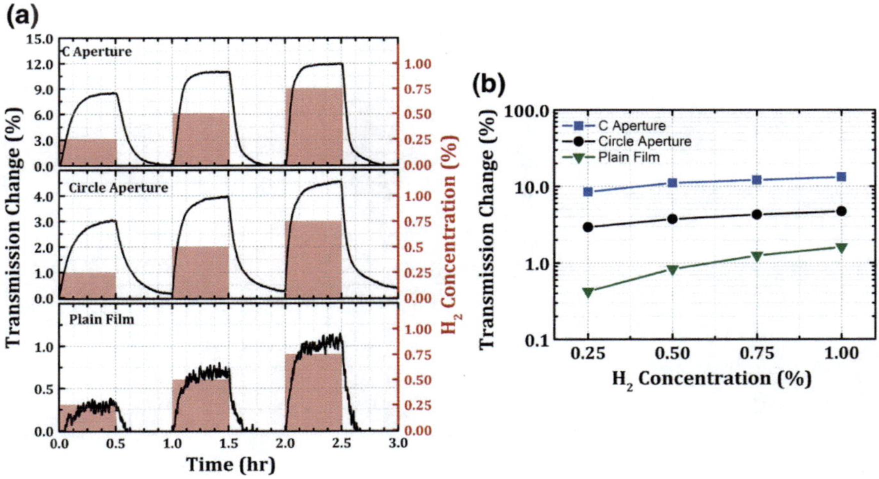

Fig. 9.14 **a** Pulse measurements showing the time dependent response of the three sensor types for various hydrogen concentrations. **b** Semi-log plot of the response of each sensor type versus hydrogen concentration. Adapted from [21]

polarization controller, and the sensor is exposed to increasing hydrogen concentration pulses of 30 min, separated by 30 min of nitrogen until the β phase transition becomes apparent.

In this experiment, several sensor types are tested for comparison purposes. These include the same C aperture as before, a circular nano-aperture with equal area, and a simple plain film facet mirror sensor. The sensors were tested down to 0.25 % hydrogen in nitrogen and the change in transmitted power was measured. A time-domain plot of the pulses and the associated change in transmission for each sensor are shown in Fig. 9.14a. The response of the sensor is then defined as the average over the last 5 min of the pulse, averaged across several pulses. These results are shown in Fig. 9.14b for each type of sensor on a semi-log plot.

The response times for each type of sensor were the same, varying from 4 to 10 min depending on the hydrogen concentration. This is because the response time is determined by the diffusion and adsorption rates of the Pd film. The C aperture sensor however, showed a much larger magnitude of response to hydrogen than the plain film, and the circle aperture was in-between. This is attributed to the extraordinary transmission resulting from guided modes in the C aperture. The circle aperture lacks the MDM or DMD waveguide structure present in the C aperture and therefore its modes are less confined than the C aperture modes. Thus, the circle aperture is less affected by changes in the refractive index or dimensions of the aperture walls. The C aperture showed a maximum response of 13.2 % at 1 % hydrogen and a response of 8.5 % at 0.25 % hydrogen whereas the circle aperture only reached a 4.7 % power change, and the plain film only had a 1.6 % power change at 1 % hydrogen according to Fig. 9.14b.

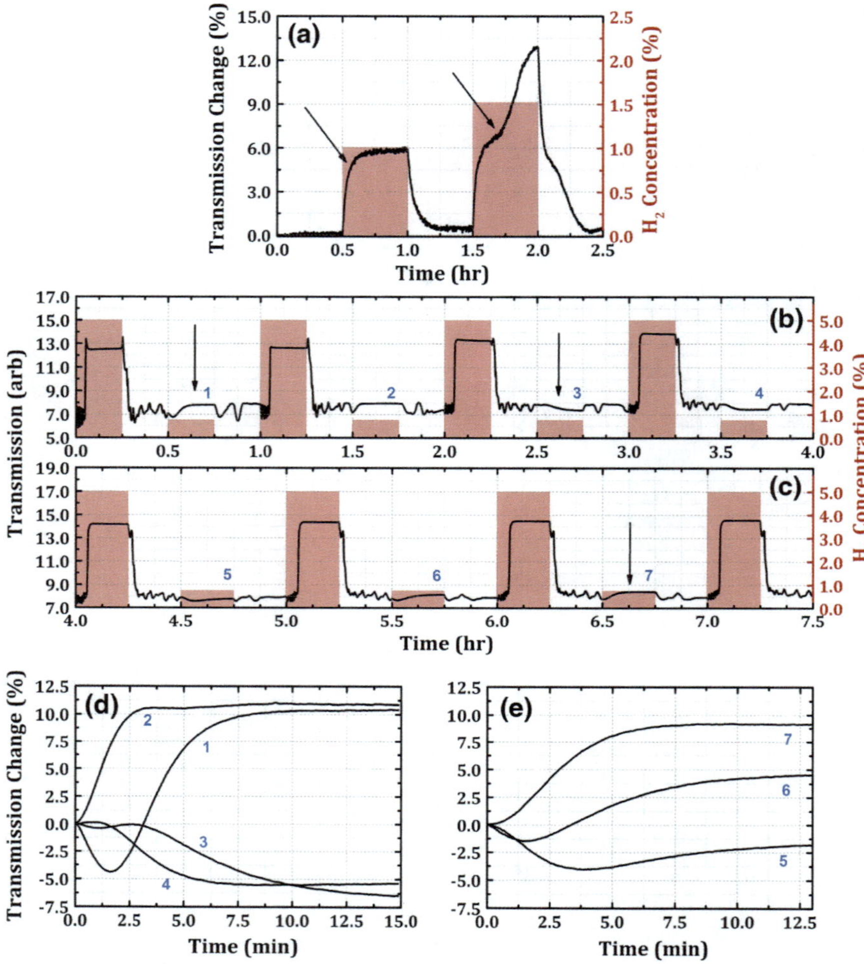

Fig. 9.15 **a** Sensor response at high concentrations; the kink in the transmitted power denotes the onset of the β phase. **b**, **c** Pulse test for the sensor alternating between 5 % hydrogen and 0.75 % hydrogen; arrows highlight the unusual effect of the 5 % hydrogen on the subsequent low concentration pulse. **d**, **e** Traces of the labeled low concentration pulses show the change in temporal response as a result of exposure to the 5 % hydrogen. Adapted from [21]

9.9 β Phase Transmission Measurements

Another area of interest is the behavior of the sensor at high hydrogen concentrations, above the α-β phase transition. Here, only a C aperture sensor is tested. It is again exposed for 30 min at the target concentration and 30 min of nitrogen to purge. The phase change can be seen starting at concentrations of around 1.5 % in Fig. 9.15a, as evidenced by a second increase in the measured transmission. Up until a certain point in time, determined by the diffusion coefficient and adsorption, the film will remain in the α phase even though the ambient concentration is high enough for a transition to the β phase. After enough hydrogen diffuses into the lattice, the film will enter the β phase and the optical and mechanical properties will begin to rapidly change again. It is also worth noting the relatively large change in transmission after the film enters the β phase. The sensitivity to hydrogen in the β phase is much larger than in the α phase, suggesting a larger change in the film's properties. In general, the concentration at which this occurs is a property of the film and is highly dependent on the deposition conditions and thickness.

It has been suggested that exposure to high concentrations of hydrogen can act to "work harden" the material [59]. However, it is also known that prolonged exposure to high concentrations can also degrade film quality [7]. To evaluate this, the sensor was also exposed to alternating 15 min pulses of 5 % hydrogen in nitrogen and 0.75 % hydrogen in nitrogen with 15 min nitrogen purges interleaved. The 5 % pulse is above the α-β phase transition threshold. In this regime, the PdH film enters the β phase and undergoes plastic deformation. The results of this can be seen in Fig. 9.15b and c. Overall this process slightly improved the response magnitudes and response times; however, an interesting side effect was the behavior of subsequent low concentration pulses. The response there seemed to cycle between a power increase and a power decrease, with each 5 % pulse advancing the location of the low concentration response in the cycle. This can be seen in Fig. 9.15d and e which plots the 0.75 % traces labeled incrementally. The response to 5 % hydrogen was always a large power increase whereas the low concentrations alternated. This behavior was repeatable in that the response at low concentration always decreased or always increased until it was again exposed to a large hydrogen concentration. The exact cause of this is unknown, but it is most likely attributable to plastic deformation and very slow relaxation of changes induced during the β phase.

9.10 Polarization Dependent Transmission Measurements

A final property of the C aperture sensor to measure in transmission is the polarization dependence and in particular the PDL, which as discussed is a result of the two asymmetric modes of the aperture. These measurements were carried out in the same way as the prior pulse tests, except that a polarization scrambler was used to continuously vary the input polarization and thereby sample all possible states of input polarization. The scrambler cycles through the entire Poincaré sphere at 10 Hz. Data was acquired at 15 kHz for 1 s so that the polarizations of minimum and maximum transmission could be properly sampled. A median filter was applied to remove spikes arising from the non-uniform sampling of the polarization states. Since the transmission change is monotonic over the test concentration range for any polarization state, the two orthogonal

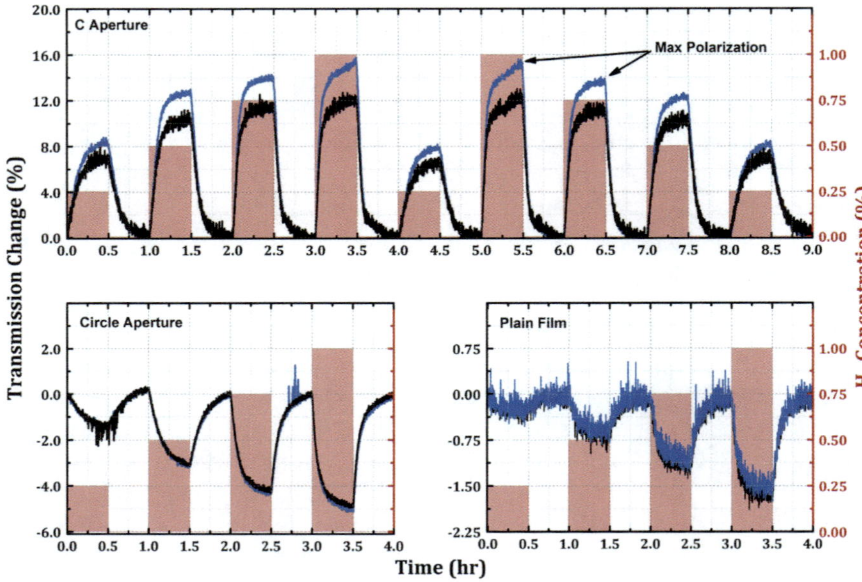

Fig. 9.16 Response of the minimum (*black*) and maximum (*blue*) transmission polarization for each sensor type versus time. Adapted from [21]

modes can be extracted by measuring the minimum and maximum of the transmitted power over some period as long as the response curves for the two polarizations do not cross. Comparisons of measured power to simulated data are difficult to make because of how sensitive the quantity is to detector alignment. However, as seen in Fig. 9.12, there is significant overlap in the radiation patterns for the two modes; hence, the measured and simulated PDL i.e., TM/TE power ratio, should be close. For the fabricated aperture, the measured PDL was 2.9, while the simulated value gave a ratio of 3.3. This discrepancy is easily within errors introduced by the uncertainties in the film thickness and its refractive index, which can vary for thin films, as well as fabrication imperfections such as over-etching.

The polarization response for the C aperture in transmission, as well as the circle aperture and plain film sensors can be seen in Fig. 9.16. As expected, the circle aperture and plain film sensors show no polarization dependence. Response curves similar to Fig. 9.14b can be found for the two states and roughly fit to a line in the 0.5–1 % H_2 region. From this, an approximate sensitivity can be calculated. This corresponds to a 3.8 % increase in power per percent hydrogen for the minimum transmission polarization and a 5.7 % power increase per percent hydrogen for the maximum state.

The C aperture, however, shows both a difference in transmitted power for both polarizations as well as a dependence of the PDL on hydrogen as shown in Fig. 9.17. A 0.07 dB increase in the PDL was measured at 0.25 % hydrogen, and a

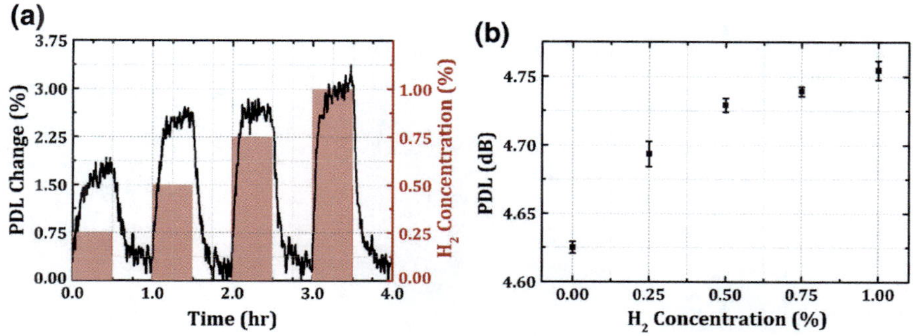

Fig. 9.17 **a** PDL of the C aperture versus time during a hydrogen pulse test. **b** Summary of the PDL response versus hydrogen concentration. Adapted from [21]

0.14 dB increase was measured at 1 % hydrogen. Since both modes must propagate through the same fiber, and radiate to the same detector over the same path, as long as a sufficiently high NA lens is used, the PDL is a self-referencing sensing metric that can be used to reduce common mode noise from mechanical disturbances to the fiber or changes in detector alignment.

9.11 Reflection Measurements

From an implementation standpoint, reflection based sensing is preferred to transmission based. A reflective sensor is less invasive than a transmissive sensor in that only a single tiny entry point to the gas volume is needed. This entry point only needs to be wide enough for the tip of the fiber sensor to fit. This results in an extremely small sensor footprint and allows for true point-based detection. In comparison, a transmissive sensor needs both an entry and an exit point and possibly also free space optics in the gas flow volume to couple between the two. This can possibly disrupt the gas flow. Further, the output from the reflective sensor is not affected by detector misalignment since the light is intrinsically coupled back into the fiber. This direct coupling allows the reflected light to then be directed to the measurement and control system, which can be co-located with the laser source, e.g. at a remote facility.

In reflection mode, the presence of a nano-aperture would not be expected to produce a noticeable effect in the overall reflection since the reflection from the surrounding film should be dominant. However, Fig. 9.18 shows that this is not necessarily the case for the nano-aperture sensor types when tested with a fixed polarization that maximizes the reflection. These results also show a much smaller difference between the C and circle apertures than the transmission mode results. This suggests that the physical mechanism for the enhanced sensitivity in reflection mode might not be an aperture resonance effect since the resonances would be

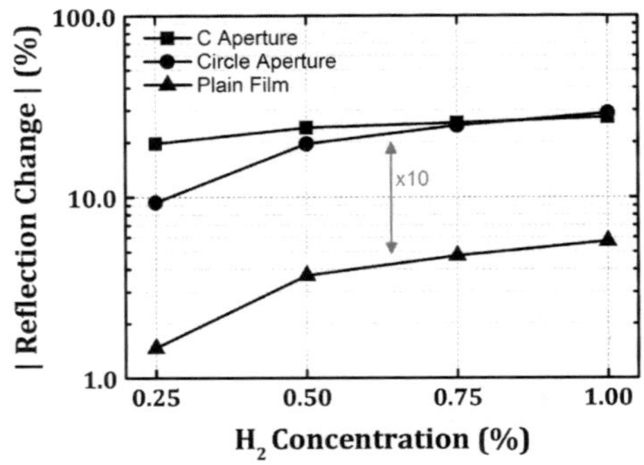

Fig. 9.18 Summary of the response in reflection for the three sensor types

different for the two aperture types. Instead, simply the presence of an aperture seems to be enhancing the sensitivity. Such an effect could arise if the surrounding film has very low reflectivity, which can happen if the film is sufficiently rough. Given the poor adhesion of Pd to oxide and the expansion of the film, this could result from localized buckling or a micro-blistering effect similar to that demonstrated by Butler.

Since the reflections from the aperture and film are coherent, there will be interference between the two fields. Furthermore, the change in refractive index and structure will affect the phase and amplitude of the reflection from the two in different ways since hydrogen more strongly incorporates at the surfaces of the aperture than in the bulk film. This difference will then manifest itself through a change in the interference, allowing the relatively small aperture area to have a large effect. In a current investigation of this physical explanation, the aperture is being etched beyond the Pd layer to several different depths into the SiO_2 fiber underneath in order to tune the phase difference between these two fields in the absence of hydrogen [60]. It is expected that the etch depth can be optimized to maximize the sensitivity of the reflectance to hydrogen.

9.12 Conclusions and Outlook

The optical fiber platform offers several unique advantages for sensing applications, such as the low cost and large supply of optical fiber and optical components around the telecom band. Furthermore, optical fiber has extremely low loss, allowing sensor deployment in many areas spread over long distances, even areas that may be far from the signal generation and measurement equipment. This also facilitates the creation of distributed systems and multiplexing for sensor networks. Finally, facet-based optical fiber sensors have a small form factor and have the ability to localize the measurement to an area a fraction of a square millimeter in size. Optical fiber is also easily integrated into systems, allowing for minimally disruptive sensing.

Table 9.1 Nano-aperture sensor performance

Mode	Sensor	ΔP at 0.25 % H_2	ΔP at 1 % H_2	Sensitivity: ΔP/ΔH_2 for H_2 between 0.5 and 1 %
Transmission	C Aperture, TE (Max)	8.5 %	15.3 %	5.6
	C Aperture, TE (Min)	6.9 %	12.0 %	3.8
	Circle	2.9 %	4.7 %	1.9
	Plain film	0.4 %	1.6 %	1.5
	PDL	0.07 dB	0.14 dB	0.09 dB per % H_2
Reflection	C Aperture	19.7 %	27.4 %	6.8
	Circle	9.3 %	28.9 %	18.7
	Plain film	1.5 %	5.7 %	4.0

For trace gas detection, specifically the detection of hydrogen, nanostructures such as apertures can be used to enhance the change in optical properties of the functional layer. A fiber sensor consisting of a palladium film with a single C-shaped aperture etched over the core was fabricated and compared experimentally to a simple plain film structure to show the increase in response due to the presence of the aperture. The C aperture was also compared to a circle aperture to show that aperture shape, and not just aperture size, can greatly affect the sensitivity of the device. Such a device offers a substantial improvement in hydrogen sensitivity over plain film sensors with minimal added fabrication complexity. The sensor was tested in several ways. The response to hydrogen in both the α and β phases of PdH was measured, showing a sharp increase at the onset of β phase and evidence of deformation of the film. The polarization dependence of the aperture was also investigated, revealing a hydrogen dependence in the PDL of the sensor. In reflection mode, both the C and the circle apertures showed increased sensitivity compared to transmission mode. In all modes and for all response variables, the C aperture showed high sensitivity well below the lower explosive limit of hydrogen in air of 4 %. A summary of the results for the various sensors and configurations can be found in Table 9.1.

To more accurately model, and thus optimize the sensor structure, future studies should be conducted to determine the exact changes in refractive index for PdH, both as a function of wavelength and as a function of film structure. Furthermore, the mechanics of hydrogen induced lattice expansion at the nanoscale for structured films needs to be more thoroughly investigated. Aside from improving the accuracy of the theoretical model, different film compositions and layer structures also need to be pursued to increase the sensitivity, decrease the response time, and improve the long term reliability and repeatability of the sensor. Platforms that are self-referencing and easy to calibrate or self-calibrating should also be explored. Despite these technical hurdles, there is still growing interest in developing robust, sensitive, low-cost, and accurate optical sensors. The unique advantages of the fiber optic platform combined with the development of inexpensive non-planar fabrication techniques such as nano-replica molding [61], superionic solid state stamping [62], pattern transfer [63], or nano-imprint lithography [64, 65] make it an attractive system for future sensor designs.

References

1. T. Wei, X. Lan, H. Xiao, Y. Han, H.-L. Tsai, Optical fiber sensors for high temperature harsh environment sensing. in *2011 IEEE Instrumentation and Measurement Technology Conference* (*I2MTC*) (2011), pp. 1–4
2. T.N. Veziroğlu, S. Şahin, 21st Century's energy: hydrogen energy system. Energy Convers. Manage. **49**, 1820 (2008)
3. C. Christofides, A. Mandelis, Solid-state sensors for trace hydrogen gas detection. J. Appl. Phys. **68**, R1 (1990)
4. Z. Zhao, Y. Sevryugina, M.A. Carpenter, D. Welch, H. Xia, All-optical hydrogen-sensing materials based on tailored palladium alloy thin films. Anal. Chem. **76**, 6321 (2004)
5. G. Choi, G. Jin, S.-H. Park, W. Lee, J. Park, Material and sensing properties of Pd-deposited WO3 thin films. J. Nanosci. Nanotechnol. **7**, 3841 (2007)
6. J. Dai, M. Yang, Y. Chen, K. Cao, H. Liao, P. Zhang, Side-polished fiber Bragg grating hydrogen sensor with WO3-Pd composite film as sensing materials. Opt. Express **19**, 6141 (2011)
7. T.B. Flanagan, W.A. Oates, The palladium-hydrogen system. Annu. Rev. Mater. Sci. **21**, 269–304 (1991)
8. Y. Morita, K. Nakamura, C. Kim, Langmuir analysis on hydrogen gas response of palladium-gate FET. Sens. Actuators B: Chem. **33**, 96 (1996)
9. J.F. Lynch, T.B. Flanagan, Dynamic equilibrium between chemisorbed and absorbed hydrogen in the palladium/hydrogen system. J. Phys. Chem. **77**, 2628 (1973)
10. D.D. Eley, Molecular hydrogen and metallic surfaces. J. Phys. Chem. **55**, 1017 (1951)
11. E. Wicke, H. Brodowsky, G. Alefeld, J. Völkl, Hydrogen in metals II. Top. Appl. Phys. **29**, 73 (1978)
12. A. Mandelis, J.A. Garcia, Pd/PVDF thin film hydrogen sensor based on laser-amplitude-modulated optical-transmittance: dependence on H2 concentration and device physics. Sens. Actuators B: Chem. **49**, 258–267 (1998)
13. C. Edwards, A. Arbabi, S.J. McKeown, R. Zhou, G. Popescu, L.L. Goddard, Optical Inspection and Metrology Using Diffraction Phase Microscopy. in *Surface Analysis Symposium 2013* (Urbana-Champaign, IL, 2013)
14. C. Edwards, S.J. McKeown, J. Zhou, G. Popescu, L.L. Goddard, Observing hydrogen induced deformations in palladium thin-films. in *IEEE Photonics Conference* (Bellevue, WA, 2013), pp. 612–613
15. M. Wang, Y. Feng, Palladium–silver thin film for hydrogen sensing. Sens. Actuators B: Chem. **123**, 101–106 (2007)
16. M. Raval, S. McKeown, A. Arbabi, L.L. Goddard, *Palladium Based Fabry–Pérot Etalons for Hydrogen Sensing*, in Imaging and Applied Optics Technical Papers, OSA Technical Digest (Optical Society of America, 2012), p. STh2B.5. http://www.opticsinfobase.org/abstract.cfm?uri=Sensors-2012-STh2B.5
17. B.G. Griffin, A. Arbabi, A.M. Kasten, K.D. Choquette, L.L. Goddard, Hydrogen detection using a functionalized photonic crystal vertical cavity laser. IEEE J. Quantum Electron. **48**, 160–168 (2012)
18. B.G. Griffin, A. Arbabi, L.L. Goddard, Engineering the sensitivity and response time of edge-emitting laser hydrogen sensors. IEEE Sens. J. **13**, 3098–3105 (2013)
19. B.G. Griffin, A. Arbabi, L.L. Goddard, Functionalized Distributed Feedback Lasers for Hydrogen Sensing Applications. Submitted, (n.d.)

20. B. Griffin, A. Arbabi, L. Goddard, *Coupled Mode Analysis of a Distributed Bragg Reflector Laser for Hydrogen Detection*, in Imaging and Applied Optics Technical Papers, OSA Technical Digest (Optical Society of America, 2012), p. STh1B.6. http://www.opticsinfobase.org/abstract.cfm?uri=Sensors-2012-STh1B.6
21. S.J. McKeown, L.L. Goddard, Hydrogen detection using polarization diversity via a subwavelength fiber aperture. IEEE Photonics J. **4**, 1752 (2012)
22. N. Liu, M.L. Tang, M. Hentschel, H. Giessen, A.P. Alivisatos, Nanoantenna-enhanced gas sensing in a single tailored nanofocus. Nat. Mater. **10**, 631–636 (2011)
23. M.A. Butler, Optical fiber hydrogen sensor. Appl. Phys. Lett. **45**, 1007–1009 (1984)
24. M.A. Butler, Micromirror optical-fiber hydrogen sensor. Sens. Actuators B: Chem. **22**, 155 (1994)
25. E. Maciak, Z. Opilski, Hydrogen gas detection by means of a fiber optic interferometer sensor. J. Phys. IV **137** 135–140 (2006) (Proceedings)
26. K.S. Park, Y.H. Kim, J.B. Eom, S.J. Park, M.-S. Park, J.-H. Jang, B.H. Lee, Compact and multiplexible hydrogen gas sensor assisted by self-referencing technique. Opt. Express **19**, 18190 (2011)
27. J. Villatoro, D. Luna-Moreno, D. Monzon-Hernandez, Optical fiber hydrogen sensor for concentrations below the lower explosive limit. Sens. Actuators B: Chem. **110**, 23–27 (2005)
28. J. Villatoro, A. Diez, J.L. Cruz, M.V. Andres, Highly sensitive optical hydrogen sensor using circular Pdcoated singlemode tapered fibre. Electronics Letters **37**, 1011–1012 (2001). http://ieeexplore.ieee.org/stamp/stamp.jsp?arnumber=00941800
29. J. Villatoro, D. Monzón-Hernández, Fast detection of hydrogen with nano fiber tapers coated with ultra thin palladium layers. Opt. Express **13**, 5087–5092 (2005). http://www.opticsinfobase.org/oe/abstract.cfm?uri=oe-13-13-5087
30. K.T. Kim, H.S. Song, J.P. Mah, K.B. Hong, K. Im, S.-J. Baik, Y.-I. Yoon, Hydrogen sensor based on palladium coated side-polished single-mode fiber. IEEE Sens. J. **7**, 1767 (2007)
31. Y.O. Barmenkov, A. Ortigosa-Blanch, A. Diez, J.L. Cruz, M.V. Andrés, Time-domain fiber laser hydrogen sensor. Opt. Lett. **29**, 2461 (2004)
32. D. Luna-Moreno, D. Monzón-Hernández, J. Villatoro, G. Badenes, Optical fiber hydrogen sensor based on core diameter mismatch and annealed Pd–Au thin films. Sens. Actuators B: Chem. **125**, 66 (2007)
33. B. Sutapun, M. Tabib-Azar, A. Kazemi, Pd-coated elastooptic fiber optic Bragg grating sensors for multiplexed hydrogen sensing. Sens. Actuators B: Chem. **60**, 27 (1999)
34. M. Buric, K.P. Chen, M. Bhattarai, P.R. Swinehart, M. Maklad, Active fiber bragg grating hydrogen sensors for all-temperature operation. IEEE Photonics Technol. Lett. **19**, 255 (2007)
35. A. Trouillet, E. Marin, C. Veillas, Fibre gratings for hydrogen sensing. Meas. Sci. Technol. **17**, 1124 (2006)
36. X. Wei, T. Wei, H. Xiao, Y.S. Lin, Nano-structured Pd-long period fiber gratings integrated optical sensor for hydrogen detection. Sens. Actuators B: Chem. **134**, 687 (2008)
37. C. Caucheteur, M. Debliquy, D. Lahem, P. Megret, Hybrid fiber gratings coated with a catalytic sensitive layer for hydrogen sensing in air. Opt. Express **16**, 16854 (2008)
38. T. Hübert, L. Boon-Brett, G. Black, U. Banach, Hydrogen sensors—a review. Sens. Actuators B: Chem. **157**, 329 (2011)
39. S.F. Silva, L. Coelho, O. Frazão, J.L. Santos, F.X. Malcata, A Review of palladium-based fiber-optic sensors for molecular hydrogen detection. IEEE Sens. J. **12**, 93–102 (2012)
40. C. Genet, T.W. Ebbesen, Light in tiny holes. Nature **445**, 39 (2007)
41. H.A. Bethe, Theory of diffraction by small holes. Phys. Rev. **66**, 163 (1944)
42. W.L. Barnes, A. Dereux, T.W. Ebbesen, Surface plasmon subwavelength optics. Nature **424**, 824 (2003)
43. R.F. Oulton, V.J. Sorger, D.A. Genov, D.F.P. Pile, X. Zhang, A hybrid plasmonic waveguide for subwavelength confinement and long-range propagation. Nat. Photonics **2**, 496 (2008)
44. A. Chandran, E.S. Barnard, J.S. White, M.L. Brongersma, Metal-dielectric-metal surface plasmon-polariton resonators. Phys. Rev. B **85**, 085416 (2012)

45. X. Shi, L. Hesselink, Design of a C aperture to achieve λ/10 resolution and resonant transmission. Josa B **21**, 1305–1317 (2004)
46. X. Shi, L. Hesselink, R.L. Thornton, Ultrahigh light transmission through a C-shaped nanoaperture. Opt. Lett. **28**, 1320–1322 (2003)
47. Y. Xie, A. Zakharian, J. Moloney, M. Mansuripur, Transmission of light through slit apertures in metallic films. Opt. Express **12**, 6106 (2004)
48. G. Veronis, S. Fan, Modes of subwavelength plasmonic slot waveguides. J. Lightwave Technol. **25**, 2511 (2007)
49. E.X. Jin, X. Xu, Finite-difference time-domain studies on optical transmission through planar nano-apertures in a metal film. Jpn. J. Appl. Phys. **43**, 407 (2004)
50. L. Goddard, S.J. McKeown, Sub-wavelength patterned fiber tips for hydrogen detection. Paper presented at 1st annual world congress of nano science and technology, Dalian, China, 2011
51. S.J. McKeown, L. Goddard, Nano-aperture fiber hydrogen sensors. Paper presented at nanoelectronic devices for defense and security conference New York, NY, 2011
52. S.J. McKeown, Fiber optic hydrogen sensing utilizing facet etched palladium nano-apertures. Fiber Optic Hydrogen Sensing Utilizing Facet Etched Palladium Nano-apertures, Thesis, University of Illinois at Urbana-Champaign, 2011
53. S.J. McKeown, B.G. Griffin, L.L. Goddard, Fiber optic hydrogen sensor utilizing facet-etched metal nano-apertures. in *23rd Annual Meeting of The IEEE Photonics Society, 2010* (Denver, CO, 2010), pp. 730–731
54. S.L. Chuang, in *Physics of Photonic Devices,* 2nd edn. (Wiley, New York, 2012)
55. A.D. Rakic, A.B. Djurisic, J.M. Elazar, M.L. Majewski, Optical properties of metallic films for vertical-cavity optoelectronic devices. Appl. Opt. **37**, 5271 (1998)
56. L.L. Goddard, K. Wong, A. Garg, E. Behymer, G. Cole, T. Bond, Measurements of the complex refractive index of Pd and Pt films in air and upon adsorption of H2 gas. Paper presented at IEEE LEOS annual meeting Newport Beach, CA, 2008
57. P. Tobiška, O. Hugon, A. Trouillet, H. Gagnaire, An integrated optic hydrogen sensor based on SPR on palladium. Sens. Actuators B: Chem. **74**, 168–172 (2001)
58. Y. Fu, N.K.A. Bryan, Investigation of physical properties of quartz after focused ion beam bombardment. Appl. Phys. B **80**, 581 (2005)
59. R.J. Smith, D.A. Otterson, The effect of hydrogen on the tensile properties of palladium. J Less Common Met. **24**, 419 (1971)
60. S.J. Mckeown, L.L. Goddard, Reflective palladium nanoapertures on fiber for wide dynamic range hydrogen sensing. Submitted, (n.d.)
61. I.D. Block, L.L. Chan, B.T. Cunningham, Large-area submicron replica molding of porous low-k dielectric films and application to photonic crystal biosensor fabrication. Microelectron. Eng. **84**, 603 (2007)
62. K.H. Hsu, P.L. Schultz, P.M. Ferreira, N.X. Fang, Electrochemical nanoimprinting with solid-state superionic stamps. Nano Lett. **7**, 446 (2007)
63. E.J. Smythe, M.D. Dickey, G.M. Whitesides, F. Capasso, A technique to transfer metallic nanoscale patterns to small and non-planar surfaces. ACS Nano **3**, 59 (2009)
64. J.A. Rogers, H.H. Lee (eds.), in *Unconventional Nanopatterning Techniques and Applications* (Wiley, New York, 2008)
65. Z. Li, Y. Gu, L. Wang, H. Ge, W. Wu, Q. Xia, C. Yuan, Y. Chen, B. Cui, R.S. Williams, Hybrid nanoimprint—soft lithography with sub-15 nm resolution. Nano Lett. **9**, 2306 (2009)

Chapter 10
Lab-in-a-Microfibre

John Canning

Abstract The laboratory-in-a-fibre was originally envisaged around structured optical fibres and their two dimensional, potentially three-dimensional (3-D), variation in structure they enabled. In contrast to D-shaped fibres which have long been used as a test bed for novel optical components and more recently as the substrate for lab-on-a-chip technologies, or lab-on-fibre, it was a proposal which looked into the future of optical fibres well beyond optical transport conduits, imagining fibres as complex 3-D instruments with multiple waveguides and waveguide devices, both along and across the fibre to enable unprecedented multiple functionality and density. Whereas the D-fibre platform has since evolved into an enormously attractive "lab-on-a-fibre" stage, the vision remains largely unfulfilled but technological advances and new ideas will make its realisation inevitable, if slow. This included the merger of 3-D optical fibre innovations with growing lab-on-fibre innovations. Although the two-dimensional structuring of the optical fibres was central to the proposal of lab-in-a-fibre, it nonetheless has some key restrictions defined by the top-down approach to optical fibre production. Macroscopic drawing down of fibres places a directional impost that gives rise to the two-dimensional layout of micro and nano structured fibres; i.e. control is only in the cross-section of the fibre. As well there are huge thermodynamic imposts integrating materials into silica itself providing motivation for different approaches to be explored. Here, recently reported bottom-up self-assembly approaches to fabricate waveguides that have the potential of enabling total control of the nanostructure are reviewed. These can be done at room temperature demonstrating in principle how the problems can be solved and simultaneously legitimising a new concept—the lab-in-a-microfibre.

J. Canning (✉)
Interdisciplinary Photonics Laboratories (IPL), The School of Chemistry, The University of Sydney, 222 Madsen Building F09, Sydney, NSW 2006, Australia
e-mail: John.canning@sydney.edu.au

A. Cusano et al. (eds.), *Lab-on-Fiber Technology*,
Springer Series in Surface Sciences 56, DOI: 10.1007/978-3-319-06998-2_10,

10.1 Introduction

The concept of laboratory-in-a-fibre (or lab-in-a-fibre), analogous to the popular lab-on-a-chip terminology, was proposed in 2006 during an invited talk at the 2nd International Symposium on Advances and Trends in Fiber Optics and Applications in Chengdu, China [1]. It was, at the time, a starting point for a long term vision of advanced and complex instrumentation within an optical fibre right down to the nano-scale. Its major theme builds on taking one of the most important technology platforms in history, the optical fibre, well beyond its core duty of optical transport. Optical fibres were seen less of an optical analogue to electronics and more as an infinitely powerful and distinct medium in its own right, one that can enable integration and creation of new, distributed advanced instrumentation, sensors and other devices as well as bring on board entirely new approaches to device physics, even quantum communications and sensing. Some novel examples include optophoresis, an optical analogue of electrophoresis where evanescent fields contribute and probe not only the electrical double layer but potentially the extended charge regions thought to be associated with dipole alignment and molecular assembly of polar solvents [2–4], and optical chromatography using multichannel capillaries in structured optical fibre. These will also contain electrodes, unique fibre Bragg gratings in each channel for combined spatial, diffusive and spectral diagnostics and much, much more. Despite significant demonstrations, beginning with the enhanced measurement of channel zeta potential by scaling with multiple channels [5] through to the concept of selective mixing of structured holes to create novel material composites seen by a traveling mode such as a white light combination of RGB emitting dyes that could not work by mixing [6], and a number of important post-processing tools using various lasers and focussed ion beam milling to enable individual hole access [7–14], the 3-D tailoring of optical fibres remains a considerable practical challenge. An attractive, alternative approach has been to reduce the problem to a 2-D deposition system, or lab-on-a-fibre, using the well-established D-shaped fibre platform as the substrate for advanced lithography and other deposition methods [15, 16]. The immediacy of this approach is apparent. The D-fibre [17, 18] has withstood the test of time, primarily in optical sensing, because it has an open, flat-side accessible core whilst retaining some of the robustness of standard fibres. A possible variant that could be demonstrated is the structured optical D-fibre where channels fill the other solid half of the D. The flat D-fibre is ideal as a standalone substrate onto which devices and materials can be integrated using standard lithographic technologies. Although other interesting variants focus on opening up structured fibres by drawing down with one preform side open [19], it is the D-flat which is crucial for successful application of lithographic technologies. Overall, the volume production of optical fibres (at $\sim$4 cents a meter for standard telecom fibre) potentially makes these systems cost competitive with chip analogues for volume markets and, in biosensing, disposability has the advantage of significantly reducing requirements on longevity and repeated cycling of devices.

Nonetheless, the 3-D challenge across more complex fibres remains the long term goal for useful multiplexing of laboratories and instruments within a single fibre. It was apparent even back then that whilst advances in controlling structured optical fibres were pointing to a promising future, there were some inherent challenges in realising this across disciplines. Consider for example that for many processing applications in biology, catalysis and chemistry, especially those using diffusive methods such as optophoresis and electrophoresis and optical chromatography, pore (rather than simply channel) selectivity and filtering can be critical. In general, parallel operation both transversely and longitudinally, also allows reduced lengths for all devices and in doing so can often overcome some of the problems associated with lower signals and higher bend loss over longer lengths. This is particularly true for electrokinetic techniques that might be used in biological applications where the double layer potential between silica and water can play an important role. In one example, the direct measurement of the zeta potential of a single channel of a photonic crystal fibre was possible by scaling the sensitivity through the measurement of more than twenty parallel channels simultaneously, reducing the required length and boosting the signal-to-noise to ensure a signal can be detected [5]. The general concept of scaling to overcome sensitivity limitations is a common tool across disciplines, including for example enhanced growth and detection of viruses and microorganisms through DNA amplification [20, 21].

Analysing the lab-in-a-fibre implementation problem there are two fundamental concerns that need to be addressed. In the first instance, the top-down drawing of preforms into optical fibres, and indeed tapers, is a macroscopic process that by necessity imposes directional impost on all features. That is, even nanostructured fibres today have their nanostructure across the fibre cross-section and so are only partially three dimensional (3D). Longitudinal variation and design of multiple sensors remains challenging. Whilst through various redrawing and tapering processes, nanoscale features are readily implemented [10, 22–24] there are some inherent difficulties still to be overcome to produce these controllably as well as sufficiently small for applications such as molecular sieves or other selective filtering methods on the molecular scale. Tapering structured optical fibres can reduce hole sizes <10 nm, which is beginning to approach the natural nanocavities dimensions within conventional glass. As well as the internal glass strain at the molecular level, the fibre design still plays a crucial role in controlling the final hole size and distribution. Such small hole sizes are extremely difficult to measure and have in practice required the use of dual focussed ion beam milling or cleaving (FIB) and scanning electron microscopy (SEM) [10, 23]. For many applications this may be fine but greater control over the microporous (<2 nm) and mesoporous (between 2 and 50 nm) ranges is desirable. Further, on the nanoscale, optimal molecular sieve applications require nanoscale variation in periodically distributed pores rather than long lengths of channels despite their widths. Patterned pores can also lead to selectivity using larger holes because the gaps in-between the 3D structure twist and turn adding to density-based selectivity of molecules where clogging up the channels can take place. An approximate 2D analogue (ultimately insufficient but nonetheless useful for a variant of optical chromatography for example) is to exploit the still relatively new technology of spun-structured optical fibres [24–28].

The second major concern is that whilst silica is the preferred host material in nearly all applications (much of new materials research has tended to focus on integration within silica [29]), its robust stability and chemical properties, including amorphous structure, is in part due to its very high thermodynamic processing conditions—these unfortunately prohibit the integration of most other materials (the rare earth oxides being a well-known exception). The need for post-addition methods raise the significant challenge of finding practical and robust stable attachment of arbitrary species within the channels, something that is evidently proving to be remarkably difficult despite the numerous reports and claims in the literature. For example, self-assembling titania was deposited inside the holes of a photonic crystal fibre to enhance evanescent field detection four fold through a novel concept of extended travelling whispering gallery modes [30]. But significant questions remain on the thermal stability and lifetime of such layers along with the integration of materials to silica more broadly. The challenges with chemical attachment, it seems, may be a driver towards low cost disposable sensors where issues such as expense, durability and recyclability can be avoided. On the other hand, silica is an outstanding substrate host because, in addition to its trademark transparency and long distance robustness, it is also chemically tough and pristine. Its general separation from other materials makes it a potential candidate for locking in bioactivity and cellular storage, ideal for long lasting diagnostics of "frozen in" biomedical samples, as well as allowing for long lived nonlinear properties such as has been demonstrated with polymer filled nanoporous glasses [31].

It would seem that in a world without today's fundamental constraints the ideal structure would be something that could be assembled into any desired format and be integrated with any desired species, and to do this truly in three dimensions over short (nanoscale) dimensions; i.e. neither the physical impost of fibre fabrication nor the thermodynamic barrier involved with silica integration would exist. Is it realistic to imagine that it would be possible to move in such a direction?

To address this concern, two key approaches have recently drawn attention—the first is based on exploring whether existing sol-gel glass methods [32] could be used at room temperature. It was found that the critical annealing step used to make uniform and stable layers could be partially circumvented through higher speed centrifugal compression [33]. These promising results led to useful metal porphyrin doped films on glass for sensor applications. A second and potentially more exciting approach is the use of evaporative self-assembly of nanoparticles, a bottom-up fabrication approach. In contrast to past work using surfactant based self-assembly at liquid interfaces to achieve silica wires through nanoparticle assembly [34] or sol-gel derived long fibre lengths [35], a physical approach based solely on topological packing through intermolecular attractive forces, replacing complex chemistry with mechanical fracturing, was demonstrated [1, 2]. This approach opened up the possibility of self-assembly of mixed nanoparticles of almost any sort. What's more, the crystal analogues mean porous structures with considerable room for control of the size and distribution of pores are possible. This latter approach finally addresses the question raised above and lays the basis for a new technology—the lab-in-a-microfibre. In this chapter, we briefly introduce and review the basis for this technology to date.

10.2 Spilt Coffee and Silica Microfibres

With the same certainty spilt coffee forms coffee rings on most hydrophilic tables through microfluidic flow during evaporation [36] (see Fig. 10.1), using nanoparticles in a drop of water and combining (almost) universal dispersion forces with strain induced fracturing, pure silica waveguides (or slabs depending on the pH of the solution and the presence of certain anions and cations) can be readily fabricated [1]. On superhydrophobic tables one expects nice spheres instead [37] but this review focuses on microfibres. Figure 10.2 summarises the fabrication of microfibres in steps. At its heart, the same convective flow observed in evaporating coffee drops occurs with drops containing nanoparticles. The added feature, though, is if the nanoparticles are of very high uniformity attractive intermolecular forces will come into play—agitated by convective flow, Marangoni reverse flow and thermal motion, these drive packing of spheres to their densest state, either hcp or fcc packing if all spheres are identical. The intermolecular "glue" leads to resistance against the radial inwards recession of the evaporating drop (which typically begins to recede when the contact angle, $\alpha < 5°$ [38]). Stresses build up as the assembling "disc" dries and cracks, from which fractures propagate inwards along the direction of recession. So high are these stresses, that cracking not only occurs radially inwards but the tapered shards begin to curl and strain upwards due to rising circumferential and interfacial disk stress (not unlike those stresses that lead to warping of silicon wafers with silica top layers) until finally shattering once the drop is completely dried. It is this dominance of a largely mechanical effect that is central to the success of the self-assembly process. It is also likely that it plays a role in previous works involving surfactant film formation but has largely been unrecognised—yet, it clearly is an integral component of packing and evaporation generally.

The significance of shape in determining convective flow and therefore overall self-assembly can be seen when observing the evaporation of a non-circular coffee stain. In the case of an ellipsoidal caplet, it is observed that there is preferential and faster flow along the longer axis of the ellipsoid shape and the brown coffee edge forms here first, before the shorter separated sides and thicker—this is an indicator that pure geometric considerations can allow total control over flow and therefore in principle self-assembly. Given the physical difficulty in controlling shapes of drops during their deposition, one elegant solution to control the effective shape of a drop is to pre-pattern the surface hydrophilicity and/or hydrophobicity of a substrate [39]. This was demonstrated using UV laser processing of silicate glass

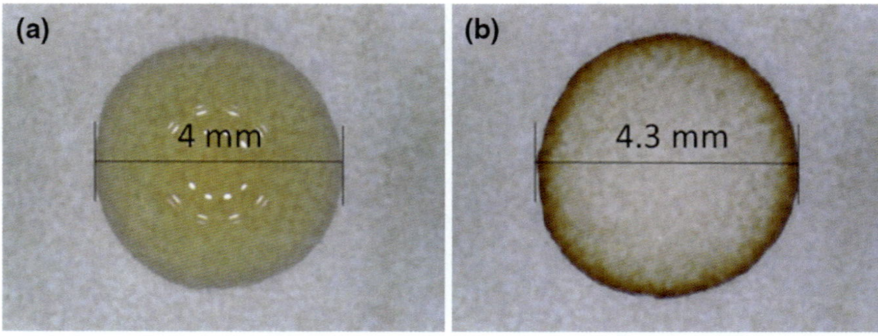

Fig. 10.1 A drop of coffee (decaffeinated) on a kitchen table **a** spreads as the water evaporates fastest where the surface area to volume ratio is highest near the boundary. The coffee particles are pushed radially outwards along the surface as flow circulates from the centre to the end, some returning via the water-air interface. The hydrophilic nature of the water kitchen table alone leads to initial spreading of the drop and the height comes down before the drop begins receding inwards. The final dry image **b** shows most of the particles accumulated at distance larger than the initial drop size with a thin, almost clear region inside

slides prior to a drop being put on. The advantages of lasers over other techniques are that they can allow sub-micron patterning of the glass surface. This processing leads to changes in the solvent (in this case water) contact angle—changes ranging more than 40° was possible by this approach, far more than is required to affect convective flow. In the simplest demonstration, the contact angle of a drop of water could be changed from one side to the other only by a few degrees ($\alpha \sim 6°$) to affect ellipsoidal-like flow within a small spherical drop. This was sufficient to reduce the taper aspect ratio to TAR = 1. Figure 10.3 summarises these results.

The discontinuity of larger drops and secondary fracture bifurcation, and indeed multiple cascading bifurcation similar to a one-direction fractal tree, has been the most effective way of producing large quantities of uniform slab waveguide microfibres. Clearly, the drops are essentially self-contained microfluidic laboratories in their own right—convective flow is generated by evaporation which can be controlled by a number of parameters, such as temperature and atmosphere pressure. The addition of salt and other materials can also affect flow. More interestingly, the shape of the drop also affects the internal microfluidic flow—there is strong geometric dependence which can be exploited to offer unparalleled control. For larger drops, adding a slight tilt to the substrate allows gravity to determine preferential flow and therefore the direction and length of alignment. In this way, microfibres longer than 11 cm have been generated, in large quantities, from silica nanoparticles [40] with diameters $\phi \sim$ 20–30 nm (measured by dynamic light scattering, DLS). It would seem there is no obvious limit to their potential length and a dynamic method which feeds solution into the process whilst slowly drawing the forming wire could in principle lead to infinitely long wires.

An example where drop shape was taken to extreme was the deposition of a drop of silica nanoparticles in water onto a D-shaped optical fibre [41]. Capillary action saw this drop extend over a significant portion of the fibre (centimetres) so the shape

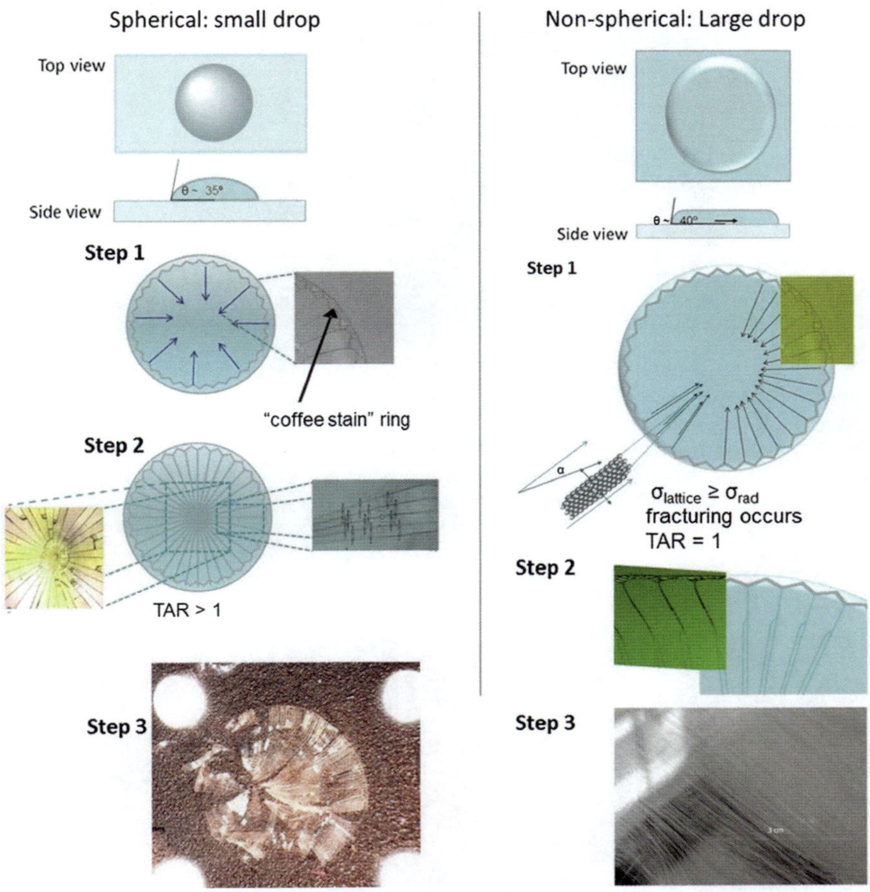

Fig. 10.2 Self-assembly of tapers and wires from silica nanoparticles ($\phi \sim$ 20–30 nm). In the small drop (spherical caplet limit—LEFT), evaporation from the ends generates convective flow within the drop and nanoparticles migrate to the edges generating the classical coffee stain ring (*Step 1*). The particles pack towards the densest configuration for spheres, a hcp/fcc lattice when all spheres are identical. As dispersive attractive forces build, stresses build up pointing radially inwards for a circular system as the drop begins receding when the contact angle is $<5°$. Experimentally, the stresses lead to a growth in potential energy and the process is observed to generate tapered shards (*Step 2*—with taper aspect ratio defined as the width of the front to the end, TAR $>$ 1) which physically curl and strain upwards with the drying process before shattering. Some of the shards are displaced away from the drop, indicating how quickly the stresses build and how violently they can be released—*bottom* left image (*Step 3*). When the drop is large (*Right*), the spherical caplet limit no longer holds and there is a discontinuity in the Marangoni induced return flow. At this point, secondary and further cascading fractures (depending on drop size and evaporation rates) are triggered (*Step 2*). These secondary fractures are parallel to the main radial fractures towards the centre, producing extremely uniform slab waveguide structures with TAR = 1. Assisted by gravity using a small tilt on the substrate, microfibres longer than 11 cm have been produced (*Step 3*) with cross sections of $\sim(10 \times 40)$ μm depending on the concentration of silica nanoparticles and evaporation rates

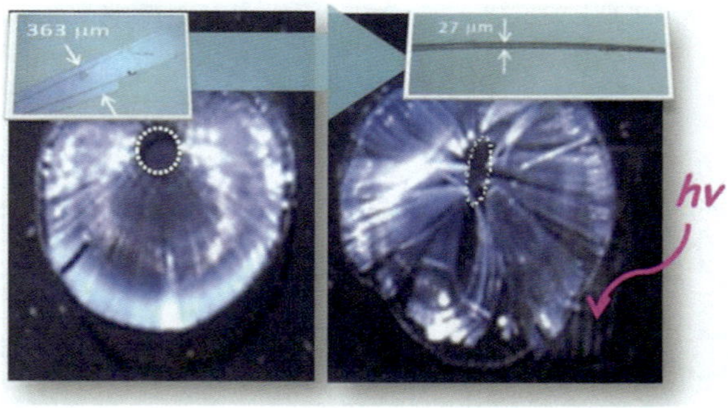

Fig. 10.3 Self-assembly of two identical drops containing silica nanoparticles ($\phi \sim$ 20–30 nm) onto borosilicate glass slides. On the *left* the drop breaks into tapered shards pointing to the *centre*. The measured contact angle ($\alpha = 24 \pm 5°$) was uniform all around the drop. On the *right*, the slide was treated in one area with 193 nm laser irradiation. The drop in this case was deposited half on the pristine sample and half on the treated area. The measured contact angle varied from $\alpha = 22 \pm 2°$ on one side to $\alpha = 17 \pm 2°$ on the other. Ellipsoidal like drop behaviour was observed. This was sufficient to lead to uniform wires along the drop. (Modified from [39])

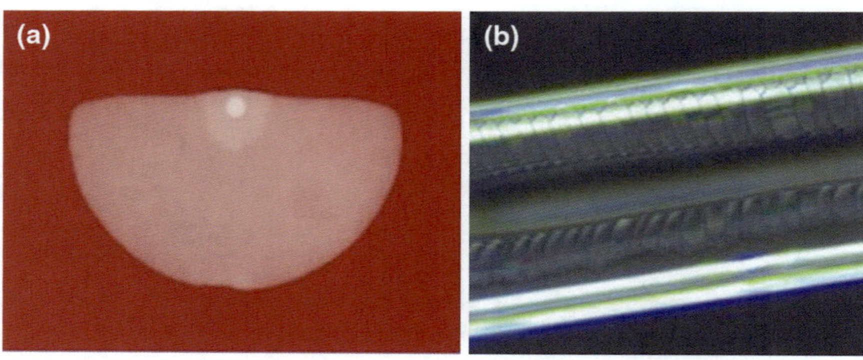

Fig. 10.4 A D-shaped fibre fabricated at the joint National Fibre Facility at the University of NSW, Sydney, Australia. When a drop of silica nanoparticle containing water ([SiO_2] $\sim$5 wt%, $\phi \sim$ 20–30 nm) is placed on the fibre, capillary action spreads it over a long length ($\sim$4 cm). Upon evaporation, cracks are observed always from the outside in generating approximately equally spaced slabs along this length. The small protruding inner cladding region stops crack propagation from the outside reaching over the core section, instead leading to a small tilt and what looks like slab scales [40]. The evanescent field propagating in the core does not see the cracked plates. Figures are modified from [40]

was essentially an extended rectangular slither. As the drop receded, evaporation was fastest from the side in because the distance was so small ($\sim$62.5 μm). Fracturing therefore occurred from the outside in generating a series of parallel silica plates $\sim$(30 × 60) μm along the entire coated section (see Fig. 10.4).

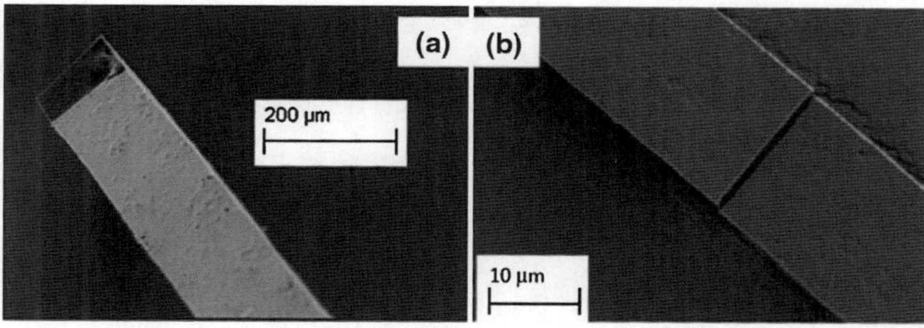

Fig. 10.5 SEM images of hand cleaved optical microfibres using an aluminosilicate tile. The cleave would appear to follow the expected fracture plane along the lattice packing; **a** A thicker wire allowed easier and more rapid handling and cleaving (nearly 100 × 100 μm square cross-section); **b** Thinner slab wires can also be cleaved but with greater difficulty. Image on the *right* is modified from [1]

10.3 The Nanostructure of Silica Microfibres

The role of intermolecular forces in driving packing towards the densest state for the given energies in the environment, and which build the stresses that give rise to uniform fracturing, suggest extremely uniform internal structure within these wires. Further circumstantial evidence can be found in the relative ease with which these microfibres can be handled (despite having a breaking point significantly less than a standard optical fibre in part due to their much thinner dimensions and higher exposure to moisture) and even manually cleaved using an aluminosilicate tile (Fig. 10.5).

To investigate these structures in more detail, high resolution microscopy methods are required [40]. Typically, scanning electron microscopy (SEM) and atomic force microscopy (AFM) are sufficient to provide high resolution characterisation of the lattice structure expected at the surface of these microfibres. Unfortunately, these and other similar methods are restricted to a local scale given the prohibitive times involved with fully mapping long wires (this was the basis for the development of Fresnel—fractal fibres to demonstrate the feasibility of real time scanning near-field optical microscopy [42]). Example images are shown in Fig. 10.6. Both these techniques are critical yet time consuming and often difficult to access given their expense and need across numerous groups highlighting a significant impediment to nanoscale research generally: the absence of sufficient and low cost, diagnostic tools for essential routine analysis. This is an area that is in need of substantial research innovation. Consequently, many of the results reviewed here were obtained across multiple institutions.

The high resolution analyses of these wires is strongly suggestive of densest packing and consistent with packing theory of identical spheres into hexagonal close packing (hcp) and face-centre cubic (fcc) lattices, both of which have the lowest free energy configurations for uniform spheres of one size. The experimental data

Fig. 10.6 A 40 × 10 μm silica microfibre **a** is examined with SEM and AFM: **b** shows a *lower* resolution close up of the *top* of the wire revealing in this case what appears to be a secondary layer. A close-up examination of the top layer with either **c** high resolution SEM (2–5 nm) or **d** high resolution AFM (2 nm) reveals well-ordered tight packing consistent with either hcp or fcc packing (Modified from [1] and [40])

suggests an overall preference towards hcp, perhaps related to the restrictive 2D ordering imposed by the surface, but this is not fully confirmed in part because the high resolution methods comes at the expense of a large field of view. This places greater emphasis on the need for more wide-field instrumental research on the nanoscale as well as different approaches—one extremely promising approach for routine analysis of larger particles has been the impressive development of holographic microscopy on a mobile phone platform [43, 44], currently reaching down to ~70 nm resolution.

Another drawback of these methods is that they are surface profile techniques and so say little about the actual volume structure although some aspects can be inferred when there is a very high level of surface uniformity. In the AFM picture of Fig. 10.7 the height variation is consistent with particle size variation within a single layer limited by the AFM tip resolution. Some 3-D information can be obtained using transmission electron microscopes (TEM) and x-ray scattering analysis for very thin samples but this too can be time consuming for marginal, if any, improvement over SEM and AFM. One alternative way to gauge the quality

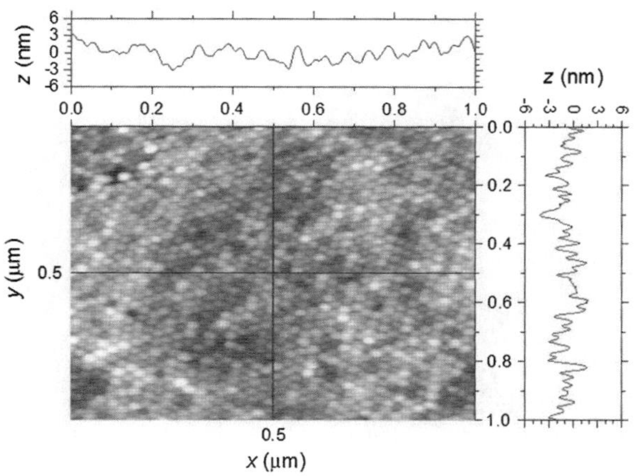

Fig. 10.7 AFM study of the surface packing of a silica microfibre self-assembled from silica nanoparticles ([SiO2] ~2.5 wt%, ϕ ~ 20–30 nm). The standard deviation of the cross-section surface profile in *x* and *y* is <25 nm over 1 μm showing good uniformity. Modified from [40]

of the internal structure of these wires is to compare an analysis of the internal pore size and distribution calculated from the unit cell of an ideal structure with those obtained by established, despite being indirect, gas adsorption approaches. If the structure is assumed to be either a hcp or fcc lattice of classical spheres, then it can be shown from the unit cell that the pore volume expected is ~26 % of the total volume [45]. This is an intriguing figure because it is close to that of a commercially porous glass manufactured by conventional glass methods—Vycor glass has an interstitial volume of 28 % and is the standard glass material used in glass-based chromatography [46]. The pore size and distribution in Vycor glass is typically ~6 nm but randomly distributed. In the self-assembled silica wires the pore sizes must be uniformly distributed. In fact, for the hcp and fcc lattice, there ought to be two pore sizes present corresponding with tetrahedral and octahedral pore sites (Fig. 10.8). From dynamic light scattering measurements (DLS), the size (ϕ) of most of our nanoparticles lies between 20 and 30 nm. Therefore, for an octahedral interstice $r_{oct} = 0.41R \sim (4.1–6.2)$ nm and for a tetrahedral interstice this is $r_{tet} = 0.22R \sim (2.2–3.3)$ nm—these sizes are much smaller than the smallest capillary channels produced using top down drawing of optical fibres [10, 23].

Unlike the direct surface characterisation methods described earlier, gas adsorption provides statistical volume and surface area information distributed over a macroscopic scale by relying on the formation of single gas molecule (picoscopic) layers on the inner surfaces of the pores of the structures. For example, N_2 adsorption studies are routinely used to quantify internal surface area, total interstitial volume and pore size and area distributions of high surface area molecular structures including mesoporous silica—such data has been cross-correlated with other less convenient methods such as H NMR studies [47]. Liquid nitrogen within the pores is slowly released by reducing pressure (keeping temperature constant)—that at the molecular level experiences van der Waals forces which give rise to a change in the desorption rate. From the experimental isotherms, the pore size and distribution is calculated most commonly by the method of Barret, Joyner and

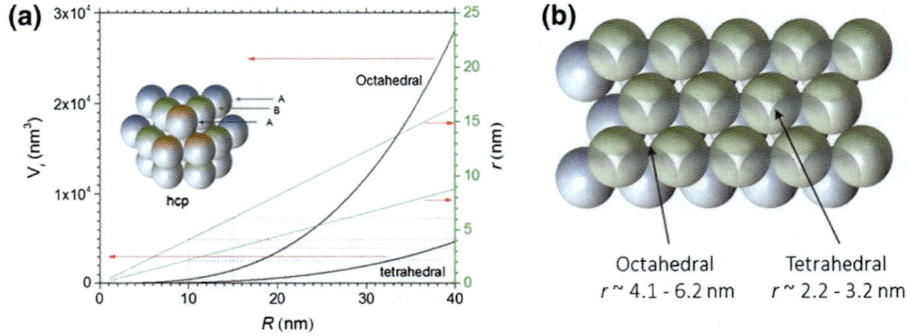

Fig. 10.8 Equally sized spheres form simple hcp and fcc lattices: **a** Simple unit cell calculation of both pore radius, *r*, and volume, *V*, against the radius of the spheres, *R*, used, and **b** a schematic illustrating the two types of pores formed and their calculated distribution for the 20–30 nm sized spheres used in this work. The size of a structure that can be supported by each interstitial region is defined by the radius of a sphere, *r*, that can fit into the gap over the radius of the nanoparticle, R (= $\phi/2$). Modified from [40]

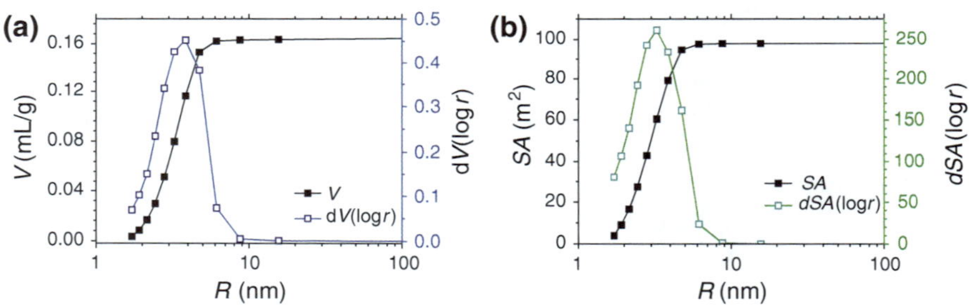

Fig. 10.9 Interstitial volume, V_i, and pore size, *r*, for hcp lattice N_2 adsorption measurements: **a** the measured volume (*V*) and pore size distribution (*dV*log*r*) as a function of particle radius, *R*, **b** surface area (SA) and distributed SA (*dSA*log*r*) within each pore as a function of *R*. Modified from [40]

Halenda (BJH) [48] which works well for mesoporous structures. As the pressure is dropped, the calculation of surface area is made using Brunauer, Emmett and Teller (BET) [49] analysis. Typical surface areas, volumes and distribution of each versus pore radii are shown in Fig. 10.9. From these values, the total interstitial pore volume is $V_i \sim 1.7$ mL/g. Similarly, the total interstitial surface area is $SA \sim 101\ \text{m}^2/\text{g}$. The distribution in pore radius is measured to be $r = (2–6)$ nm, shown in Fig. 10.4b. This size distribution is in excellent agreement with the calculated interstices, $r \sim (2.2–6.2)$ nm, assuming most of our nanoparticles lie within 20 and 30 nm and is a strong indication that the surface lattice structure is as every bit as well ordered within the bulk of the volume of the wires.

Occasionally, there are some wires in which there exists what appears to be a higher free energy packing bcc-like configuration—in this case these values become $r_{oct} = 0.155R \sim (1.6\text{–}2.6)$ nm and $r_{tet} = 0.291R \sim (2.9\text{–}4.4)$ nm respectively. Figure 10.8 shows the trend in individual interstice volume calculated from the unit cell properties of hcp (or fcc). Clearly, there is considerable scope to tune the volume size using different sized nanoparticles or adjusting the crystal topology. Along with nanoparticle type and size, the addition of temperature to the fabrication process should pave the way for accessing and freezing in a range of crystal states that have stable potential wells at different energies, some which may be more suitable for various applications than others. More interestingly, this self-assembly approach may allow the generation of completely new crystal configurations that have no analogy in nature, an exciting area for basic materials research.

Given the aspect ratios involved and the length of the wires, microfluidic flow enabled self-assembly driven by packing towards a densest state proves to be a powerful tool and it has ramifications that extend to all microfluidic systems down to the molecular level—for example, how such flow drives self-assembly is likely to be a crucial aspect in the evolution of biological systems and their functionality, particularly where structured water [2–4] facilitates lubricant-like transport. From a materials perspective, the overall assembled structure is much more crystalline than familiar bulk silica glass despite the fact that the nanoparticles themselves may be less so. This in of itself is a very interesting question—what is the internal structure of these nanoparticles like given the very constricted volumes in which they are contained? In normal glass for example interstitial random vacancy sites can have ring structures approaching several nanometers in size. This might be expected to perhaps affect surface quality and ordering of 20 nm sized nanoparticles—the volume constraints almost certainly play an important role in determining the nature of silica fabrication and the ability to produce these particles. Are there, for example, any analogous cristobalite ordering as observed by x-ray distribution profiles of normal glass which suggest highly ordered cristobalite-like local structure (<2 nm) [50, 51]. These may be made up of tessellated non-additive covalent bonding of tetrahedral units, disappearing over intermediate length scales. For the wires themselves, in the first instance the nanoparticle can be considered closer to a classical hard sphere atom (still used to describe crystal structures). But more significantly, the dispersion forces holding the nanoparticles together are additive in nature and therefore have a long range impact compared to covalent bonds, one that might lead to longer range order and other interesting size-dependent effects over simple tessellation. All-in-all the self-assembled wires pose an interesting, and as yet unclassified, material bringing together covalent glass nanoparticles into a crystal lattice bound by dispersion forces—a new type of glass in its own right?

The ease of fabricating crystal lattices with such tiny pore sizes, almost commensurate in scale with existing nanocavities in conventional glass, is a striking difference between top down and bottom-up fabrication processes. Dispersion forces have allowed glass fabrication in a low temperature regime with all the significant implications that accompany that. It raises fundamentally intriguing questions about the nature of the glass structure within the nanoparticles themselves and whether internal nanoparticle strain contributes to other phenomena, possibly eventually covalent bonding between particles. As these approaches are integrated increasingly into practical systems, research into these differences will need to grow both in fundamental and applied domains (they are in of themselves incredibly important fundamental subjects).

The remarkable agreement between adsorption and calculation suggest the estimated pore volume is $\sim 26\ \%$ (with some variation from this ideal single sized sphere lattice), opening up the way for controlled diffusion-based applications. It should be noted, as referenced earlier, these observations and inferences are applicable in two and three dimensions such as the self-assembled nanoparticle films on D-fibres. The control of the diffusion is one essential part to all micro and nano laboratories, whether they are on a chip or on a fibre or within a microwire. Good optical transparency, for example, might lead to novel optical chromatography both in microfibre and optical fibre form. Further, the two different pore sizes of the hcp lattice structure potentially allows very selective filtering to be undertaken (in addition to that at the surface) so that it can be analysed directly, using optical means [1, 2] as well as other existing techniques. Given the uniformity of the process, it is possible to envisage a stage where individual nanoreactions or processes can be undertaken in different pores in a controlled manner, potentially feeding into each other – characteristics of future lab-in-a-wire, or lab-in-a-microfibre, technologies. Multiple-site simultaneous repetition of processes that are close to identical offers the unique potential of scaling up nanoreaction products, improving the signal-to-noise of various detection means. This may also reduce significantly both sensor detection and diagnostic tolerances required for processes such as single molecule detection. Finally the nanoscale integration of materials into silica, once considered almost impossible, opens an alternative approach to the fabrication of metamaterials and other novel composites.

10.4 Doped Microwires

Clearly, one of the most significant motivations for the development of cold-fabrication of silica glasses is the drive to circumvent what can be described as the great thermodynamic challenge: overcoming the intrinsic gap in temperature required to process glass (>1,900 C) versus almost any other material. Silica is perhaps the leading platform technology for communications and increasingly sensing, especially as technology moves towards integration through Smartgrids and Smart (or intelligent) networks fuelled by optical fibre. The thermodynamic barrier between silica and most other materials is in part what has made it successful: both low loss and robust operation require this. And yet, this success may be what limits the next stage of pervasiveness—for many applications the functionality required demands other materials and in sensing and biodiagnostics these are often carbon or organic based. Cold convective self-assembly using dispersion forces which are almost ubiquitous and attractive for all materials, overcomes the thermodynamic barrier to material integration—it has only now just become technically possible to integrate directly almost anything into silica, including contrasting carbon and organic species. This does not only mean simple post-diffusion, or percolation, into the microwires but direct integration during fabrication.

Figure 10.10 shows examples of various organic species introduced into the self-assembled microwires during their fabrication that were not previously possible [52]. These species can enter the system by either chemical or physical adsorption; even if chemical attachment is generally low on silica, physical adsorption is sufficient enough to see strong integration into the self-assembled structure. In contrast to simple percolation, these species can be trapped within interstitial regions of the self-assembled structure if not become part of that structure through their own dispersive forces – these are true hybrid/composite material systems of the kind that have not been previously reported. Given that the self-assembled wires shown in Fig. 10.10 do not show appreciable degradation in mechanical integrity for relatively high concentrations of dopants, it is likely that the molecules are adsorbed onto the surface of the silica nanoparticles and themselves contribute to attractive dispersive forces. Some support for this was found using photon correlation spectroscopy, or dynamic light scattering (DLS), of the solutions containing both nanoparticles and dopants [53]. The role of dipole-dipole attractions and surrounding structured water need detailed consideration as it almost certainly plays a critical role. Generally, in basic solution, an increase in particle size up to a few nanometres was observed for both Rhodamine B and hydrated Zn-acetate containing solutions, though the broadening of the Zn acetate distribution suggests much less homogenous adsorption between particles. The slightly narrower distribution of Rhodamine B suggests even adsorption and smoothing of surface roughness—results are shown in Fig. 10.11.

This physical difference between percolation and integration has significant implications because it results in different trapping within the interstitial regions. In addition to controlled accessibility by controlling pore size, many of the adsorbed species, in contrast to simple percolation, will be more difficult to remove, or access, by subsequent solvent and chemical attack. This is schematically illustrated in Fig. 10.12a. Thus a prediction is that the functionality of an active species can be isolated and retained as a consequence of inaccessibility (not withstanding further electrostatic, conformational and chemical interactions with the silica surface). Preliminary results in trapping species such as Rhodamine 6G (Fig. 10.13), and other species, support this argument. Figure 10.13 shows the fluorescence of the laser dye measured along the wires and quantified using a fluorescence microscope and camera. Integration allows higher concentrations to be introduced; these in turn are much more difficult to extract using ethanol, which tends to remove the in-diffused dye readily [56].

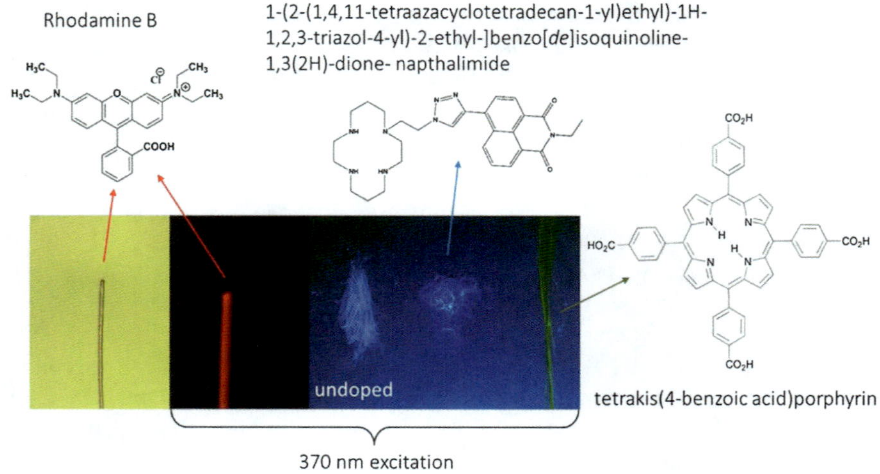

Fig. 10.10 Examples of dopants introduced during self-assembly. These organic dyes would be obliterated under normal silica processing conditions. Here, under UV irradiation they fluoresce across the RGB spectrum. Modified from [52]

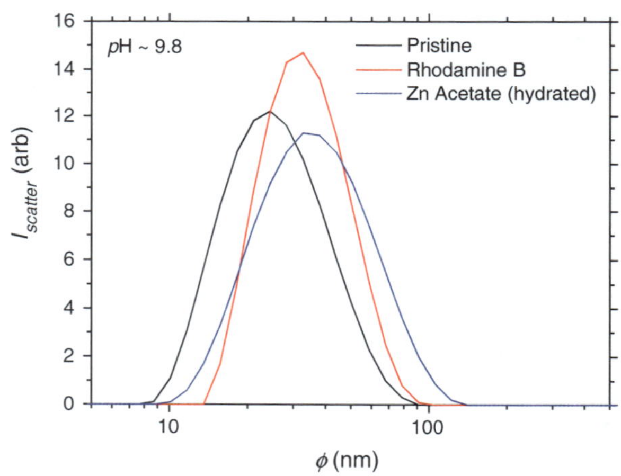

Fig. 10.11 Dynamic light scattering (DLS) size distribution profiles showing a shift to larger nanoparticle size upon mixing with either Rhodamine B or hydrated Zn-acetate. The broadening of the hydrated Zn acetate distribution indicates an uneven size increase compared to the much more uniform layering of particles with Rhodamine B. In both cases the peak shift is about 8 nm, suggesting several layers are formed [53]. The molecular dimensions are similar

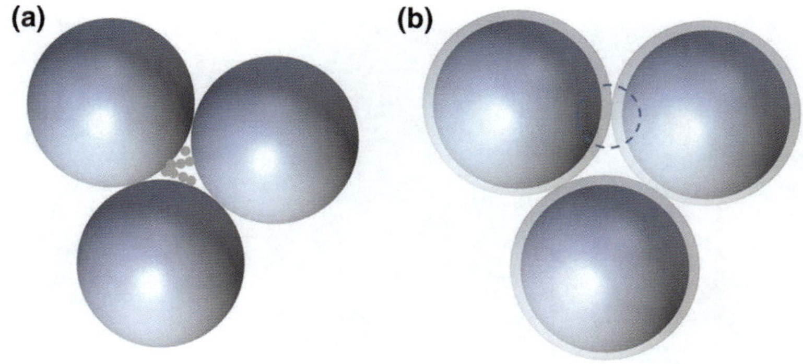

Fig. 10.12 Simple schematic highlighting differences between adsorbed species through percolation (**a**) and through integration (**b**), following a simplistic interpretation of DLS measured particle size increases shown in Fig. 10.11. The possible role of adsorbed species in electrostatic binding between particles can affect both the quantity and accessibility of attached species; for the latter case the prediction of stronger trapping of some of the integrated species is reasonable and from this the potential of the silica lattice to act as a preservative buffer, isolating and insulating molecular species not unlike that previously reported for silica sol gels [54, 55]

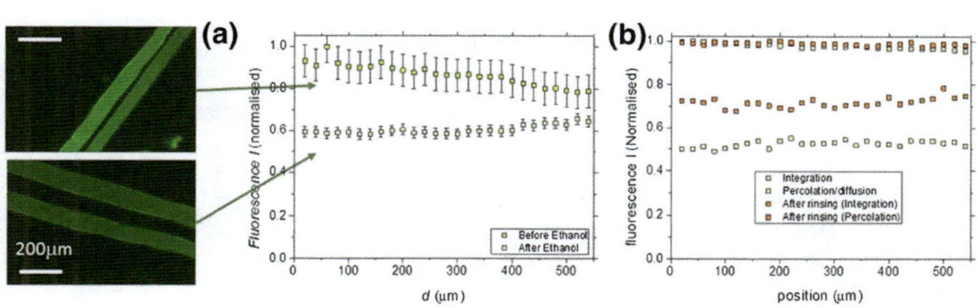

Fig. 10.13 Self-assembled pairs of microfibres containing Rhodamine 6G laser dye, introduce by percolation (or in-diffusion) and by integration prior to self-assembly. The fluorescence intensity along the microfibres is less for the in-diffused case. Washing the dye out with ethanol **b** shows that integration generally retains a higher percentage of material, consistent with some of the species tightly trapped within the structure. Modified from [56]

10.5 Complex and Mixed Nanoparticle Self-Assembly

An alternative to simple doping (where the added molecule is generally adsorbed onto the silica nanoparticle surfaces and then are integrated throughout the self-assembly process), is the mixing of nanoparticles not only of different size but of different materials. Intermolecular forces are ubiquitous and generally attractive between most materials although the strongest forces are between like materials. Whilst the intermolecular attraction is strongest between similar sized, similar materials, it is possible to generate different lattice structures by mixing particle

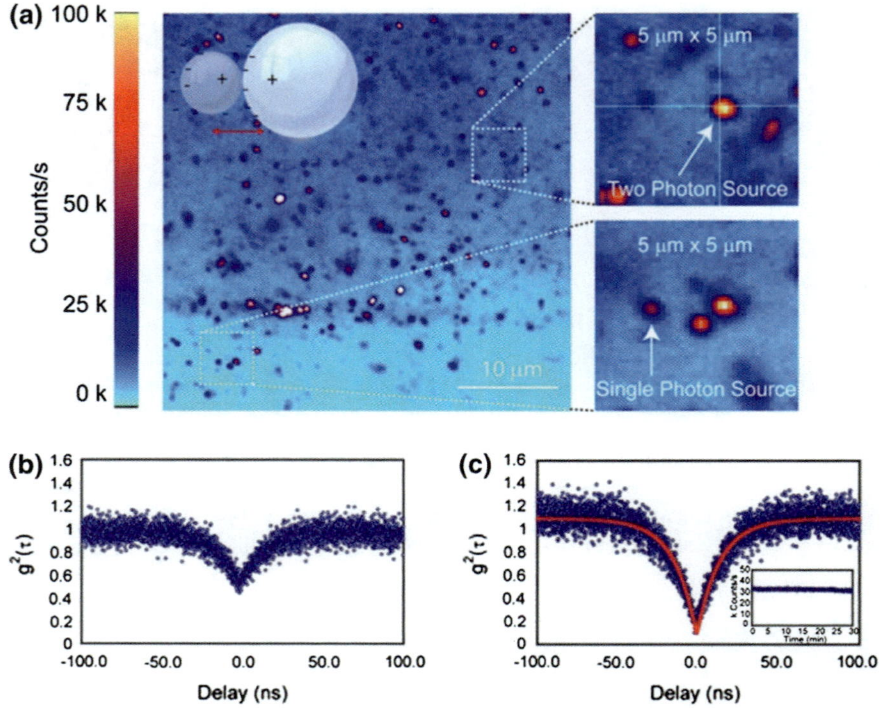

Fig. 10.14 Confocal image of fluorescent signal from nanodiamonds embedded in silica nanoparticles is shown in (**a**). Approximate size differences are shown (nanodiamond:nanosilica ~50:24 nm). Depending on the centre examined, the correlation function, $g^2(\tau)$ of 0.5 and near 0 corresponding to two and one photon emitting sites respectively are observed, (**b**) and (**c**). Modified from [1, 65, 66]

sizes and generate new composites by using different material nanoparticles. In this way the physical and chemical properties of these new hybrid materials can be artificially tailored for almost any potential use (even when the basic material is identical in all particles). An interesting proposition is that metamaterials [57–61], both hybrid metal and dielectric systems and in some cases all dielectric, could be manufactured this way through nanoparticle combinations, establishing the necessary periodic structure through generating a desired crystal lattice configuration.

The advantage of this approach is that all manners of novel composites are possible as are the integration and controlled distribution of specifically functionalised nanoparticles. A particularly innovative example is the nanodiamond which contains a single nitrogen vacancy (NV) defect site that emit single photons (in the red) when excited by green light [61]. Recent innovations promise to extend nanodiamond defect emission into the near IR [62]. The red emitters are (at the time of writing) the most robust single photon emitters known, making them highly attractive and practical sources for future quantum communication and sensor sources—if they could be integrated into key material platforms such as silicon or silica. Like most materials, the NV centres anneal out at temperatures lower than typical silica processing conditions ($\sim$700 °C) compared to the processing temperature required for silica. Thus demonstrations comparable in significance to much of the quantum work done using cold ion traps [63] have been in low melting point systems such as tellurite glass [64]. Unfortunately, these demonstrations remain largely incompatible with the need for quantum communication technologies amenable to integration with existing telecommunications and computer technologies based on key platform materials such as silicon or silica. Overcoming this is arguably the single most important area in quantum research. Thus there is a critical impetus for demonstrating the potential of realistic material systems—integrating these single emitting nanodiamonds into silica is one milestone towards that. This was achieved using self-assembly of a mixture of silica and nanodiamond nanoparticles [1, 65, 66]. Figure 10.14 shows single photon emission from such a hybrid system. Also observed are two photon emitters, the interference of which is often preferred in quantum communication demonstrators [67–71].

10.6 Conclusions

In summary, the ability to "bottom-up" self-assemble waveguide-like silica structures at room temperature using nanoparticles springs open a hitherto impossible domain of novel material structures, ones that are actually quite important in expanding the functionality of silica, as indeed any material in general. By extension, it has been shown that 2-D structures can be coated on the lab-on-a-fibre choice platform—the D-fibre. These self-assembled structures are by design porous and the pore size, surface area and volume determined by the size and range of particles which in turn determine the lattice packing. For uniform spheres, there are two pore sites—tetrahedral and octahedral—uniformly distributed throughout the structure. The ability to fabricate and control such features, which are already smaller than the smallest features produce by top-down drawing methods and are commensurate with inherent nanocavities sizes in normal bulk glass, provides interesting fundamental research at near molecular level as well as introducing new engineering tools for the next generation of components and functionalised waveguides. Molecular sieves and other filtration designs are possible, making novel optical chromatography and optophoresis real. Percolation and post integration is feasible and has been demonstrated with various dyes. From a material perspective, the ability to control silica dielectric properties is of both fundamental interest, exploring novel new silica structures for example, and practical interest: for example, low k values are of central importance to solving the need for both thinner and thinner insulator layers and reduced dielectric constant in the next generation of smaller and smaller electronic and optical chips [72, 73]. The possibility of preserving and isolating molecular functionality is at the heart of the lab-in-a-microfibre concept in which potentially hundreds, if not more, of selectively quarantined nanoscale laboratories both for diagnostics and/or processing are tailored and distributed along and across the microfibre (or film platform for that matter) allowing controlled and deterministic accessibility. Perhaps more interesting is the doping of the wires during fabrication where high concentrations are possible as result of adsorption on the silica nanoparticle surface. This leads to less out-diffusion from the final wires compared to simple diffusion or percolation.

By mixing nanoparticles of different composition, a new approach to complex, and novel, composites is possible—one clear example was the creation of a nanodiamond and silica hybrid material allowing for the first time single photon emitters to be demonstrated within a robust silica framework, a milestone that signifies it is possible to integrate such emitters into silica and indeed any key material platform using this new approach to material engineering based on complex, mixed nanoparticle self-assembly.

Bringing all the technologies together in waveguide, or microfibre, form and recognising that these individual steps can be brought together multiple times to add extensive functionality within the one structure, opens up the possibility of true bottom up engineering of functional devices and brings substance to the concept of lab-in-a-microfibre.

Acknowledgments As with any entry into a broad new field, it is only possible with the contributions of many people. The following people, by institution and in alphabetical order, are acknowledged for various contributions that were carried out to make the lab-in-a-microfibre potentially realisable:

The University of Sydney—Fahad Khaksar Ali, Dr. Mattias Aslund, Jean-Gabriell Brisset, Dr. Kevin Cook, Prof. Maxwell J. Crossley, Stephanie Eid, Oscar Fawkes, Dr. George Huyang, Hari Athitha Jeyaseelan, Lachlan Lindoy, Nicolas Losio, Lucas Moura, Sarah Murphy, Miles Ma, Masood Naqshbandi, Melissa Nash, Dr. Liyang Shao, Dr. Jeff Shi, Dr. Patrick Trimby, Elijah Tyedmers, Hadrien Weil; *University of NSW* – Dr. Yanhua Liu, Prof. Gang-Ding Peng; *The University of Technology*—Dr. Alison Beavis, Dr. David Bishop, Prof. Andrew McDonagh, *The Royal Melbourne Institute of Technology (RMIT)* – Dr. Brant Gibson; *Wollongong University* – Dr. Donqi Shi; *Universite de Paris Sud, France* – Dr. Matthieu Lancry (Marie Curie Exchange, EC FP7 Program).

Various funding sources are aslo acknowledged, particularly from the Australian Research Council (ARC). Both Marie Curie Exchange and EC FP7 Program funding helped to support exchange visits by Hadrien Weil, Jean-Gabriel Brisset and Elodie Medeiros from the Universite de Paris Sud, France. Lucas Moura was supported by the "Conselho Nacional de Desevolvimento Científico e Tecnológico" (CNPq) Brazil. M. Naqshbandi received a Gritton Student Scholarship and Liyang Shao an Australian Endeavour Research Fellowship. Hari Athitha Jeyaseelan, Melissa Nash, Lachlan Lindoy and Miles Ma were supported by IPL Summer Scholarships. Oscar Fawkes was supported with some project funding from the Foundation of Inorganic Chemistry, at The University of Sydney.

References

1. J. Canning, New trends in structured optical fibres for telecommunications and sensing (Invited). in *5th International Conference on Optical Communications and Networks and the 2nd International Symposium on Advances and Trends in Fiber Optics and Applications (ICOCN/ATFO 2006)* (Chengdu, China, 2006)
2. P. Fenter, L. Cheng, S. Rihs, M. Machesky, M.J. Bedzyk, N.C. Sturchio, Electrical double-layer structure at the rutile-water interface as observed in situ with small-period X-ray standing waves. J. Colloid Interface Sci. **225**, 154 (2000)
3. *Water in Biomaterials Surface Science*, ed. by M. Morra (Wiley, New York, 2001)
4. I.-F.W. Kuo, C.J. Mundy, An ab Initio molecular dynamics study of the aqueous liquid-vapor interface. Science **303**(5658), 658–660 (2004)
5. J. Canning, E. Buckley, S. Huntington, K. Lyytikäinen, Using multi-microchannel capillaries for determination of the zeta potential of a microfluidic channel. Electrochim. Acta **49**, 3581–3586 (2004)
6. J. Canning, M. Stevenson, T.K. Yip, S.K. Lim, C. Martelli, White light sources based on multiple precision selective micro-filling of structured optical waveguides. Opt. Express **16**(20), 15700–15708 (2008)
7. S. Pilevar, K. Edinger, W. Atia, I. Smolyaninov, C. Davis, Focused ion-beam fabrication of fiber probes with well-defined apertures for use in near-field scanning optical microscopy. Appl. Phys. Lett. **72**(24), 3133–3135 (1998)
8. J. Canning, M. Åslund, Applications of CO_2 laser processing of optical waveguides. in *Australian Conference on Optical Fibre Technology (ACOFT'99)* (The University of Sydney, NSW, Australia 1999)
9. H. Lehmann, S. Bruckner, J. Kobelke, G. Schwotzer, K. Schuster, R. Willsch, Toward photonic crystal fiber based distributed chemosensors. Proc. SPIE **5855**, 419 (2005)
10. B.C. Gibson, S.T. Huntington, S. Rubanov, P. Olivero, K. Digweed, J. Canning, J. Love, Exposure and characterization of nanostructured hole arrays in tapered photonic crystal fibers using a combined FIB/SEM technique. Opt. Express **13**(22), 9023–9028 (2005)
11. Y. Lai, K. Zhou, L. Zhang, I. Bennion, Microchannels in conventional single-mode fibers. Opt. Lett. **31**, 2559–2561 (2006)
12. C. Martelli, P. Olivero, J. Canning, N. Groothoff, B. Gibson, S. Huntington, Micromachining structured optical fibres using focussed ion beam (FIB) milling. Opt. Lett. **32**(12), 1575–1577 (2007)
13. C.M.B. Cordeiro, C.J.S. de Matos, E.M. dos Santos, A. Bozolan, J.S.K. Ong, T. Facincani, G. Chesini, A.R. Vaz, C.H.B. Cruz, "Towards practical liquid and gas sensing with photonic crystal fibres: side access to the fibre microstructure and single-mode liquid-core fibre", Meas. Sci. Technol. **18**, 3075 (2007)
14. Y-J. Rao, M. Deng, D-W. Duan, X.-C. Yang, T. Zhu, G.-H. Cheng, Micro Fabry-Perot interferometers in silica fibers machined by femtosecond laser. Opt. Express **15**, 14123–14128 (2007)
15. M. Consales, A. Ricciardi, A. Crescitelli, E. Esposito, A. Cutolo, A. Cusano, Lab-on-Fiber Technology: Toward Multifunctional Optical Nanoprobes. Nano **6**(4), 3163–3170 (2012)
16. M. Consales, M. Pisco, A. Cusano, Lab-on-fiber technology: a new avenue for optical nanosensors. Phot. Sens. **2**(4), 289–314 (2012)
17. R.B. Dyott, J. Bellow, Polarisation-holding directional coupler made from elliptically cored fibre having a D section. Electron. Lett. **19**(16), 601 (1983)
18. R.B. Dyott, J. Bello, V.A. Handerek, Indium-coated D-shaped-fiber polarizer. Opt. Lett. **12**, 287–289 (1987)
19. S.C. Warren-Smith, H. Ebendorff-Heidepriem, T.C. Foo, R. Moore, C. Davis, T.M. Monro, Exposed-core microstructured optical fibers for real-time fluorescence sensing. Opt. Express **17**, 18533–18542 (2009)

20. J.W. Krebs, P. Feorino, D. Warfield, G. Schochetman, DNA amplification for direct detection of HIV-1 in DNA of peripheral blood mononuclear cells. Science **239**, 292–295 (1988)
21. R.J. Steffan, R.M. Atlas, DNA amplification to enhance detection of genetically engineered bacteria in environmental samples. Appl. Environ. Microbiol. **54**(9), 2185–2191 (1988)
22. Y. Ruan, H. Ebendorff-Heidepriem, S. Afshar, T.M. Monro, Light confinement within nanoholes in nanostructured optical fibers. Opt. Express **18**, 26018–26026 (2010)
23. C.M. Rollinson, S.T. Huntington, B.C. Gibson, S. Rubanov, J. Canning, Characterization of nanoscale features in tapered fractal and photonic crystal fibers. Opt. Express **19**(3), 1860–1865 (2011)
24. M. Fuochi, J.R. Hayes, K. Furusawa, W. Belardi, J.C. Baggett, T.M. Monro, D.J. Richardson, Polarization mode dispersion reduction in spun large mode area silica holey fibres. Opt. Express **12**(9), 1972–1977 (2004)
25. A. Michie, J. Canning, I. Bassett, J. Haywood, K. Digweed, A. Lau, D. Scandurra, M. Aslund, B. Ashton, M. Stevenson, J. Digweed, Spun elliptically birefringent photonic crystal fibre. Opt. Express **15**(4), 1811–1816 (2007)
26. Andrew Michie, John Canning, Ian Bassett, John Haywood, Katja Digweed, Brian Ashton, Michael Stevenson, Justin Digweed, Alfred Lau, Daniel Scandurra, Spun elliptically birefringent photonic crystal fibre for current sensing. Meas. Sci. Tech. **18**(10), 3070–3075 (2007)
27. A. Argyros, J. Pla, F. Ladouceur, L. Poladian, Circular and elliptical birefringence in spun microstructured optical fibres. Opt. Express **17**, 15983–15990 (2009)
28. Y.K. Chamorovsky, N.I. Starostin, S.K. Morshnev, V.P. Gubin, M.V. Ryabko, A.I. Sazonov, I.'L. Vorob'ev, Spun microstructured optical fibres for Faraday effect current sensors. Quant. Electron. **39,** 1074 (2009)
29. J. Ballato, P. Dragic, Rethinking Optical Fiber: New Demands, Old Glasses. J. Am. Chem. Soc. **96**(9), 2675–2692 (2013)
30. J. Canning, W. Padden, D. Boskovic, M. Naqshbandi, H. de Bruyn, M.J. Crossley, Manipulating and controlling the evanescent field within optical waveguides using high index nanolayers. Opt. Mat. Express **1**(2), 192–200 (2011)
31. M.F. Koldunov, A.A. Manenkov, Polymer-filled nanoporous glass: a new material for solid-state dye lasers and nonlinear optical elements. in *Proceedings of SPIE 6054, International Conference on Lasers, Applications, and Technol. 2005: Advanced Lasers and Systems, 605401* (2006); doi:10.1117/12.660490
32. R. Ciriminna, A. Fidalgo, V. Pandarus, F. Béland, L.M. Ilharco, M. Pagliaro, The Sol-Gel route to advanced silica-based materials and recent applications. Chem. Rev. **113**(8), 6592–6620 (2013)
33. G. Huyang, J. Canning, I. Petermann, D. Bishop, A. McDonagh, M.J. Crossley, "Room temperature sol-gel fabrication and functionalisation for sensor applications", Photonic Sensors. Springerlink (2012). doi:10.1007/s13320-012-0075-2
34. Q. Huo, D. Zhao, J. Fang, K. Weston, S.K. Buratto, G.D. Stucky, S. Schacht, F. Schuth, Room temperature growth of mesoporous silica fibers: A new high-surface-area optical waveguide. Adv. Mater. **9**(121), 974–978 (1997)
35. K. Matsuzaki, D. Arai, N. Taneda, T. Mukaaiyama, M. Ikemura, Continuous silica glass fiber produced by sol-gel process. J. Non-Cryst. Sol. **112**(1–3), 437–441 (1989)
36. R.D. Deegan, O. Bakajin, T. F. Dupont, G. Huber, S.R. Nagel, Capillary flow as the cause of ring stains from dried liquid drops. Nature **389**, 827–829 (1997)
37. M. Naqshbandi, J. Canning, A. Lau, M.J. Crossley, Controlled fabrication of macroscopic mesostructured silica spheres for potential diagnostics and sensing applications. in *The International Quantum Electronics Conference (IQEC)/Conference on Lasers & Electro-Optics (CLEO) Pacific Rim, (IQEC/CLEO-Pacific Rim 2011), Syd.* (Australia, 2011)
38. H. Hu, R.G. Larson, Evaporation of a sessile drop. J. Phys. Chem. B. **106**(6), 1334–1344 (2002)

39. J. Canning, H. Weil, M. Naqshbandi, K. Cook, M. Lancry, Laser tailoring surface interactions, contact angles, drop topologies and the self-assembly of optical microwires. Opt. Mat. Express **3**(2), 284–294 (2013)
40. J. Canning, M. Ma, B. Gibson, J. Shi, K. Cook, M.J. Crossley, Highly ordered mesoporous silica microfibres produced by evaporative self-assembly and fracturing. Opt. Mat. Express **3**(2), 284–294 (2013)
41. L. Moura, J. Canning, L. Lindoy, K. Cook, M.J. Crossley, Y. Luo, G-D. Peng, L. Glavind, M. Kristensen, A fluorescence study of self-assembled silica layers on D-shaped optical fibre. in *4th Asia Pacific Optical Sensors (APOS 2013)* (Wuhan, China, 2013)
42. C.M. Rollinson, S.M. Orbons, S.T. Huntington, B.C. Gibson, J. Canning, J.D. Love, A. Roberts, D.N. Jamieson, Metal-free scanning optical microscopy with a fractal fibre probe. Opt. Express **17**(3), 1772–1780 (2009)
43. A. Greenbaum, W. Luo, B. Khademhosseinieh, T-W. Su, A.F. Coskun, A. Ozcan, Increased space-bandwidth product in pixel super resolved lens free on-chip microscopy. Sci. Rep. **3**, 1717 doi:10.1038/srep01717 (2013)
44. O. Mudanyali, E. McLeod, W. Luo, A. Greenbaum, A.F. Coskun, Y. Hennequin, C.P. Allier, A. Ozcan, Wide-field optical detection of nanoparticles using on-chip microscopy and self-assembled nanolenses. Nat. Photon. **7**, 247–254 (2013)
45. D.M. Adams, *Inorganic Solids* (Wiley, Great Britain, 1974)
46. P. Levitz, G. Ehret, S.K. Sinha, J.M. Drake, Porous Vycor glass: the microstructure as probed by electron microscopy, direct energy transfer, small angle scattering, and molecular adsorption. J. Chem. Phys. **95**(8), 6151 (1991)
47. R. Schmidt, E.W. Hansen, M. Sticker, D. Akporiaye, O.H. Ellestad, Pore size determination of MCM-41 Mesoporous materials by means of ‘H NMR spectroscopy, N2 adsorption, and HREM. A preliminary study. J. Am. Chem. Soc. **117**, 4049–4056 (1995)
48. E.P. Barrett, L.G. Joyner, P.P. Halenda, The determination of pore volume and area distributions in porous substances. I. computations from nitrogen isotherms. J. Am. Chem. Soc. **73**(1), 373–380 (1951)
49. S. Brunauer, P.H. Emmett, E. Teller, Adsorption of Gases in Multimolecular Layers. J. Am. Chem. Soc. **60**, 309 (1938)
50. J.T. Randall, H.P. Rooksby, B.S. Cooper, Structure of glasses: the evidence of X-ray diffraction. J. Soc. Glass Tech. **14**, 219 (1930)
51. E. Bourova, S.C. Parker, P. Richet, Atomistic simulation of cristobalite at high temperature. Phys. Rev. B **62**, 12052 (2000)
52. E. Tyedmers, F. Ali, N. van K. Losio, S. Murphy, S. Eid, C. Shen, M. Naqshbandi, J. Canning, M.J. Crossly, P. Rutledge, Chemical sensing with optical fibres and waveguides: diagnostics for a sustainable environment. in *Talented Students Program (TSP) Report, (Staff Supervisor: J. Canning) Undergraduate Faculty Program, School of Chemistry* (The University of Sydney, 2013)
53. N. Lai, K. Li, T. Jegathees, J. Canning, The detection of dopant adsorption onto silica nanoparticles through photon correlation spectroscopy. in *Submitted to the Australia-New Zealand Conference on Photonics (ANZ-COP)* (Perth, Australia, 2013)
54. D.-M. Liu, I.-W. Chen, Encapsulation of protein molecules in transparent porous silica matrices via an aqueous colloidal sol–gel process. Acta Materiala **47**(18), 4535–4544 (1999)
55. E.H. Lan, B.C. Dave, J.M. Fukuto, B. Dunn, J.I. Zink, J.S. Valentine, Synthesis of sol-gel encapsulated heme proteins with chemical sensing properties. J. Mater. Chem. **9**, 45–53 (1999)
56. O. Fawkes, A comparison of methods for the effective integration of dye into self-assembling silica nanowires. A Report to the School of Chemistry, The University of Sydney in partial fulfilment of the Michael Bishop Award Program for High School Students (2013)
57. D.R. Smith, J.B. Pendry, M.C.K. Wiltshire, Metamaterials and negative refractive index. Science **305**, 788–792 (2004)

58. E. Ozbay, Science Meets Magic: Photonic Metamaterials. in ed. by A. Adibi, S.-Y. Lin, A. Scherer, *Photonic and Phononic Properties of Engineered Nanostructures II, Proceedings of SPIE* vol. 8269, 82690L (2012)
59. K.V. Do, X. Le Roux, C. Caer, D. Morini, L. Vivien, E. Cassan, All-dielectric photonic metamaterials operating beyond the homogenization regime. Adv. Electromagnetics **1**(1), 1–10 (2012)
60. F. von Cube, S. Irsen, R. Diehl, J. Niegemann, K. Busch, S. Linden, From Isolated Metaatoms to Photonic Metamaterials: Evolution of the Plasmonic Near-Field. Nano Lett. **13**(2), 703–708 (2013)
61. I. Aharonovich, A.D. Greentree, S. Prawer, Diamond photonics. Nat. Photon. **5**, 397–405 (2011)
62. D.W.M. Lau, T.J. Karle, B.C. Johnson, B.C. Gibson, S. Tomljenovic-Hanic, A.D. Greentree, S. Prawer, Very bright, near-infrared single photon emitters in diamond. APL Mater. **1**, 032120 (2013)
63. D. Kielpinski, C. Monroe, D.J. Wineland, Architecture for a large-scale ion-trap quantum computer. Nature **417**, 709–711 (2002)
64. M.R. Henderson, B.C. Gibson, H. Ebendorff-Heidepriem, K. Kuan, S. Afshar, V.J.O. Orwa, I. Aharonovich, S. Tomljenovic-Hanic, A.D. Greentree, S. Prawer, T.M. Monro, Diamond in tellurite glass: a new medium for quantum information. Adv. Mat. **23**(25), 2806–2810 (2011)
65. J. Canning, M. Naqshbandi, B.C. Gibson, M. Nash, H. Jeyaseelan, M.J. Crossley, *"Nanoparticle self-assembly: a new approach to fabricating optical interconnects, single photon sources and more", Integrated Photonics Research, Silicon and Nano-Photonics (IPR)* (OSA's Advanced Photonics Congress, Cheyenne Mountain Resort, Colorado Spring, Colorado United States, 2012)
66. B.C. Gibson, J. Canning, M. Naqshbandi, M. Ma, M.M. Nash, M.J. Crossley, *Room temperature self-assembly of diamond and silica nanoparticles* (Int. Conf. on Diamond and Carbon Materials, Riva del Garda, Italy, 2013)
67. H.P. Yuen, J.H. Shapiro, Optical communication with two-photon coherent states–Part I: Quantum-state propagation and quantum-noise. IEEE Trans. Inf. Theory **24**(6), 657–668 (1978)
68. M. Fiorentino, P.L. Voss, J.E. Sharping, P. Kumar, All-fiber photon-pair source for quantum communications. IEEE Photn. Technol. Lett. **14**(7), 983–985 (2002)
69. X. Li, J. Chen, P. Voss, J. Sharping, P. Kumar, All-fiber photon-pair source for quantum communications: Improved generation of correlated photons. Opt. Express **12**(16), 3737–3744 (2004)
70. A. Politi, M.J. Cryan, J.G. Rarity, S. Yu, J.L. O'Brien, Silica-on-silicon waveguide quantum circuits. Science **320**(5876), 646–649 (2008)
71. M. Kristensen, T. Balle, J. Selchau, K.B. Sigvardt, N. Groothoff, Generating and sensing signals for quantum cryptography using phase encoding in compact silica-on-silicon Mach-Zehnder circuits with Bragg gratings. in *Proceedings of SPIE 8351, Third Asia Pacific Optical Sensors Conference*, (Sydney, Australia, 83510X, 2012); doi:10.1117/12.914310
72. J. Canning, M. Ma, B.C. Gibson, J. Shi, K. Cook, M.J. Crossley, The nanostructure of self-assembled silica microwires: a crystalline pure silica zeolite? in *Australia New Zealand Conference on Optics and Photonics (ANZCOP)* (Perth, Australia, 2013)
73. M. Ree, J. Yoon, K. Heo, *Low-K nanoporous interdielectrics: Materials, Thin Film Fabrications, Structures and Properties* (NOVA Science Publishers Inc., New York, 2010)

Chapter 11
Lab on Fiber by Using the Breath Figure Technique

Marco Pisco, Giuseppe Quero, Agostino Iadicicco, Michele Giordano, Francesco Galeotti and Andrea Cusano

Abstract In this chapter, we report on a novel fabrication process developed to realize metal-dielectric crystals onto optical fibers by a self-assembly technique. Breath figures methodology is selected as a technological tool to operate directly onto non-conventional substrates like optical fibers. Regular and ordered metallo-dielectric crystals are easily integrated onto the optical fiber tip, providing the basis for the rapid and cost effective prototyping of photonic-plasmonic nanoprobes for advanced sensing applications. In order to validate the proposed fabrication route, we develop a first technological platform capable of supporting interferometric effects assisted by surface plasmon excitation at the metallo-dielectric interface. We investigated the sensing properties of the realized optical fiber probes, the results of which revealed an exceptional sensitivity with respect to the refractive index, as high as 2300 nm/RIU, enabling the employment of the optical fiber platform to be used as optical nano-probe for label-free chemical and biological sensing.

M. Pisco · G. Quero · A. Cusano (✉)
Department of Optoelectronic Division—Engineering, University of Sannio, Corso Garibaldi 107, 82100 BN, Italy
e-mail: a.cusano@unisannio.it

A. Iadicicco
Department of Engineering, University of Naples “Parthenope”, Centro Direzionale Isola C4, 80143 NA, Italy

M. Giordano
Institute for Composite and Biomedical Materials, Consiglio Nazionale delle Ricerche, piazzale E Fermi1, 80055 Portici, NA, Italy

F. Galeotti
Istituto per lo Studio delle Macromolecole, Consiglio Nazionale delle Ricerche, via Bassini 15, 20133 Milan, Italy

A. Cusano et al. (eds.), *Lab-on-Fiber Technology*,
Springer Series in Surface Sciences 56, DOI: 10.1007/978-3-319-06998-2_11,

11.1 Introduction

In 1966, the Nobel Prize winner C. K. Kao concluded his study on “Dielectric-Fibre Surface Waveguides for optical frequencies” [1] envisaging that “*a fibre of glassy material (…) represents a possible practical optical waveguide with important potential as a new form of communication medium*” [1, 2]. In the following years, the fiber optics field has undergone a tremendous growth and advancement. Initially conceived as a medium to carry light and images for medical endoscopic applications, optical fibers were later proposed as an adequate information-carrying medium for telecommunication applications. As optical fibers established their position in the telecommunications industry and its commercial market matured, parallel efforts were carried out by a number of different research groups around the world to exploit them also in sensing applications. Initially, fiber sensors were lab curiosities and simple proof-of-concept demonstrations but in a few years their operation and instrumentation have become well understood and developed. Over the years, optical sensors enjoyed an increased acceptance as well as a widespread use for structural sensing and monitoring applications in civil engineering, aerospace, marine, oil & gas, composites, smart structures, bio-medical devices, the electric power industry and many others [3–5].

Although fiber optic sensors have, in many cases, completely replaced and outperformed their counterparts based on conventional technologies, it is an established fact that optical fibers are still mainly conceived as a communication medium. For this reason, the fiber optic industry focused its efforts on a small set of materials and structures able to provide light guidance in the transparency range of silica glass.

A significant technological breakthrough would be the development of components and devices “all in fiber” through the integration of advanced functional materials at micro and nano-scale, exhibiting the more disparate properties, combined with suitable light matter interaction mechanisms. This achievement would be the cornerstone of a new photonics technological revolution that would lead to the definition of a novel generation of micro- and nano- photonic devices “all-in-fiber”. An additional driving force pushing in this direction is the increasing market demand for highly integrated and multifunctional sensor probes with advanced performances and unrivaled features.

In this context, the growth of the life science field, has driven photonics technology into important interdisciplinary fields such as biophotonics and nanophotonics. In particular the optical biosensor field is experiencing a new surge in activity, seeking to exploit novel optical structures and bio-coating materials capable of detecting and discriminating among large classes of molecules. Thanks to rapid advances in this research field, the highly bio-sensitive/selective optical sensors are becoming a viable alternative to traditional “solution based” assay biosensors for applications in genomics, proteomics and drug discovery, as well as in the food industry, homeland security and environmental monitoring applications.

Optical fibers are well suited to support this revolution also by virtue of their dynamicity and versatility offered by using this unique technology. Ten years ago, first demonstrations of microstructured optical fibers [6, 7] opened up new avenues for the development of advanced light waveguides, adding new fiber functionalities, new insights in light manipulation and control [8, 9]. Microstructured optical fibers, because they have many holes, can be easily filled by fluids and have therefore also provided the basis for novel optofluidic devices and components [8–12].

This impressive driving action combined with an ever increasing market demand for a continuous technological innovation, has determined a strong involvement of the optical fibers research community in looking for new fiber materials to use [13] as well as silica fibers with sub-micro diameters to achieve ultimate light matter interaction levels [14].

All the above considerations are leading an envision of the "Lab-on-Fiber" concept [15, 16] as a concrete technological solution to actual market demand; this new technology essentially deals with the development of novel and highly functionalized materials, devices and components, completely integrated in a single optical fiber and ready to be incorporated in modern optical systems for communication and sensing applications. This new vision, thus, relies on the development of a technological platform based on a single optical fiber where several materials, macro- and nano-structures are constructed, embedded and connected all together to provide the necessary physical connections and light-matter interactions. The Lab-on-Fiber technology could definitely open the way for the implementation of sophisticated autonomous multifunction sensing and actuating systems "all in fiber", meaning the existence of adjacent multi operational labs which are able to analyze and exchange sensorial data, providing unique auto diagnostics properties.

11.2 Lab on Fiber Technologies

In recent years, remarkable theoretical, numerical and experimental studies on novel optical fibers and novel optical fiber components have been conducted with the aim of concurring to launch the "Lab on Fiber" technologies [15–17]. From this perspective, we refer to "Lab on Fiber" technology as the basis for the development of a technological world completely contained in a single optical fiber where several structures and materials at sub or wavelength scale are constructed, embedded and connected all together to provide the necessary physical connections and light-matter interactions useful in providing a wide range of functionalities and unparalleled performances. Optical fiber technology combined with the new concept of "Lab on Fiber" has the potential to enable the implementation of sophisticated autonomous multifunctioning sensing and actuating systems all of them integrated in optical fibers with unique advantages in terms of miniaturization, light weight, cost effectiveness, robustness and power consumption.

The most obvious technological strategy to produce multi-material optical fiber with new functionalities is to create a new fiber fabrication process. The idea relies on the preparation of a multi-material pre-form suitable for the fiber drawing technique. This technological approach was successfully adopted to produce novel optical fibers acting as photodetectors to visible and infrared light [18] or optical fibers with chemiluminescent features for sensing applications [19] as well as piezoelectric fibers for acoustic transduction [20].

Alternative approaches recently proposed are focused on the integration of photonic crystal structures and/or plasmonic nanostructures on the end facet of optical fibers. In this scenario, the creation of periodic micro and nanostructures on the end facet of optical fibers is of great interest because it may yield versatile optical devices well-suited to serve as miniaturized probes for remote sensing applications. The reduction of feature sizes in micrometer and nanometer scales permits the possibility of obtaining intriguing spectral features, even by lighting the structured pattern out of the plane of periodicity [16, 21–23]. Several approaches have been recently introduced to fabricate metallic and dielectric structures on the optical fiber end facet [2]. Some approaches rely on technique of transferring planar nano-scale structures, fabricated on a planar wafer by means of standard lithographic techniques, onto the optical fiber end facet. These methods exploit well-assessed fabrication processes developed for planar substrates, but they are limited by the final transferring step that plays a fundamental role in determining both the fabrication yield and the performance of the final device. For example it is worth focusing attention on [21–25] where a transfer technique was demonstrated with regards to various metallic patterns. They were first defined by electron-beam lithography (EBL) and then moved to the optical fiber facets by using a sacrificial polymeric film to strip and transfer the metallic patterns. Successively, in order to increase the production throughput, the same authors developed an alternative procedure to realize metallic nanostructures by embedding and sectioning a metallized array of epoxy nanoposts using an ultramicrotome [26]. After the sectioning stage, the thin epoxy slabs are manually transferred to the optical fibers tips and then the epoxy matrix is etched, leaving the metallic pattern on the optical fiber.

Following this basic approach, featured by the two distinct phases of generation and transfer of the nanostructures, the fabrication of a silicon photonic crystal (PC) on the fiber tip was also recently proposed [27]. Similarly, Shambat et al. [28] realized a photonic crystal exhibiting resonant cavities via EBL, transferred it onto the fiber by micromanipulator and bonded the PC on the fiber by depositing small amounts of epoxy, acting as adhesive, on the cladding surface. Scheerlinck et al. [29] developed a procedure in which the realization and transferring of metallic gratings on the optical fiber occurred in a single step by means of UV-based nano imprint and transfer lithography. By following these approaches, multifunctional optical fiber nanoprobes composed of a variety of gold and silver patterns on the facet of an optical fiber were proposed and experimentally demonstrated by highlighting the potential impact of fiber tip probes for sensing applications.

Alternative approaches are based on direct-write patterning of the fiber tip. These methods, based on conventional lithographic techniques adapted to operate on unconventional substrates such as the optical fiber tip, are able to efficiently provide nanostructured devices on the optical fiber, but they require complex and expensive fabrication procedures with a relatively low throughput.

By following direct-writing approaches, fiber based probes composed of a gold layer with sub-wavelength apertures [30] and periodic gold dots arrayed [31] on the facet of an optical fiber were already proposed and experimentally demonstrated by highlighting the potential impact of fiber tip nanoprobes for chemical and biological sensing applications. Such prototypes were possible via Focused Ion Beam (FIB) milling and Electron Beam Lithography (EBL). On this line, recently, spin coating combined with EBL nanotechnology, has been proposed for creating hybrid metallo-dielectric nanostructures, giving rise to resonant effects exploitable for label-free chemical and biological sensing [16].

The direct writing techniques have the merit to be able to sculpt the desired micro and nanostructures with high definition on the optical fiber. So arbitrary sizes and exotic shapes can be defined, inspiring novel configurations for completely new functionalities. The choice of materials is also increased insofar as diversified deposition techniques are developed and assessed on the optical fiber substrates. Nevertheless, all these methods are evidently serial because each device must be written individually by using uneconomical technological equipment. Overall the direct writing approaches, while they are able to efficiently provide nanostructured devices on the optical fiber, are still very challenging by requiring complex and expensive fabrication procedures with a relatively low throughput.

In order to create patterned structures onto the optical fiber tip usinga low cost fabrication approach, we propose the exploitation of self-assembly processes, to be specialized for operating on the optical fiber substrates [32–34]. The autonomous organization of micro and nano structures in regular ensembles [35–37] on the optical fiber end facet is the key tool in obtaining ordered patterns on a high number of optical fibers simultaneously by using easy and low cost fabrication strategy, suitably employed in the mass production of technologically advanced devices.

In the research effort of developing a direct and low-cost method that enables the fabrication of self-assembled periodic structures, the Breath Figures (BFs) are a valid option [34]. In fact, by exploiting this matter auto-organization phenomenon, thin polymeric films having a distribution of micrometric cavities arranged in hexagonal fashion are readily obtainable: (i) through a single step approach, (ii) with no need for lithographic processing, (iii) in a few seconds, and (iv) by means of very simple and accessible laboratory equipment. Indeed, the BF process is a templating method in which the template consists of an ordered array of water droplets that can be removed by simple evaporation. This is an undisputable advantage with respect to most of the other known templating approaches, where the templates need to be removed after the fabrication of the porous films and in most of the cases they are not easily prepared or eliminated. For these reasons, BF is one of the most widely employed methods for the fabrication of porous polymer films [35–37].

On this line, in this chapter, we present recent results on the research activities aimed to use self-assembly to create ordered metallo-dielectric structures on the optical fiber tip. In particular, we report details on the fabrication process based on the BF technique, supported by a wide morphological analysis, and discuss the related principle of operation through numerical and experimental analysis. Finally, the functionality of the realized probes for sensing applications is also demonstrated.

11.3 Breath Figure Structures onto the Optical Fiber Tip

The strategy for the fabrication of periodic structures onto the optical fibers initially involves the coating of the fiber facet by a microporous polymer film, realized via the BF technique. This self-assembly approach, in fact, allows for the preparation of honeycomb-like arrays at micro scale with a high degree of order despite the simplicity of the process. The mechanism of the BF formation has been described in detail by different authors [38–40], and will be, here, briefly recalled. The whole process relies on the precipitation of a polymer around condensed water droplets, initiated by rapid evaporation of a polymer solution in a humid environment. Its salient stages are depicted in Fig. 11.1. In the initial stage, the endothermic evaporation of the solvent results in a decrease of the system temperature, thus triggering the water condensation. Once a droplet of water has nucleated on the polymer solution, it grows at the expense of the vapor of the surrounding atmosphere. During this stage, the growing droplets auto-organize at the polymer solution/air interface into a close hexagonal arrangement. Once the film returns to ambient temperature, the condensed water and residual solvent evaporate, leaving behind a honeycomb structure. When the process is adequately controlled and the key parameters (polymer concentration, kind of solvent, evaporation rate and relative humidity) accurately adjusted, the manipulation of some morphology features in the final film, i.e. degree of order, distribution and size of the cavities, is easily feasible.

Even though the applicability of the BF approach to the fabrication of both metallic disk arrays and metallic meshes has already been established [41, 42], this is the first study, to the best of our knowledge, that makes use of BFs as a tool for realizing self-assembled metallo-dielectric periodic structures on the end facet of a standard single mode optical fiber.

To reach this goal, the standard setup, which is normally utilized for making honeycomb films onto glass or silicon substrates, must be adapted to correctly work onto non-conventional substrates as is the case with optical fibers. The main issue to be addressed deals with the fiber tip size: 125 μm of diameter for a standard single mode fiber. Typically, in fact, highly regular BF arrays are obtained by spreading a few droplets of the polymer solution over a flat surface of 1–5 cm^2, hence the fiber facet is definitely too small for such an approach. To overcome this issue, the optical fiber was embedded in a ceramic ferule with a diameter of 2.5 mm (similar to the ferule used in standard fiber connectors FC/PC) and then accurately polished. This modified fiber tip offers a larger surface as compared to the bare fiber, therefore resulting in easier handling during both the fabrication process and the operative stage of the final device.

Even though we have experimented with the possibility of preparing BF films directly onto the fiber tip, owing to the little casting volume (4 μL) and consequently quicker evaporation of the polymer solution, we found the reproducibility of this approach rather poor. Therefore, the use of a holder that enlarges the deposition surface, is recommendable. To this purpose, an Al holder of 20 × 20 × 8 mm was

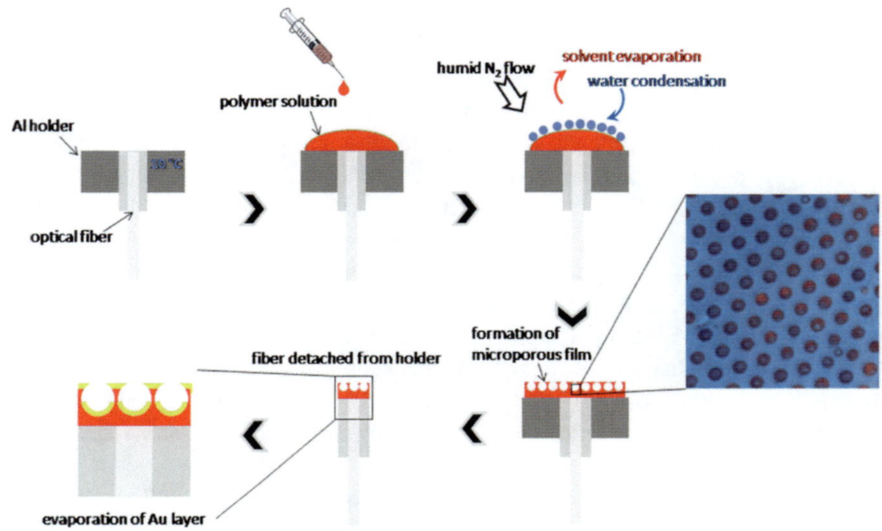

Fig. 11.1 Schematic overview of the fabrication process. The detail of the microporous film is a real view taken by a confocal microscope, in the central area of the fiber face

shaped in order to exactly fit the ceramic ferule external diameter, thus preventing any solvent from leaking, so that the polymer solution can be easily drop cast over an area of about 1 cm^2 (see initial sketch in Fig. 11.1). Since the fiber is placed exactly in the center of the holder, only the central portion of the film is deposited on it, which overcomes also the problem of poor pattern homogeneity that is usually encountered on the edges of BF films [43]. Moreover, this approach is in principle suitable for patterning at the same time more than one fiber per time, simply by arranging multi-slots in the Al holder and by casting a wider film. The Al holder was pre-cooled by placing it on a cold stage which was maintained at a temperature of +10 °C, and removed from there just before starting the film deposition. In fact, given the high thermal conductibility of the metal substrate, as compared to glass, the evaporation-induced cooling of the polymer solution is mainly dispersed through the Al substrate, resulting in poor water condensation. For this reason, such a pre-cooling step is required to promote an efficient water condensation on the evaporating film and, consequently, the BF formation.

Films were prepared by drop casting on the fiber/holder assembly a 4 mg/mL CS_2 solution of a fluorinated fluorescent dye-terminated linear polystyrene, which we had already described as a good candidate for producing highly ordered BF structures [44]. To speed up the solvent evaporation and promote water condensation, a flux of moist nitrogen (75 % R.H. at 25 °C) was directed towards the fiber facet. The solvent completely evaporated within 20 s, leaving on the holder and the fiber surface an opaque film, which shone in bright iridescent colors indicating a periodic variation in the refractive index (air holes/polymer walls) throughout the surface of the film. Thanks to the marked fluorescent emissions of the polymer, we were able to rapidly check the quality of patterns formed on the fiber by fluorescence microscopy. Once the BF process was complete, the fiber was carefully detached from the holder, retaining on its tip a portion of the honeycomb film.

Since all other external parameters (temperature and air humidity) as well as those related with the preparation procedure (solvent, polymer concentration and casting volume) were maintained constant, the control on the pore size was achieved by varying the flow of humid nitrogen directed at the fiber tip. Even though a flow rate/size dependence was not accurately determined at this stage of the study, by varying the air flow between 100 and 300 L/h it was possible to obtain honeycomb arrays with cavities ranging from 2.5 to 1.0 μm (external diameter).

Once the optical fiber facet was covered with the polymeric honeycomb structure, the quality of the pattern in the center of the fiber was checked by fluorescence microscopy, so that well-ordered and defect-free arrays with the desired dimensions were selected for the following step. The second fabrication step consisted in the vacuum evaporation of a thin layer of Au over it. To this purpose, the fiber was placed in a vacuum evaporation chamber and 30–40 nm of Au were deposited on top. By this two-step procedure, prototypes consisting of a single mode optical fiber end-coated with ordered metal-dielectric crystals were successfully produced (Fig. 11.1, final sketch).

11.4 Morphological Characterization

A complete morphological characterization of a representative sample achieved by means of a flow of humid nitrogen of 300 L/h is, here, reported and discussed.

Figure 11.2a shows a scanning electron microscope (SEM) top view image of the sample in which the ceramic ferule (diameter 2.5 mm) with smoothed edge can be seen while the pattern is poorly visible. In Figs. 11.2b, c, we show a magnified SEM image and atomic force microscope (AFM) image of the structure, respectively, allowing to accurately study the topography of the patterned region. By analyzing the images, the key parameters of the periodic pattern of holes have been measured revealing a lattice period and hole diameter of 2.67 μm and 0.95 μm, respectively. Relative standard deviation (%RSD) of both geometrical parameters has been estimated to be 1.3 and 3.2%, respectively. Additionally, the AFM profile, shown in Fig. 11.2d, made it possible to measure the mean pores depth, the result being 1.78 μm (RSD of 3.5 %). Finally, in order to investigate the overall structure's thickness and uniformity, we firstly removed the polystyrene on half the area of the ceramic ferule by using an excimer laser (operating at the wavelength of 248 nm), and then measured the structure's height with AFM analysis. Laser fluence and repetition rate were opportunely selected in order to ablate the polystyrene layer without damage to the ceramic surface. A SEM top view image of the carved sample is shown in Fig. 11.3a where the treated region is clearly evident. In Fig. 11.3b, we report an AFM profile of the treated region edge, revealing a

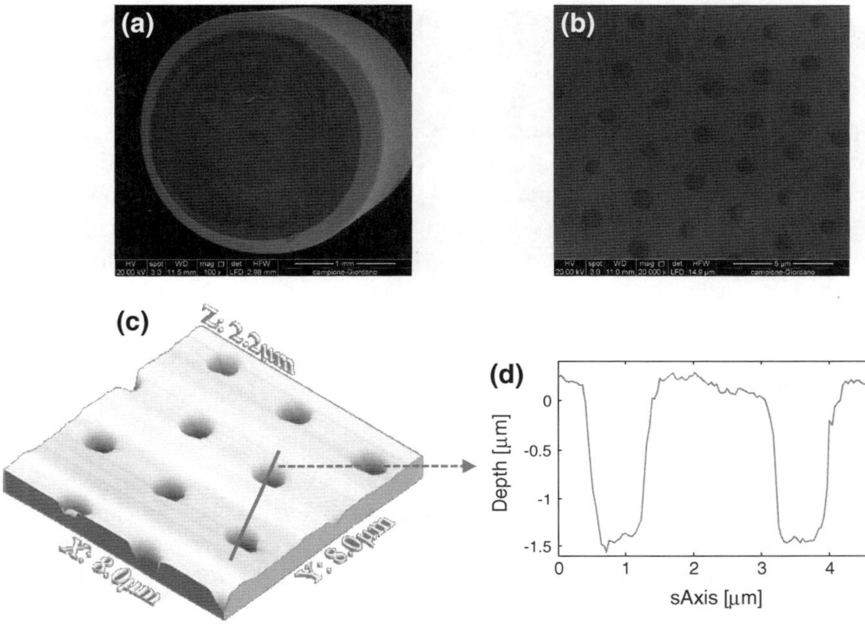

Fig. 11.2 Morphological characterization via SEM and AFM images: **a** SEM top view image; **b** magnified SEM image; **c** AFM image of patterned region; and **d** AFM profile

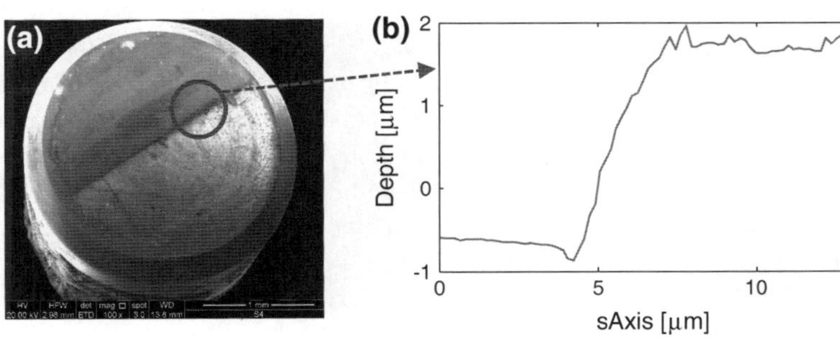

Fig. 11.3 Morphological characterization via SEM and AFM images after laser treatment: **a** SEM *top view* image; **b** AFM profile of the pattern edge

polystyrene height of 2.5 µm with a relative standard deviation of 2.3 %. The polystyrene thickness is evidently higher than the holes height, revealing the presence of a polystyrene uniform undercoating estimated to be 0.72 µm thick.

From these results it is evident that the breath figure technique enables us to realize metallo-dielectric structures directly self-assembled on a fiber optic tip with a highly regular pattern.

11.5 Spectral Reflectance via Numerical and Experimental Analysis

Spectral reflectance measurements were carried out by means of an easy measurement setup involving a 3 dB 1 × 2 coupler connected to a broadband light source (covering the wavelength range 1,250–1,650 nm), the fiber probe and the optical spectrum analyser. In addition, to compensate for intensity fluctuations of the source vs wavelength, the sample reflectance was normalized by using a fiber-optic reference mirror, fabricated by depositing a 160 nm-thick gold film on the tip of a standard single-mode fiber. The schematic of the characterization setup is reported in the Fig. 11.4a. In Fig. 11.4b, we show the experimental reflectance spectrum (blue line) of the selected sample, revealing a broadband peak, centred around 1400 nm.

In order to investigate the physics underlying the observed spectra, we performed a numerical analysis on the basis of the results obtained from the morphological characterization reported in the previous section. The structure realized is numerically reconstructed in the form of a thin polystyrene uniform layer supporting the BF pattern as schematized in Fig. 11.5a. The whole structure is thus assumed to be covered by a conformal thin film of gold.

The simulations were carried out by the finite-element method (FEM) using the commercial modeling tool COMSOL Multiphysics (RF Module). In order to numerically retrieve the reflectance of photonic crystals by FEM, a common approach requires the calculation of the scattering parameters of a slab composed of the unit cell under periodic boundary conditions. Rather than considering a (honeycomb) primitive unit cell replicated by the Blok periodic boundary conditions, we selected a non-primitive rectangular cell of the hexagonal tiling and, to reduce the computation burden, we restricted the computational domain to one quarter of this rectangular cell (see Fig. 11.5b). At the bottom side of the slab, we assumed an homogenous glass substrate (having the role of an optical fiber), while at the top side we posed an air thickness (having the role of the surrounding medium). By exploiting the crystal symmetry, the quarter of the cell is transversely terminated with two horizontal perfectly-electric-conducting and two vertical perfectly-magnetic-conducting walls, so as to simulate a normal incident plane-wave with a vertically-polarized electric field [21, 45, 46]. The input power is set to 1 W. The structural parameters used for the simulations are retrieved from the morphological characterizations. Refractive index of glass and polystyrene are supposed 1.45 and 1.58 respectively. Both losses and dispersion of gold have been taken into account [16].

The dimensional parameters used for the simulation are: holes diameter of 0.95 μm, pitch of 2.67 μm, depth of 1.78 μm, basement height of 0.72 μm whereas the gold thickness is assumed to be 33 nm.

In Fig. 11.4b, the numerical reflectance spectrum (green line) is reported and compared with the experimental results (blue line) obtained in the wavelength range 1,250–1,650 nm.

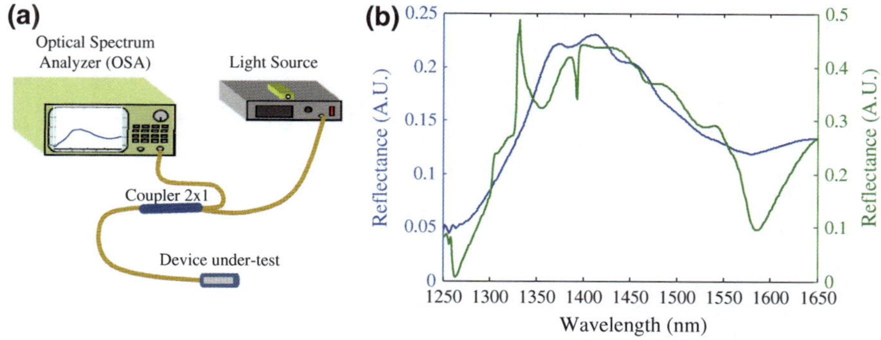

Fig. 11.4 **a** Schematic optoelectronic setup; and **b** Numerical and experimental reflectance spectra

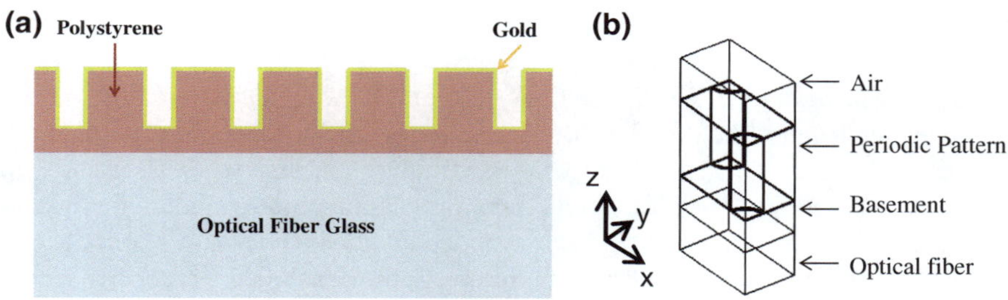

Fig. 11.5 **a** Schematic side view of the hybrid metallo-dielectric structures; and **b** Computational domain

From the reported data, it can be appreciated that numerical and experimental data agree with each other, even if two main differences can be easily observed. The first is related to the reflectance values. Maximum reflectance of 25 and 50 % have been found for the experimental and numerical data, respectively. The observed mismatch can be explained taking into account the unavoidable fabrication defects in respect to the ideal structure considered in the numerical analysis. The second difference relies on the presence in the numerical spectra of narrow peaks and dips superimposed on the broadband reflectance peak, such as those located at 1,260, 1,330 and 1392 nm. By means of a careful analysis of the electric field distributions, it can be easily recognized that these spectral features are associated to the excitation of surface plasmonic resonances.

Obviously the presence of sharp spectral features, erroneously foreseen by the numerical analysis but not observable in the experimental spectra, limits the prediction capability of the unit cell based model to study and design advanced structures conceived on the optical fiber tip.

In order to improve the numerical prediction capability, we developed a numerical model offering a more faithful representation of the realty with respect to the mostly used unit cell model. The major discrepancies between the experimental

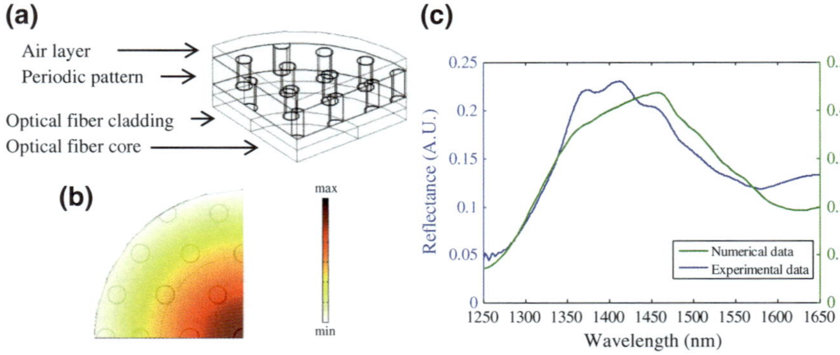

Fig. 11.6 **a** Computational domain of the optical fiber based model; **b** Normalized electric field distribution of the input light source superimposed to the periodic pattern onto the optical fiber; **c** Experimental reflectance spectrum in comparison with the numerical reflectance retrieved by the optical fiber based model

scenario and the numerical simulations rely on the waveform impinging on the structure under investigation and on the finiteness of the periodic crystal. In the model, using the unit cell of the periodic structure and its symmetries, the light source is a plane wave with the direction of propagation orthogonal to the plane of periodicity and the periodic structure is assumed infinite in the plane of periodicity in virtue of the imposed boundary conditions. The developed numerical model, instead, admits a Gaussian-like mode as input light source and it is characterized by a space-limited periodic pattern. The corresponding computational domain, shown in Fig. 11.6a, involves a thin slice of optical fiber, composed of a core with a diameter of 9 μm and a reduced cladding with diameter of 16 μm terminated by a perfect electric conductor (PEC) condition. This optical fiber supports a single mode clearly resembling the fundamental mode of the standard single mode fibers. Consequently, as evident in Fig. 11.6b, showing the input mode field distribution superimposed on the computational domain, in the numerical simulations, most light content impinges on a restricted area close to the optical fiber core which is the only one effectively illuminated in the experimental scenario.

In Fig. 11.6c, we show the numerical reflectance spectrum (green line) retrieved using the "optical fiber based model" in comparison with the experimental one (blue line) demonstrating a better agreement when compared with the unit cell model.

Having improved the numerical prediction capability, in Fig. 11.7 we show the electric field distribution at different wavelengths, representative of the investigated spectral range. At a first glance, the normalized electric field appears quite similar to that observed in a simple interferometric structure, featured by sinusoidal behavior along the propagation direction. However, remarkable differences are evident in correspondence to the metal-dielectric interfaces and in correspondence to the holes pattern, where a significant field localization can be clearly appreciated. In this regard, the Fig. 11.7 also shows the top view of a slice

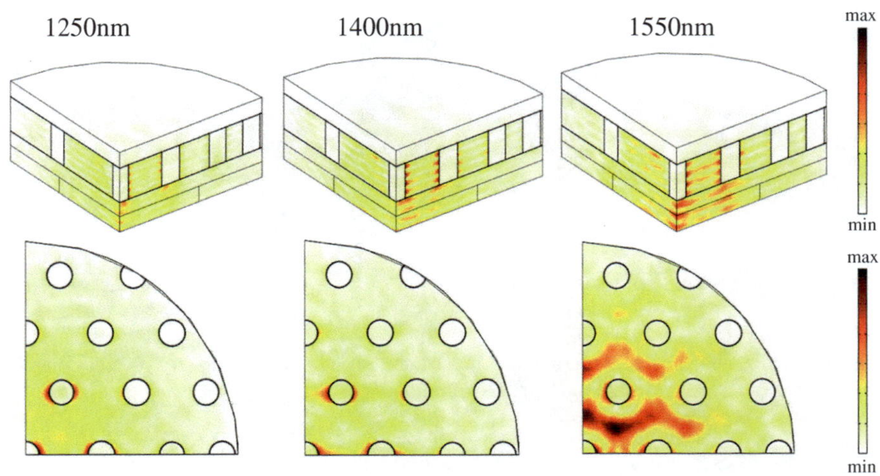

Fig. 11.7 Normalized electric field distributions at the wavelengths 1,250, 1,400 and 1,550 nm and top view of the normalized electric field distributions in a middle slice at the same wavelengths

positioned in the middle of the patterned region, where field localization can be easily noticed especially close to the gold disks and at the interfaces between the thin layer of gold and the surrounding dielectric.

These considerations lead us to envision the interplay taking place between the classical interference mechanism and the plasmonic excitation occurring at the metal interfaces.

11.6 Sensing Characteristics

All the results so far reported open up very intriguing scenarios for the development of a novel generation of miniaturized and cost-effective fiber optic nanoprobes useful in many applications including physical, chemical, and biological sensing. In this section we report on experimental results aimed at investigating the refractive index sensitivity of the realized sample. Since the physical mechanism at the basis of the realized probes relies on the interplay between interferometric and plasmonic effects, a significant refractive index sensitivity is expected especially due to the local field enhancement occurring at the metal interface.

In order to assess the sensitivity of the fabricated samples to the Surrounding Refractive Index (SRI), the optical fiber probes were immersed in different liquids with known refractive indices and the corresponding reflectance spectra were thus measured. To this aim, water ($n = 1.333$), ethanol ($n = 1.362$), isopropyl alcohol ($n = 1.378$), and some solutions with different concentration of water and ethanol were used as external medium. In Fig. 11.8a, the spectra of the investigated sample

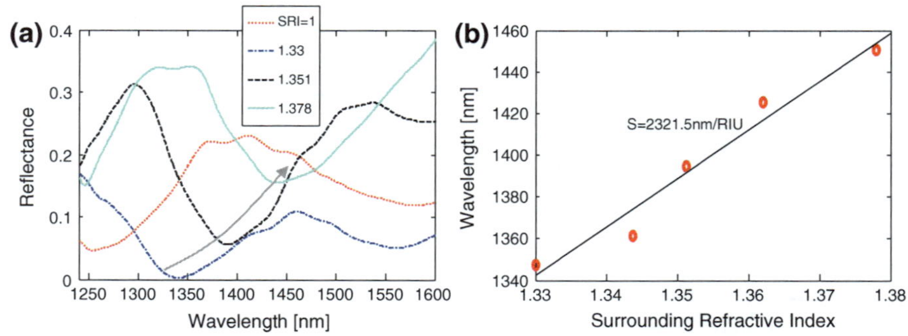

Fig. 11.8 SRI analysis: **a** Experimental reflectance spectra as function of the surrounding refractive index; **b** wavelength of the reflection dip versus SRI

for different SRIs are shown. It can be noted that the maximum reflectance value keeps quasi unchanged, when the external medium changes from air to water, while it monotonically increases for higher SRIs. A red-shift of the spectral curves with increasing values of the SRI can be clearly appreciated. In Fig. 11.8b we plot the wavelength of the reflection minimum as a function of the SRI, revealing an impressive sensitivity of 2,300 nm/RIU in the refractive index range of 1.33–1.378.

Despite the enhanced sensitivity for SRI slightly higher than 1.33, the SRI sensitivity results are quite low in correspondence to the air as a surrounding medium. This behavior resembles the typical sensitivity curves exhibited by most of the evanescent wave sensors and thus is consistent with the plasmonic nature of the field enhancement occurring at the metal interface, while the interplay between interferometric and plasmonic effects can explain the exceptional refractive index sensitivity here observed.

It is worth highlighting that the reported sensitivity is significantly higher than that demonstrated in the past concerning Local Surface Plasmon Resonance (LSPR)-based fiber tip probes (typically hundreds of nm/RIU) [16, 27, 30, 31]. Moreover, the results obtained are significant higher than those reported in an excellent review by Van Duyne et al. [47] concerning LSPR sensors implemented in "planar" configurations (reporting sensitivities ranging from 200 nm/RIU to 538 nm/RIU).

11.7 Conclusions

We have demonstrated a fabrication process which enables the integration of metallo-dielectric crystals directly onto optical fiber tips by using a self-assembly approach. The presented fabrication technique relies on the use of the Breath Figure methodology directly operating onto the optical fiber tip by adopting a judicious fiber preparation. In order to adapt the self-assembly procedure, allowing to correctly operate onto non-conventional substrates as the case of optical fibers, we specifically prepared a customized Aluminum holder, designed to house the optical fiber by exactly fitting the ceramic ferule external diameter so as to enlarge the available deposition surface, facilitating the formation of regular and ordered lattice structures in controlled conditions. Once created the holes pattern in the polystyrene layer can be used also as a template for successive depositions of functional materials, able to confer advanced functionalities to the final device.

It is remarkable that, for the first time to the best of our knowledge, the proposed approach enables the creation of periodic structures on an unconventional substrate such as the optical fiber end facets by using a self-assembling approach offering a high production throughput without using sophisticated and expensive technologies.

The effectiveness of the proposed technological process has been confirmed through the realization of specific optical fiber platforms based on the integration of hybrid metallo-dielectric crystals onto the optical fiber tip. The miniaturized fiber-tip devices are composed of a thin polystyrene layer patterned with a hexagonal lattice of holes, then further covered by a conformal layer of gold at nanoscale. Specifically, we selected gold as functional material in order to confer plasmonic features and to judiciously exploit them for sensing applications.

A deep morphological analysis has been carried out by means of SEM and AFM tools, in order to retrieve detailed information on the geometrical composition of the realized samples as well as to demonstrate the successful creation of well-ordered self-assembled lattice structures onto optical fibers.

Paralleling the experimental activities, we also developed a numerical tool able to design arbitrary structures conceived to be integrated onto optical fiber end facets, by taking into account both the field distribution of the fundamental mode in single mode optical fiber as well as the finite size of the crystal structure. As far as the production of structures on the optical fiber tip is finding increasing interest in the scientific community, the availability of a numerical tool able to faithfully describe and predict the spectral features of the realized samples can be considered as a beneficial side effect of this activity, useful also for further development in the design of all fiber devices based on the integration of functional materials at nano scale.

The results of the full-wave numerical analyses are in agreement with the experimental data and also demonstrated that the spectral features of the realized sensing probes are essentially dominated by interferometric effects significantly assisted and modified by plasmonic interactions occurring at the metal interfaces.

In order to show the functionalities of the realized optical probes, we focused our attention on investigating the sensing performances. Taking advantage of the plasmonic character of the proposed nanoprobes, we investigated the sensing performances in terms of sensitivity to the surrounding refractive index, revealing unprecedented performances (as high as about 2,000 nm/RIU) when compared to other plasmonic nanoprobes realized on the optical fiber tip using traditional nanotechnologies. The excellent sensitivities versus SRI changes make these structures promising candidates for novel miniaturized label free biological nanoprobes.

We point out that no attempts at this stage have been made to optimize the platform performances in terms of refractive index sensitivity. However, by exploiting the large set of degrees of freedom offered by the composite metallo-dielectric structures, combined with the versatility of the proposed fabrication technique, further optimization margins exist.

The achieved experimental results can be considered as a significant step forward in the Lab on Fiber technology roadmap, demonstrating the feasibility of the proposed technological approach to attain advanced miniaturized sensors by exploiting an easy and low cost fabrication process, suitable to be employed in rapid prototyping and mass production. Additionally, since the technique is intrinsically suitable to create periodic pattern on multiple fibers simultaneously, further efforts will be devoted to:

- enlarging the set of employable materials in order to add new functionalities to the final device
- demonstrating the effectiveness of this method for high throughput fabrication
- changing the pattern features to obtain regular cavities patterns with pitches down to hundreds of nanometers, enabling a better use of enhanced light matter interaction occurring in confined domains.

References

1. K.C. Kao, G.A. Hockham, Dielectric-fibre surface waveguides for optical frequencies. Proc. IEE **113**(3), 191–198 (1966)
2. A. Cusano, M. Consales, M. Pisco, A. Crescitelli, A. Ricciardi, E. Esposito, A. Cutolo, Lab on fiber technology and related devices, part I: a new technological scenario; Lab on fiber technology and related devices, part II: the impact of the nanotechnologies. Proc. SPIE **8001**, 800122 (2011)
3. B. Culshaw, A. Kersey, Fiber-optic sensing: a historical perspective. J. Lightwave Technol. **26**(9), 1064–1078 (2008)
4. A. Cusano, A. Cutolo, J. Albert, (eds.), *Fiber Bragg Grating Sensors : Recent Advancements, Industrial Applications and Market Exploitation* (Bentham Science Publishers, Oak Park, 2011), pp. 197–217
5. A. Cusano, M. Giordano, A. Cutolo, M. Pisco, M. Consales, Integrated development of chemoptical fiber nanosensors. Curr. Anal. Chem., Bentham Sci. Publ. **4**(4), 296–315 (2008)
6. P. Russell, Photonic crystal fibers. Science **299**, 358–362 (2003)
7. J.C. Knight, Photonic crystal fibers. Nature **424**, 847–851 (2003)
8. A. Cusano, D. Paladino, A. Iadicicco, Microstructured fiber bragg gratings. J. of Lightwave Technol. **27**(11), 1663–1697, ISSN: 0733-8724 (2009)
9. A. Iadicicco, S. Campopiano, A. Cusano, Long period gratings in hollow core fibers by pressure assisted arc discharge technique. Photonics Technol. Lett. **23**(21), 1567–1569 (2011)
10. D. Psaltis, S.R. Quake, C. Yang, Developing optofluidic technology through the fusion of microfluidics and optics. Nature **442**, 381–386 (2006)
11. J. Canning, M. Stevenson, T.K. Yip, S.K. Lim, C. Martelli, White light sources based on multiple precision selective micro-filling of structured optical waveguides. Opt. Express **16**(20), 15700–15708 (2008)
12. M. Pisco, A. Iadicicco, S. Campopiano, A. Cutolo, A. Cusano, Structured chirped fiber bragg gratings. J. Lightwave Technol. **26**(12) 1613–1625 (15 June 2008)

13. F. Poli, A. Cucinotta, S. Selleri, *Photonic crystal fibers, properties and applications* (Material Science Springer-Verlag, Dordrecht, 2007)
14. G. Brambilla, Optical fibre nanowires and microwires: a review. J. Opt. **12**(4), 043001 (2010)
15. M. Consales, M. Pisco, A. Cusano, Lab-on-fiber technology: a new avenue for optical nanosensors. Photonic Sens. **2**(4), 289–314 (2012)
16. M. Consales, A. Ricciardi, A. Crescitelli, E. Esposito, A. Cutolo, A. Cusano, Lab-on-fiber technology: toward multifunctional optical nanoprobes. ACS Nano **6**(4), 3163–3170 (2012)
17. A.F. Abouraddy, M. Bayindir, G. Benoit, S.D. Hart, K. Kuriki, N. Orf, O. Shapira, F. Sorin, B. Temelkuran, Y. Fink, Towards multimaterial multifunctional fibres that see, hear, sense and communicate. Nat. Mater. **6**, 336–347 (2007)
18. M. Bayindir, F. Sorin, S. Hart, O. Shapira, J.D. Joannopoulos, Y. Fink, Metal–insulator–semiconductor optoelectronic fibres. Nature **431**, 826–829 (2004)
19. Alexander Gumennik, Alexander M. Stolyarov, Brent R. Schell, Chong Hou, Guillaume Lestoquoy, Fabien Sorin, William McDaniel, Aimee Rose, John D. Joannopoulos, Yoel Fink, All-in-fiber chemical sensing. Adv. Mater. **24**(45), 6005–6009 (2012)
20. S. Egusa, Z. Wang, N. Chocat, Z.M. Ruff, A.M. Stolyarov, D. Shemuly, F. Sorin, P.T. Rakich, J.D. Joannopoulos, Y. Fink, Multimaterial piezoelectric fibres. Nat. Mater. **9**(8), 643–648 (2010)
21. S. Fan, J.D. Joannopoulos, Analysis of guided resonances in photonic crystal slabs. Phys. Rev. B **65**, 235112 (2002)
22. A. Ricciardi, I. Gallina, S. Campopiano, G. Castaldi, M. Pisco, V. Galdi, A. Cusano, Guided resonances in photonic quasicrystals. Opt. Express **17**(8), 6335–6346 (2009)
23. M. Pisco, A. Ricciardi, I. Gallina, G. Castaldi, S. Campopiano, A. Cutolo, A. Cusano, V. Galdi, Tuning efficiency and sensitivity of guided resonances in photonic crystals and quasi-crystals: a comparative study. Opt. Express **18**(16), 17280–17293 (2010)
24. E.J. Smythe, M.D. Dickey, G.M. Whitesides, F.A. Capasso, A technique to transfer metallic nanoscale patterns to small and non-planar surfaces. ACS Nano **3**, 59–65 (2009)
25. E.J. Smythe, M.D. Dickey, J. Bao, G.M. Whitesides, F. Capasso, Optical antenna arrays on a fiber facet for in situ surface-enhanced Raman scattering detection. Nano Lett. **9**(3), 1132–1138 (2009)
26. D.J. Lipomi, R.V. Martinez, M.A. Kats, S.H. Kang, P. Kim, J. Aizenberg, F. Capasso, G.M. Whitesides, Patterning the tips of optical fibers with metallic nanostructures using nanoskiving. Nano Lett. **11**(2), 632–636 (2011)
27. W. Jung, B. Park, J. Provine, R.T. Howe, O. Solgaard, Highly sensitive monolithic silicon photonic crystal fiber tip sensor for simultaneous measurement of refractive index and temperature. J. Lightwave Technol. **29,** 1367–1374 (2011)
28. G. Shambat, J. Provine, K. Rivoire, T. Sarmiento, J. Harris, J. Vuckovic, Optical fiber tips functionalized with semiconductor photonic crystal cavities. Appl. Phys. Lett. **99**, 191102 (2011)
29. S. Scheerlinck, P. Dubruel, P. Bienstman, E. Schacht, D. Van Thourhout, R. Baets, Metal grating patterning on fiber facets by UV-based nano imprint and transfer lithography using optical alignment. J. Lightwave Technol. **27**(10), 1415–1420 (2009)
30. A. Dhawan, M.D. Gerhold, J.F. Muth, Plasmonic structures based on subwavelength apertures for chemical and biological sensing applications. Sens. J. IEEE **8**(6), 942–950 (2008)
31. Y. Lin, Y. Zou, R.G. Lindquist, A reflection-based localized surface plasmon resonance fiber-optic probe for biochemical sensing. Biomed. Opt. Express **2**, 478–484 (2011)
32. M. Pisco, G. Quero, A. Iadicicco, M. Giordano, F. Galeotti, A. Cusano, Lab on fiber using self-assembly technique: a preliminary study. Proceedings of SPIE 8421, OFS2012 22nd International Conference on Optical Fiber Sensors, 842188, 2012
33. M. Pisco, G. Quero, A. Iadicicco, M. Giordano, F. Galeotti, A. Cusano, Lab on fiber by using the breath figure technique. Proceedings of SPIE 8774, 87740R, Optical Sensors 2013

34. M. Pisco, G. Quero, A. Iadicicco, M. Giordano, F. Galeotti, A. Cusano, Ultrasensitive nanoprobes based on metallo-dielectric crystals integrated onto optical fiber tips using the breath figures technique. Proceedings of the SPIE, Vol 8794, 87942P, Fifth European Workshop on Optical Fibre Sensors, 2013
35. George M. Whitesides, Bartosz Grzybowski, Self-assembly at all scales. Science **295**(5564), 2418–2421 (2002)
36. J.F. Galisteo-López, M. Ibisate, R. Sapienza, L.S. Froufe-Pérez, Á. Blanco, C. López, Self-assembled photonic structures. Adv. Mater. **23**(1), 30–69 (2011)
37. Y.A. Vlasov, X.-Z. Bo, J.C. Sturm, D.J. Norris, On-chip natural assembly of silicon photonic bandgap crystals. Nature **414**, 289–293 (2001)
38. A. Bolognesi, C. Mercogliano, S. Yunus, M. Civardi, D. Comoretto, A. Turturro, Self-organization of polystyrenes into ordered microstructured films and their replication by soft lithography. Langmuir **21**(8), 3480–3485 (2005)
39. M. Srinivasarao, D. Collings, A. Philips, S. Patel, Three-dimensionally ordered array of air bubbles in a polymer film. Science **292**(5514), 79–83 (2001)
40. M.H. Stenzel, C. Barner-Kowollik, T.P. Davis, Formation of honeycomb-structured, porous films via breath figures with different polymer architectures. J. Polym. Sci. Polym. Chem. **44**, 2363 (2006)
41. M. Haupt, S. Miller, R. Sauer, K. Thonke, A. Mourran, M. Moeller, Breath figures: self-organizing masks for the fabrication of photonic crystals and dichroic filters. J. Appl. Phys. **96**, 3065 (2004)
42. P. Escalé, L. Rubatat, L. Billon, M. Save, Recent advances in honeycomb-structured porous polymer films prepared via breath figures. Eur. Polymer J. **48**(6), 1001–1025 (2012)
43. M. Hernández-Guerrero, M.H. Stenzel, Honeycomb structured polymer films via breath figures. Polym. Chem. **3**, 563–577 (2012)
44. F. Galeotti, V. Calabrese, M. Cavazzini, S. Quici, C. Poleunis, S. Yunus, A. Bolognesi, Self-functionalizing polymer film surfaces assisted by specific polystyrene end-tagging. Chem. Mater. **22**, 2764–2769 (2010)
45. A. Ricciardi, M. Pisco, I. Gallina, S. Campopiano, V. Galdi, L. O' Faolain, T.F. Krauss, A. Cusano, Experimental evidence of guided-resonances in photonic crystals with aperiodically ordered supercells. Opt. Lett. **35**, 3946–3948 (2010) http://www.opticsinfobase.org/ol/abstract.cfm?URI=ol-35-23-3946
46. A. Ricciardi, M. Pisco, A. Cutolo, A. Cusano, L.O' Faolain, T.F. Krauss, G. Castaldi, V. Galdi, Evidence of guided resonances in photonic quasicrystal slabs. Phys. Rev. B **84**, 085135 (2011)
47. J.N. Anker, W.P. Hall, O. Lyandres, N.C Shah, J. Zhao, R.P. Van Duyne, Biosensing with plasmonic nanosensors. Nat. Mater. **7**, 442–453 (2008)

Chapter 12
Electrohydrodynamic Dispenser for Delivering Multiphase Samples at Nanoscale

Sara Coppola, Veronica Vespini, Francesco Merola, Melania Paturzo, Lisa Miccio, Oriella Gennari, Simonetta Grilli and Pietro Ferraro

Abstract Manipulation of liquids on micro- and nanoscale is a key issue in many fields of technology and biotechnology. Electric field induced formation of microliter and nanoliter droplets is very useful in lab-on-chip applications and would represent a new and contact-less way for functionalizing smart materials [1–3]. Ink-jet printing, manipulation of biomolecules, deposition of inorganic, organic and biological inks [4, 5], dispensing of small amounts of material into well-defined areas would be a further possibility for functionalizing sensing area for lab-on-a-fiber devices and related applications.

S. Coppola (✉) · V. Vespini · F. Merola · M. Paturzo · L. Miccio · O. Gennari · S. Grilli · P. Ferraro
INO – CNR, Via Campi Flegrei 34, 80078 Pozzuoli, NA, Italy
e-mail: sara.coppola@ino.it

V. Vespini
e-mail: veronica.vespini@ino.it

F. Merola
e-mail: francesco.merola@ino.it

M. Paturzo
e-mail: melania.paturzo@ino.it

L. Miccio
e-mail: lisa.miccio@ino.it

O. Gennari
e-mail: oriella.gennari@ino.it

S. Grilli
e-mail: simonetta.grilli@ino.it

P. Ferraro
e-mail: pietro.ferraro@ino.it

A. Cusano et al. (eds.), *Lab-on-Fiber Technology*,
Springer Series in Surface Sciences 56, DOI: 10.1007/978-3-319-06998-2_12,

12.1 Introduction

In this chapter we combine the capabilities of the pyrofluidic platform here described with the possibility of working in condition of very high resolution, showing amazing perspectives for fibre integration into a chip. The fabrication of smart and personalized materials requires the ability of high precision "on-demand" material dispensing. In order to satisfy this claim the multi purposed properties of the pyrofluidic platform are described. This innovative technology is based on the use of pyroelectric fields for generating an electro-hydrodynamic effect able to manipulate and dispense liquid compartments with very low volumes (down to attolitres) and radii down to around 300 nm from a sample reservoir [6]. A pyroelectric crystal (e.g. Lithium Niobate, LN) is used for creating a spatial distribution of surface electric charges. The phenomenon exploited in the process is the electro-hydrodynamic effect [7] where a liquid reservoir, for example in the form of a sessile drop, is deformed and successively small volumes of liquid are drawn under the action of the electric field. This technique, compared to traditional electro-hydrodynamic methods [8], is able to generate electric fields (typical 10 kV/mm) and is free from electrodes, high-voltage circuits and/or special capillary nozzles, thus extending the capability of use to high viscous polymer materials. The ejected liquid is released onto the receiving substrate of interest. For the dispensing application this effect is used for transfering materials between two substrates and for manipulating droplets and nanoparticles threedimensionally [9, 10]. The simplicity of the technique suggests potential uses for printing mutliphase materials providing the flexibility desired in direct patterning on a fibre of fragile organics or biological materials that are incompatible with conventional patterning methods. This system is easy to use and non-invasive, and is therefore potentially useful also as an instrument for in situ drawing and delivering of liquid biological samples within a chip. Different methods of pyro-electro-hydrodynamic manipulation, dispensing, delivery and patterning of multiphase materials, potentially useful for lab-on-a-fibre applications, are described.

12.2 Manipulation of Polymer- and Oil-Based Materials for Patterning Micro Lenses

Microlenses are key components for optical devices and are widely applied in several application fields, such as communications, three-dimensional displays, optical data storage and photodetectors, playing a fundamental role in many optical systems. Numerous classes of microlenses exist, depending on the embedding technology and the specific applications. At now, microlenses with variable focusing have been obtained by actuation of liquid crystals or other liquids through electro-wetting, electrophoresis or hydrodynamic pressure [11, 12]. On the other hand, a large variety of fabrication processes have been developed for plastic/polymer based microlenses [13, 14], such as embossing [15], soft-lithography [16], micro-molding [17], photolithography [18], electron beam lithography [19] reactive ion etching [20], laser assisted technique [16] and printing techniques [21, 22]. In particular, these last techniques were advised as direct methods for high-quality and high-precision processes making the fabrication short time-consuming and cost-effective.

The ability to pattern PDMS reliably in the form of both thick substrates and thin membranes or films is critical to expanding the scope of its applications, especially in the fields of microfluidics and bioengineering. Polydimethylsiloxane (PDMS) is an elastomer material widely used in different fields of application, such as micro/nanofluidics [23], electrical insulation, micro/nanoelectromechanical (MEMS/NEMS) devices [24], soft lithography [25], quantum dots [26]. The PDMS material offers many advantages . It is optically transparent, electrically insulating, mechanically elastic, and gas permeable. PDMS is also biocompatible [27], thus finding application in the field of bioengineering where the position of cells on a substrate is important for different purposes. These include biosensor fabrication for drug toxicity and environmental monitoring [28], tissue engineering [29], patterning of active proteins [30], patterning of animal cells [31], and basic biology studies where the role of cell adhesion, shape, proliferation, and differentiation are studied as a function of cell–cell and cell–extracellular matrix interactions [32]. In this section we propose a different approach for direct PDMS patterning. A lithography-based technique for realizing periodic patterning of thin-film PDMS structures onto lithium niobate (LN) substrates is presented and discussed. The PDMS film is microstructured according to the wettability variation of the LN substrate induced by the pyroelectric effect. The appropriate thermal treatment applied to the crystal induces the polymerization of the PDMS film, leading to a stable and reliable PDMS pattern. The wettability patterning and its periodicity is achieved by the exploitation of the periodic reversed domains fabricated in congruent z-cut LN wafers. Both sides polished and 500 μm thick LN crystals (from Crystal Technology, Inc.) were subject to standard electric field poling [33] in order to achieve a square array of hexagonal reversed domains. The samples were first resist patterned (photoresist Shipley S1813-J2, around 1.3 μm thick) by conventional mask lithography in order to achieve a square array of circular resist openings. The subsequent application of high voltage pulses, exceeding the coercive field of the material (around 21 kV/mm), allowed to fabricate the periodic domain reversed sample. The period of the structure was around 200 μm along both x and y crystal axis direction. Figure 12.1 shows the optical microscope image of the Periodically Poled Lithium Niobate (PPLN) samples fabricated by the electric field poling and used for the described experiments.

It is well known that LN is a rhombohedral crystal belonging to the point group 3 m at room temperature [34]. The lack of inversion symmetry induces different effects including the pyroelectricity. This is the manifestation of the spontaneous polarization change ΔP following to a temperature variation ΔT, according to the equation

$$\Delta P = P_S \Delta T$$

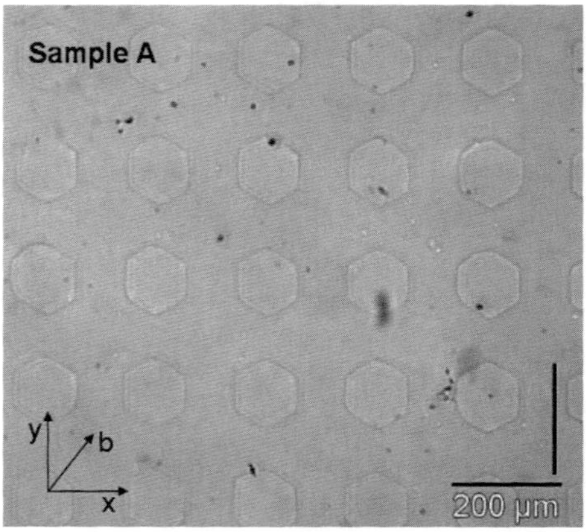

Fig. 12.1 Optical microscope image of a PPLN sample with a square array of reversed domains. The period of the structures is around 200 μm [50]

where ΔP is the coefficient of the polarization vector and P_S (Ps = 78 μC/cm^2) is the pyroelectric coefficient. At equilibrium, all Ps in the crystal are fully screened by the external screening charge and no electric field exists [35]. The change of the polarization, occurring with temperature variation, perturbs such equilibrium, causing a lack or excess of surface screening charge. Consequently, an electrostatic state appears and generates a high electric field at the crystal surface [36, 37]. Figure 12.2a, b show the schematic view of the PPLN sample cross section with the charge distribution occurring at the equilibrium state and in case of heating/cooling treatment, respectively. The arrows indicate the orientation of the ferroelectric domains.

According to the pyroelectric effect [35] the heating process makes the polarization magnitude to decrease, thus leaving surface screening charges uncompensated (see Fig. 12.2b top). These generate a net electric charge distribution depending on the inverted domain structure, with positive and negative sign onto the reversed and un-reversed domain regions, respectively. The screening charges in excess, continuously produced during the heating process, are no more attracted by the polarization charge and consequently are free to diffuse into the PDMS film.

The lens-like array topography exhibited by the PDMS film can be considered as the result of the equilibrium condition between the surface tensions and the electric forces related to the charge redistribution on the substrate. Figure 12.3 shows schematically the process steps implemented for the direct patterning of PDMS films. A layer of PDMS prepolymer solution (Dow Corning Sylgard 184, 10:1 mixing ratio base to curing agent) was spun onto the original z- face of the PPLN substrate at 6000 RPM for 2 min. The PDMS coated sample was then placed onto an hot-plate at a temperature of 170 °C for 30 s, thus inducing a rapid heating of the sample.

Here we show how microlens arrays made of PDMS can be self-formed on a polar dielectric substrate due to the pyroelectric effect. This self-patterning of PDMS by PyroElectroWetting (PEW) approach has been recently discovered [38]. We demonstrate here that the discovered process reported in [38] can be exploited

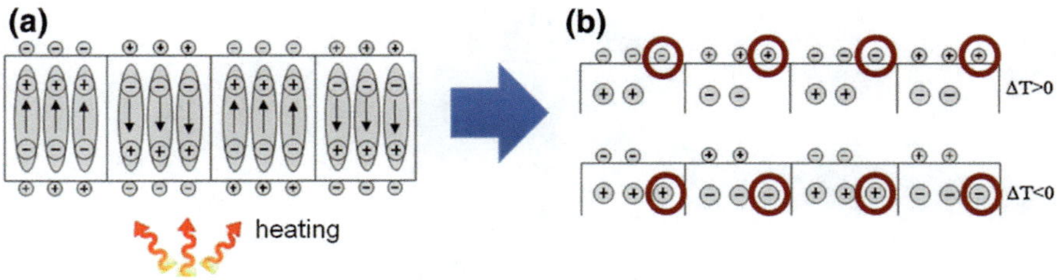

Fig. 12.2 Schematic view of the PPLN sample cross section with the charge distribution exhibited **a** at the equilibrium state; **b** in case of heating (*top*) and (*bottom*) cooling process [50]

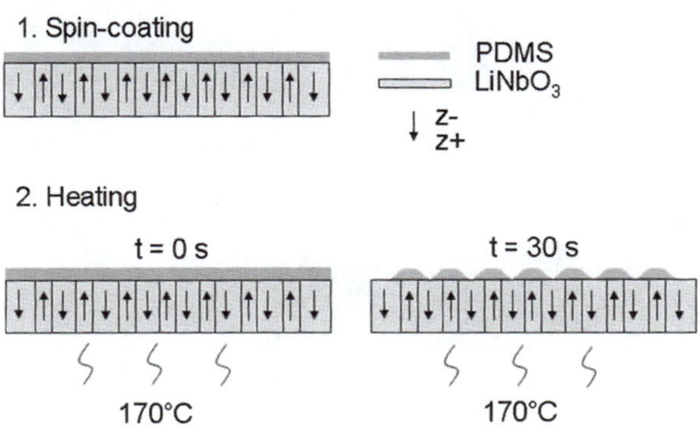

Fig. 12.3 Schematic view of the process steps

in several different ways to produce polymeric microlens arrays with different focal lengths, depending on the parameters adopted in the process of functionalization of the substrate. In fact, through this effective lithography-based technique, we obtained PDMS microlens arrays made of thousand of lenses as shown in Fig. 12.4 having micrometric size (100 mm of diameter) and focal length in the range of 300–1100 μm. To the best of our knowledge such kind of method for fabricating polymeric microlens arrays has never been performed before.

Different surface-charge lithography experiments were performed by varying both the temperature and the time length of the thermal treatment and the values at 170 °C and 30 s appeared to be the best settings for activating the pyroelectric effect, through the abrupt temperature increase of the substrate, as well as to insure the cure of the PDMS layer with the modified profile. The PDMS layer assumes a wave-like profile as shown in the schematic 3D and profile views in Fig. 12.5a, b, as a result of the equilibrium condition between the surface tension and those electric forces.

The lithography method presented here appears to exhibit various differences and advantages compared to the conventional PDMS patterning techniques. This method can be considered cost-effective and relatively quick, since no sophisticated process steps, such as atmosphere and pressure control or UV treatment are required by the procedure. In fact, nowadays, standard PPLN crystals are routinely

Fig. 12.4 Optical microscope image of the PDMS microlens array

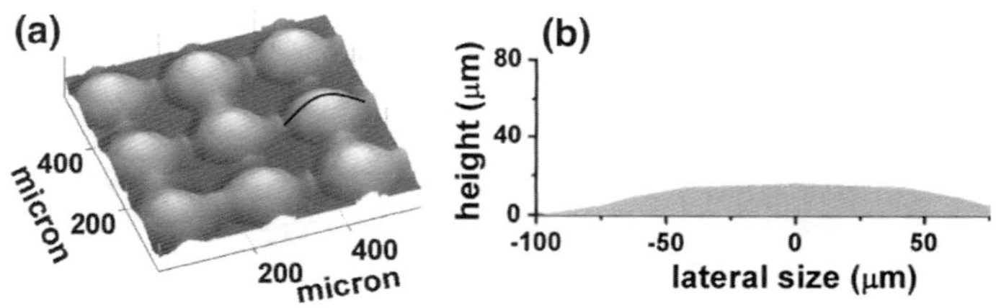

Fig. 12.5 Schematic view of **a** the micro-bump array and of **b** the profile along the black line in correspondence of a single micro-bump

fabricated through relatively cheap processes, while PDMS coating and thermal treatments can be performed by conventional spin-coaters and hot-plates. The PDMS patterns obtained here could be used also as alternative moulds for non-conventional soft lithography applications. The extensive homogeneity of the PPLN crystals available nowadays, and the inherent flexibility in the choice of the desired geometries, makes this technique interesting for applications in the field of biology, such as for cell patterning, where high throughput screens into single integrated microfluidic devices are desirable and the additional chemical patterning of the substrate could be avoided. Moreover, the possibility to pattern PDMS layers directly onto LN substrates could open the way for a variety of applications in microfluidics and biology, where the chemical stability and the specific properties (piezoelectricity; electro-optics; pyroelectricity; nonlinearity; etc.) of the bare LN regions, could be exploited for innovative operational functions. Other possible applications of this technique include microlens arrays and liquid-filled microlens arrays for integrated optical devices. Periodic square arrays of PDMS micro-bumps, with intermediate wells, have been fabricated by an appropriate thermal treatment providing both surface modification and thermal curing of the PDMS layer. This lithographic technique could be a quick and cheap alternative to the conventional PDMS patterning methods which generally use external forces, load/pressure or sophisticated surface treatments. Potential applications of the technique are in the field of biology and biomedicine for fabricating large areas adhesion/growth site arrays for cells analysis.

The same physical principle exploited to fabricate PDMS microlenses on LN substrate is used to generate an array of liquid micro-lenses. The most popular and well established approach to obtain liquid lenses is based on the electrowetting effect [39]. The electrode-based liquid lenses usually consist of two immiscible liquids manipulated applying an electric voltage that cause a reshaping and a re-arrangement of the liquid meniscus.

The change of curvature, in modulus and sign, has direct effect on the refraction of through-transmitted light, thus allowing to switch between a converging and a diverging lens with flexible focal lengths [40–46]. A sessile liquid drop, free standing on a flat electrode surface, can be also manipulated by a second needle-like electrode immersed into the drop [47]. Another configuration are the pressure-driven liquid lenses that make use of a small liquid reservoir having flexible and transparent membranes. The lens shape, and thus the focal power of the micro-fluidic lens, can be changed by varying the hydrostatic pressure of the liquid volume [48, 49].

The PPLN sample was mounted onto a digitally controlled hot-plate to ensure a reliable control of the substrate temperature during the experiments. The liquid used in this work was a carboxylic acid (pentanoic acid—$C_5H_{10}O_2$) in the form of an oily substance. The values of the dielectric constant and of the refractive index are 2.66 and 1.407–1.41, respectively. The PPLN sample was coated with a thin film of this oil (oil thickness is about 200 μm) and subject first to an heating process up to 100 °C at a rate of around 20 °C/min, and then let cooling down to room temperature.

The heating process was performed by increasing the temperature from 40 °C up to 100 °C. The cooling was achieved by letting the temperature to decrease from 100 °C down to 40 °C, thus with an approximate rate of 12 °C/min. The whole process took about 5 min and the liquid lens array kept stable topography for about 30 min at 40 °C, after its formation.

The liquid microlenses were formed in correspondence of the hexagonal domains and thus with a lateral dimension of about 100 μm. Moreover, the lens formation followed faithfully the domain grating, thus exhibiting perfect homogeneity over the whole patterned region [50]. Figure 12.6 shows the optical microscope image of the formed microlenses array and the corresponding schematic views of the sample cross section in two different cases, that we name *wave-like lenses regime* and *separated lenses regime,* by using a thick or a thin layer of liquid onto the crystal surface, respectively [12] . More complex geometries, such as hexagonal or circular arrays, could be used for further investigations, by simply providing the appropriate photolithographic mask for the pattern generation [51, 52].

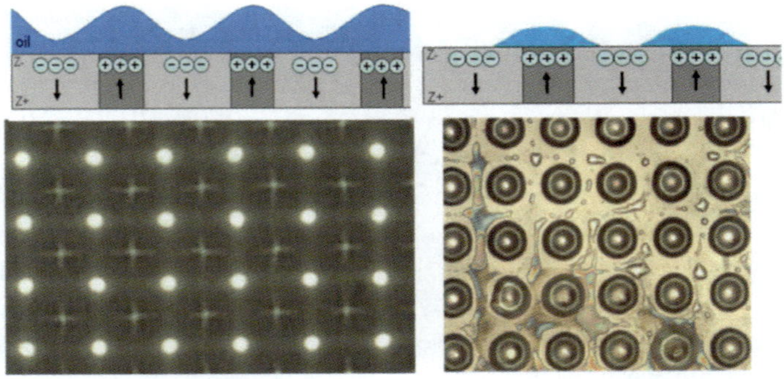

Fig. 12.6 *Top* schematic views of the sample cross section corresponding to the wave-like lenses regime and to the separated lenses regime of the microlens array, respectively. *Bottom* optical microscope image of the formed microlenses array The difference between the two regimes stands in the liquid thickness. The *black arrows* indicate the orientation of the spontaneous polarization

12.3 Pyro-Electrohydrodynamic Platform for Manipulating Multiphase Liquids at Nanoscale

The ability to manipulate tiny amounts of liquid is a soaring desire in many fields of technology, biotechnology and combinatorial chemistry [5–33]. A variety of functionalities such as manipulation [53] and droplet dispensing, also very useful in microfluidics and optofluidics systems, can be performed by different approaches, for example by electrowetting (EW) or by thermocapillarity [53, 54]. In fact, the development of such techniques could have impact in biological assays or functional devices technologies. Various hypothesis and methods are being followed in order to accomplish this demand for a reliable liquid manipulation on micro-nano scale, such as the adaptation of the shape of liquid menisci, patterning, displacement and drop-formation [5]. Standard microfluidic systems relied essentially on hydrodynamic forces while the electric field assisted approaches, have become increasingly popular only recently, owing to their enhanced flexibility and additional functionalities [55, 56]. Electric field induced formation of microliter and nanoliter droplets is useful in electrospray mass spectroscopy, ink-jet printing, manipulation of biomolecules [57] providing the so called "on-demand" material dispensing . In the past, researchers have used thermal treatments to induce specific surface tension gradients able to eject droplets from nozzles [58]. On-demand drops much smaller than the delivery nozzle have been produced by pulsed EHD jets [59]. Recently, the resolution of the EHD approach has been improved through fine microcapillary nozzles in order to pattern at sub-micrometric scale i.e. below the limit of resolution provided by commercial thermal or piezoelectric inkjet printers. Anyway it is important to note that a method, providing an order of magnitude reduction in drop volume by using the same nozzle, has been reported in literature [60]. Water droplets of few attoliters have been dispensed by nanopipetting [61] and very recently a nanoscale dispensing technique based on the use of an hollow atomic force microscope probe modified by focus ion beam lithography has been presented in literature [62]. Nonetheless, even though commercial EHD jet printing systems are not yet available, the future development could satisfy the growing demand in the field of nanotechnology for ever more sophisticated lithographic methods able to realize ever finer and more complex patterning. Such EHD printing methods can also provide the flexibility desired in direct patterning of fragile organics or biological materials that are incompatible with conventional patterning methods such as photolithography.

Recently a novel concept of EW was developed, where the actuation of liquids in contact with a polar dielectric crystal was achieved through an electrode-less configuration. In particular, the method made use of the pyroelectric effect induced onto lithium niobate (LN) crystals functionalized with appropriate micro-engineered ferroelectric domains [63]. Wettability patterning, formation of tuneable liquid microlens arrays and self-assembling lithography were demonstrated to be possible by that approach [38, 50]. In this chapter we propose an approach that, differently from the standard EHD techniques [8, 64], adopts an electrode-less configuration providing high versatility. Electric forces are activated pyroelectrically by scanning a heated-tip on a functionalized substrate and the pyroelectric functionality of a LN substrate is used for non-contact manipulation of liquids leading to the formation of a smart "dispensing gun". The technique presented here allows one to avoid the use of high-voltage power supplies and electrical circuits, and moreover there is no need to design and fabricate nanocapillary nozzles. The flexibility offered for manipulating the "dispensing-gun" is intriguing, demonstrating that various functionalities can be engineered for a smart manipulation of the liquid reservoir and of the dispensing process.

Figure 12.7 shows the schematic view of our electrohydrodinamic system. A liquid drop or a liquid film is deposited on a microscope glass slide, while the upper plate is a z-cut LN wafer. The experiment made use of a non-contact thermal stimulus applied through a IR source (CO2 laser) emitting at a wavelength of $\lambda = 10.6$ μm and of a hot tip source in contact with the LN crystal. The laser and the tip can be scanned to induce point-wise thermal-stimuli. The laser beam irradiates the upper functionalized substrate (z-cut LN, 500 μm thick) and the separation distance between the two substrates is D = 1 mm.

LN reacts to the thermal-stimuli by building-up an electric potential across the z-cut LN crystal's surfaces because of the pyroelectric effect, that consists in the spontaneous polarization change ΔPs following to a temperature variation ΔT. At equilibrium, the crystal Ps is fully screened by the external screening charge and no electric field exists [36]. When the tip-source or the laser beam locally heats the crystal, a sudden surface charge density σ immediately appears given by $\Delta\sigma = p\Delta T$neglecting losses, where Pc is the material-specific pyroelectric coefficient (Pc $= -8.3 \times 10^{-5}$ C/°C/m^2 for LN at 25 °C). The electric field exerts an attractive force on the liquid as shown in Fig. 12.7b. In the case of a sufficiently strong electric field, thin liquid jets can be released from conical tip structures

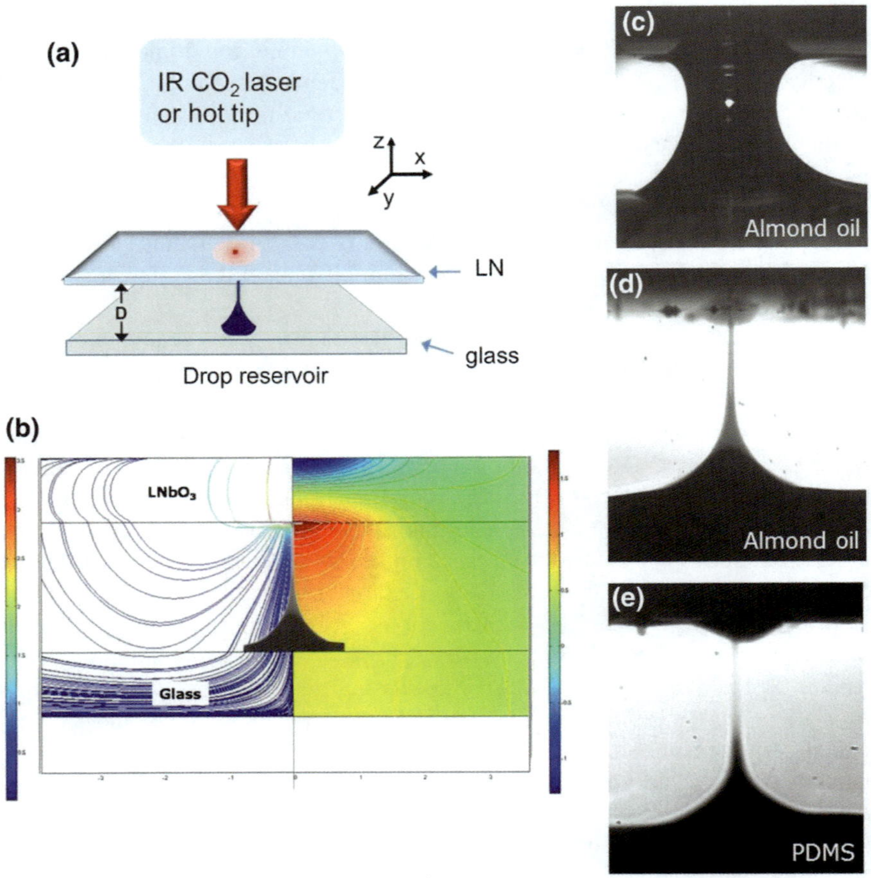

Fig. 12.7 **a** Schematic diagram of the dispensing microfluidic system by using the IR laser (CO2) as a thermal-stimuli tool or a resistive heated-tip in contact with the LN sample. A sessile drop or liquid film is deposited onto the *bottom* substrate (glass). **b** 3D axis-symmetric plot of the electric field lines (*left*) and electric potential (*right*) obtained by a finite element method simulation. The electric potential is generated by the pyroelectric effect induced onto the functionalized substrate through the thermal stimuli of the heated-tip. **c** Image of the typical liquid bridge obtained when D is shorter than the critical distance (see (12.1)). The different contact angles onto the up and *bottom* solid-liquid interfaces are clearly visible. **d** Image of the typical liquid shooting effect during the dispensation of almond oil through the heated-tip thermal excitation. **e** Same as e, but in this case the liquid is PDMS. Due to the higher viscosity the blasting liquid cone is continuous

(similar to Taylor's cone usual in electro-spray). On the other hand in our case the liquid is not conductive, consequently we are not exactly in the Taylor's cone regime [55].

When the liquid (either the sessile droplets or a film) starts to deform under the action of the electric field, two evolutions are possible. Case (I): if the liquid volume and the separation distance D between the two plates are appropriate then a stable liquid bridge can be formed (see Fig. 12.7c). For a given volume, the critical distance below which a bridge can be established is expressed by [65].

$$Dc = (1 + \vartheta/4)\ V^{1/3} \tag{12.1}$$

where θ is the contact angle and V is the volume.

A typical liquid bridge is shown in Fig. 12.7c. Nevertheless the most relevant case for us is the Case (II): if the separation D is above the critical value, a stable liquid bridge cannot be established between the plates. We designed our experiments with the aim at using such instability for dosing and dispensing liquid drops. The typical situation of liquid shooting in case of two different liquids, almonds oil and polydimethylsiloxane (PDMS), are shown in Fig. 12.7d, e respectively.

Results shown in Fig. 12.8a, b demonstrate the possibility to dispense nL almond-oil drops from a liquid film and a sessile droplet reservoir. A heated tip was used in case of the liquid film (see Fig. 12.8a), while for the sessile drop the thermal excitation was induced by an IR laser. The dynamic evolutions show the deformation of the liquid in both situations and it is possible to note from the sequences how either the height of the film and the drop increases under the action of the EHD force.

Figure 12.8c, d present the temporal behaviour of the liquid volume transferred to the substrate (insets) and the height's change (main plots) during shooting. For the case of Fig. 12.8a the rate is approximately of ~30 nL/ms. After 5 shots the total volume transferred to the pendant droplet is 160 nL. The drop acts like a dispensing-gun with a repetition rate that depends on the liquid response to EHD force. The experiment clearly shows that, after the Taylor's cone has been established to allow the first shot, the dispensing-gun blasts liquid droplets periodically till the electric filed is on. The period of the shooting was of 50 ms. The cycle is repeated many times during cooling. Further studies should address a deeper investigation on periodicity of the process and the volume regularity of the sequential shoot drops.

Figure 12.8b shows how the CO2 laser beam can be used to extract completely the liquid from the reservoir-drop, probably thanks to the higher heating efficiency. In particular, the irradiation of the substrate by five laser pulses (power of 10 W and 100 ms long) made the droplet reservoir to dispense up to 55 shots. The initial volume was 180 nL while the final total volume transferred at last frame in Fig. 12.8b was 164 nL and the remaining drop was 16 nL. The shooting period was ~200 ms while the volume transferred in each shot was estimated to be 3 nL.

The flexibility of our approach is demonstrated by the following experiments where various functionalities have been proofed. The movement of either the heated-tip or the laser beam allows for example one to change the firing direction of droplets in a wide solid angle as shown in Fig. 12.9a. In fact, the electric field distribution can be changed rapidly (~2 s) by varying the heated-tip position, because the regions with highest electric field follow the tip's movement. The off-axis shooting angles reach values up to ~20 °C, thus giving the possibility for dispensing liquids on an area of ~23 mm^2 even though the droplet reservoir

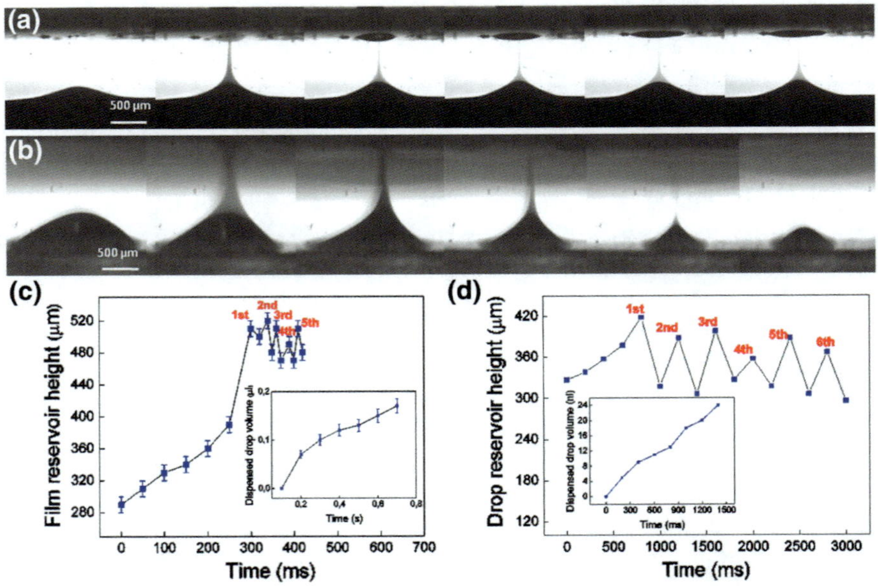

Fig. 12.8 **a** Sequence of different dispensing shots in case of liquid film and thermal stimuli induced by the heated-tip. **b** Sequence of several dispensing shots in the case of a sessile droplet and thermal stimuli induced by the IR laser. **c** (*main*) Plot of the temporal height variation of the droplet in the case of Fig. 12.8a and (*inset*) plot of the corresponding volume transfer rate. The single shots are detectable. The period of the shooting cycle is 50 ms. **d** (*main*) Plot of the temporal height variation of the droplet in the case of Fig. 12.8b and (*inset*) plot of the corresponding volume transfer rate. The single shots are detectable. The period of the shooting cycle is 200 ms

maintains a fixed position. Larger angles induce displacements of the drop reservoir, as described below. Another functionality of the "dispensing-gun" allows to deliver liquid in different locations along a line by the continuous shooting effect occurring during an appropriate movement of the gun. In this case we reports two different experiments: sessile drop (Fig. 12.9b) and a film layer (Fig. 12.9c). In both cases the "dispensing-gun" can be moved easily, just by moving the heated-tip. However in the first case the sessile droplet starts to move only at a critical angle. In fact the asymmetrical deformation experienced by the drop, under the action of no-axial electric force, creates the unbalance of the solid-liquid interface tensions with a net force resultant (Fig. 12.9b) that push forward the drop. This unbalance causes the drops displacement in analogy with what happens for thermocapillarity where instead the thermal gradient causes the unbalance of the solid-liquid interface tensions. Figure 12.9b shows the possibility to dispense the liquid in three different locations along a single scanning line, by moving the drop. In the case of a liquid layer the "dispensing-gun" can be moved easier because no solid-liquid interface tension prevents the movement of the blasting cone. A sequence of images show (Fig. 12.9c) lateral movement (x-axis) up to 1.6 mm with no interruption of dispensing action and 1 mm along the y-axis.

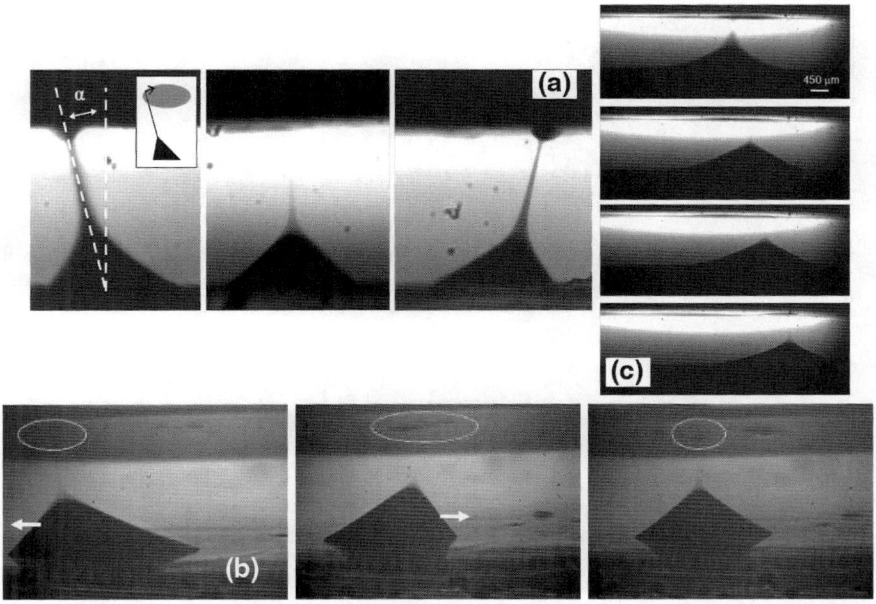

Fig. 12.9 **a** Dispensing nL liquid droplets along angled directions ($\sim 20°$) from a sessile drop reservoir. **b** Sessile droplet reservoir on PDMS coated glass. The activation and the successive lateral displacement of the "dispensing gun" allowed one to deliver liquid along relatively short liquid lines at different locations. **c** In case of the liquid film, the "dispensing gun" can be easily moved according to a 2D geometry

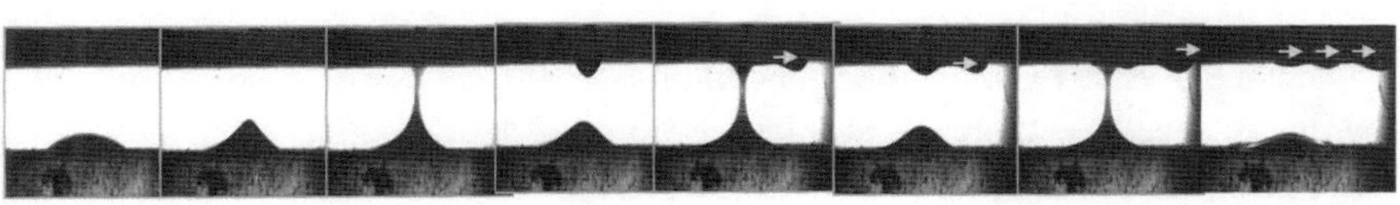

Fig. 12.10 Combination of the dispensing function with the transportation of the droplets. The angle in this case is clearly visible in the picture and is $\sim 11°$ with respect to the normal to the substrate

One more fascinating function that can be engineered is the harmonic combination of the dispensing function synchronized with the transportation of droplets, while they are continuously formed shown in the sequence of Fig. 12.10. The sequence of captured images in Fig. 12.10 shows clearly the formation and synchronized carrying of three distinct droplets in a row on right side. Such drops could be easily collected and managed into a microfluidic system. This function can be successfully implemented by choosing in an opportune way the position of the thermal stimuli (heated-tip in this case). The beauty of the physics is that the process seems to be self-organized based on two different physics effects: EHD and thermocapillarity but activated by a single external stimuli. The lateral

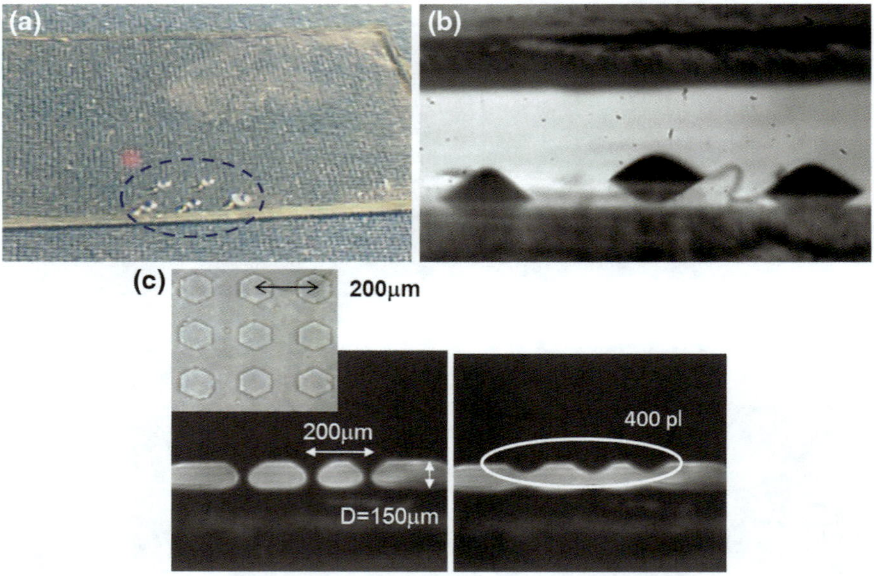

Fig. 12.11 **a** Five sessile drop reservoirs on the PDMS coated glass. **b** Sessile drops activated in **c** for multiple dispensing. The liquid deposition from each sessile drop arranged according to a 2D geometry on the *bottom* glass substrate was possible by scanning the IR laser on the substrate in three different locations. **c** Since D was shorter than the critical distance, the breaking of the liquid bridges gave the possibility of dispensing 400 pL pendant droplets onto the functionalized substrate corresponding to the desired locations driven by the geometry of the PPLN sample

displacement is driven by thermocapillarity that pushes the droplets in colder regions (in our case the right and left sides) (Fig. 12.10).

It is important to note that the thermal stimulus generated by a laser provides different advantages. The thermal excitation can be addressed easily in different locations (i.e. in correspondence of the different sessile drop reservoirs) by the precise movement of the beam. The heating energy can be varied by modulating the beam power. The beam focusing by a lens allows to better restrict the area on which the thermal stimulus should be applied. As illustrated in Fig. 12.11a the laser beam was addressed to the three different drop reservoirs deposited on the glass substrate as shown in the perspective view in Fig. 12.11b. The drops were activated sequentially. High throughput dispensing is possible by splitting the laser beam to get parallel blasting from multiple "dispensing-guns".

One more experiment was performed with a LN sample functionalized through a periodically poled structure, in order to demonstrate the possibility of dispensing pico-liter droplets in specific locations. The structure consisted of a square array of hexagonal regions with reversed polarization at a lateral distance of about 200 μm. A PDMS liquid layer was spin coated onto the glass substrate. The thermal stimulus enabled the formation of three droplets in correspondence of the

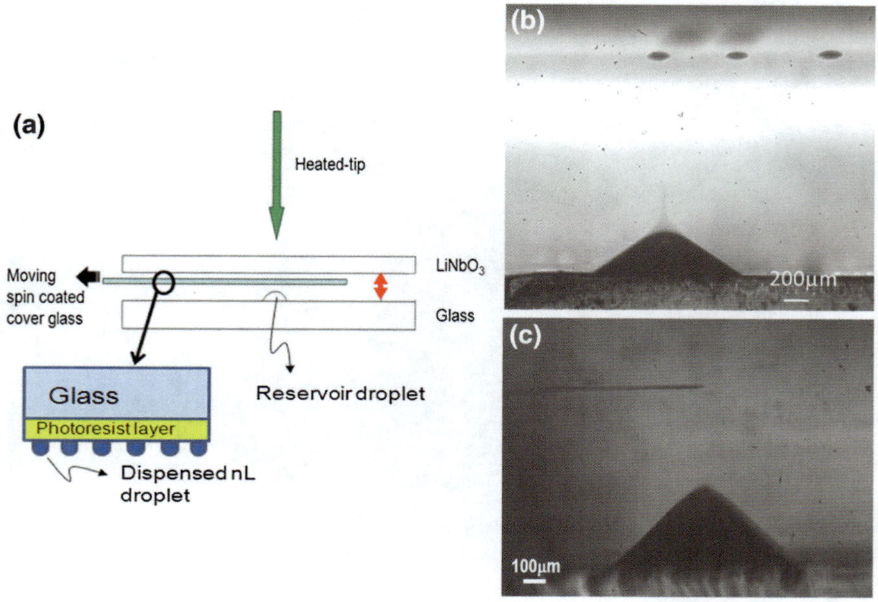

Fig. 12.12 **a** Arrangement for liquid dispensing onto a dielectric substrate. A photoresist spin coated cover glass is inserted in between the LN *upper* substrate and the glass plate in order to avoid the droplet spreading and to dispense droplets on different substrates. **b** Side view of the droplet dispensing onto the moving substrate. **c** Side view of the typical line printing onto the substrate

hexagons. Liquid filaments and unstable liquid bridges were formed due to the higher viscosity of the PDMS layer, as clearly shown in Fig. 12.11c. In fact, the separation distance D = 150 µm was again shorter than the critical distance given by (12.1). After the bridges collapse, 400 pL pendant droplets have been clearly dispensed.

The PEHD dispenser was implemented under a different configuration in order to improve the performance. All of the above experiments clearly show the possibility of drawing and dispensing liquid samples from a drop or film reservoir onto a functionalized substrate such as LN through intriguing functionalities. Anyway some drawbacks exist that prevent the application of the technique for developing a reliable tool for dispensing liquid droplets or lines with stable and defined volumes. In fact, the deposition of the droplets directly onto the warm LN wafer causes the spreading of the liquid due to the EW effect. In other words, the uncompensated charges generated pyroelectrically on the crystal surface reduce the surface tension of the dispensed droplets, causing spreading. Moreover the deposition of droplets onto chips of interest which are different from the LN plate would be desirable. In order to overcome these disadvantages the set-up was improved according to the scheme illustrated in Fig. 12.12a. The new set-up

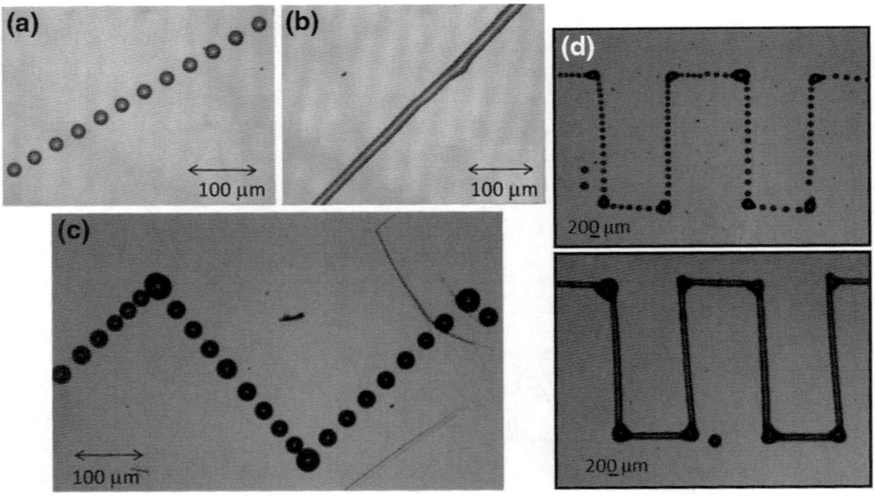

Fig. 12.13 **a** Simple liquid patterns consisting of periodic separate droplets, (diameters around 40 μm) and **b** continuous straight line (width around 40 μm). **c** *Dotted* patterned Greek fret. **d** Greek fret with combined continuous and *dotted lines*

allows one to avoid the spreading of the dispensed droplets, and gives the possibility of dispensing the liquid droplets according to specific geometries, such as aligned separate droplets or continuous lines. A liquid drop was deposited onto a glass slide over which a z-cut LN wafer was positioned at a changeable distance. A photoresist coated cover glass was inserted in between and mounted onto a computer controlled x-y translation stage. A hot tip was locally in contact with the LN crystal and induced a point-wise thermal-stimulus to dispense separate droplets or lines in case of shorter distance between the reservoir and the resist coated substrate, as shown respectively in Fig. 12.12b, c.

The control of translation direction and speed of the substrate permits to obtain droplets aligned along straight or lines, respectively. Moreover the volume variation of the drop reservoir allows one to dose the printed droplets with different volumes and sizes, Fig. 12.13.

This innovative technology reported here could open a new route in microfluidics and nanofluidics where manipulation, dispensing and dosage of nano-pico droplets is in high-demand in different fields of science and technology such as biotechnology, combinatory chemistry, pharmacology, deposition of inorganic, organic and biological matter. Different methods of pyro-electro-hydrodinamic dispensing and patterning of nL-pL droplets can be performed by a completely new approach without nozzles and electrodes but with much higher versatility to be integrated in microdevices and can be used for lab-on-a-fibre applications.

12.4 Self-assembling of Polymeric Liquids for Fabricating Single or Arrays of 3D Microstructures

Surface tension and capillarity are the key forces driving the formation of a wide variety of liquid shapes pervading the nature. The steady dew drops appearing on plant leaves and spider webs result from the minimization of the overall surface energy [66]. Break-up of viscoelastic filaments passes through the formation of temporary pearls interconnected by a thin thread, the so called beads-on-a string (BOAS) structures [67]. Thanks to the surface tension, the interfaces of such spontaneous structures exhibit extremely good optical quality. Therefore the ability of manipulating tiny amounts of liquids on surfaces has become of great interest for replicating those structures created by Mother Nature, which find applications in micro/optofluidics, nanotechnology and biotechnology. In particular, electrowetting [47] has become one of the most widely used tools for micro/nanoscale manipulation of liquids, leading to the development of a variety of applications ranging from adjustable lenses [46, 68] to lab-on-a-chip devices. Moreover, new techniques for manipulating surface wettability have been presented recently [25] and a wide variety of techniques have been developed for fabricating 3D structures of interest to the optics, photonics and nanotechnology fields. The so called soft lithography is able to transfer microscale patterns to a substrate by using an elastomeric stamp or mold, typically made of polydimethylsiloxane (PDMS) [25]. This technique allows one to develop biocompatible lab-on-chip devices for a wide variety of applications ranging from organic LED to cell and tissue engineering [69] and biomolecular analysis [70]. Among others, the so called 'capillary force lithography' is able to pattern polymers at nano/microscale in a single step avoiding the use of external forces [71]. Moreover, structures have been fabricated by using different procedures or effects such as the pyroelectric effect [72], the bond-detaching [73], the dip-pen lithography [74], laser-induced thermoplastic method [75]. More recently, other approaches have been developed for generating self-patterned structures by using destabilizing forces produced by electric fields, namely electrohydrodynamic (EHD) lithography [73], or by temperature gradients [76], or also by mechanical stress [77]. These methods are characterized by the challenging implementation of a fabrication procedure based on the critical control of liquid film instabilities. In fact, even little perturbations could drag the nanofluidic system towards non-fully predictable configurations. Nevertheless amazing results about polymeric patterns have been reported, demonstrating the possibility of controlling the process with high accuracy. The EHD lithography is performed usually at temperatures above the glass transition of the polymer film (typically PS or PMMA) [78] and permanent microstructures are obtained by a successive cooling process. Electrodes with relief structures have also been used or hierarchical structures have been obtained by using multiple materials. So far EHD lithography is very effective for fabricating polymeric structures with high accuracy at micro- and nano-scale, however it suffers different limitations such as the low aspect ratio and the time

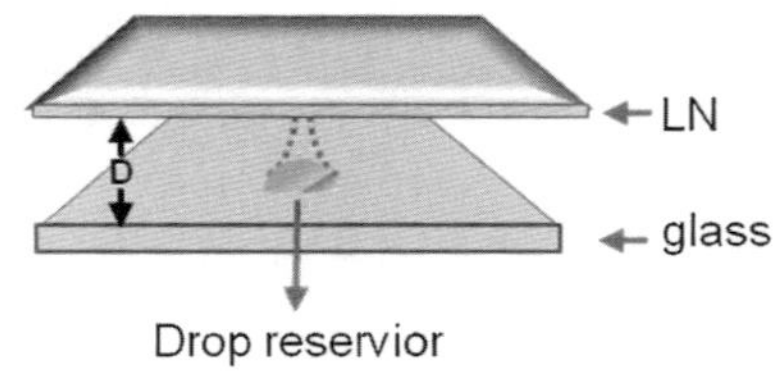

Fig. 12.14 Polymer drop reservoir subjected to the pyro-electric field

consuming of the annealing and cooling processes. Moreover, only a few classes of structures can be realized such as pillars, dots and lines [6].

Here we show a novel approach which exploits instabilities and self-assembling of polymeric liquids for fabricating single or arrays of unique 3D microstructures useful as micro-photonic components. The fascinating aspect is that the shape of such microstructures is first induced spontaneously by liquid instabilities through an EHD pressure and successively frozen, i.e. quickly solidified, by appropriate thermal treatments [79]. Therefore, the technique is able to convert fluidic configurations, which are otherwise intrinsically unstable and short-lived, into cross-linked and permanent 3D structures. Under appropriate conditions, discussed later, single or multiple ordered structures are obtained here without resorting to expensive and time consuming nanotechnology processes required for structuring the electrode. Moreover, multiphase or biological structures are fabricated through the appropriate dispersion of solid particles into the PDMS layer. Furthermore, the photonic application of one frozen microstructure is fully demonstrated.

The experimental configuration adopted in this work consists basically of a glass slide supporting a sessile polymer drop as liquid reservoir, facing a lithium niobate (LN) substrate at distance D (Fig. 12.14). Any temperature change of LN builds up electric charges through pyroelectric effect.

At the equilibrium, the spontaneous polarization Ps of the LN crystal is fully compensated by the external screening charge and no electric field exists. According to the pyroelectric effect the temperature change ΔT, obtained heating the LN substrate, causes a variation ΔPs which builds-up an electric potential across the z surfaces. Therefore, neglecting the losses, a surface charge density $\sigma = Pc\Delta T$ immediately appears. Pc is the material specific pyroelectric coefficient ($Pc = -8.3 \times 10^{-5}$ C/°C/m^2 for LN at 25 °C). The electric field generated by the surface charge σ causes an electrohydrodynamic (EHD) instability on the polymer film. Heating the sample, capillary film instability in term of surface waves begin to develop and in turn leads to a hexagonally ordered film fragmentation. The resulting strong electric field is able to exert a force on the liquid PDMS creating a bridge across the two substrates, as shown in Fig. 12.15. The fluid dynamics causes liquid depletion from the unstable bridge and thus the formation of various temporary liquid silhouettes.

The pyroelectric effect is activated by different sources depending on the specific experiment: hot tip of a conventional soldering iron; jet of hot air from a conventional hot air gun; hotplate. Once the system is activated by the thermal excitation, the liquid PDMS is subjected to an EHD pressure that causes the

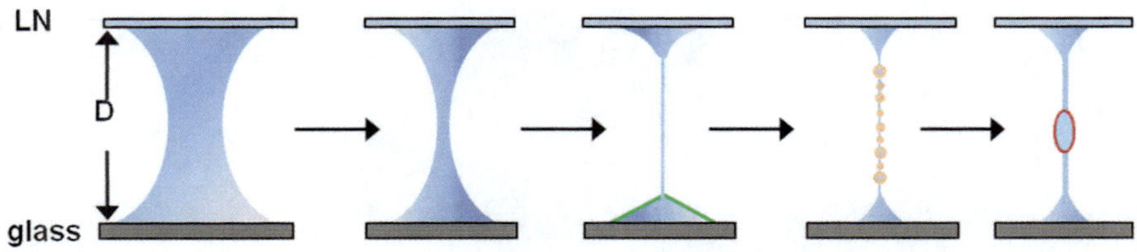

Fig. 12.15 Scheme of the temporal evolution of the unstable liquid bridge during the PEHD process, due to the depletion of the liquid and leading to unstable polymer bridges and thinner polymer columns

formation of various liquid shapes that are rapidly cross-linked. The temperature of the operation used in the experiment is about 120 °C.

The dynamic evolutions were observed by a high-speed (CMOS) camera imaging system. Despite the final stage of such unstable liquid column is the collapse and breaking, we are able to freeze the microstructures by a rapid-curing approach. A hot-air jet or an halogen lamp could be used for both stimulating the generation of the structures and for their rapid curing (typically in about 60 s at 200 °C).

The micro-sized structures of Fig. 12.16 are high aspect ratio structures fabricated using this novel technique. For the fabrication of this structures a PDMS polymer was used (Sylgard 184, 10:1 mixing ratio base to curing agent). The reservoir drops were placed onto the glass base while the Lithium Niobate LN wafer, lid of the package, was stimulated thermally by the hot-air-jet for 15 s at 150 °C. Successively the package was turned around to put the LN wafer in contact with a cold surface in order to induce a rapid temperature gradient. The package was then observed by an optical microscope. While the nanofluidic instability took place with continuous formation of beads-on-a-string BOAS, the same hot-air-jet was used for curing such structures (typically in about 60 s at 200 °C).

This is a typical case where the same thermal source was used for stimulating the pyroelectric effect as well as for the rapid curing of the structures. In this case the structures were formed by the combination of the horizontal and vertical PEHD instabilities. The partial curing sustained by the PDMS during the first hot-air-jet (used as pyroelectric stimulus) appears to favour the formation of BOAS. The images refer to different samples to demonstrate the repeatability of the results, Fig. 12.17. One more interesting aspect is the possibility of regulating the position of the beads along the wire prior to the rapid curing of the structure, by modulating appropriately the pyroelectric field.

The frozen microstructures, represented in Figs. 12.16 and 12.17, such as soft solid-like bridges with different aspect ratios and BOAS structures, could find application in different fields. For example, the wires are potential optical waveguides similar to optical fibers and could be used for collecting or distributing light signals while microspheres may be used as resonators where the resonance modes, known as whispering gallery modes (WGMs), arise from the confinement of light

Fig. 12.16 High-aspect ratio microstructures

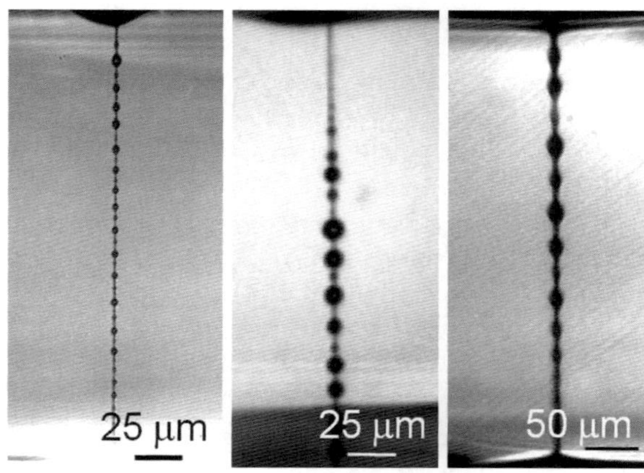

Fig. 12.17 Microstructures with BOAS

by total internal reflection. In particular, BOAS may be used as passive or active optical microresonators (i.e. by exciting WGMs). As active elements, we successfully embedded CdSe microspheres inside BOAS, and excited them by an UV laser.

In Fig. 12.18 is reported a microaxicon structure of PDMS fabricated using the 3D lithography technique. A layer of PDMS polymer diluted with hexane (mixing ratio 3:1 PDMS/hexane) was spin coated on a microscope slide at 3000 RPM (Round Per Minute) for 1 min. Thin spacers of 100 μm were glued on the edges of the base to superimpose the LN plate at a fixed distance. The sample with the film of PDMS was then placed on a hot-plate at a temperature of 115 °C for 10 min, generating a series of PDMS cones. By heating the sample, a film instability arises and the rapid curing process of the polymer generates the formation of microaxicon structure. The lid of LN was then removed, leaving the microaxicon deposited on glass.

The cone of Fig. 12.18 is an "axicon" lens. We have used this structure, in combination with a laser source, as an "optical tweezer", in order to trap and move micrometric latex particles. It is able to produce Bessel beams having high

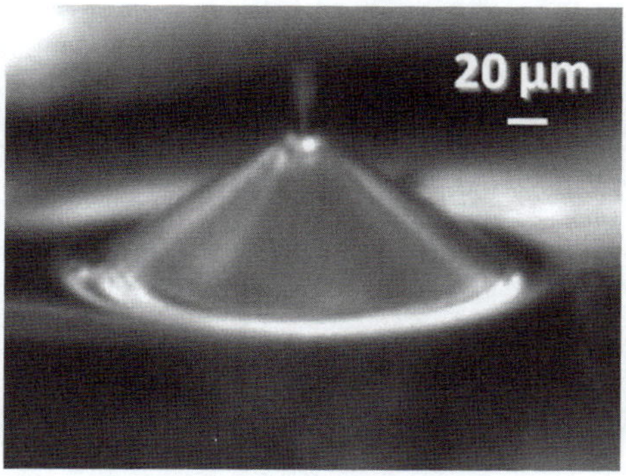

Fig. 12.18 Microscope image of the fabricated PDMS microaxicon

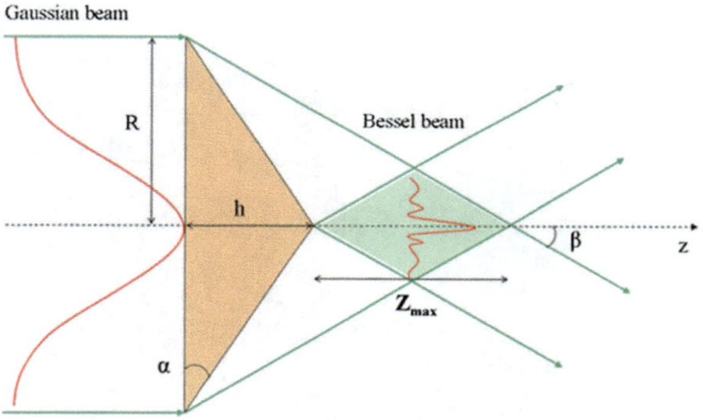

Fig. 12.19 A gaussian beam passing through the axicon becomes a Bessel beam

depth-of-focus when compared to focussed Gaussian beams obtained by high numerical aperture microscope objectives, Fig. 12.19.

This is a clear advantage for optical trapping (i.e. by the light forces) compared to the use of microscope objectives. In fact, an axicon is very useful as optical tweezers for trapping and sorting microparticles or live biological cells. Small dimensions of such axicons (hundreds or tens of microns) could make these elements suitable for embedding them into lab-on-chip devices to image or trap micro-objects in microfluidic channels. This would be unrealizable with a cumbersome microscope objective having limited depth of field. We performed optical tweezers experiments that demonstrate real application of the frozen PDMS axicons, Fig. 12.20. In a first experiment a 10 µm latex microsphere, dispersed in water, was trapped and moved, while a second experiment demonstrates the high depth-of-focus of the axicon by trapping up to six particles simultaneously in multiple planes along the z direction. Such multiple trapping of microscopic objects may be of great interest for biology and/or medicine applications.

As said before, BOAS may be used as microresonators. Refractive index changes can induce a resonance shift that is used for label-free detection of molecules or viruses with unsurpassed sensitivity [71]. The PDMS BOAS fabricated here may be used for sensing applications since PDMS has demonstrated

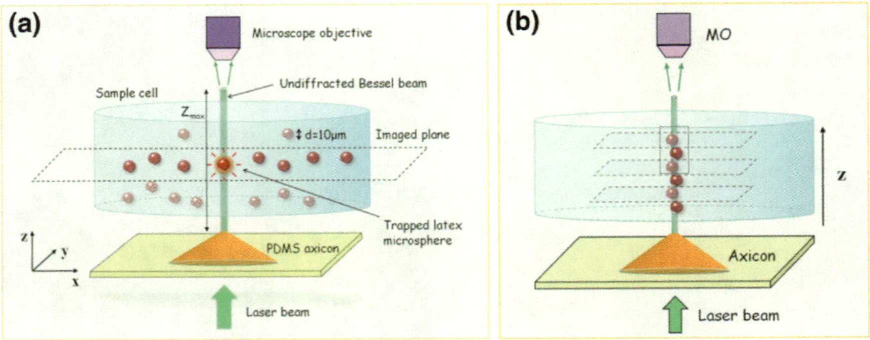

Fig. 12.20 Sketch of the set-up for single (**a**) and multiple (**b**) trapping

Fig. 12.21 QDs-embedded microstructures

quality factors of 10^6. Such microresonators do not require complex fabrication processes and the smooth surfaces are spontaneously given by the surface tension. Moreover, we have demonstrated the fabrication of BOAS by embedding CdSe/Zns core/shell QDs, Fig. 12.21.

The well known softness of PDMS would provide additional functionality for tuning the cavity. The fabrication process is easier to accomplish compared to other methods where the QDs are embedded into the periphery of polymer microspheres through complicated chemical procedures and enables the inclusion of the QDs into deeper regions of the structures. To obtain the structures of Fig. 12.21 samples of degassed PDMS with CdSe QDs nanocrystals were prepared by mixing 3 ml of PDMS (Sylgard 184) with 300 μL of Lumidots CdSe 590 nm (5 mg/mL concentration in toluene, fluorescence emission maxima spanning the visible spectrum, Sigma Aldrich). The mixtures were obtained by stirring and subsequent vacuum drying in order to vaporize the solvent (toluene). A mixed droplet was placed on the glass slide and the auxiliary LN plate was mounted on independent vertical translation stages (separate configuration) to control the reciprocal distance D while PDMS wire was obtained. The activation of the process leading to the formation of the microstructures and the rapid curing were obtained using a conventional halogen lamp (beam power around 150 W) with a parabolic spectrum focused onto the microstructures. In conclusion, we have demonstrated a rapid way to freeze nanoliter liquid instabilities producing microphotonics elements, a sort of natural nanofluidic "optical foundry". We believe that further theoretical investigations, in combination with the simple rapid-curing approach demonstrated here, will allow in the future the development of a previously undescribed "3D lithography concept/platform," in which nano-liquid instabilities could be "fluidynamically" designed a priori with the aim of fabricating even more complex shapes exploitable in many fields.

12.5 Conclusion

In this chapter we combine the capabilities of the pyrofluidic platform with the possibility of working in condition of very high resolution, showing amazing perspectives for fibre integration into a chip. A variety of functionalities of the pyro-electrohydrodynamic platform, such as droplet self assembling and dispensing, have been described for manipulating multiphase liquids at micro as well as nanoscale. Pyro-self-assembling of polymer material for patterning single or array of microlenses appears to exhibit various differences and advantages compared to the conventional lithography techniques. In fact, the simplicity of the method proposed associated with the flexibility of the process for fabricating 3D polymer microstructures demonstrates the great potentiality of the pyro platform functionalities exploitable in many fields, from optics to biosensing. In particular, the manipulation of polymer in combination with the high resolution of the pyro dispensing at nanoscale suggests different innovative and potential uses for in situ and non-invasive instruments, potentially useful for different kind of lab-on-chip and lab-on-fibre applications.

References

1. T. Taniguchi, T. Torii, T. Higuchi, Chemical reactions in microdroplets by electrostatic manipulation of droplets in liquid media. Lab. Chip. **2**, 19–23 (2002)
2. M.G. Pollack, A.D. Shenderov, R.B. Fair, Electrowetting-based actuation of droplets for integrated microfluidics. Lab. Chip. **2**, 96–101 (2002)
3. G. Marchand, C. Delattre, R. Campagnolo, P. Pouteau, F. Ginot, Electrical detection of DNA hybridization based on enzymatic accumulation confined in nanodroplets. Anal. Chem. **77**, 5189–5195 (2005)
4. V. Taly, B.T. Kelly, A.D. Griffiths, Droplets as microreactors for high-throughput biology. ChemBioChem **8**, 263–272 (2007)
5. T.M. Squires, S.R. Quake, Microfluidics: Fluid physics at the nanoliter scale. S.R. Rev. Mod. Phys. **77**, 977–1026 (2005)

6. P. Ferraro, S. Coppola, S. Grilli, M. Paturzo, V Vespini, Dispensing nano-pico droplets and liquid patterning by pyroelectrodynamic shooting, Nat. Nanotechnol. **5**, 429–435 (2010)
7. R.T. Collins, J.J. Jones, M.T. Harris, O.A. Basaran, Electrohydrodynamic tip streaming and emission of charged drops from liquid cones. Nat. Phys. **4**, 149–154 (2008)
8. J.U. Park, M. Hardy, S.J. Kang, K. Barton, K. Adair, D.K. Mukhopadhyay, C.Y. Lee, M.S. Strano, A.G. Alleyne, J.G. Georgiadis, P.M. Ferreira, J.A. Rogers, High-resolution electrohydrodynamic jet printing. Nat. Mater. **6**, 782–789 (2007)
9. S. Coppola, V. Vespini, S. Grilli, P. Ferraro, Self-assembling of multi-jets by pyro-electrohydrodynamic effect for high throughput liquid nanodrops transfer. Lab. Chip. **11**, 3294–3298 (2011)
10. V. Vespini, S. Coppola, S. Grilli, M. Paturzo, P. Ferraro, Pyroelectric Adaptive Nanodispenser (PYRANA) microrobot for liquid delivery on a target. Lab. Chip. **11**, 3148–3152 (2011)
11. S. Xu, Y.-J. Lin, S.-T. Wu,, Dielectric liquid microlens with well-shaped electrode. Opt. Express **17**, 10499–10505 (2009)
12. L. Miccio, A. Finizio, S. Grilli, V. Vespini, M. Paturzo, S. De Nicola, P. Ferraro, Tunable liquid microlens arrays in electrode-less configuration and their accurate characterization by interference microscopy. Opt. Express **17**, 2487–2499 (2009)
13. H. Ottevaere, B. Volckaerts, J. Lamprecht, J. Schwider, A. Hermanne, I. Veretennicoff, H. Thienpont,, Two-dimensional plastic microlens arrays by deep lithography with protons: fabrication and characterization. J. Opt. A: Pure Appl. Opt. **4**, S22–S28 (2002)
14. M. He, X.-C. Yuan, N.Q. Ngo, J. Bu, S.H. Tao,, Single-step fabrication of a microlens array in sol-gel material by direct laser writing and its application in optical coupling. J. Opt. A: Pure Appl. Opt. **6**, 94–97 (2004)
15. C.Y. Chang, S.Y. Yang, J.L. Sheh, A roller embossing process for rapid fabrication of microlens arrays on glass substrates. Microsyst. Technol. **12**, 754–759 (2006)
16. J. Shi, Z. Stratton, S-C.S. Lin, H. Huang, T.J. Huang, Tunable optofluidic microlens through active pressure control of an air-liquid interface. Microfluid. Nanofluid. **9**, 313–318 (2010)
17. J.-H. Zhu, J.-xia Shi, Y. Wang, P.-sheng He, Spherical micro-lens array of PMMA produced by micro-molding. Chin. J. Chem. Phys. **19**, 443–446 (2006)
18. A. Schilling, R. Merz, C. Ossmann, H.P. Herzig, Surface profiles of reflow microlenses under the influence of surface tension and gravity. Opt. Eng. **9**, 2171–2176 (2000)
19. W. Cheong, L. Yuan, V. Koudriachov, W. Yu, Single-step fabrication of continuous surface relief micro-optical elements in hybrid sol-gel glass by laser direct writing. Opt. Express **10**, 443–448 (2002)
20. D.W.D. Monteiro, O. Akhzar-Mehr, P.M. Sarro, G. Vdovin, Single-mask microfabrication of aspherical optics using KOH anisotropic etching of Si. Opt. Express **11**, 2244–2252 (2003)
21. I.A. Grimaldi, A. De Girolamo Del Mauro, G. Nenna, F. Loffredo, C. Minarini, F. Villani, Microstructuring of polymer films by inkjet etching. J. App. Polym. Sci. **122**, 3637–3643 (2011)
22. J.Y. Kim, N.B. Brauer, V. Fakhfouri, D.L. Boiko, E. Charbon, G. Grutzner, J. Brugger, Hybrid polymer microlens arrays with high numerical apertures fabricated using simple ink-jet printing technique. Opt. Mater. Expr. **1**, 259–269 (2011)
23. B.H. Jo, L.M. Van Lerberghe K.M. Motsegood, D.J. Beebe, Three-dimensional micro-channel fabrication in polydimethylsiloxane (PDMS) elastomer. J. Microelectromech. Syst. **9**, 76–81 (2000)
24. T. Sulchek, R. Hsieh, J.D. Adams, S.C. Minne, C.F. Quate, D.M. Adderton, High-speed atomic force microscopy in liquid. Rev. Sci. Inst. **71**, 2097–2099 (2000)
25. Y. Xia, G.M. Whitesides, Soft lithography. Annu. Rev. Mater. Sci. **28**, 153–184 (1998)
26. D. Bodas, C. Khan-Malek, Direct patterning of quantum dots on structured PDMS surface. Sens. Actuators B **128**, 168–172 (2007)
27. S.L. Peterson, A. McDonald, P.L. Gourley, D.Y. Sasaki, Poly(dimethylsiloxane) thin films as biocompatible coatings for microfluidic devices: Cell culture and flow studies with glial cells. J. Biomed. Mater. Res. **72A**, 10–18 (2005)
28. S.I. Morefield, E.W. Keefer, K.D. Chapman, G.W. Gross, Drug evaluations using neuronal networks cultured on microelectrode arrays. Biosen. Bioelectron. **15**, 383–396 (2000)

29. H. Andersson, A. van den Berg, Microfabrication and microfluidics for tissue engineering: state of the art and future opportunities. Lab .Chip. **4**, 98–103 (2004)
30. K. Atsuta, H. Noji, S. Takeuchi, Micro patterning of active proteins with perforated PDMS sheets (PDMS sheets). Lab. Chip. **4**, 333–336 (2004)
31. M.N. De Silva, R. Desai, D.J. Odde, Micro-patterning of animal cells on PDMS substrates in the presence of serum without use of adhesion inhibitors. Biomed. Microdev. **6**, 219–222 (2004)
32. C.S. Chen, M. Mrksich, S. Huang, G.M. Whitesides, D.E. Ingber, Geometric control of cell life and death. Science **276**, 1425–1428 (1997)
33. S. Grilli, M. Paturzo, L. Miccio, P. Ferraro, In situ investigation of periodic poling in congruent LiNbO3 by quantitative interference microscopy. Meas. Sci. Technol. **19** (2008)
34. R.S. Weis, T.K. Gaylord, Lithium niobate: summary of physical properties and crystal structure. Appl. Phys. A **37**, 191–203 (1985)
35. E.M. Bourim, C.W. Moon, S.-W. Lee, I.K. Yoo, Investigation of pyroelectric electron emission from monodomain lithium niobate single crystals. Phys. B Conden. Matter **383**, 171–182 (2006)
36. B. Rosenblum, P. Bräunlich, J.P. Carrico, Thermally stimulated field emission from pyroelectric LiNbO3. Appl. Phys. Lett. **25**, 17–19 (1974)
37. G. Rosenman, D. Shur, Y.E. Krasik, A. Dunaevsky, Electron emission from ferroelectrics. J. Appl. Phys. **88**, 6109–6161 (2000)
38. P. Ferraro, S. Grilli, L. Miccio, V. Vespini, Wettability patterning of lithium niobate substrate by modulating pyroelectric effect to form microarray of sessile droplets. Appl. Phys. Lett. **92**, 213107 (2008)
39. B. Berge, J. Peseux, Variable focal lens controlled by an external voltage: an application of electrowetting. Eur. Phys. J. E **3**, 159–163 (2000)
40. D. Grahan-Rowen, Liquid lenses make a splash. Nat. Photonics Vol Sample **2–4** (2006)
41. G. Beni, M.A. Tenan, Dynamics of electrowetting displays. J. Appl. Phys. **52**, 6011 (1981)
42. S. Hayes, D.J. Feenstra, Video-Speed electronic paper based on electrowetting. Nature **5**, 383–385 (2003)
43. B.H.W. Kuiper, Hendriks, Variable–focus liquid lens for miniature cameras. Appl. Phys. Lett. **85**, 1128–1130 (2004)
44. L. Dong, A.K. Agrawal, D.J. Beebe, H.R. Jiang, Adaptive liquid microlenses activated by stimuli-responsive hydrogels. Nature **2**, 551–554 (2006)
45. C.C. Cheng, J.A. Yeh, Dieletrically actuated liquid lens. Opt. Express **15**, 7140–7145 (2007)
46. D. Psaltis, S.R. Quache, C.H. Yang, Developing optofluidic technology through the fusion of microfluidics and optics. Nature **442**, 381–386 (2006)
47. F. Mugele, S. Herminghaus, Electrostatic stabilization of fluid microstructures. Appl. Phys. Lett. **81**, 2303–2305 (2002)
48. P.M. Moran, S. Dharmatilleke, A.H. Khaw, K.W. Tan, M.L. Chan, I. Rodriguez, Fluidic lenses with variable focal length. Appl. Phys. Lett. **88**, 041120 (2006)
49. H.W. Ren, D. Fox, P.A. Anderson, B. Wu, S.T. Wu, Tunable-focus liquid lens controlled using a servo motor. Opt. Express **14**, 8031–8036 (2006)
50. S. Grilli, L. Miccio, V. Vespini, A. Finizio, S. De Nicola, P. Ferraro, Liquid micro-lens array activated by selective electrowetting on lithium niobate substrates. Opt. Express **16**, 8084–8093 (2008)
51. L. Miccio, M. Paturzo, S. Grilli, V. Vespini, P. Ferraro, Hemicylindrical and toroidal liquid microlens formed by pyro-electro-wetting. Opt. Lett. **34**, 1075–1077 (2009)
52. B.S. Gallardo, V.K. Gupta, F.D. Eagerton, L.I. Jong, V.S. Craig, R.R. Shah, N.L. Abbot, ELectrochemical principles for active control of liquids on submillimeter scales. Science **283**, 57–60 (1999)
53. W.K. Choi, E. Lebrasseur, M.I. Al-Haq, H. Tsuchiya, T. Torii, H. Yamazaki, E. Shinohara, T. Higuchi, Nano-liter size droplet dispenser using electrostatic manipulation technique. Sens. Actuators A Phys. **136**, 484–490 (2007)
54. F. Mugele, J.C. Baret, Electrowetting: from basics to applications. Phys. Condens. Matter **17**, R705–R774 (2005)
55. A.M. Gañán-Calvo, Electro-flow focusing: the high-conductivity low-viscosity limit. Phys. Rev. Lett. **98**, 239904 (2007)

56. F. Malloggi, S.A. Vanapalli, H. Gu, D. van den Ende, F. Mugele, Electrowetting-controlled droplet generation in a microfluidic flow-focusing device. J. Phys. Condens. Matter **19**, 462101 (2007)
57. O.Y. Loh, A.M. Ho, J.E. Rim, P. Kohli, N.A. Patankar, H.D. Espinosa, Electric field-induced direct delivery of proteins by a nanofountain probe. PNAS **105**, 16438–16443 (2008)
58. R. Suryo, O.A. Basaran, Dripping of a liquid from a tube in the absence of gravity. Phys. Rev. Lett. **96**, 034504 (2006)
59. C.H. Chen, D.A. Saville, I.A. Aksay, Scaling laws for pulsed electrohydrodynamic drop formation. Appl. Phys. Lett. **89**, 124103 (2006)
60. A.U. Chen, O.A. Basaran, A new method for significantly reducing drop radius without reducing nozzle radius in drop-on-demand drop production. Phys. Fluids **14**, L1–L4 (2002)
61. K.T. Rodolfa, A. Bruckbauer, D.J. Zhou, A.I. Schevchuk, Y.E. Korchev, D. Klenerman, Nanoscale pipetting for controlled chemistry in small arrayed water droplets using a double-barrel pipet. Nano Lett. **6**, 252–257 (2006)
62. T. Ondarçuhu, J. Arcamone, A. Fang, H. Durou, E. Dujardin, G. Rius, Pérez-Murano, F. Controlled deposition of nanodroplets on a surface by liquid nanodispensing: Application to the study of the evaporation of femtoliter sessile droplets. Eur. Phys. J. Spec. Top. **166**, 15–20 (2009)
63. Ferroelectric Crystals for Photonic Applications, Including Nanoscale Fabrication and Characterization Techniques, Series in Materials Science, vol. 91, ed. by P. Ferraro, S. Grilli, P. De Natale, Springer (Germany) 2008
64. A. Barrero, I.G. Loscertales, Micro- and nanoparticles via capillary flows. Annu. Rev. Fluid Mech. **39**, 89–106 (2007)
65. N. Maeda, J.N. Israelachvili, M.M. Kohonen, Evaporation and instabilities of microscopic capillary bridges. PNAS **100**,803–808 (2003)
66. Y.M. Zheng, H. Bai, Z.B. Huang, X.L. Tian, Y. Zhao, J. Zhai, L. Jiang, Directional water collection on wetted spider silk. Nature **463**, 640–643 (2010)
67. P.P. Bhat, S. Appathurai, M.T. Harris, M. Pasquali, G.H. McKinley, O.A. Basaran, Formation of beads-on-a-string structures during break-up of viscoelastic filaments. Nat. Phys. **6**, 625–631 (2010)
68. N.A. Malvadkar, M.J. Hancock, K. Sekeroglu, W.J. Dressick, M.C. Demirel, An engineered anisotropic nanofilm with unidirectional wetting properties. Nat. Mater. **9**, 1023–1028 (2010)
69. J.C. McDonald, D.C. Duffy, J.R. Anderson, D.T. Chiu, H.K. Wu, O.J.A. Schueller, G.M. Whitesides, Fabrication of microfluidic systems in poly(dimethylsiloxane). Electrophoresis **21**, 27–40 (2000)
70. S. Park, Y.S. Huh, H.G. Craighead, D. Erickson, A method for nanofluidic device prototyping using elastomeric collapse. Pnas **106**, 15546–15554 (2009)
71. K.Y. Suh, Y.S. Kim, H.H. Lee, Capillary force lithography. Adv. Mater. **13**, 1386–1389 (2001)
72. S. Grilli, V. Vespini, P. Ferraro, Surface-charge lithography for direct PDMS micro-patterning. Langmuir **24**, 13262–13265 (2008)
73. A.L. Thangawng, M.A. Swartz, M.R. Glucksberg, R.S. Ruoff, Bond-detach lithography: a method for micro/nanolithography by precision PDMS patterning. Small **3**, 132–138 (2007)
74. L. Huang, A.B. Braunschweig, W. Shim, L.D. Qin, J.K. Lim, S.J. Hurst, F.W. Huo, C. Xue, J.W. Jong, C.A. Mirkin, Matrix-assisted dip-pen nanolithography and polymer pen lithography. Small **6**, 1077–1081 (2010)
75. L. Wang, D. Zhang, H. Zhang, J.Z. Jiang, Fabrication of micropillars by laser-induced thermoplastic method. Appl. Phys. Lett. **97**, 137905 (2010)
76. E. Schaffer, S. Harkema, R. Blossey, U. Steiner, Temperature-gradient-induced instability in polymer films. Europhys. Lett. **60**, 255–261 (2002)
77. W. Monch, S. Herminghaus, Elastic instability of rubber films between solid bodies. Europhys. Lett. **53**, 525–531 (2001)
78. S. Harkema, Capillary Replication Ch. 3,5 (Department of Polymer Physics, University of Groningen, The Netherlands) (2005)
79. S. Grilli, S. Coppola, V. Vespini, F. Merola, A. Finizio, P. Ferraro, 3D lithography by rapid curing of the liquid instabilities at nanoscale. Proc. Natl. Acad. Sci. **108**, 15106–15111 (2011)

Chapter 13
Fiber Optic Sensors Based on Nanostructured Materials

Cesar Elosua, Miguel Hernaez, Ignacio R. Matias and Francisco J. Arregui

Abstract Fiber optic sensors have been developed taking advantage on the synergy between the properties of nanostructured materials and the ones that characterize an optical fiber. The mechanical properties of optical fiber introduce some restrictions to the techniques used for the deposition of materials. As an alternative to the classical deposition procedures, wet coating techniques have been successfully applied in these cases. The current chapter put emphasis on materials that can be incorporated using wet coating techniques. The first one presented is the multilayer based nanostructures: among the different alternatives, we have focused on materials prepared with the Layer-by-Layer technique. Another type of products used for the fabrication of optical fiber sensors is sol-gel matrices, which are made of silica, so that its optical properties are similar to the ones of an optical fiber. The other two described type of products have focused the attention of many researchers in the recent years. Firstly, materials with an enhanced selectivity are presented: the molecularly imprinted polymers (MIPs). Finally, sensors based on metallic nanolayers and particles are presented. All these materials and techniques have acquired a great importance in the field of optical fiber sensors due to their versatility and the good features that offer.

C. Elosua (✉) · M. Hernaez · I. R. Matias · F. J. Arregui
Department of Electric and Electronic Engineering,
Public University of Navarre, Madrid, Spain
e-mail: cesar.elosua@unavarra.es

M. Hernaez
e-mail: miguel.hernaez@unavarra.es

I. R. Matias
e-mail: natxo@unavarra.es

F. J. Arregui
e-mail: parregui@unavarra.es

A. Cusano et al. (eds.), *Lab-on-Fiber Technology*,
Springer Series in Surface Sciences 56, DOI: 10.1007/978-3-319-06998-2_13,

Fiber optic sensors have been developed taking advantage on the synergy between the properties of nanostructured materials and the ones that characterize an optical fiber. In this chapter, four types of materials employed to develop this kind of devices will be described: multilayer nanostructures, sol-gel derived materials, molecularly imprinted polymers and metallic thin films and nanoparticles.

13.1 Introduction

Compared to electronic sensors, optical fiber sensors offer some interesting features such as immunity to electromagnetic interferences, small size or multiplexing capabilities [1]. The apparition of new types of fibers and optical devices, together with the use of nanostructured materials, has made possible the development of optical fiber sensors based on this type of membranes.

There are diverse techniques such as those used in the semiconductor industry optimized for the deposition on flat substrates. Here, the mechanical properties of optical fiber (reduced dimensions and flexibility) and especially its shape (cylindrical or even conical in the case of tapered optical fibers) introduce some restrictions to the techniques used for the deposition of materials if the goal is to achieve homogeneous coatings onto these cylindrical or conical optical waveguides. As an alternative to the classical deposition techniques, wet coating techniques have been successfully applied in these cases because they make possible the deposition of the sensing material incorporated in a homogeneous and azimuthally symmetric coating. Moreover, diverse nanostructured materials are compatible with this kind of deposition methods where the fiber is dipped into a solution. For this reason, the current chapter put emphasis on in materials that can be incorporated using wet coating techniques. More specifically, four nanostructured types of materials are described in this chapter. The first one presented is the multilayer based nanostructures: among the different alternatives, we have focused on materials prepared with the Layer-by-Layer technique. This procedure allows controlling relevant parameters about the material morphology such as porosity or roughness just by adjusting some fabrication parameters such as the ionic strength, pH or temperature of the precursor solutions. These nanostructures can be deposited on any substrate independently on their shape, which is very important as it was cited earlier in the case of optical fiber sensors [2, 3].

Another type of materials used for the fabrication of optical fiber sensors is sol-gel matrices: this type of membranes is made of silica, so that its optical properties are similar to the ones of an optical fiber. In this manner, sol-gel membranes could be considered like an extension of the substrate doped with a certain sensing agent. On the other hand, the development of sol-gel has reached the nano structural level in the way that the porous size can be controlled depending on the construction conditions, allowing the discrimination of analytes in terms of their molecular size. Even in some cases, the sol-gel matrix can act itself as the sensing material [4].

The other two described type of materials have focused the attention of many researchers in the recent years due to their great potential. Firstly, materials with an enhanced selectivity are presented: the molecularly imprinted polymers (MIPs). These polymers are prepared to emulate the behavior of enzymes: they memorize the shape of the target analyte and react only in its presence. Synthesis process can be designed to match the requirements of a certain application achieving a high selectivity, and furthermore, the transduction can be also set to measure luminescent emission changes if the proper fluorophore is included along the synthesis of the polymer [5, 6].

Finally, sensors based on metallic nanolayers and particles are presented. In this case, transduction is based on optical resonances that vary their spectral response in the presence of the target molecules. This kind of responses is more robust that the ones based on intensity variations. The nature of the resonance depends on some physical properties of the nanostructure such as thickness or refractive index [7–9].

All these materials and techniques have acquired a great importance in the field of optical fiber sensors due to their versatility and the good features that offer.

13.2 Multilayer Based Nanostructures

Although there are another coating methods based on multilayer deposition, such as Langmuir–Blodgett, that can be used for fabricating optical fiber sensors [10–14], this section is focused on the Layer-by-Layer technique due to its simplicity and importance.

The conventional Layer-by-Layer (LbL) method, also known as Electrostatic Self-Assembly Multilayer (ESAM), consists of the alternate immersion of the substrate into aqueous solutions with opposite electric charge (see Fig. 13.1). This way, the particles are adsorbed by electrostatic attraction. It was reported for the first time by Iler in 1966 [15] and in the first nineties, Decher et al. rediscovered the technique [2, 16, 17]. Since then, the number of publications has increased exponentially [18–22]. Moreover, in addition to the electrostatic driven LbL assembly or conventional LbL, many different intermolecular interactions, such as hydrogen bonding, charge transfer interaction, molecular recognition, coordination interactions, can be also used as driving force for the multilayer fabrication by means of the LbL technique.

One of the main advantages that this technique offers is its simplicity. In addition, the LbL method allows controlling with high precision the thickness of the film by just selecting the number of deposited bilayers. It is a very important feature because this parameter plays a key role in the behavior of optical fiber sensors. Other benefits of LbL are that, since it is a wet technique, it can be applied on surfaces with any shape (conical, cylindrical, plain, etc.). It is important to note that the polyanions and polycations overlap each other at the molecular level, which produces a homogeneous material. The composition and thickness of an

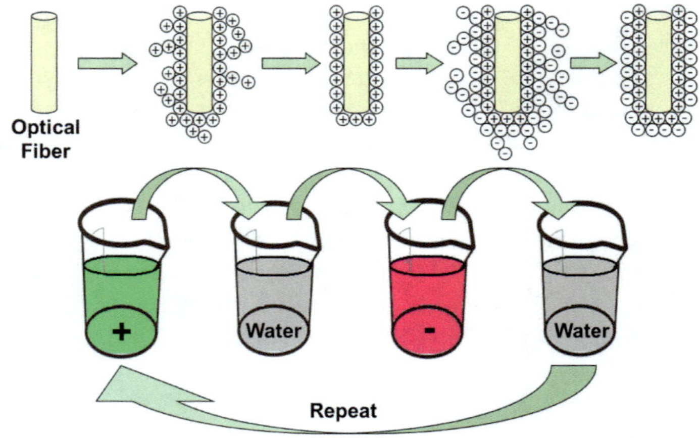

Fig. 13.1 Schematic representation of the different steps involved in the deposition of a nanocoating by using the Layer by Layer method

individual bilayer can be controlled by adjusting the deposition parameters (concentration of solutions, pH, temperature, immersion times, etc.). This way, the LbL technique allows the deposition of different kind of substances, such as polymers [23, 24], nanoparticles [25], fluorescent indicators [26], quantum dots [27] or colorimetric dyes [28]. In addition, it has been used as well to deposit biofilms including enzymes [29, 30], antibodies [31] or viruses [32].

This versatility makes the LbL technique a suitable method for fabricating optical fiber sensors based on nanostructured coatings. Many different approaches have been developed in the last years [3, 33] and, in this section, some representative examples will be presented.

One of the most used optical fiber topologies is the nanoFabry-Perot interferometer. This simple scheme consists on a nanocoating deposited on a perpendicularly cleaved optical fiber tip. A double optical mirror is created by the change of refractive index at the two interfaces (optical fiber core–nanocoating and nanocoating–external medium). This double reflection of light produces a variation of the total reflected optical power as a function of the effective refractive index of the coating. Several examples of these interferometers can be found in bibliography. The first optical fiber nanoFabry–Perot fabricated by means of the LbL technique was reported in the late nineties by Arregui et al. [34, 35]. Different groups have contributed to this technology since then. For instance, Jiang et al. have developed Fabry–Perot refractometers using LbL nanocoatings made of poly(diallyldimethylammonium chloride) (PDDA) and poly(styrenesulfonate sodium salt) (PSS) [36, 37]. In addition, humidity sensors [38], pH sensors [39], chemical compounds or volatile organic compounds (VOCs) detectors [35, 40] and optical fiber biosensors [41, 42] based on this kind of devices have been also fabricated using the LbL method. This simple structure allows the monitoring of the target parameter using a narrow optical band, or in other words, just by measuring intensity changes. Sensors

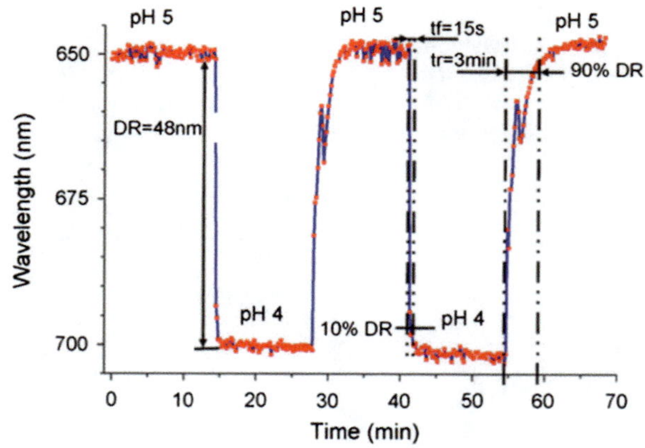

Fig. 13.2 Wavelength shift of a nanoFabry–Perot optical fiber pH sensor. Reprinted from [39] with permission from Elsevier

based on intensity have a high cross-sensitivity to undesired artifacts such as accidental bending of the optical fiber, light source fluctuations or just aging of the optoelectronic components. It can be overcome because nanoFabry–Perots also present spectral response depending of the target. Therefore, just by measuring the wavelength shift of the spectral characteristic curve of the nanoFabry-Perot the target can be determined. As an example of that, in Fig. 13.2 the response of a nano-Fabry–Perot pH sensor when it is immersed into pH 5 and 4 solutions is shown [39].

Another sensing topology widely used in optical fiber is based on long period gratings (LPG). This structure consists of a periodic index modulation of the re-fractive index of the core of a single mode fiber, with a period typically in the range 100 μm–1 mm. This modulation induces attenuation bands in the transmission spectrum. The wavelength of these bands is affected by temperature, strain, curvature or the refractive index of the surrounding medium. This way, if a LPG is coated by a material whose refractive index is sensitive to a certain magnitude, the attenuation bands of this LPG will suffer a spectral shift when this magnitude changes [3, 33]. The contribution of James et al. to this field has reached a particular relevance in the last years, with different approaches of optical fiber sensors based on coated LPGs [43–45]. Among other examples, they have created VOC detectors [46–48], chemical sensors [49] or interferometers [50] based on this technology. In Fig. 13.3 the response of an ammonia detector developed by this group is plotted [47]. The shift of the attenuation bands to higher wavelengths as long as the concentration of ammonia is increasing can be clearly appreciated. Additionally, a great variety of coated-LPG-based sensors developed by other groups can be found in the bibliography [46–48, 51–53].

Other sensing structures that have been widely used are tapers. A taper consists of a fragment of single mode fiber that has been stretched, reducing dramatically the diameter of the core and the cladding, in order to maximize the interaction between the guided light and the external medium. In the tapered zone, light becomes guided through the cladding of the fiber and the surrounding medium acts as a cladding. This way, the fiber is very sensitive to changes of the surrounding

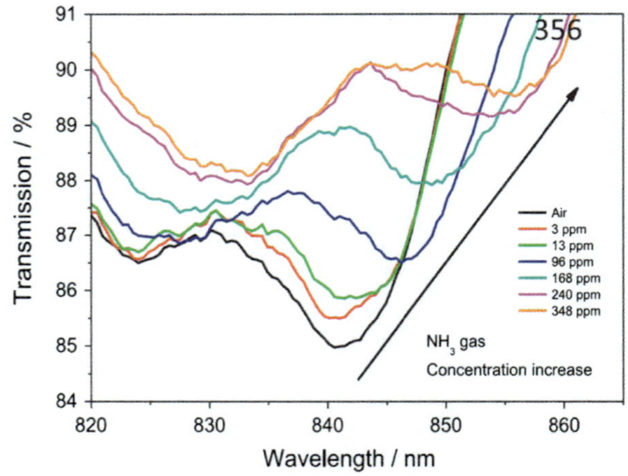

Fig. 13.3 Characterization of an optical fiber ammonia detector based on a coated LPG. Reprinted from [47] with permission from Elsevier

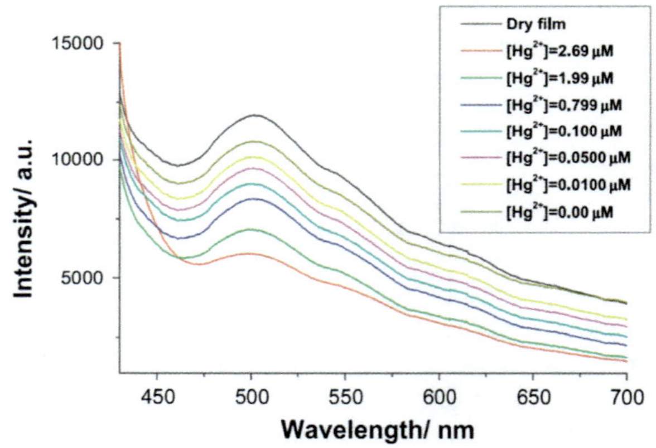

Fig. 13.4 Response of an optical fiber sensor based on fluorescence to variations in the concentration of ions Hg^+ in the surrounding medium$^+$. Reprinted from [64] with permission from Elsevier

refractive index [3]. A great variety of optical fiber sensors based on tapered single-mode fiber (SMF) has been developed in the last decades [54, 55].

Other optical fiber sensors fabricated by means of the LbL method includes devices based on colorimetric dyes [28], microstructured fibers [55, 56], fluorescence [57, 58], lossy mode resonances (LMR) [59–63], etc. As an example of a fluorescent based optical fiber sensor, in Fig. 13.4 the response of the fluorescence emission peak as a function of the concentration of Hg^+ ions is shown [64].

13.3 Sol-Gel Matrices

Sol-gel technique is a wet deposition method with the special feature that allows porous matrices whose composition is similar to glass to be prepared at room temperature. It is very relevant in the case of optical fiber sensors because the substrate is made of glass, so that the losses due to refractive index change are reduced. On the other hand, the porous size can be adjusted to satisfy the specifications required by the final application depending on the reaction conditions. The method is based on the polymerization of metallic or semi metallic hydroxides. The chemical species involved are typically a precursor, water, a solvent and a catalyzer.

13.3.1 Chemical Reaction and Morphology

The whole sol-gel process is divided in six stages: hydrolysis, condensation, gelation, curing, drying and densification. In the first step, water molecules replace alkoxy groups (–OR, where R is an alkyl group) by hydroxyl ones (–OH). The alkoxides most used are tetraethyl orthosilicate (TEOS) and thetramethyl or-thosilicate (TMOS). This reaction is followed by condensation between silanols (Si–OH) to conform siloxane bonds (Si–O–Si), which generates water or alcohol molecules. The kinetics of both reactions depends on factors such as pH, in a way that condensation can take place before hydrolysis occurs [65]. Once this step has concluded, colloidal Si–O–Si particles are formed. Gelation occurs when the colloids are so linked that the viscosity increases abruptly. At this point, there are still no bonded particles remaining and the gel shows a high viscosity and low elasticity, which is known as the gel point: at this moment, the optical fiber is typically dipped into the solution [66, 67].

Along the curing stage, the gel is immersed into a liquid in order to enhance the crosslinking reactions that improves the structure of the final matrix, and so, its mechanical properties. Curing conditions allow the porous size to be controlled: specifically, the used solvent, temperature and pH, among others [68–70].

If the curing is performed under atmospheric conditions, a xerogel is obtained [71], whereas under supercritical conditions (high temperature and pressure), the final matrix is called aerogel. In the first case, the gel suffers significant morphological changes due to the evaporation of the solvents; this process produces mechanical stress in the structure due to the vaporization of the solvents through the pores. On the other hand, with supercritical conditions, solvent removal is made at high temperature and pressure to eliminate capillarity forces, and so, cracking. This aerogel matrix shows a higher porous density (around 80 and 99 %) [71–73]. Figure 13.5 shows a comparison between a xerogel and an aerogel [74].

Final stage is densification, and it is needed if mechanically stable materials are required. In most optical applications, both water molecules and silanols groups have to be removed. Moreover, when the process is over, the resulting surface shows a high hydrophobicity [75]. This is another way to control the size of the porous by the holes formed when water and silanols molecules are removed. Figure 13.6 summarizes the main steps of the sol-gel process.

The morphology of the resulting matrix is described in terms of the porosity, specifically, the porous size and distribution. Regarding to the first parameter,

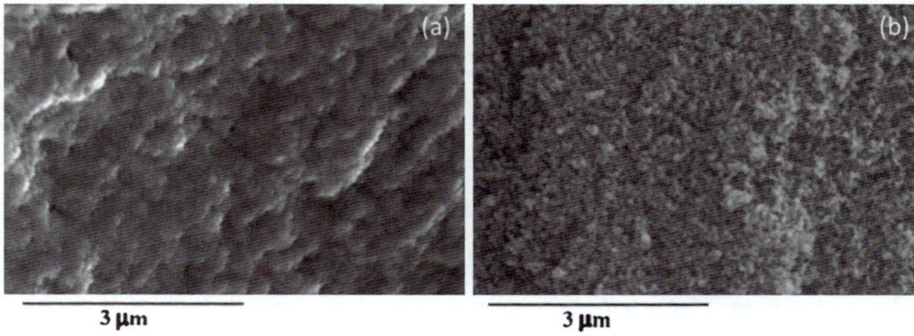

Fig. 13.5 SEM micrographs of **a** a xerogel aged and an **b** aerogel aged in 2.0 M NH_3 (aq.). Reprinted from [73] with permission from Elsevier

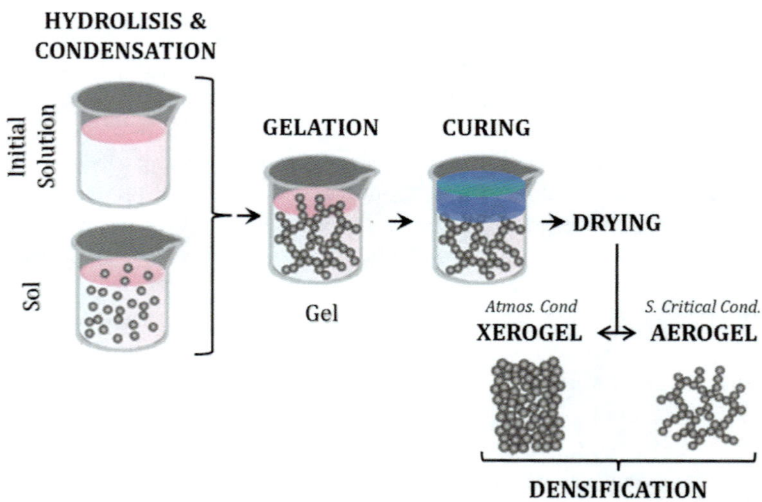

Fig. 13.6 Scheme with the main steps of the sol-gel process

there are three types of materials: macroporous (diameter over 50 nm), mesoporous (diameter between 2 and 50 nm) and microporous (below 2 nm). As it will be exposed later, this feature can be used to discriminate compounds depending on their molecular size. Other important quantitative parameters to characterize the morphology are the superficial area (expressed in $m^2\ g^{-1}$) and the fraction of porosity (which relates the bulk density with the skeletal one in %). In most cases, a high area implies a greater interface, and so, a better sensitivity.

13.3.2 Applications

Several optical sensors have been developed using sol-gel as membrane to encapsulate the sensing material. The membrane can be deposited along the fiber (evanescent field or LPG sensors), onto a cleaved ended pigtail (reflection sensors) or even act as a segment of light conducting guide [76]. Typically, the fiber is dipped in the first stages when the solution can be attached to the fiber and the following steps can take place onto this substrate.

As it has been said before its optical properties are similar to the ones of the optical fiber [77]: this feature allows luminescent materials to be used. Moreover, the sol-gel also minimizes the leaking effect of the reagent even in the case of sensors designed to work with water solutions. The chemical dye is included at the initial step of the process so the final matrix is doped with it [4]. Thanks to the room ambient conditions required in the process, the reagent suffers little variations, whereas the sensing properties remain unaltered [78]. Furthermore, in some cases the membrane reduces the degradation produced by ambient conditions such as external light, which is another benefit of this kind of matrices [4] .

One of the most relevant applications where sol-gel based sensors is focused on is pH monitoring. In these devices, a pH indicator is immobilized into the sol-gel matrix, which can also enhance the pH range of the chemical indicator [79]. The resulting membrane is uniform and with certain morphology: in the case of pH sensors, porous size is typically high to ensure the diffusion of the solution through the membrane.

The measurements can be based on optical absorbance or luminescence emission. This kind of chemical dyes can be also used to prepare sensors for gas detection such as ammonia [82]. Another relevant application is oxygen detection, in which metallic salts are employed (ruthenium and platinum ones mainly) [83]. Recent works are focused on coating silica nanoparticles with the sensing material, and then, add this mixture to the sol-gel initial solution [84].

Thanks to the densification step, the resulting sol-gel shows a hydrophobic surface, which is very useful in the case of volatile organic compounds detection: many of these solvents are nonpolar, and therefore, would get diffused better into a hydrophobic matrix. Taking advantage of this feature, the sol-gel matrix can be doped with materials sensitive to organic vapors: as an example, a vapochromic compound was dissolved in ethanol and then added into the initial sol-gel solution [81]. The diffusion of the organic vapor is affected by its polarity, but also by its molecular size. Actually, the porous size can be used to filter vapors depending on the size of their molecules. In this manner, xerogel films can be designed to act themselves as the sensing material: depending on the densification stage, a humidity sensor can be developed if the resulting surface of the membrane is not hydrophobic [85] (Fig. 13.7a). On the contrary, if the resulting matrix is nonpolar, the porous size can be designed to discriminate between VOCs and their polarity [82]. Figure 13.8 shows the comparative response of reflection sensors with no

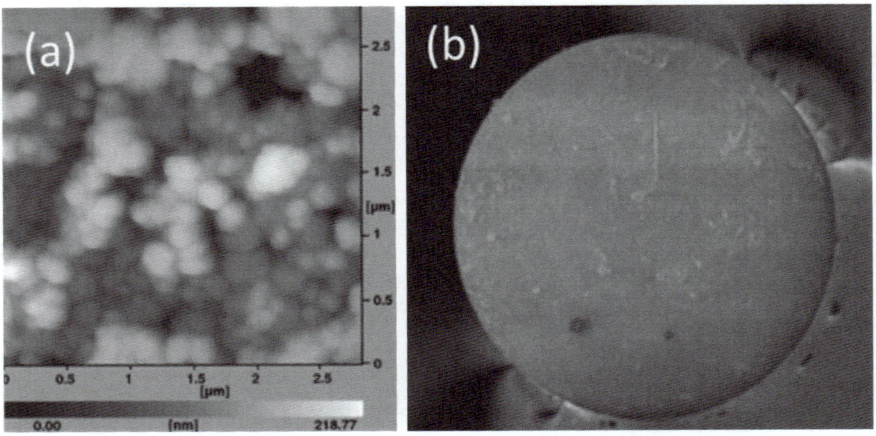

Fig. 13.7 **a** AFM image from a macro porous xerogel prepared doped with a pH indicator based on Eu^{+3}. **b** Field emission scanning electron microscopy (FESEM) micrograph of the xerogel film deposited on the tip of the optical fiber. Reprinted from [80, 81] with permission from Elsevier

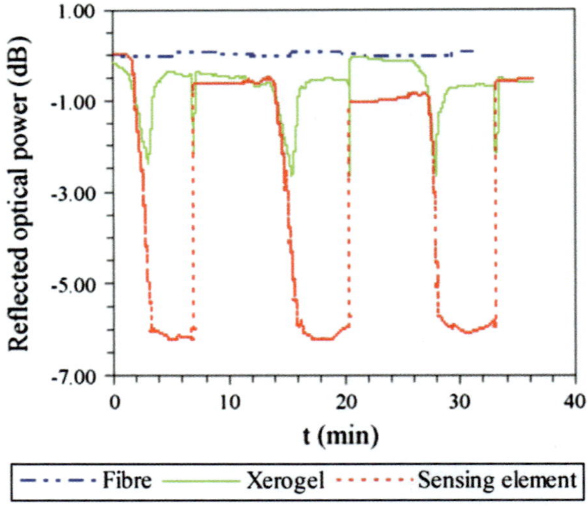

Fig. 13.8 Response signal from a bare fiber alone (*blue line*), a fiber with a xerogel matrix (*green line*) and a doped xerogel with a vapochromic material (*red line*) in the presence of a 5 mM ethanol vapor. Reprinted from [84] with permission from Elsevier

deposition (bare fiber), with a xerogel membrane and one doped with a material sensitive to organic vapors: although the non-doped membrane is also sensitive to organic vapors, it is clear that the response is better in the last case.

13.4 Molecularly Imprinted Polymers

This type of sensing materials is based on molecular recognition. In nature, the best examples for this reaction are enzymes: these proteins are able to react only to a certain molecule depending on its shape and chemical structure with a high selectivity and in a reversible way [86]. Actually, several optical fiber biosensors use enzymes taking advantage of these features (high specificity), specifically, the ones developed to detect glucose [54]. The main drawback of biosensors is the low stability and short lifetime of the sensing material. In this context, the molecularly imprinted polymers (MIPs) technique proposed at the beginning of the seventies [87] is an artificial method to obtain molecules that mimic the behavior of enzymes with the advantages of a higher stability and lifetime. These materials are designed at the molecular level and many researchers are focused currently on this technology due to its great potential applications [5].

13.4.1 Molecular Imprinting Basic Concepts

The selectivity of MIPs is based on their ability of "memorizing" the target analyte, known in the synthesis process as the imprinted molecule. The method consists on preparing a polymer scaffold in presence of the target, in a way the target acts as a template. Once the reaction ends, the template is removed: in this manner, the resulting polymer shows a cavity where only the target molecule fits. Functional monomers and copolymers, together with the target reagent, are required to obtain a MIP: the main stages of the reaction are explained in the following paragraphs.

In the first step, the imprinted molecule and the functional monomer are connected by a covalent link or through non covalent interactions, forming an adduct. The type of linkage determines the final features of the MIP: the first approach is called covalent imprinting whereas the second one is known as non-covalent imprinting. The generated adduct can be considered as the monomer unit of the final MIP. Thereafter, a polymerization reaction takes place by copolymer agents: as a result, the adducts are immobilized in a three dimensional polymer structure. Finally, template molecules are removed from the network, leaving cavities with the similar shape to the target molecule. If the process is successful, these cavities memorize the size and chemical structure of the template, and therefore, the target molecules can get bind to the MIP selectively [6]. The sensing principle is based on physicochemical detectable changes (optical absorption, luminesce) when the template is bound to the polymer structure. The process is summarized in Fig. 13.9.

In the case of the covalent imprinting, the covalent bond generated in the first step is formed when a guest molecule is bind to the structure. The obtained adducts are chemically stable thanks to the nature of the linkage: therefore, different

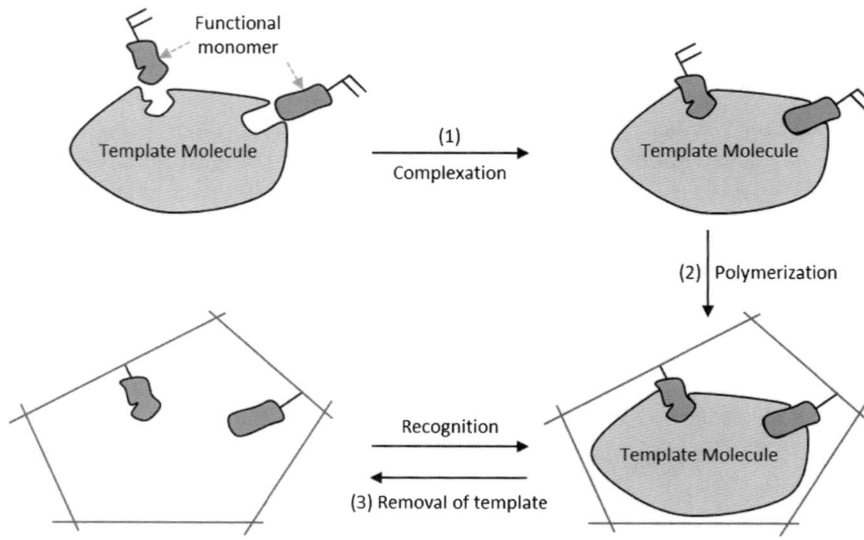

Fig. 13.9 Main steps of the molecular imprinting process

polymerization conditions (temperature or pH) can be used. On the other hand, the reversibility of covalent linkages is limited, and also, the binding and release processes are slow [88].

Non covalent imprinting is based on interactions such as hydrogen bonding, electrostatic interaction or coordination-bond formation, which easies the first stage. The removal of the template can be performed with the solvents that extract the template from the MIP without damaging its structure. Due to the weaker nature of the non-covalent linkages, polymerization conditions have to be carefully defined to maximize this kind of interactions; moreover, the selectivity could be worsen in the case there are several nonspecific bounding sites. However, the faster reaction time and the versatility of non-covalent imprinting have made this option more popular compared to its covalent counterpart [89].

13.4.2 Sensors Based on MIPs

The structural changes suffered by the imprinted polymer when the target reagent is present can induce variations on its optical properties, such as the refractive index: in the case the MIP is deposited onto a cleaved ended fiber because this change can be transduced into an optical signal by measuring the reflected optical power. An extrinsic sensor inspired on this approach to detect formaldehydes is reported in [90]. In this type of studies, polymers imprinted and non-imprinted with the target molecule are prepared just to check the performance of the sensing material. Moreover, the response of the sensor to reagents similar to the target

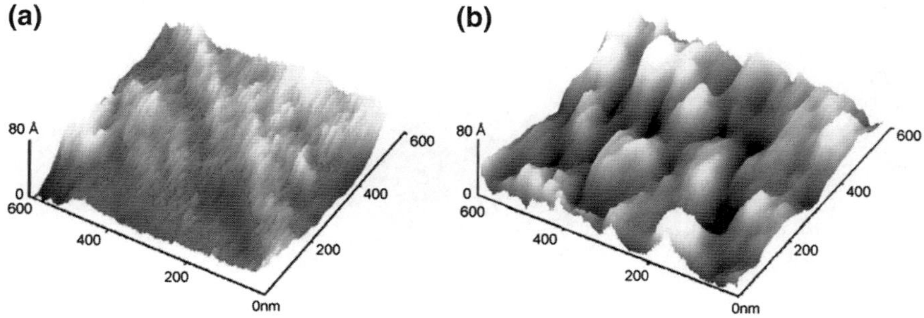

Fig. 13.10 AFM surface micrograph of the MIP a before removing the template and b once it is eliminated. Reprinted from [90] with permission from Elsevier

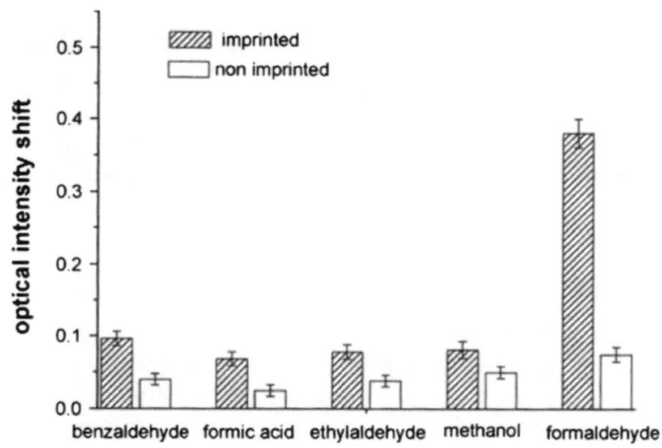

Fig. 13.11 Optical response of imprinted and non-imprinted polymers to different organic solvents. Reprinted from [90] with permission from Elsevier

analyte (in terms of size or functional groups) is also studied to check the selectivity of the device. AFM images obtained from the printed MIP described in this work show the cavities in the case of the removal of the target molecule, as can be observed in Fig. 13.10. The first image is obtained once the MIP is prepared (polymerization step in Fig. 13.10), whereas the second one is registered when the template is removed (last step in Fig. 13.10). It is evident that free binding sites are available.

Regarding to the sensor response, one of the most important features is the selectivity of the imprinted polymer. In the case of formaldehyde, solvents with –OH groups can be bound to the MIP decreasing its selectivity. Figure 13.11 shows the polymer response in the case of some organic solvents, together with the intensity change observed in the non-imprinted polymer. The highest response is registered with formaldehyde, and in every case, the response observed with the other polymer is lower.

There is another approach based on standard single mode fiber where the MIP is deposited onto a cleaved ended pigtail, forming an interferometer [91]. Using a broadband source, the reflected spectrum is shifted in presence of the target molecule, in this case, the toxin Microcystin-LR (see Fig. 13.12).

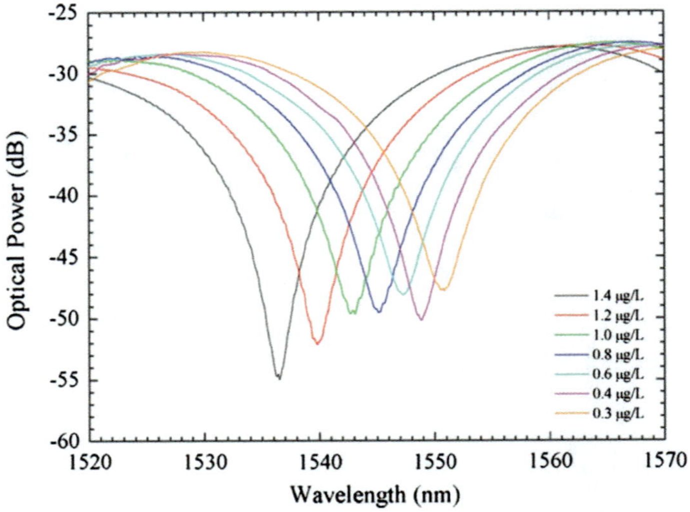

Fig. 13.12 Wavelength shift observed in the optical spectrum depending on the concentration of the target analyte. Reprinted from [91] with permission from Elsevier

There is also another sensing mechanism based on MIPs: luminescent molecules can be included in the polymerization process so that they get attached to the final scaffold. In this manner, the emission of the fluorophore varies depending on whether the target molecule is attached or not to the MIP. Following this idea, optical sensors have been developed to detect anthracene [92] and aluminum ions in aqueous media [93]. This idea has been applied to develop optical fiber sensors able to detect chemical solvents [94] and cocaine [95].

MIPs technology is a recent area in chemistry, but due to its versatility, it has been studied in detail along the last years. Its use in lab on fiber applications is very relevant thanks to the high sensitivity and selectivity that can be obtained. Its polymeric nature, together with its chemical stability, makes it combinable with other nanostructured materials such as sol-gel [92], electrospun [96] and even with LbL technique in the case the MIP also shows an electrolyte behavior. Therefore, this promising technology is supposed to focus the attention of optical fiber sensor researchers along the following years.

13.5 Metallic Nano Thin Films

A different kind of nanostructured coatings that have been included in the fabrication of optical fiber sensors are thin films based on metals and metal oxides.

Among other applications, the surface plasmon resonance (SPR) phenomenon has supposed an important breakthrough in chemical and biological optical fiber sensing due to the high sensitivities that the devices based on this effect can reach.

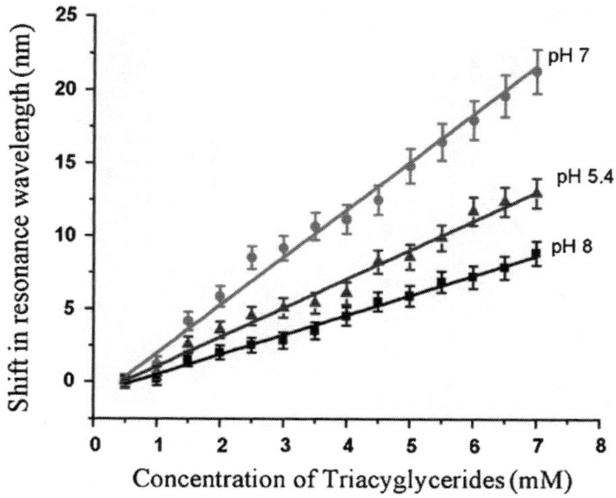

Fig. 13.13 Comparison of the wavelength shift as a function of triacyglycerides concentration at different-pH mediums. Reprinted from [124] with permission from Elsevier

Different phenomena can occur when an optical waveguide is coated by a metallic thin film. SPR consists of the excitation of a surface plasmon wave at the metal-dielectric interface. When some particular conditions of the incident and the surface plasmon wave match, a resonance is generated and a sharp absorption peak can be observed in the transmitted light. The resonance wavelength depends on the refractive index of the material in contact with the metal. In other words, a change in this refractive index will produce a wavelength shift of the resonance peak [8, 97, 98] . This fact has been applied in the design and fabrication of sensors for different applications [99–101].

The optical fiber configuration developed by Jorgenson and Yee [102] overcomes the disadvantages of the traditional sensing configuration and allows the development of optical fiber SPR devices. This way, the well-known advantages of optical fiber sensing are added to the previously mentioned high sensitivity of SPR-based devices. For these reasons, optical fiber sensors based on SPR have acquired a great relevance in the last decades [7, 9, 103–107].

Different optical fiber structures, symmetrical or not, have been applied to perform SPR-based sensing, from a simple metal-coated optical fiber [108, 109] to tapered fibers, D-shaped fibers [110–112], gratings [113, 114], etc.

Although several examples of optical fiber SPR-based sensors can be found in literature, biosensors and chemical sensor are of particular interest [115–123]. As an example, in Fig. 13.13, the response of an optical fiber SPR sensor to variations in the concentration of triacyglycerides [124]. It can be clearly appreciated in this graph that a shift in the resonance wavelength is produced when the concentration of the target varies.

Recent advances in nano-fabrication and nanoparticle synthesis technology have made possible to achieve the patterning of metallic nano-structures. These

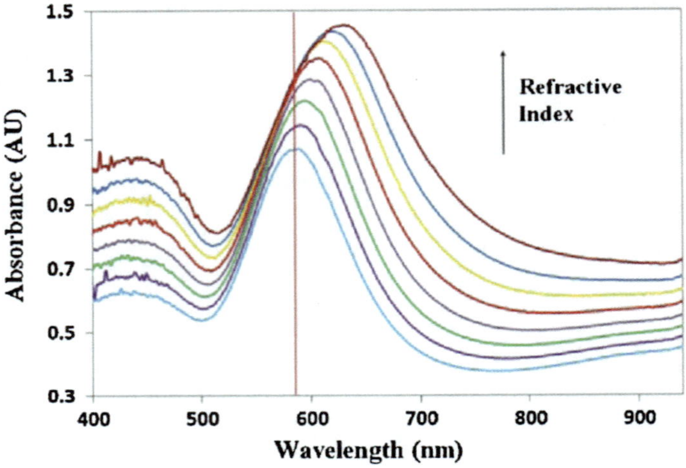

Fig. 13.14 Absorbance spectra of a 60 nm Au nanospheres based LSPR sensor when the refractive index of the testing solutions increase. Reprinted from [130] with permission from Elsevier

advances have allowed obtaining optical fiber sensors based on localized surface plasmon resonances (LSPR) from metallic-nanoparticles thin films [9, 107]. This kind of sensors is acquiring particular relevance in the field of optical fiber sensing and biosensing [120, 122, 125–129]. As an example, in Fig. 13.14 the response to external refractive index variations of a LSPR-based sensor made of Au nanospheres is shown [130].

It is important to remark that SPR-based sensors show much higher wavelength sensitivity than LSPR-based sensors. However, sensors based on LSPR allow controlling the operation wavelength and, in general, present a higher linearity [131–133]. For these reasons, LSPR sensors are considered a really promising technology and a lot of research is being done in this field.

13.6 Conclusions

The remarkable rise of nanotechnology in the last decades has provoked the apparition of new materials and techniques that offer properties not available until now. The combination of these materials with optical fiber has given a push to the continuous development of new optical fiber sensors based on these materials.

This chapter has cited some of these optical fiber sensors based on nanostructured coatings. For the sake of brevity, it has focused on four different families of sensors, according to the nature of the coatings. Thus, materials based on multilayered nanostructures, sol-gel matrices, molecularly imprinted polymers and metallic thin films or nanoparticles have been described, highlighting their advantages and the reasons that make them suitable for the fabrication of optical fiber sensors.

Along this chapter, some typical optical fiber sensing architectures have been explained and analyzed, presenting a brief review of optical fiber sensors based on nanocoatings. In addition, some examples of the most representative sensing mechanisms and applications have been shown. This way, a variety of samples of optical fiber devices designed for sensing humidity, pH, presence of VOCs, concentration of chemical compounds in aqueous solution, proteins or drugs have been described.

References

1. B. Culshaw, Optical fiber sensor technologies: opportunities and – perhaps—pitfalls. J. Lightwave Technol. **22**(1), 39–50 (2004)
2. G. Decher, M. Eckle, J. Schmitt, B. Struth, Layer-by-layer assembled multicomposite films. Curr. Opin. Colloid Interface Sci. **3**(1), 32–39 (1998)
3. F. Arregui, I. Matias, J. Goicoechea, I. Villar, in *Optical Fiber Sensors Based on Nanostructured Coatings*, ed. by F.J. Arregui (Springer, US, 2009), pp. 275–301
4. C. McDonagh, C.S. Burke, B.D. MacCraith, Optical chemical sensors. Chem. Rev. **108**(2), 400–422 (2008)
5. S.A. Piletsky, A.P.E. Turner, Imprinted polymers and their application in optical sensors, in *Optical Biosensors*: *Today and Tomorrow* (Elsevier Science B.V., Amsterdam, 2008)
6. C. Branger, W. Meouche, A. Margaillan, Recent advances on ion-imprinted polymers. React. Funct. Polym. **73**(6), 859–875 (2013)
7. X. Wang, O.S. Wolfbeis, Fiber-optic chemical sensors and biosensors (2008–2012). Anal. Chem. **85**(2), 487–508 (2013)
8. B.D. Gupta, R.K. Verma, Surface plasmon resonance-based fiber optic sensors: principle, probe designs, and some applications. J. Sens. **2009**, 1 (2009)
9. Y. Chen, H. Ming, Review of surface plasmon resonance and localized surface plasmon resonance sensor. Photonic Sens. **2**(1), 37–49 (2012)
10. K. Kudo, Optical fiber bio-sensor using adsorption LB films, IEICE Trans. Electron. **E87-C**(2), 185–187 (2004)
11. H. Matsuo, S. Kuniyoshi, K. Kudo, K. Tanaka, Evanescent wave optical fiber sensor using adsorption LB films. Synth. Met. **115**(1), 37–39 (2000)
12. M. Consales, A. Crescitelli, M. Penza, P. Aversa, P.D. Veneri, M. Giordano, A. Cusano, SWCNT nano-composite optical sensors for VOC and gas trace detection. Sens Actuators, B **138**(1), 351–361 (2009)
13. M. Penza, G. Cassano, P. Aversa, A. Cusano, A. Cutolo, M. Giordano, L. Nicolais, Carbon nanotube acoustic and optical sensors for volatile organic compound detection. Nanotechnology **16**(11), 2536–2547 (2005)
14. S.W. James, I. Ishaq, G.J. Ashwell, R.P. Tatam, Fibre optic sensing using Langmuir Blodgett thin film overlays, in *Proceedings of SPIE—The International Society for Optical Engineering* (2004), p. 308
15. R.K. Iler, Multilayers of colloidal particles. J. Colloid Interface Sci. **21**(6), 569–594 (1966)
16. J. Schmitt, T. Gruenewald, G. Decher, P.S. Pershan, K. Kjaer, M. Loesche, Internal structure of layer-by-layer adsorbed polyelectrolyte films: a neutron and X-ray reflectivity study. Macromolecules **26**(25), 7058–7063 (1993)
17. G. Decher, Fuzzy nanoassemblies: toward layered polymeric multicomposites. Science **277**(5330), 1232–1237 (1997)

18. P.T. Hammond, T.R. Farhat, Designing a new generation of fuel cells using layer-by-layer deposition of polyelectrolytes, in *ACS National Meeting Book of Abstracts* (2004)
19. P.T. Hammond, Form and function in multilayer assembly: new applications at the nanoscale. Adv. Mater. **16**(15), Spl. Issue, 1271–1293 (2004)
20. D. Yoo, S.S. Shiratori, M.F. Rubner, Controlling bilayer composition and surface wettability of sequentially adsorbed multilayers of weak polyelectrolytes. Macromolecules **31**(13), 4309–4318 (1998)
21. S.S. Shiratori, M.F. Rubner, pH-dependent thickness behavior of sequentially adsorbed layers of weak polyelectrolytes. Macromolecules **33**(11), 4213–4219 (2000)
22. J.D. Mendelsohn, C.J. Barrett, V.V. Chan, A.J. Pal, A.M. Mayes, M.F. Rubner, Fabrication of microporous thin films from polyelectrolyte multilayers. Langmuir **16**(11), 5017–5023 (2000)
23. Y. Lvov, G. Decher, H. Möhwald, Assembly, structural characterization, and thermal behavior of layer-by-layer deposited ultrathin films of poly(vinyl sulfate) and poly(allylamine). Langmuir **9**(2), 481–486 (1993)
24. W.B. Stockton, M.F. Rubner, Molecular-level processing of conjugated polymers. 4. Layer-by-layer manipulation of polyaniline via hydrogen-bonding interactions. Macromolecules **30**(9), 2717–2725 (1997)
25. J. Bravo, L. Zhai, Z. Wu, R.E. Cohen, M.F. Rubner, Transparent superhydrophobic films based on silica nanoparticles. Langmuir **23**(13), 7293–7298 (2007)
26. P. Schuetz, F. Caruso, Electrostatically assembled fluorescent thin films of rare-earth-doped lanthanum phosphate nanoparticles. Chem. Mater. **14**(11), 4509–4516 (2002)
27. Y. Jin, X. Gao, Plasmonic fluorescent quantum dots. Nat. Nanotechnol. **4**(9), 571–576 (2009)
28. J. Goicoechea, C.R. Zamarreño, I.R. Matías, F.J. Arregui, Optical fiber pH sensors based on layer-by-layer electrostatic self-assembled Neutral Red. Sens. Actuators, B **132**(1), 305–311 (2008)
29. F. Caruso, D. Trau, H. Möhwald, R. Renneberg, Enzyme encapsulation in layer-by-layer engineered polymer multilayer capsules. Langmuir **16**(4), 1485–1488 (2000)
30. J. Hodak, R. Etchenique, E.J. Calvo, K. Singhal, P.N. Bartlett, Layer-by-layer self-assembly of glucose oxidase with a poly(allylamine)ferrocene redox mediator. Langmuir **13**(10), 2708–2716 (1997)
31. X. Cui, R. Pei, Z. Wang, F. Yang, Y. Ma, S. Dong, X. Yang, Layer-by-layer assembly of multilayer films composed of avidin and biotin-labeled antibody for immunosensing. Biosens. Bioelectron. **18**(1), 59–67 (2003)
32. P.J. Yoo, K.T. Nam, J. Qi, S. Lee, J. Park, A.M. Belcher, P.T. Hammond, Spontaneous assembly of viruses on multilayered polymer surfaces. Nat. Mater. **5**(3), 234–240 (2006)
33. I.D. Villar, I.R. Matias, F.J. Arregui, Fiber-optic chemical nanosensors by electrostatic molecular self- assembly. Curr. Anal. Chem., **4**(4), 341–355 (2008)
34. F.J. Arregui, I.R. Matias, Y. Liu, K.M. Lenahan, R.O. Claus, Optical fiber nanometer-scale Fabry-Perot interferometer formed by the ionic self-assembly monolayer process. Opt. Lett. **24**(9), 596–598 (1999)
35. F.J. Arregui, Y. Liu, I.R. Matias, R.O. Claus, Optical fiber humidity sensor using a nano Fabry-Perot cavity formed by the ionic self-assembly method. Sens. Actuators, B **59**(1), 54–59 (1999)
36. M. Jiang, Q. Li, J. Wang, W. Yao, Z. Jin, Q. Sui, J. Shi, F. Zhang, L. Jia, W. Dong, Optical response of fiber-optic Fabry-Perot refractive-index tip sensor coated with polyelectrolyte multilayer ultra-thin films. J. Lightwave Technol. **31**(14), 2321–2326 (2013)
37. M. Jiang, Q. Li, Q. Sui, L. Jia, P. Peng, Spectral characteristics of refractive index based on nanocoated optical fiber F-P sensor. Guang Pu Xue Yu Guang Pu Fen Xi/Spectr. Spectral Anal. **33**(1), 261–265 (2013)
38. J.M. Corres, I.R. Matias, M. Hernaez, J. Bravo, F.J. Arregui, Optical fiber humidity sensors using nanostructured coatings of SiO nanoparticles. Sens. J. IEEE **8**(3), 281–285 (2008)

39. J. Goicoechea, C.R. Zamarreño, I.R. Matias, F.J. Arregui, Utilization of white light interferometry in pH sensing applications by mean of the fabrication of nanostructured cavities. Sens. Actuators, B **138**(2), 613–618 (2009)
40. C. Elosúa, C. Bariáin, I.R. Matías, F.J. Arregui, A. Luquin, E. Vergara, M. Laguna, Indicator immobilization on Fabry-Perot nanocavities towards development of fiber optic sensors. Sens. Actuators B **130**(1), 158–163 (2008)
41. X. Zhang, Y. Guan, Y. Zhang, Ultrathin hydrogel films for rapid optical biosensing. Biomacromolecules **13**(1), 92–97 (2012)
42. L.H. Chen, X.M. Ang, C.C. Chan, M. Shaillender, B. Neu, W.C. Wong, P. Zu, K.C. Leong, Layer-by-layer (chitosan/polystyrene sulfonate) membrane-based fabry-perot interferometric fiber optic biosensor. IEEE J. Sel. Top. Quantum Electron. **18**(4), 1457–1464 (2012)
43. S.W. James, R.P. Tatam, Optical fibre long-period grating sensors: characteristics and application. Meas. Sci. Technol. **14**(5), R49–R61 (2003)
44. I.M. Ishaq, A. Quintela, S.W. James, G.J. Ashwell, J.M. Lopez-Higuera, R.P. Tatam, Modification of the refractive index response of long period gratings using thin film overlays. Sens. Actuators, B **107**(2), 738–741 (2005)
45. S. Korposh, S.W. James, S. Lee, S. Topliss, S.C. Cheung, W.J. Batty, R.P. Tatam, Fiber optic long period grating sensors with a nanoassembled mesoporous film of SiO_2 nanoparticles. Opt. Express **18**(12), 13227–13238 (2010)
46. T. Wang, S. Korposh, R. Wong, S. James, R. Tatam, S. Lee, A novel ammonia gas sensor using a nanoassembled polyelectrolyte thin film on fiber-optic long-period gratings. Chem. Lett. **41**(10), 1297–1299 (2012)
47. T. Wang, S. Korposh, S. James, R. Tatam, S. Lee, Optical fiber long period grating sensor with a polyelectrolyte alternate thin film for gas sensing of amine odors. Sens. Actuators, B **185**, 117–124 (2013)
48. S.M. Topliss, S.W. James, F. Davis, S.P.J. Higson, R.P. Tatam, Optical fibre long period grating based selective vapour sensing of volatile organic compounds. Sens. Actuators, B **143**(2), 629–634 (2010)
49. S. Korposh, R. Selyanchyn, W. Yasukochi, S. Lee, S.W. James, R.P. Tatam, Optical fibre long period grating with a nanoporous coating formed from silica nanoparticles for ammonia sensing in water. Mater. Chem. Phys. **133**(2–3), 784–792 (2012)
50. S. Korposh, S. Lee, S.W. James, R.P. Tatam, Refractive index sensitivity of fibre-optic long period gratings coated with SiO_2 nanoparticle mesoporous thin films. Meas. Sci. Technol. **22**(7), 075208 (2011)
51. D. Viegas, M. Hernaez, J. Goicoechea, J.L. Santos, F.M. Araújo, F. Arregui, I.R. Matias, Simultaneous measurement of humidity and temperature based on an SiO_2-nanospheres film deposited on a long-period grating in-line with a fiber Bragg grating. IEEE Sens. J. **11**(1), 162–166 (2011)
52. D. Viegas, J. Goicoechea, J.M. Corres, J.L. Santos, L.A. Ferreira, F.M. Arajo, I.R. Matias, A fibre optic humidity sensor based on a long-period fibre grating coated with a thin film of SiO_2 nanospheres. Meas. Sci. Technol. **20**(3), 034002 (2009)
53. D. Viegas, J. Goicoechea, J.L. Santos, F.M. Araújo, L.A. Ferreira, F.J. Arregui, I.R. Matias, Sensitivity improvement of a humidity sensor based on silica nanospheres on a long-period fiber grating. Sensors **9**(1), 519–527 (2009)
54. J.M. Corres, A. Sanz, F.J. Arregui, I.R. Matías, J. Roca, Fiber optic glucose sensor based on bionanofilms. Sens. Actuators, B **131**(2), 633–639 (2008)
55. I.R. Matias, F.J. Arregui, J.M. Corres, J. Bravo, Evanescent field fiber-optic sensors for humidity monitoring based on nanocoatings. IEEE Sens. J. **7**(1), 89–95 (2007)
56. B. Larrió.n, M. Hernáez, F.J. Arregui, J. Goicoechea, J. Bravo, I.R. Matías, Photonic crystal fiber temperature sensor based on quantum dot nanocoatings. J. Sens. **2009**, 932471 (2009)
57. J. Bravo, I.R. Matías, I. Del Villar, J.M. Corres, F.J. Arregui, Nanofilms on hollow core fiber-based structures: an optical study. J. Lightwave Technol. **24**(5), 2100–2107 (2006)
58. I.R. Matias, J. Bravo, F.J. Arregui, J.M. Corres, Nanofilms on a hollow core fiber. Opt. Eng. **45**(5), 050503 (2006)

59. A.B. Socorro, I.D. Villar, J.M. Corres, F.J. Arregui, I.R. Matias, Tapered single-mode optical fiber pH sensor based on lossy mode resonances generated by a polymeric thin-film. IEEE Sens. J. **12**(8), 2598–2603 (2012)
60. A.B. Socorro, J.M. Corres, I. Del Villar, F.J. Arregui, I.R. Matias, Fiber-optic biosensor based on lossy mode resonances. Sens. Actuators, B **174**, 263–269 (2012)
61. M. Hernaez, C.R. Zamarreño, I. del Villar, I.R. Matias, F.J. Arregui, Lossy mode resonances supported by TiO_2-coated optical fibers. Procedia Eng., **5**, 1099 (2010)
62. I. Del Villar, C.R. Zamarreño, M. Hernaez, F.J. Arregui, I.R. Matias, Generation of lossy mode resonances with absorbing thin-films. J. Lightwave Technol. **28**(23), 3351–3357 (2010)
63. I. del Villar, M. Hernaez, C.R. Zamarreno, P. Sánchez, C. Fernández-Valdivielso, F.J. Arregui, I.R. Matias, Design rules for lossy mode resonance based sensors. Appl. Opt. **51**(19), 4298–4307 (2012)
64. H.V.R. Gonçalves, A.J. Duarte, F. Davis, S.P.J. Higson, J.C.G. Esteves da Silva, Layer-by-layer immobilization of carbon dots fluorescent nanomaterials on single optical fiber. Anal. Chim. Acta **735**, 90–95 (2012)
65. C.J. Brinker, Hydrolysis and condensation of silicates: effects on structure. J. Non-Cryst. Solids **100**(1–3), 31–50 (1988)
66. K. Hwang, J. Kwon, J. Oh, J. An, B. Kim, Surface morphological properties of sol-gel derived SiO_2 fiber. J. Mater. Sci. **39**(5), 1683–1687 (2004)
67. C. Brinker, S. Wallace, N. Raman, R. Sehgal, J. Samuel, S. Contakes, *Sol-Gel Processing of Amorphous Nanoporous Silicas: Thin Films and Bulk* (2002), pp. 123-139
68. P.J. Davis, R. Deshpande, D.M. Smith, C.J. Brinker, R.A. Assink, Pore structure evolution in silica gel during aging/drying. IV. Varying pore fluid pH. J. Non-Cryst. Solids **167**(3), 295–306 (1994)
69. S. Liu, L.L. Hench, Control of the texture of gel-silica monoliths by aging treatments, in *Proceedings of SPIE—The International Society for Optical Engineering* (1992), p. 14
70. T. Mizuno, H. Nagata, S. Manabe, Attempts to avoid cracks during drying. J. Non-Cryst. Solids **100**(1–3), 236–240 (1988)
71. J. Estella, J.C. Echeverría, M. Laguna, J.J. Garrido, Silica xerogels of tailored porosity as support matrix for optical chemical sensors. Simultaneous effect of pH, ethanol:TEOS and water:TEOS molar ratios, and synthesis temperature on gelation time, and textural and structural properties. J. Non-Cryst. Solids **353**(3), 286–294 (2007)
72. J. Estella, J.C. Echeverría, M. Laguna, J.J. Garrido, Effect of supercritical drying conditions in ethanol on the structural and textural properties of silica aerogels. J. Porous Mater. **15**(6), 705–713 (2008)
73. J. Estella, J.C. Echeverría, M. Laguna, J.J. Garrido, Effects of aging and drying conditions on the structural and textural properties of silica gels. Microporous Mesoporous Mater. **102**(1–3), 274–282 (2007)
74. M. Kruk, M. Jaroniec, R. Ryoo, S.H. Joo, Characterization of MCM-48 silicas with tailored pore sizes synthesized via a highly efficient procedure. Chem. Mater. **12**(5), 1414–1421 (2000)
75. J.C. Echeverría, P. de Vicente, J. Estella, J.J. Garrido, A fiber-optic sensor to detect volatile organic compounds based on a porous silica xerogel film. Talanta **99**, 433–440 (2012)
76. S. Tao, C.B. Winstead, R. Jindal, J.P. Singh, Optical-fiber sensor using tailored porous sol-gel fiber core. IEEE Sens. J. **4**(3), 322–328 (2004)
77. V. Matejec, J. Mrázek, J. Skokánková, M. Chomát, I. Kašík, Optical properties and sensitivity of xerogel detection layers for fiber-optic hydrocarbon sensors. J. Sol-Gel. Sci. Technol. **32**(1–3), 237–242 (2004)
78. A. Lobnik, N. Majcen, K. Niederreiter, G. Uray, Optical pH sensor based on the absorption of antenna generated europium luminescence by bromothymolblue in a sol-gel membrane. Sens. Actuators, B **74**(1–3), 200–206 (2001)
79. W. Cao, Y. Duan, Optical fiber-based evanescent ammonia sensor. Sens. Actuators, B **110**(2), 252–259 (2005)

80. S. Dong, M. Luo, G. Peng, W. Cheng, Broad range pH sensor based on sol-gel entrapped indicators on fibre optic. Sens. Actuators, B **129**(1), 94–98 (2008)
81. J. Estella, P. De Vicente, J.C. Echeverría, J.J. Garrido, A fibre-optic humidity sensor based on a porous silica xerogel film as the sensing element. Sens. Actuators, B **149**(1), 122–128 (2010)
82. C. Chu, Y. Lo, T. Sung, Review on recent developments of fluorescent oxygen and carbon dioxide optical fiber sensors. Photonic Sens. **1**(3), 234–250 (2011)
83. C. Chu, T. Sung, Y. Lo, Enhanced optical oxygen sensing property based on Pt(II) complex and metal-coated silica nanoparticles embedded in sol-gel matrix. Sens. Actuators, B **185**, 287–292 (2013)
84. M. Bezunartea, J. Estella, J.C. Echeverría, C. Elosúa, C. Bariáin, M. Laguna, A. Luquin, J.J. Garrido, Optical fibre sensing element based on xerogel-supported $[Au_2Ag_2(C_6F_5)_4(C_{14}H_{10})]n$ for the detection of methanol and ethanol in the vapour phase. Sens. Actuators, B **134**(2), 966–973 (2008)
85. J. Musgo, J.C. Echeverría, J. Estella, M. Laguna, J.J. Garrido, Ammonia-catalyzed silica xerogels: Simultaneous effects of pH, synthesis temperature, and ethanol:TEOS and water:TEOS molar ratios on textural and structural properties. Microporous Mesoporous Mater. **118**(1–3), 280–287 (2009)
86. A.P.F. Turner, Biosensors. Curr. Opin. Biotechnol. **5**(1), 49–53 (1994)
87. G. Wulff, R. Grobe-Einsler, R. Vesper, A. Sarhan, On the specificity distribution of chiral cavities prepared in synthetic polymers. Die Makromolekulare Chemie **178**, 2817 (1977)
88. T.A. Sergeyeva, Molecularly imprinted polymers as synthetic mimics of bioreceptors. 1. General principles of molecular imprinting. Biopolymers Cell **25**(4), 253–265 (2009)
89. S.A. Piletsky, E.V. Piletskaya, T.L. Panasyuk, A.V. El'skaya, R. Levi, I. Karube, G. Wulff, Imprinted membranes for sensor technology: opposite behavior of covalently and noncovalently imprinted membranes. Macromolecules **31**(7), 2137–2140 (1998)
90. N. Wu, L. Feng, Y. Tan, J. Hu, An optical reflected device using a molecularly imprinted polymer film sensor. Anal. Chim. Acta **653**(1), 103–108 (2009)
91. R.B. Queirós, S.O. Silva, J.P. Noronha, O. Frazão, P. Jorge, G. Aguilar, P.V.S. Marques, M.G.F. Sales, Microcystin-LR detection in water by the Fabry-Pérot interferometer using an optical fibre coated with a sol-gel imprinted sensing membrane. Biosens. Bioelectron. **26**(9), 3932–3937 (2011)
92. Y. Chen, J.J. Brazier, M. Yan, P.R. Bargo, S.A. Prahl, Fluorescence-based optical sensor design for molecularly imprinted polymers. Sens. Actuators, B **102**(1), 107–116 (2004)
93. S.M. Ng, R. Narayanaswamy, Fluorescence sensor using a molecularly imprinted polymer as a recognition receptor for the detection of aluminium ions in aqueous media. Anal. Bioanal. Chem. **386**(5), 1235–1244 (2006)
94. X. Ton, B. Tse Sum Bui, M. Resmini, P. Bonomi, I. Dika, O. Soppera, K. Haupt, A versatile fiber-optic fluorescence sensor based on molecularly imprinted microstructures polymerized in situ. Angew. Chem. Int. Ed. **52**(32), 8317–8321 (2013)
95. T.H. Nguyen, S.A. Hardwick, T. Sun, K.T.V. Grattan, Intrinsic fluorescence-based optical fiber sensor for cocaine using a molecularly imprinted polymer as the recognition element. IEEE Sens. J. **12**(1), 255–260 (2012)
96. S. Piperno, B. Tse Sum Bui, K. Haupt, L.A. Gheber, Immobilization of molecularly imprinted polymer nanoparticles in electrospun poly(vinyl alcohol) nanofibers. Langmuir **27**(5), 1547–1550 (2011)
97. B. Liedberg, C. Nylander, I. Lunström, Surface plasmon resonance for gas detection and biosensing. Sens. Actuators **4**(C), 299–304 (1983)
98. J. Homola, Electromagnetic theory of surface plasmons, in *Springer Series on Chemical Sensors and Biosensors*, vol. 4 (2006), pp. 3–44
99. G. Dougherty, Compact optoelectronic instrument with a disposable sensor based on surface plasmon resonance. Meas. Sci. Technol. **4**(6), 697–699 (1993)

100. S. Ekgasit, A. Tangcharoenbumrungsuk, F. Yu, A. Baba, W. Knoll, Resonance shifts in SPR curves of nonabsorbing, weakly absorbing, and strongly absorbing dielectrics. Sens. Actuators, B **105**(2), 532–541 (2005)
101. E. Stenberg, B. Persson, H. Roos, C. Urbaniczky, Quantitative determination of surface concentration of protein with surface plasmon resonance using radiolabeled proteins. J. Colloid Interface Sci. **143**(2), 513–526 (1991)
102. R.C. Jorgenson, S.S. Yee, A fiber-optic chemical sensor based on surface plasmon resonance. Sens. Actuators, B **12**(3), 213–220 (1993)
103. B. Culshaw, A. Kersey, Fiber-optic sensing: a historical perspective. J. Lightwave Technol. **26**(9), 1064–1078 (2008)
104. A. Cusano, J.M. López-Higuera, I.R. Matias, B. Culshaw, Editorial optical fiber sensor technology and applications. IEEE Sens. J. **8**(7), 1052–1054 (2008)
105. B. Lee, Review of the present status of optical fiber sensors. Opt. Fiber Technol. **9**(2), 57–79 (2003)
106. B. Lee, S. Roh, J. Park, Current status of micro- and nano-structured optical fiber sensors. Optical Fiber Technology **15**(3), 209–221 (2009)
107. S. Roh, T. Chung, B. Lee, Overview of the characteristics of micro- and nano-structured surface plasmon resonance sensors. Sensors **11**(2), 1565–1588 (2011)
108. S. Singh, S.K. Mishra, B.D. Gupta, SPR based fibre optic biosensor for phenolic compounds using immobilization of tyrosinase in polyacrylamide gel. Sens. Actuators, B **186**, 388–395 (2013)
109. R. Slavík, J. Homola, J. Ctyroký, Single-mode optical fiber surface plasmon resonance sensor. Sens. Actuators, B **54**(1), 74–79 (1999)
110. T. Allsop, R. Neal, C. Mou, P. Brown, S. Rehman, K. Kalli, D.J. Webb, D. Mapps, I. Bennion, Multilayered coated infra-red surface plasmon resonance fibre sensors for aqueous chemical sensing. Opt. Fiber Technol. **15**(5–6), 477–482 (2009)
111. T.D.P. Allsop, R. Neal, C. Mou, K. Kalli, S. Saied, S. Rehman, D.J. Webb, P.F. Culverhouse, J.L. Sullivan, I. Bennion, Formation and characterization of ultra-sensitive surface plasmon resonance sensor based upon a nano-scale corrugated multi-layered coated D-shaped optical fiber. IEEE J. Quantum Electron. **48**(3), 394–405 (2012)
112. X. Yu, S. Zhang, Y. Zhang, H. Ho, P. Shum, H. Liu, D. Liu, An efficient approach for investigating surface plasmon resonance in asymmetric optical fibers based on birefringence analysis. Opt. Express, **18**(17), 17950–17957 (2010)
113. J. Albert, L. Shao, C. Caucheteur, Tilted fiber Bragg grating sensors. Laser Photonics Rev. **7**(1), 83–108 (2013)
114. T. Schuster, R. Herschel, N. Neumann, C.G. Schäffer, Miniaturized long-period fiber grating assisted surface plasmon resonance sensor. J. Lightwave Technol. **30**(8), 1003–1008 (2012)
115. S. Cheng, L. Chau, Colloidal gold-modified optical fiber for chemical and biochemical sensing. Anal. Chem. **75**(1), 16–21 (2003)
116. H.S. Jang, K.N. Park, C.D. Kang, J.P. Kim, S.J. Sim, K.S. Lee, Optical fiber SPR biosensor with sandwich assay for the detection of prostate specific antigen. Opt. Commun. **282**(14), 2827–2830 (2009)
117. N. Lai, C. Wang, H. Chiang, L. Chau, Detection of antinuclear antibodies by a colloidal gold modified optical fiber: comparison with ELISA. Anal. Bioanal. Chem. **388**(4), 901–907 (2007)
118. Lin, T.-. & Chung, M.-. 2008, "Using monoclonal antibody to determine lead ions with a localized surface plasmon resonance fiber-optic biosensor", *Sensors,* vol. 8, no. 1, pp. 582-593
119. M. Piliarik, J. Homola, Z. Maníková, J. Ctyroký, Surface plasmon resonance sensor based on a single-mode polarization-maintaining optical fiber. Sens. Actuators, B **90**(1–3), 236–242 (2003)
120. V.V.R. Sai, T. Kundu, S. Mukherji, Novel U-bent fiber optic probe for localized surface plasmon resonance based biosensor. Biosens. Bioelectron. **24**(9), 2804–2809 (2009)

121. R. Slavík, J. Homola, E. Brynda, A miniature fiber optic surface plasmon resonance sensor for fast detection of staphylococcal enterotoxin B. Biosens. Bioelectron. **17**(6–7), 591–595 (2002)
122. J. Tang, S. Cheng, W. Hsu, T. Chiang, L. Chau, Fiber-optic biochemical sensing with a colloidal gold-modified long period fiber grating. Sens. Actuators, B **119**(1), 105–109 (2006)
123. Y. Yanase, A. Araki, H. Suzuki, T. Tsutsui, T. Kimura, K. Okamoto, T. Nakatani, T. Hiragun, M. Hide, Development of an optical fiber SPR sensor for living cell activation. Biosens. Bioelectron. **25**(5), 1244–1247 (2010)
124. A. Baliyan, P. Bhatia, B.D. Gupta, E.K. Sharma, A. Kumari, R. Gupta, Surface plasmon resonance based fiber optic sensor for the detection of triacylglycerides using gel entrapment technique. Sens. Actuators, B **188**, 917–922 (2013)
125. J. Cao, M.H. Tu, T. Sun, K.T.V. Grattan, Wavelength-based localized surface plasmon resonance optical fiber biosensor. Sens. Actuators, B **181**, 611–619 (2013)
126. L. Chau, Y. Lin, S. Cheng, T. Lin, Fiber-optic chemical and biochemical probes based on localized surface plasmon resonance. Sens. Actuators, B **113**(1), 100–105 (2006)
127. T. Lin, M. Chung, Detection of cadmium by a fiber-optic biosensor based on localized surface plasmon resonance. Biosens. Bioelectron. **24**(5), 1213–1218 (2009)
128. K. Mitsui, Y. Handa, K. Kajikawa, Optical fiber affinity biosensor based on localized surface plasmon resonance. Appl. Phys. Lett. **85**(18), 4231–4233 (2004)
129. P.J. Rivero, A. Urrutia, J. Goicoechea, F.J. Arregui, Optical fiber humidity sensors based on Localized Surface Plasmon Resonance (LSPR) and Lossy-mode resonance (LMR) in overlays loaded with silver nanoparticles. Sens. Actuators, B **173**, 244–249 (2012)
130. J. Cao, M.H. Tu, T. Sun, K.T.V. Grattan, Wavelength-based localized surface plasmon resonance optical fiber biosensor. Sens. Actuators B **181**, 611–619 (2013)
131. J. Cao, E.K. Galbraith, T. Sun, K.T.V. Grattan, Cross-comparison of surface plasmon resonance-based optical fiber sensors with different coating structures. IEEE Sens. J. **12**(7), 2355–2361 (2012)
132. E. Petryayeva, U.J. Krull, Localized surface plasmon resonance: nanostructures, bioassays and biosensing—a review. Anal. Chim. Acta **706**(1), 8–24 (2011)
133. B. Sepúlveda, P.C. Angelomé, L.M. Lechuga, L.M. Liz-Marzán, LSPR-based nanobiosensors. Nano Today **4**(3), 244–251 (2009)

Chapter 14
Sensitive and Selective Lab-on-a-Fiber Sensor for Bacteria Detection in Water

Wojtek J. Bock, Saurabh Mani Tripathi and Mateusz Smietana

Abstract Pathogenic and toxic bacteria are one of the most dangerous agents of food-borne disease. In the suitable environment (24–70 °C) they can quickly multiply leading to widespread, severe outbreak, deadly especially for children and the elderly. In this chapter we will first, briefly, revise various existing technologies for the detection of pathogenic *Escherichia coli* (*E. coli*) and hepatotoxic *micro-cystin-LR* (MC-LR) toxin. We will then present, in detail, our recently developed long-period-fiber-grating (LPFG) based, ultra stable, label free, and highly accurate methods for the specific detection (both quantitative and qualitative) of *E. coli* and MC-LR. We will draw the protocols to achieve the exceptionally accurate (>99 % for *E. coli* concentrations of 10^3 cfu/ml) detection and, using the scan electron microscopy and atomic force topography, will also show that the results are based on the real bacterial/toxin binding to the optical-fiber probe and not merely due to the sample-refractive-index variations. The chapter should be of particular interest for the students and researchers both in the industry and in academia.

14.1 Introduction

Pathogenic *Escherichia coli* (*E. coli*) is one of the most dangerous agents of food-borne disease. Consumption of contaminated food or water can be deadly, especially for children and the elderly. Although *E. coli* infection is most common in

W. J. Bock (✉)
University of Quebec in Outaouais, Gatineau, QC, Canada
e-mail: Wojtek.Bock@uqo.ca

S. M. Tripathi
Indian Institute of Technology Kanpur, Kanpur, Uttar Pradesh, India
e-mail: smt@iitk.ac.in

M. Smietana
Warsaw University of Technology, Warsaw, Poland
e-mail: smiema01@uqo.ca

A. Cusano et al. (eds.), *Lab-on-Fiber Technology*,
Springer Series in Surface Sciences 56, DOI: 10.1007/978-3-319-06998-2_14,

developing countries, many recent outbreaks in Europe and Northern America have been attributed to a strain of *E. coli* which has been identified among the most common causes of diseases related to food safety [1] (http://www.cdc.gov/ecoli/2011/ecolio104/). Accurate routine testing is crucial for outbreak prevention. Nevertheless, currently available tests require time-consuming amplification of samples. The standard detection process of *E. coli* bacteria takes about 24 h to obtain results from culturing methods. Although more recent detection techniques such as PCR, ELISA and IMS offer a more rapid detection, analysis time of several hours is still required.

For the specific and rapid detection of such pathogens, bio-recognition elements such as antibodies [3, 18], nucleic acids (DNA/RNA) [5, 8] and bacteriophages [6, 19] have widely been used for the specific capturing of the target bacteria. Their binding can be detected by fluorescence labeling methods [17, 24] or by label-free methods [15, 22, 26]. Each of these recognition elements has its own advantages and disadvantages. For example, recognition based on nucleic acid, though offering high specificity, suffers from the inability to discriminate between viable and non-viable cells [8]. For antibody-based recognition elements, the drawbacks are high price, stability and cross-binding to other bacteria which may result in false positives [6]. Bacteriophage, on the other hand, offers a superior alternative to other recognition elements in terms of their high specificity, fast binding, easy/low-cost production and stability. Furthermore, the bacteriophage has a unique feature to discriminate between the viable and non-viable cells. In principle the bacteriophage binds to both viable and non-viable bacteria as long as the surface receptor (lipopolysaccharides, teichoic acids, proteins) is intact and recognized by the bacteria. To see whether the bacteria are viable, the experiment can be extended past the latent period (the time from infection to bacterial lysis by the bacteriophage progeny). The signal should change since the intact bacteria are now broken (lysed).

In the fluorescent-label method, the recognition elements are labeled with specific dyes, making this procedure complex, time-consuming and allowing potential alteration of the properties of the analyte [6]. Label-free methods, on the other hand, are based on changes occurring at the sensor surface, on changes in the analyte refractive index (ARI) or on changes in the thickness of the bio-film—all these changes could modify the optical properties of a sensor. This method is not only very fast but also allows real-time monitoring of the biomolecular interactions occurring while the process of detection takes place.

Several schemes based on the label-free methods, like evanescent wave coupling in grating couplers [13], leaky optical waveguides [26] and surface Plasmon resonances (SPR) [2, 4, 12, 16, 23] for the detection of *E. coli* have been proposed in recent past. Out of these, the SPR based sensors, both in the angular interrogation regime [4, 16, 23] and spectral interrogation regime [2, 12], have received much attention due to their extremely high sensitivity which is necessary for highly accurate bio-measurements. However, drawbacks associated with the SPR based sensors are numerous, such as the incorporation of bulky moving part (the high index prism used in angular interrogation regime) [4, 16], the expensive and elaborated metal deposition requirements to support the surface Plasmon wave, the limitation of operation to the highly lossy visible spectral regime ($\sim$400–800 nm) [2, 12], and the possibility of false positives due to input power fluctuations and/or connector losses associated with the intensity based measurements.

In this chapter, we discuss stable, label-free, bacteriophage-based detection of *E. coli* using ultra sensitive long-period fiber gratings (LPFGs). We follow two approaches (i) physical adsorption based and (ii) covalent chemical binding based, involving bacteriophage T4 immobilization on the LPFG surface and analysing the *E. coli* binding using the highly accurate spectral interrogation mechanism. In contrast to the widely used SPR based sensors, no moving part or metal deposition is required in our sensor, making the present sensor extremely accurate, fast, very compact and cost effective. We demonstrated that our detection mechanism is capable of reliable detection of *E. coli* concentrations as low as 10^3 cfu/ml (cfu: colony forming units) with an experimental accuracy greater than 99 %.

14.2 Materials and Methods

The LPFGs are periodic refractive index variations created within the core region of optical fiber (diameter $\sim$9 μm) to redirect part of the optical field from the fiber core region (known as the core mode) to the cladding region (diameter $\sim$125 μm) known as the cladding mode of the optical fiber. The specific wavelength at which the field is redirected is known as the resonance wavelength (λ_R) and is expressed as [14],

$$\lambda_R = \Lambda\left(n_{eff}^{c} - n_{eff}^{cl} + \frac{\kappa_{c-c} - \kappa_{cl-cl}}{k_0}\right) \tag{14.1}$$

where n_{eff}^{c} and n_{eff}^{cl} are the effective refractive indices of the core and the cladding mode; Λ is the grating period; and κ_{c-c} and κ_{cl-cl} are the self-coupling coefficients of the core mode and the cladding mode, respectively. A small fraction of the cladding mode field, evanescent field, travels outside of the optical fiber, interacting with the outer region, changing the n_{eff}^{cl} and thus changing the λ_R. The LPFG sensing principle relies on measuring the changes in λ_R, the smaller the period of refractive index variation (Λ) the larger would be cladding mode evanescent field leading to high sensitivity.

14.2.1 LPFG Fabrication

To carry out the experiments we fabricated several highly sensitive LPFGs at our lab using single mode optical fiber SMF-28^{TM} (Corning: NY, 14831 USA). For ease of fabrication the photosensitivity of the fibers was increased by hydrogen loading at 150 bars in hydrogen chamber for 15 days. Photosensitive LPFGs were then inscribed into the fiber core using a chromium amplitude mask ($\Lambda = 226.8$ μm) and a high-power KrF excimer laser (Lumonics™ Lasers: Pulse Master®-840) emitting at 248 nm at a pulse repetition rate of 100 Hz, pulse duration of 12 ns and peak pulse energy of 10 mJ. The LPFGs were then annealed at 150 °C for 3 h to release the excess hydrogen and thus to stabilize their optical properties. The LPFGs so fabricated have λ_R in the low loss window of telecommunication wavelengths (~ 1.56 μm) for air as surrounding medium. Since the measurements were to be performed in aqueous solutions the λ_R would shift by ±350 nm shifting away from the low loss regime. To overcome it we slowly etched the cladding of the fiber in 4 % HF for ~3 h. This shifted the λ_R close to its turning point ~1.58 μm, enhancing the sensitivity of the LPFGs [20].

14.2.2 *E. coli* Culturing

Frozen stock of *E. coli* B was used to seed Luria-Bertani (LB) media overnight. Bacteria were harvested by centrifugation at 3,000 g for 10 min, followed by washing in phosphate-buffered saline (PBS) buffer. Dilutions of the overnight culture were plated on LB agar to determine the titer, expressed in cfu. Appropriate dilutions of bacteria stock were made in the PBS buffer from the overnight culture stock.

14.2.3 T4 Bacteriophage Preparation

To produce a stock of T4-bacteriophage solution, 100 μL of *E.coli* log-phase culture was added to 3 ml of cooled top agar and the mixture was poured onto an LB agar plate until it solidified. Then 100 μL of phage stock (T4 phage) specific to *E. coli* B was added to the solidified top agar and incubated at 37 °C overnight. A macro-plaque was developed on the lawn of bacteria after incubation. The top agar was scrapped off with 3 ml of lambda buffer, the suspension was collected in a 50 ml centrifuge tube, followed by an additional washing of the LB agar plate with another 3 ml of lambda buffer. Three drops of chloroform were then added to the suspension, which was then vortexed and centrifuged at 3,500 g for 10 min. The phage containing a supernatant was then filtered through a 0.22 μm filter to remove bacterial debris. The phage was freshly prepared for each experiment. The titer of the phage stock was determined by serial dilution of the stock. Phage concentration was expressed in plaque forming units (pfu).

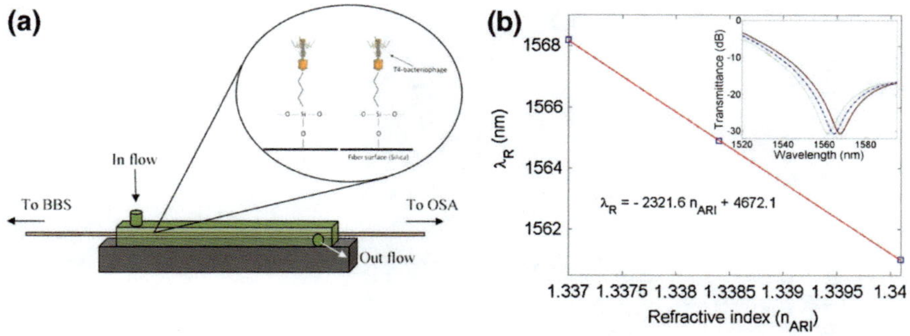

Fig. 14.1 **a** Schematic of the experimental arrangement; inset shows Si–O-Si covalent binding of the bacteriophages with the optical fiber surface (SiO_2). **b** Variation of the resonance wavelength with ARI; the transmission spectrum is shown in the inset

14.2.4 Experimental Set-up

The presence of macrobends along the LPFG region is the primary cause of measurement errors in the LPFG based sensors. In order to avoid them we maintained a constant tension along the LPFG throughout the experiments by attaching the fiber near one end of the LPFG and applying a fixed force near the other end of it (see Fig. 14.1a). Light was launched into the fiber using an *Agilent*-83437A broadband source (BBS) and the transmission spectrum was recorded using an *Agilent*-86142B optical spectrum analyzer (OSA) with a resolution of 0.02 nm. Prior to the bacteria measurements, the ambient refractive index (ARI) sensitivity of the LPFGs was determined by using analytes of known refractive indices measured by Abbe refractometer (Atago DR-M2) operating at $\lambda = 1.55$ μm. The variation in the λ_R of a typical LPFG in response to changing the ARIs is shown in (Fig. 14.1b), with the transmission spectrum shown in the inset. The slope of the λ_R with the ARI and hence the ARI sensitivity being 2321.6 nm/RIU.

14.3 Results and Discussion

14.3.1 Physical Adsorption Based Sensing

In our *E. coli* sensor development, we first study the physical adsorption based senor in which the bacteriophages stick and are physically adsorbed to the sensor surface and the changes in the resonance wavelength are monitored. To carry out this study we first cleaned the LPFG surface and balanced its pH by washing it in deionized water followed by a PBS (Phosphate Buffer Saline) wash. The LPFGs were then incubated with the bacteriophages for a sufficiently long time ($\sim$4 h) for the phages to be physically adsorbed on the sensor surface. The excess and

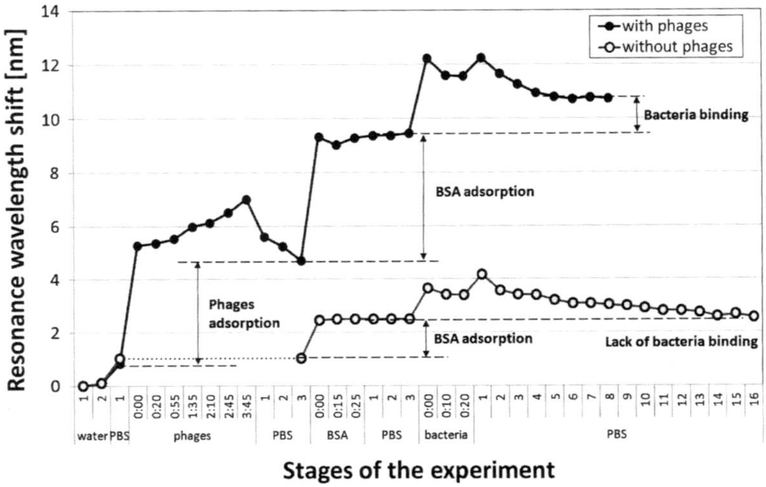

Fig. 14.2 Resonance wavelength shifts, referred to initial value for water surrounding the samples, at various stages of the bacteria experiment, compared for the samples with and without the phages adsorbed on the surface of the LPFG. In the case of phages, BSA and bacteria, the time of the appropriate procedure is given, while for the PBS the consecutive rinsing cycles are marked on the time scale

non-sticky phages were then washed away in a subsequent wash in the PBS buffer. Non specific sites on the sensor surface were blocked by incubating the sensor in BSA (Bovine serum albumin) for ~30 min. Finally the sensor was incubated with the *E. coli* bio -sample for ~20 min, and subsequently washed again in PBS buffer. The results of the experiments with bio-liquids are compared in Fig. 14.2 for the samples with and without a phage overlayer. The variation in the state of the LPFG surface, i.e., modification in the RI of the bio-overlay, as well as variation of its thickness in this case, results in a clear spectral shift of the resonances. The physical adsorption of the bacteriophages to the LPFGs increases the effective cladding radius, resulting in a resonance wavelength shift of ~6 nm during our experiment (see Fig. 14.2).

An increase in the sensor response with time during the phage incubation has been observed also by other authors using different instrumentation techniques. They attributed the effect to an increase in the density of the phage-based overlay [6, 25]. The observed effect may have been slightly amplified in our experiment by some evaporation of the solution during the 4 h incubation. Such evaporation results in an increase of the liquid density and the concentration of phages and thus an increase in the RI of the bio-overlay. In the following stage of the experiment, a decrease of the resonance wavelength is observed. This effect can be explained by the removal of the loosely bound cells from the fiber surface when the LPFG is rinsed with the PBS [6, 25]. The adsorbed cells create a thin overlayer with an external RI equal to that of the PBS. When PBS responses before and after the introduction of the phages are compared, a shift of ~3.7 nm induced by the overlayer can be seen. The RI of this kind of the overlayer (protein) is typically assumed to be from 1.56 [6] to 1.57 [25]. The height of the T4 phages is estimated to be ~200 nm [11, 21], however their orientation on the surface is random, and the surface coverage is usually low [15]. This distribution was shown to result in an average surface coverage of about 2 nm, when incubation took 2–3 h and a gold surface was used [6]. A higher thickness/density of the overlayer was obtained by Nanduri et al. [15] when a flow cell was used. Surprisingly, it was found that the differences in the surface coverage by the phages had no significant effect on the sensitivity of bacteria detection [6]. In the next step, the sample LPFGs were immersed in a BSA-containing solution so that the protein would block the non-specific sites on the surfaces of the sensor. We performed a number of control experiments where we introduced bacteria to the unmodified LPFG surfaces and after applying the BSA blocking agents. To crosscheck the reduction in non-specific adsorption due to BSA incubation, we used a green fluorescent protein marked *E. Coli* for testing the LPFG surface treated and untreated with the BSA using a fluorescence microscope. This confirmed that the BSA coated surface showed significant reduction in non-specific adsorption. Like the earlier stage where phages were adsorbed, this step in the experiment resulted in a further red shifts of the resonance of ~4.7 and ~1.5 nm for the phage overlayer sample and the sample without it, respectively. The difference in the spectral shifts is mainly due to differences in the sensitivity of the two samples; with high-RI nano- over layer (formed by the phages) modifying the RI sensitivity of the LPFG. This accounts for the fact that the shift induced by adsorption of the BSA for the sample with the phages on the surface is more than 3 times as large as for the other sample. The final step of the experimental procedure involves introduction of the bacteria followed by multiple rinses in the PBS. For both types of sample (with and without phages), the presence of the bacteria induced a further spectral shift of the resonances. This shift takes place during the first several minutes of the final experimental stage and is followed by a slight decrease and then stabilization of the response in time. The phage-bacterium interaction investigated by other authors was shown to cause a time-related structural change in the surface-bound bacteria, possibly due to collapse of the cell [10]. The effect observed in [10] was correlated to an increase in mass density, which in turn increases the RI. However, in our experiment the effect can also be seen for the samples where no phages were present, so its origins must not be due simply to the increase in the mass density.

An important point to note here is that due to the physical adsorption mechanism used in our detection procedure, the sensor is able to detect only high concentrations of *E. coli*, with no link between a given bacterial concentration and a spectral shift recorded. For a quantitative detection of *E. coli* bacterial cells in the bio-sample, we next carried out a covalent immobilization of the bacteriophages on the LPFGs surface and studied the sensors response to various concentrations of *E. coli* in the bio-sample.

14.3.2 Covalent Bacteriophage Immobilization and Quantitative E. coli Detection

To covalently immobilize the bacteriophages on the fiber surface we first silanized it by incubating the cleansed LPFG samples in 3-Aminopropyltris(trimethylsiloxy)silane (APTES) $H_2N\text{-}(CH_2)_3\text{-}Si\text{-}(OCH_3)_3$ (5 % in methanol), for 3 h. This resulted in a mono (multi) layer poly silane film, covalently bonded to the silica surface of the optical fiber with the amino functional groups on the top of the film [7]. The incubated LPFG samples were then rinsed first in methanol, and second in deionized water. Next, the amino functional groups exposed on the top of the film were functionalized by an amine-reactive homo-bifunctional cross-linker glutaraldehyde (5 % in deionized water), for 30 min. Thus, the amino groups on the top of the LPFG sensor surface were activated to bond with the bacteriophages. The LPFGs were rinsed once again with deionized water to remove the excess glutaraldehyde and then with the PBS buffer to neutralize the surface to pH 7.5. Having activated the surface, the T4 bacteriophages were covalently immobilized through their amine groups on the LPFG surface by immersing the LPFG samples in the bacteriophage solution (10^{10} pfu/ml) for 4 h. The functionalization of the fiber surface by the APTES and the eventual covalent binding of the bacteriophages are schematically shown in the inset of Fig. 14.1a. The LPFG sensors were washed copiously with the PBS buffer to rinse off the excess (unbound) phages. This step was followed by incubation of the gratings in BSA (1 mg/ml) for 30 min to block the non-specific sites on the LPFG surface. The excess BSA was removed by washing the sample LPFGs again in the PBS buffer. At this point, the sensors were all set to bind the bacteria on the LPFG appropriately fonctionalized surface, and we therefore continuously recorded the transmission spectrum right from this stage. After applying the PBS buffer wash the LPFGs were incubated in different concentrations of *E. coli* B bacteria for 20 min, and were finally rinsed again with the PBS buffer. Here, we would like to mention that since the bacteriophages are easy and cheap to produce and the long-period fiber gratings are easy to fabricate using widely available telecom-grade standard single mode optical fiber, we have always used separate LPFGs for different measurements to avoid contamination and for bio-safety reasons..

The micrographs of the optical fiber surface, obtained using a Hitachi TM-3000 N scanning electron microscope (SEM) at three different concentrations (10^3, 10^5 and 10^9 cfu/ml) of *E. coli* bacteria bound intentionally on the LPFG surface, are shown in Fig. 14.3. To clearly show the extent of *E. coli* coverage on optical fiber surface we have selected a bigger scale for higher concentration of *E. coli*, 10^9 cfu/ml (Fig. 14.2c). To increase the topographic contrast and to avoid the metallization of *E. coli* bacterium we used the back scattering electron (BSE) mode of the SEM. The high-energy electrons originating in the incident electron beam back scatter form of the *E. coli* volume due to the elastic

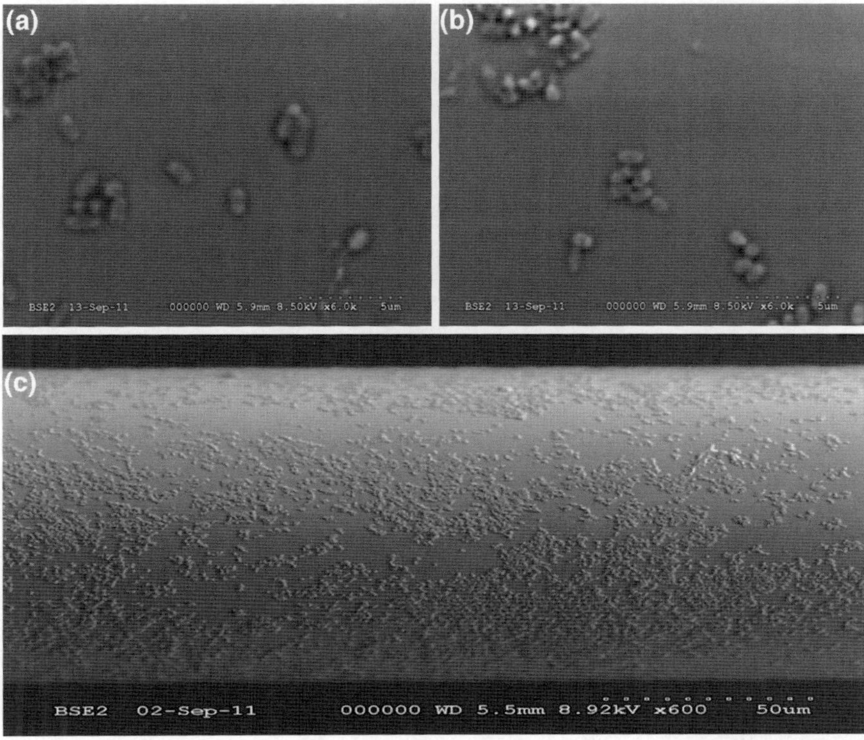

Fig. 14.3 SEM micrograph of the *E. coli* immobilized on the surface of the LPFG **a** 10^3 cfu/ml of *E. coli* **b** 10^5 cfu/ml of *E. coli* and **c** 10^9 cfu/ml of *E. coli* concentration in the PBS buffer

scattering interactions with the *E. coli* bacterium atoms, which are in turn recorded using an asymmetrical, directional BSE detector concentrically placed with the incident electron beam, above the optical fiber. As can be observed from Fig. 14.3, even for a small concentration of the *E. coli*, 10^3 cfu/ml, the bacteria are well bound on the LPFG surface, with surface coverage increasing as the *E. coli* concentration is increased. Unlike the adsorption-based measurements where the *E. coli* bacteria could easily be wiped away from the fiber surface by subsequent PBS buffer washes, in the present experiments the *E. coli* bacteria remain firmly attached to the fiber surface due to the covalently immobilized bacteriophages on it, making the measurements stable in time. This feature enables us to estimate the amount of bacteria in the solution by monitoring the spectral shifts of the resonance wavelength λ_{R} as a function of varying *E. coli* concentrations in the PBS buffer. This shift is influenced not only by the changing refractive index of the solution but also by the change in the cladding radius due to the *E. coli* bacteria attached to it: the shifts in resonance wavelength ($\Delta\lambda_R$) with respect to the changes in the ARI (Δn_{ARI}) and the cladding radius (Δr_{cl}) can be expressed as:

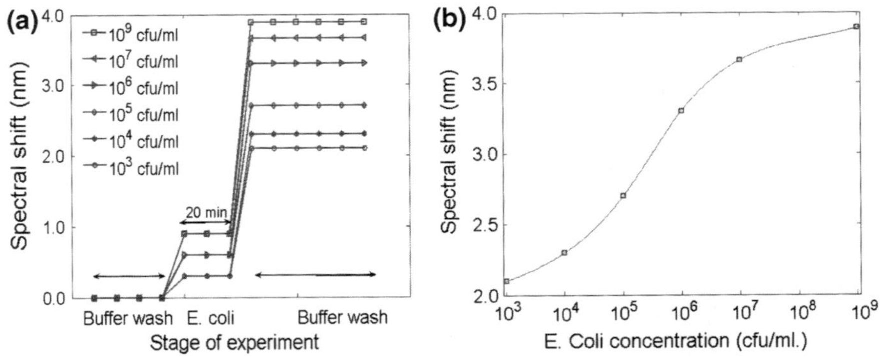

Fig. 14.4 Variation of **a** spectral shift of the resonance wavelength prior to and after the *E. coli* incubation on the LPFG, and **b** spectral shift of the resonance wavelength as a function of the *E. coli* concentration in the PBS buffer

$$\Delta\lambda_R^m = -\Lambda\left(\frac{\partial n_{eff}^{0m}}{\partial n_{ARI}}\Delta n_{ARI} + \frac{\partial n_{eff}^{0m}}{\partial r_{cl}}\Delta r_{cl}\right) \tag{14.2}$$

where we have disregarded the self-coupling terms as a first-order perturbation [14].

The spectral shifts in the λ_R corresponding to the stages of the experiment recorded at the last three steps, namely *pre* PBS buffer wash—Bacteria incubation—*post* PBS buffer wash, are shown in Fig. 14.4a. A spectral shift in the λ_R of ~2–4 nm, given the same overlay medium (*pre* and *post* PBS buffer washes) after *E. coli* incubation, is a clear indicator of various concentrations of bacteria binding on the LPFG surface. This has been shown more clearly in Fig. 14.4b where we have plotted the spectral shift of the λ_R between the *pre* buffer wash stage and *post* PBS buffer wash stages against varying *E. coli* concentrations. We observe that the wavelength shift initially increases exponentially with increasing *E. coli* concentrations and then, after 10^6–10^7 cfu/ml, it starts saturating as the concentrations of *E. coli* continue to increase in the sample. This can be easily explained by the fact that an increased concentration of *E. coli* in the PBS decreases the refractive index of the solution and at the same time increases the surface coverage of the LPFG surface (see Fig. 14.3). At high *E. coli* concentrations, however, the sensor surface is almost completely covered with the bacteria and therefore the resonance-wavelength shift starts saturating. The overall response time of the sensor, after the *E. coli* incubation, is ~20 min and the transmission spectrum can be continuously monitored using an optical spectrum analyzer. This sensor, thus, provides an attractive, reliable and very fast way to monitor the growth/decay of *E. coli* culture. The primary cause of low spectral shifts in the λ_R at lower *E. coli* concentrations is the smaller fractional surface coverage area of the LPFG by the *E. coli*. Further, owing to an experimental spectral resolution of 0.02 nm used in our measurements, the experimental accuracy associated with the minimum resonance-wavelength shift observed in our experiments (~2.1 nm, corresponding to

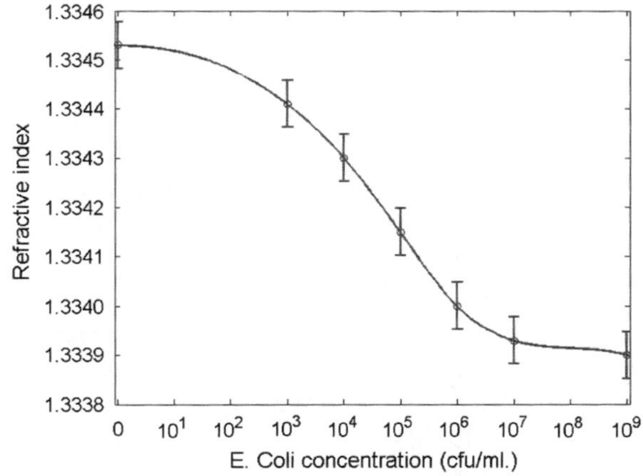

Fig. 14.5 Variation of the refractive index (measured at $\lambda = 1.55$ μm) of the bio-sample as a function of the *E. Coli* concentration in the PBS buffer

an *E. coli* concentration of 10^3 cfu/ml) comes out to be ~99.05 %. We would like to mention here that, although the present sensor is capable of measuring even lower concentrations of *E. coli*, the experimental error associated with such measurements would be greater because of the reduced spectral shifts. For example, using the present LPFGs the experimental accuracy associated with an *E. coli* concentration of 10^2 cfu/ml is 98.95 %, which drops down to 98.82 % for an *E. coli* concentration of 10 cfu/ml. Nevertheless, since the resonance wavelength is independent of the grating length [9, 14], lower concentrations of *E. coli* (0–100 cfu/ml) could be efficiently measured by increasing the fractional surface coverage by *E. coli*, which could be done by reducing the LPFG length. This, however, will result in a decreased transmission loss at the resonance wavelength [9, 14] making detection challenging. Finally, in order to estimate the influence of *E. coli* bacteria on refractive index variation of the bio-sample we measured the refractive indices of various concentrations of the *E. coli* (in PBS buffer) using an Abbe refractometer (Atago DR-M2) with an accuracy of 0.0001 RIU, shown in Fig. 14.5. The overall refractive index difference in the PBS solution, as observed form Fig. 14.4, is $\sim 6 \times 10^{-4}$ RIU for the *E. coli* concentrations varying between 0 and 10^9 cfu/ml. This implies that, based on an exclusive refractive index sensing of the bio-samples, the LPFGs used in our experiments (with a sensitivity of 2321.6 nm/RIU) should yield a net resonance-wavelength shift of ~1.3 nm. This shift is much smaller than the ~4 nm spectral shifts observed in our experiments: a further confirmation of an efficient *E. coli* binding on the optical fiber surface.

14.4 Conclusions

In this chapter we reported development of a fast, reliable, cost-effective, and label-free sensor for detection of *E. coli*. Using a highly accurate spectral interrogation mechanism we carried out a real-time monitoring of the bacteria binding on fiber surface by measuring the spectral shift in the resonance-wavelength of the LPFG. Using covalently immobilized T4 bacteriophages on the surfaces of the long-period optical fiber gratings, we observed resonance-wavelength shifts ranging from 2 to 4 nm for *E. coli* concentrations varying between 10^3 and 10^9 cfu/ml, with an experimental accuracy larger than 99 %. No bacteria dissociated during the subsequent washes of the sensor as they used to do in physical adsorption-based measurement procedures, making the measurements highly stable compared to those of adsorption-based sensors. We showed that the present sensor can efficiently detect a bacteria concentration as low as 10^3 cfu/ml. Even lower concentrations can be detected by increasing the fractional surface coverage of the LPFG by the bacteria through reducing the grating length. The extremely fast detection time, $\sim$30 minutes, makes the present sensor an ideal choice for clinical applications where a fast, precise detection of the pathogenic bacteria is often the difference between life and death.

Acknowledgments The authors gratefully acknowledge support for this work from the Natural Sciences and Engineering Research Council of Canada, the Canada Research Chairs Program, and from NanoQuébec.

References

1. WHO 2011 Press release, Investigation Update: Outbreak of Shiga toxin-producing E. *coli* O104 (STEC O104:H4) Infections Associated with Travel to Germany. http://www.cdc.gov/ecoli/2011/ecolio104/
2. P. Adam, M. Piliarik, H. Sipova, T. Springer, M. Vala, J. Homola, Surface Plasmons for Biodetection, in *Wiley Series in Microwave and Optical Engineering*, ed. by G. Xiao, W.J. Bock (Wiley, Hoboken, 2012), pp. 119–146
3. P. Arora, A. Sindhu, N. Dilbaghi, Biosensors as innovative tools for the detection of food borne pathogens. Biosens. Bioelectron. **28**, 1–12 (2011)
4. Y.M. Bae, B.K. Oh, W. Lee, W.H. Lee, J.W. Choi, Detection of insulin–antibody binding on a solid surface using imaging ellipsometry. Biosens. Bioelectron. **20**, 895–902 (2004)
5. A.J. Baeumner, R.N. Cohen, V. Miksic, J.M. Min, RNA biosensor for the rapid detection of viable *Escherichia coli* in drinking water. Biosens. Bioelectron. **18**, 405–413 (2003)
6. S. Balasubramanian, I.B. Sorokulova, V.J. Vodyanoy, A.L. Simonian, Lytic phage as a specific and selective probe for detection of *Staphylococcus aureus*—A surface plasmon resonance spectroscopic study. Biosens. Bioelectron. **22**, 948–955 (2007)
7. S.P. Chandran, S. Hotha, B.L.V. Prasad, Tunable surface modification of silica nanoparticles through 'click' chemistry. Curr. Sci. **95**, 1327–1333 (2008)
8. B.V. Dorst, J. Mehta, K. Bekaert, E. Rouah-Martin, W. De Coen, P. Dubruel, R. Blust, J. Robbens, Recent advances in recognition elements of food and environmental biosensors: a review. Biosens. Bioelectron. **26**, 1178–1194 (2010)
9. T. Erdogan, Fiber grating spectra. J. Lightwave Technol. **15**, 1277–1294 (1997)
10. C. Garcia-Aljaro, X. Munoz-Berbel, A.T.A. Jenkins, A.R. Blanch, F.X. Munoz, Surface plasmon resonance assay for real-time monitoring of somatic coliphages in wastewaters. Appl. Environ. Microbiol. **74**(13), 4054–4058 (2008)
11. L. Gervais, M. Gel, B. Allain, M. Tolba, L. Brovko, M. Zourob, R. Mandeville, M. Griffiths, S. Evoy, Immobilization of biotinylated bacteriophages on biosensor surfaces. Sens. Actuators B Chem. **125**(2), 615–621 (2007)

12. J. Homola (ed.), *Surface Plasmon Resonance Based Sensors* (Springer, Berlin, 2006)
13. R. Horváth, H.C. Pedersen, N. Skivesen, Optical waveguide sensor for on-line monitoring of bacteria. Opt. Lett. **28**, 1233–1235 (2003)
14. R. Kashyap, *Fiber Bragg Gratings* (Academic Press, Boston, 2009)
15. V. Nanduri, S. Balasubramanian, S. Sista, V.J. Vodyanoy, A.L. Simonian, Highly sensitive phage-based biosensor for the detection of β-galactosidase. Anal. Chim. Acta **589**, 166–172 (2007)
16. B.K. Oh, W. Lee, B.S. Chun, Y.M. Bae, W.H. Lee, J.W. Choi, The fabrication of protein chip based on surface plasmon resonance for detection of pathogens. Biosens. Bioelectron. **20**, 1847–1850 (2005)
17. Q. Pan, X.L. Zhang, H.Y. Wu, P.W. He, F.B. Wang, M.S. Zhang, J.M. Hu, B. Xia, J.G. Wu, Aptamers that preferentially bind type IVB pili and inhibit human monocytic-cell invasion by Salmonella enterica serovar typhi. Antimicrob. Agents **49**, 4052–4060 (2005)
18. K. Rijal, A. Leung, P.M. Shankar, R. Mutharasan, Detection of pathogen *Escherichia coli* O157:H7 AT 70 cells/mL using antibody-immobilized biconical tapered fiber sensors. Biosens. Bioelectron. **216**, 871–880 (2005)
19. A. Shabani, M. Zourob, B. Allain, C.A. Marquette, M.F. Lawrence, R. Mandeville, Bacteriophage-modified microarrays for the direct impedimetric detection of bacteria. Anal. Chem. **80**, 9475–9482 (2008)
20. X. Shu, L. Zhang, I. Bennion, Sensitivity characteristics of long-period fiber gratings. J. Lightwave Technol. **20**, 255–266 (2002)
21. A. Singh, N. Glass, M. Tolba, L. Brovko, M. Griffiths, S. Evoy, "Immobilization of bacteriophages on gold surfaces for the specific capture of pathogens. Biosens. Bioelectron. **24**(12), 3645–3651 (2009)
22. M. Smietana, W.J. Bock, P. Mikulic, A. Ng, R. Chinnappan, M. Zourob, Detection of bacteria using bacteriophages as recognition elements immobilized on long-period fiber gratings. Opt. Express **19**, 7971–7978 (2011)
23. A.D. Taylora, Q. Yua, S. Chena, J. Homola, S. Jianga, Comparison of *E. coli* O157:H7 preparation methods used for detection with surface plasmon resonance sensor. Sens. Actuators B **107**, 202–208 (2005)
24. S. Tombelli, M. Minunni, M. Mascini, Analytical applications of aptamers. Biosens. Bioelectron. **20**, 2424–2434 (2005)
25. H. Zhu, I.M. White, J.D. Suter, M. Zourob, X. Fan, Analytical applications of aptamers. Anal. (Lond.) **133**(3), 356–360 (2008)
26. M. Zourob, S. Mohr, B.J. Brown, P.R. Fielden, M.B. McDonnell, N.J. Goddard, Bacteria detection using disposable optical leaky waveguide sensors. Biosens. Bioelectron. **21**, 293–302 (2005)

Chapter 15
Photonic Crystal Fiber as a Lab-in-Fiber Optofluidic Platform

Fei Tian, Svetlana Sukhishvili and Henry Du

Abstract The ability to design and fabricate photonic crystal fiber (PCF) of seemingly unlimited, axially aligned air cladding structures for vastly different optical properties is arguably one of the most significant recent advances in more than half a century of modern fiber optics. The combined characteristics of PCF as both a waveguide for laser transmission/excitation and a microfluidic cell for gas/ liquid transport/reactions make it a unique lab-in-fiber platform for chemical and biological processes and their real-time monitoring. The easy access of the fiber air channels for surface functionalization at the molecular and nano scales and the ready incorporation of long-period gratings (LPG) as an integral part of the PCF-based lab-in-fiber platform further expand the realm of its applications. The aim of this chapter is to review the state-of-the-art advances in the science and technology of PCF lab-in-fiber optofluidics relevant to technologically important chemical and biological events and their measurements. The chapter will begin with a brief introduction of the fundamentals, fabrication, optical properties, and general areas of applications of PCF. PCF as a natural lab-in-fiber optofluidics will be highlighted and contrasted with lab-on-chip optofluidics derived from planar silicon device technology. We will then discuss the chemical or biological surface treatment of PCF air channels to impart specific functionality for molecular recognition, the immobilization of plasmonic nanostructures at the channel surface for surface-enhanced Raman scattering and the in situ laser spectroscopy measurements. Considerable focus will be placed on the design, development, and implementation of highly index-sensitive PCF-LPG lab-in-fiber optofluidics for

---

F. Tian (✉) · S. Sukhishvili · H. Du
Department of Chemical Engineering and Materials Science,
Stevens Institute of Technology, Hoboken, NJ 07030, USA
e-mail: ftian1@stevens.edu

S. Sukhishvili
e-mail: ssukhish@stevens.edu

H. Du
e-mail: hdu@stevens.edu

A. Cusano et al. (eds.), *Lab-on-Fiber Technology*,
Springer Series in Surface Sciences 56, DOI: 10.1007/978-3-319-06998-2_15,

in situ investigation of immunoassays as label-free bioreactors and biosensors as well as for real-time monitoring of layer-by-layer assembly of stimuli-responsive polyelectrolyte thin films as nano-sensors and nano-actuators. We will conclude the chapter by sharing our views on future opportunities and challenges in the exciting field of PCF lab-in-fiber optofluidics.

15.1 Introduction

Silica photonic crystal fiber (PCF, also termed microstructured fiber) represents arguably one of the most exciting recent advances in more than half of a century of fiber optic development [1, 2]. Unlike all-solid conventional optical fiber (COF), which consists typically of high-index core and low-index cladding, the solid- or air-core in PCF is surrounded by a cladding structure composed of a fine array of air channels running axially along the entire fiber length. The diameter of the air channels ranges from sub-microns to tens of microns. The endless degrees of freedom in designing and fabricating the cladding microstructure and thus the ability to realize vastly different and novel optical properties have led to the emergence of PCF as a major field of research. PCF as both a light guide and a microfluidic gas or liquid transmission cell renders it an inherent optofluidic platform, ushering in a promising lab-in-fiber paradigm for a multitude of applications ranging from the monitoring of chemical and biological processes at the molecular level to the study of controlled drug release in technologically and physiologically relevant environment [3].

15.2 PCF as a Light Guide and Its Fabrication

Light is guided in PCF via two mechanisms, depending on fiber microstructure. PCF with a high-index core and a low-index cladding, often found in the solid-core type as illustrated in Fig. 15.1a, operates similarly to that of COF: it guides light by modified total internal reflection, as a result of the index contrast between the solid core and the air-silica cladding. PCF with a low-index core, typically the hollow-type as shown in Fig. 15.1b guides light in the core via the so-called bandgap confinement, made possible by the intricately designed photonic crystal cladding structure [4–7].

Kaiser and Astle [8] reported the first PCF fabrication in 1974. There are two main steps in the fabrication process: the first one always involves preform development that assumes the structure suitable for desired properties; the second one entails fiber drawing at elevated temperature in a drawing tower. Additional steps such as doping can be introduced to meet the need for hybrid fiber types, in

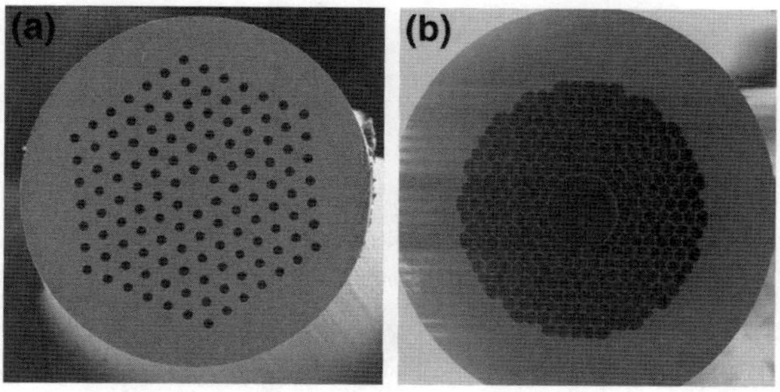

Fig. 15.1 SEM images of **a** *solid-core* PCF with hexagonally arranged cladding air channels and **b** *hollow-core* PCF with photonic bandgap cladding structure

which light guide via both index-guiding and band gap confinement can be accomplished [9].

The preform is often made either by stack and draw method or by sol-gel technique, with the former being the prevailing approach [10]. The stack and draw method has been demonstrated to be versatile to fabricate PCF with complex and diverse cladding microstructures. In this approach, silica capillary tubes and/or rods are directly stacked into a closely packed assembly. The stack is subsequently fused together to form a preform. The preform is drawn to PCF with the desired size and microstructure in a drawing tower. The sol-gel technique is another attractive preform fabrication method because of the significant flexibility in fiber design, its low cost and the ability to produce preforms in large scale [11]. A typical sol-gel casting process developed in OFS Laboratories [12] consists of several steps. First, an array of mandrel elements is arranged in a mold with the designed structure. Then a dispersion of colloidal silica at high pH is filled in the mold. After gelation at lower pH, the mandrel elements are extracted, leaving a series of air channels in the gel body. Thermochemical treatment is employed in the drying and purification processes to help remove undesired water and organic remnants. Finally the gel is sintered at 1,600 °C, resulting in a preform of vitreous glass ready for PCF drawing in the drawing tower.

15.3 PCF as Optofluidics Platform

Microfluidics is the science and technology that manipulates small volumes of fluidic samples in a confined geometry, typically under laminar flow condition [13]. Rapid advance in microfluidics, in conjunction with available micro-fabrication technology, has led to the proliferation of the field of optofluidics [14, 15]. Optofluidics combines microfluidics with propagating light for process control, process monitoring, and sensing and measurements.

The distinct features of PCF optofluidics, compared with its counterpart from planar Si technology, include light path spanning the entire fiber length, low-cost and high-volume fabrication without reliance on photolithography, and easy light coupling. Figure 15.2 shows an example of PCF optofluidics by selectively sealing the air channels of a hollow-core PCF [16]. This process allows the formation of hybrid PCF with various functional materials filled in the core. UV-curable polymer is infiltrated into the air channels of PCF using a syringe. The flow rate of the polymer is dependent on the dimension of the air channels and the larger hole of fiber core will have a much longer infiltration length. Selective sealing is then achieved by multistep injection-cure-cleave process. Such a hybrid PCF with optical confinement in low index materials is of particular interest for spectroscopy and sensing applications in aqueous solutions. The inherently high rates of mass transport and heat transfer in a microfluidic system make PCF optofluidics an efficient lab-in-fiber micro-reactor with built-in optical interrogation capability for process monitoring.

15.4 PCF as Lab-in-Fiber Microreactor

Photochemistry has been widely used in a variety of applications in photo-medicine [17], chemical synthesis [18], data storage [19] and conversion and storage of solar energy [20]. Monitoring the progress of photochemical reactions in parallel is essential. This capability can be accomplished through miniaturization and integration of sample preparation, optical excitation and detection functionalities, all combined in a single device. These microdevices act as versatile microfluidic platforms which promote the interaction between light and fluids for photochemical activation. There is a surging interest in using PCF for microreactors with nL sample volumes. The intense overlap of the guided light and reactant and product species inside the microchannels of PCF offers exciting new opportunities for in situ spectroscopy of photochemical activation.

The sensitivity and performance of conventional spectroscopic technique using cuvettes is limited by a few centimeters of the interaction path length. This challenge can be overcome using PCF microreactor with path length up to several meters. Moreover, the propagating light is mainly confined in the fiber core, which favors the photochemical activation of the photosensitive compound even with moderate laser power [21]. For example, hollow core PCF has been used as a highly efficient mircoreactor for photolytic conversion of vitamin B12 to vitamin B12b [20] and the photoisomerization of azobenzene derivatives [22].

Platinum-based drugs are well-established anticancer compounds. Traditional drugs do not differentiate cancer cell from healthy tissues, which limits their application due to severe side-effect and acquired resistance. It is of great importance to use photoactivated platinum complexes that can be localized at the tumor site [23]. A highly-controllable photochemical microreactor based on PCF has been demonstrated to simultaneously activate and monitor the reaction dynamics of photoaquation of nontoxic metal complex vitamin B12 (cyanocobalamin, CNCbl) [21]. Figure 15.3 shows the photochemical conversion of CNCbl to $[H_2OCbl]^+$.

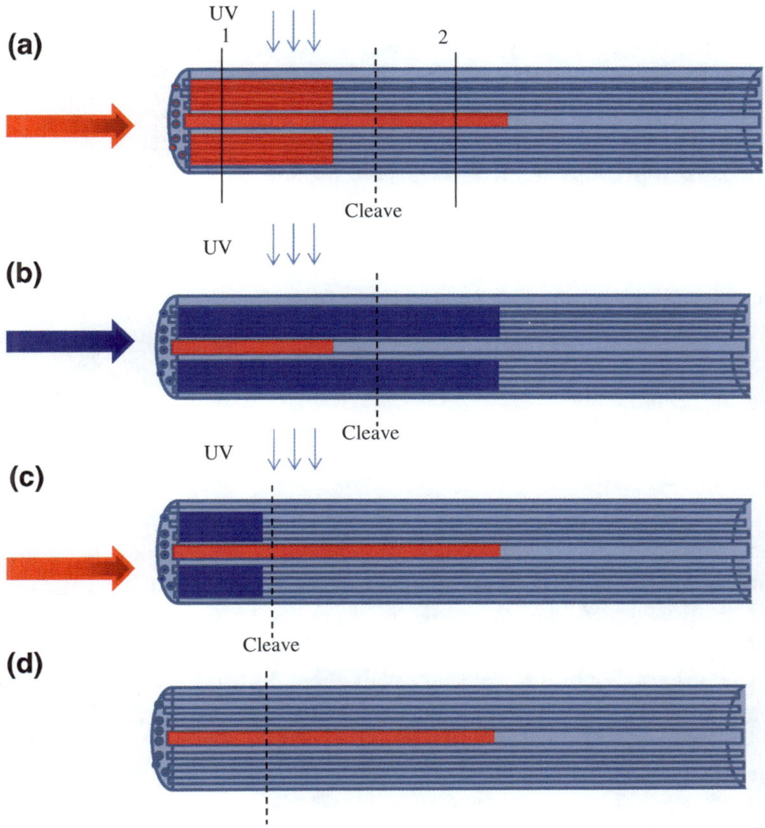

Fig. 15.2 Flow chart of selective filling of the microchannels of PCF. Reproduced with permission [16]

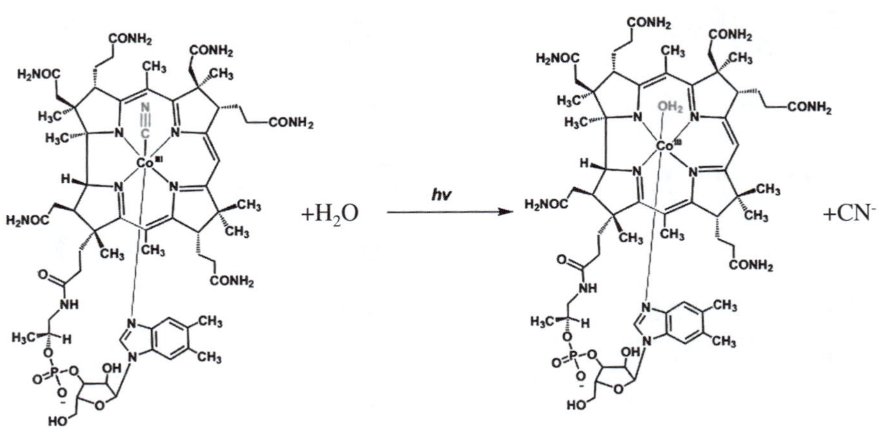

Fig. 15.3 Schematic of photochemical conversion of CNCbl to $[H_2OCbl]^+$ [21]

Because hollow-core PCF allows maximized light-sample interaction inside the hollow core, it is also highly suitable as a spectroscopic nanoreactor for studying and further understanding the catalytic reaction process [24]. For example, it has been employed both as a tubular reactor and a spectroscopic system for studying the well-known photo-Fenton chemistry in which hydrogen peroxide can be decomposed to hydroxyl radicals catalyzed by ferrous ions [24]:

$$Fe^{II} + H_2O_2 \rightarrow Fe^{III} + {}^{\cdot}OH + OH^- \tag{15.1}$$

Enhancements in orders of magnitude in both the reaction yield and detection sensitivity can be achieved with PCF, compared with conventional quartz cuvette, due to long path length and efficient overlap with the light field [24].

15.5 PCF as Optofluidic Sensor and Process Monitor

The accessibility to the air channels of PCF has opened up endless possibilities for functionalization of the channel surfaces at the molecular and the nano scales, especially for chem/bio sensing and measurements. Major efforts have been made in this regard such as the incorporation of colloidal nanoparticles, self-assembly multilayers, or nanoporous materials with PCF. Various chemical sensors have been achieved successfully in PCF with such functional coatings in both vapor and aqueous conditions [25]. With the immobilization of bio-specific species, such as specific antigens, complementary strand of DNAs and proteins in PCF, biological sensing and detection can be carried out accordingly. By integrating the fiber with proper indicators or stimuli-responsive materials, a wide range of PCF-based chem/bio sensors can be developed [26, 27]. The review below focuses mainly on two sensing schemes: PCF with built-in surface-enhanced Raman scattering (SERS) functionality for chemical sensing and PCF inscribed with long-period gratings for process monitoring, with the recognition of other applicable sensing modalities such as absorption and fluorescence spectroscopies.

15.5.1 SERS-Active PCF

SERS was discovered in 1974 by Fleischmann et al. who observed intense Raman signal on a roughened silver electrode surface with adsorbed pyridine in aqueous solution [28]. There has been an unabated interest in developing robust SERS substrates for chemical and biological detections ever since. Integrating SERS with PCF takes the advantage of the best of both worlds have to offer for ultra-high sensitivity measurements of minute solutions [29]. With easy access to the cladding air channels for molecular and nano-scale surface modification, a substantial amount of work has been carried out to impart SERS functionality in PCF with success [30–32].

Gold and silver nanoparticles are extensively used in SERS sensing due to strong surface plasmon resonance when excited by light in the visible to NIR range. There are two strategies to incorporate nanoparticles into PCF. One is to mix the SERS-active nanoparticles with the analyte solution of interest and fill the resultant solution in the cladding air channels of PCF. In one design, gold nanoparticles mixed with Rhodamine B (RhB) aqueous solution was filled in the air channels to about 1 cm long by capillary force [33]. The measurement scheme had a detection limit of 10^{-7} M of RhB, made possible via SERS enhancement. The other strategy, a preferred one, is to immobilize the gold or silver nanoparticles using polyelectrolyte adhesion layer on the air channels of PCF to provide SERS activity along the entire fiber length. A critical step to realize full-length SERS PCF is to ensure the uniformity of immobilized nanoparticles throughout the cladding air channels. Much of our related work has been devoted to achieving this objective with both solid-core and hollow-core PCF for a high-degree of control in the surface coverage density [34–36].

Figure 15.4 shows the cross sectional SEM of solid-core and hollow-core PCF used in this work. A sensitivity of 1×10^{-7} M or $\sim$48 ppb Rhodamine 6G in a minute $\sim 10^{-7}$–10^{-8} L aqueous solution using over 20 cm-long PCF has been demonstrated [34]. Figure 15.5 shows the Raman spectra of 2 μM Rhodamine 6G (R6G) solution using a SERS-active PCF with immobilized silver nanoparticles (1 particle/μm^2) and others as control and reference for the forward propagating geometry [37].

One important criterion in considering PCF for evanescent field-based sensing is the maximization of the mode-field overlap for the light-analyte interaction inside the air channels and the minimization of the confinement loss of the guided modes. By carefully designing the cladding structure of the PCF, the confinement loss decreases and the overlap increases with the increase in the number of rings and diameter of air channels, and the decrease in the web thickness between air channels [38]. For example, we have studied the design of a steering wheel patterned cladding structure of PCF with three large cladding air channels [39]. Numerical simulation of the triangular lattice of the channels demonstrated that confinement loss of less than 0.7 dB/m at 850 nm, and 29, 13.7, and 7.2 % of light intensity overlap in air channels at wavelengths 1,500, 1,000 nm, and 850 nm can be achieved with such a PCF. To assess the effect of cladding microstructure of PCF on the SERS performance of the fiber, three types of PCF have been imparted with SERS functionality and used for SERS measurement of ethanol: nonlinearity PCF (NL PCF) as shown with the core surrounded by six rings of air channels, steering-wheel PCF (SW PCF) and suspended-core PCF (SC PCF) [40]. Numerical analysis showed that the power fractions in the air channels upon filtration with ethanol for the three PCF structures were 0.708, 0.483 and 10.859 %, respectively. Raman measurements were conducted for the three types of PCF with 24 cm in length completely filled with ethanol. The experiments were carried out by laser excitation at one end of the PCF and signal collection at the other end. As shown in

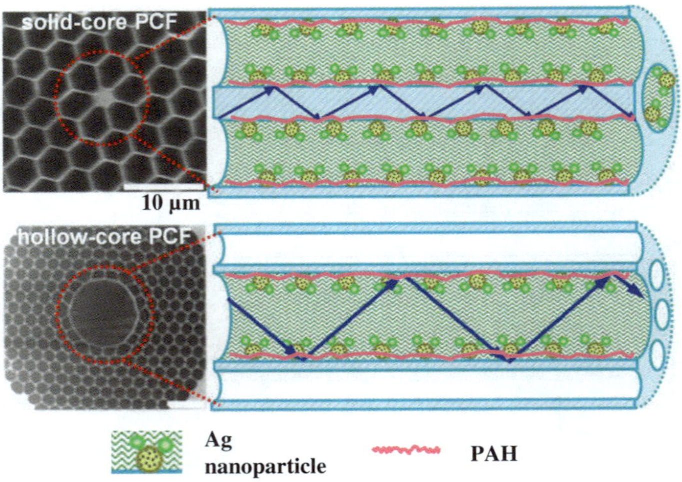

Fig. 15.4 SEM micrographs of *solid-core* PCF and *hollow-core* PCF and schematic illustration of light guiding in the corresponding liquid-filled structures with *Ag nanoparticles* immobilized by positively-charged polyallylamine hydrochloride (*PAH*). Reproduced with permission [34]

Fig. 15.6, the ratio of Raman intensities at 2,930 cm^{-1} (characteristic of C–H stretching vibration from ethanol) for SW PCF, NL PCF, and SC PCF was about $\sim$1:14:28, in qualitative agreement with the numerical results.

Different from solid-core PCF, hollow-core PCF provides a sensing platform for direct interaction between light and analyte present in the hollow core. We have demonstrated a full-length liquid-core PCF with the fiber length up to 30 cm for SERS detection in a forward propagating mode [34]. The key to achieving the full-length liquid-core PCF for SERS detection is to incorporate a uniform layer of SERS-active nanoparticles inside the hollow core along the entire length of the PCF with relatively low surface coverage density. Figure 15.7a shows the Raman spectrum obtained for hollow-core PCF with the core filled with 10^{-5} M R6G. Figure 15.7b, c and d show three hyperspectral Raman images corresponding to silica (485 cm^{-1}), R6G (1,351 cm^{-1}), and water (3,381 cm^{-1}) in liquid-core PCF. The dashed ring-like Raman distribution in Fig. 15.7b indicates that the core and two adjacent air channels allowed forward propagating of silica signal. Core mode contribution to the measured Raman intensities can also be demonstrated by the confinement of the Raman distributions for R6G and water as shown in Fig. 15.7c and d. The above results strongly suggested that the hollow-core PCF can be suitable as a highly sensitive SERS-active platform due to the robust waveguide upon infiltration with liquid, the effective coupling of R6G to SERS signal and forward-propagation of the Raman signal along the liquid core.

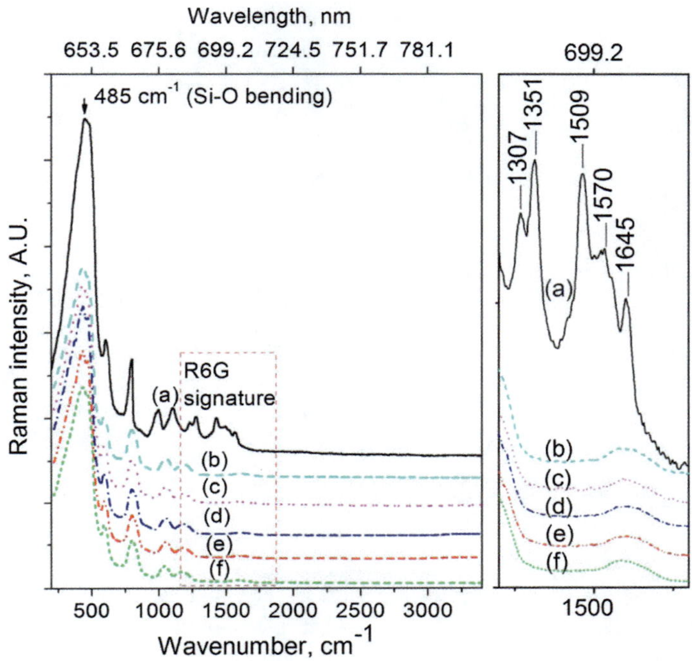

Fig. 15.5 Raman spectra for *a* SERS-active PCF filled with 2 μM R6G solution, *b* SERS-active PCF filled with water, *c* PCF with adsorbed gelatin and filled with R6G solution, *d* PCF with adsorbed gelatin and filled with water, *e* PCF filled with water, and *f* bare PCF. The right panel corresponds to spectral enlargement in the region of characteristic peaks of R6G. The forward Raman measurements were done at 632.8 nm, ~5 mW laser power, and acquisition time of 10 s. Reproduced with permission [37]

Fig. 15.6 Raman spectra from NL PCF, SW PCF, and SC PCF filled with ethanol. The Raman measurements were carried out at 632.8 nm and 5 mW power, with an acquisition time of 20 s. Reproduced with permission [40]

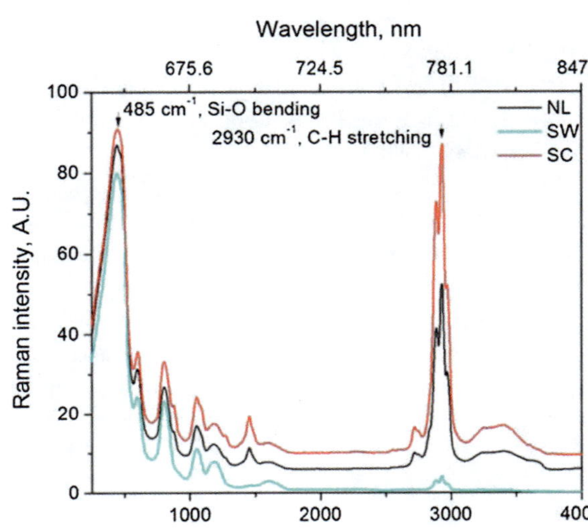

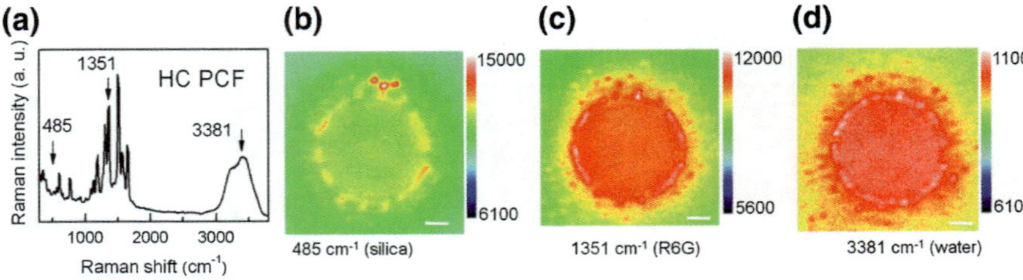

Fig. 15.7 SERS spectra of 10^{-5} M R6G (**a**) and hyperspectral Raman images of silica (**b**), R6G (**c**), and water (**d**) from hollow-core PCF with immobilized silver nanoparticles of ~0.5 particle/μm2 in coverage density. The measurements were done using forward-propagating geometry at an excitation wavelength of 632.8 nm. Reproduced with permission [34]

15.5.2 Long-Period Gratings in PCF as Index Transduction Platform

Long-period gratings (LPG) in PCF are periodical modulations of the refractive index axially along the fiber with a period in the range of 100 μm to 1 mm. The core mode of LPG is coupled to the forward propagating cladding mode at some resonance wavelength, resulting in significant attenuation in its transmission spectrum. The resonance wavelength, λ, is described by phase matching condition $\lambda = (n_{core}^{eff} - n_{clad}^{eff})\wedge$, where, $(n_{core}^{eff}$ and $(n_{clad}^{eff}$ are respective effective indices of the core and the cladding modes, and Λ is grating period. The strong dependence of λ on $(n_{clad}^{eff}$ makes LPG in PCF an optofluidic platform that is highly sensitive to the changes in $(n_{clad}^{eff}$ induced by any physical, chemical, or biological perturbations in the cladding structure. The ability to bring a measurant solution inside of the cladding air channels to directly interact with the cladding mode further boosts the refractive index sensitivity of PCF-LPG.

Periodic index perturbation in LPG can be accomplished by laser techniques such as UV, infrared femtosecond laser or CO_2 laser irradiation as well as by non-laser approaches such as arc discharge, acoustic wave coupling, focused ion beam and periodic mechanical stress loading. CO_2 laser irradiation is advantageous due to system flexibility and ease of control. While silica material itself has a very low absorption at the wavelength of UV light, CO_2 laser is readily utilized as a versatile tool for LPG fabrication in PCF due to the strong absorption of the silica material at the wavelength of 10.6 μm. Many studies have focused on the investigation of the LPG coupling mechanism associated with CO_2 laser for various types of fibers [41, 42]. LPG fabricated by CO_2 laser exhibits good temperature stability [43].

15.5.2.1 PCF-LPG as Optofluidic Platform for Sensing in Liquid Phase

Refractive index is a fundamental material property that is of great importance in science and technology. The detection of minute deviations in refractive index is essential for process control and quality assurance in pharmaceutical, beverage and food industries, for instance. There is a growing interest in the development of PCF-LPG for such applications. The sensitivity of PCF-LPG increases significantly with the refractive index of the probing medium. This correlation makes PCF-LPG especially suited for liquid phase measurements.

PCF-LPG has been shown to be sensitive to adsorption of nanometer thick biomolecules such as DNA and proteins [44]. The performance of PCF-LPG based biosensor has been further enhanced by integrating it with a flow cell [45]. The microfluidic cell can optically couple light into and out of PCF while enabling continuous liquid flow through the air channels as part of an integral surface modification/binding and *in situ* measurement scheme [45]. The negatively charged inner surface of the microchannels in the PCF is modified by electrostatic self-assembled polymer monolayer for the subsequent adsorption of antibody. Upon exposure of the modified surface to specific secondary antibody, there is a resonance wavelength shift due to the adsorption of the specific secondary antibody. The PCF-LPG can thus monitor each and every step of the multiple modification and binding sequences. This optofluidics device scheme has been shown to be sensitive to monolayer adsorption/binding event [45].

15.5.2.2 PCF-LPG as Optofluidic Platform for Layer-by-Layer Process Monitoring

Layer-by-layer (LbL) assembly, first introduced by Decher in 1997 [46], has been regarded as a robust processing method to obtain thin films with multi-functionalities. The LbL technique has been used to produce ultrathin films at interface by sequential alternating adsorption of polymers from solutions. The technique has received growing attention as a simple and environmentally benign way to create advanced surface coatings with a nanoscopically controlled structure. LbL approach is advantageous in several ways. First, it is able to incorporate with a wide range of functional molecules, such as electroactive polymers, organic dyes, semiconductor quantum dots, electrochemically active species, inorganic nanomaterials, and biologically active molecules within nanostuctured films (see Fig. 15.8) [47–49]; Second, LbL layers can be functionally designed to respond to environmental stimuli such as pH, ionic strength, temperature, and light irradiation, imparting distinct physical, chemical, and biological properties in the films and/or coated surface itself; Third, it can be deposited virtually on the surface of substrate with any shape and chemistry.

Integrating LbL assembly in the air channels of the PCF-LPG of high index sensitivity thus create another exciting opportunity to monitor and understand the LbL process in microfluidic systems and to develop similarly configured

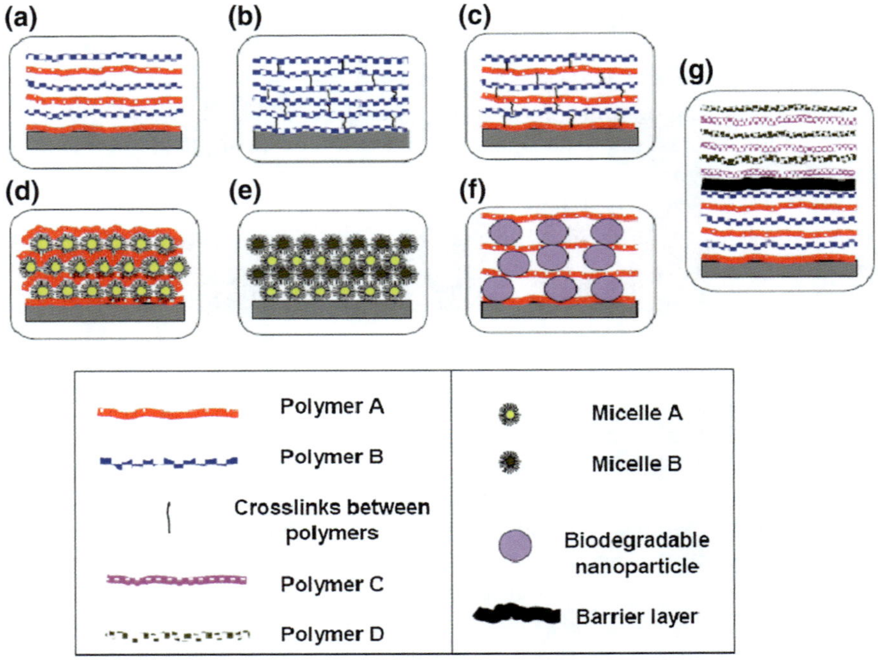

Fig. 15.8 Schematic illustration of various nanostructured materials prepared via LbL. The depicted structures represent: **a** a polymer/polymer non-crosslinked LbL film, **b** a single-component crosslinked hydrogel, **c** a two-component crosslinked hydrogel, **d** a polymer/micelle coating, **e** a micelle/micelle coating, **f** a polymer/biodegradable nanoparticle film, and **g** a stratified film with a barrier layer in between. Reproduced with permission [47]

microsensors and microactuators. Shown in Fig. 15.9 is an optically coupled flow chamber that we have used to perform real-time and in situ transmission measurements during LPG fabrication and LbL assembly under continuous flow of the liquid medium through the PCF air channels [50].

The LbL procedure basically involves the immersing of the platform surface in alternating polyelectrolyte solutions, between which thorough rinsing is employed to remove excess polymer molecules. The specific steps for LbL in PCF-LPG are depicted in Fig. 15.10.

In one set of experiments, the PCF-LPG (1.37 cm in length and 490 μm in periodicity) was precleaned with 10 wt% hydrogen peroxide and Millipore (Milli-Q system) filtered water. One bilayer of BPEI/PMAA was deposited and thermally cross-linked at 125 °C for 1 h for enhanced adhesion. PVPON/PMAA multilayers were deposited from 0.2 mg/ml polymer solution at pH 2. Each step was followed by rinsing with 0.01 M phosphate buffer solution at the same pH. After six bilayers of PVPON/PMAA were deposited in the PCF-LPG, the PMAA layers were chemically cross-linked including activation in EDC solution (5 mg/ml, pH = 5) and subsequent treatment in AADH cross-linker solution (9 mg/ml, pH = 4). PVPON

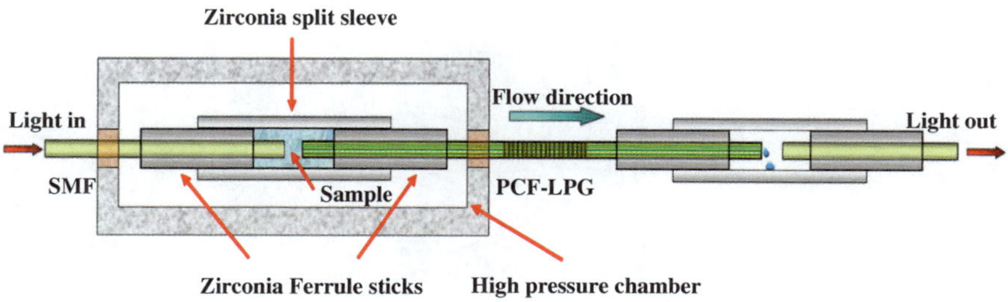

Fig. 15.9 Optically coupled microfluidic module for LbL in PCF-LPG and in situ transmission measurements. Reproduced with permission [50]

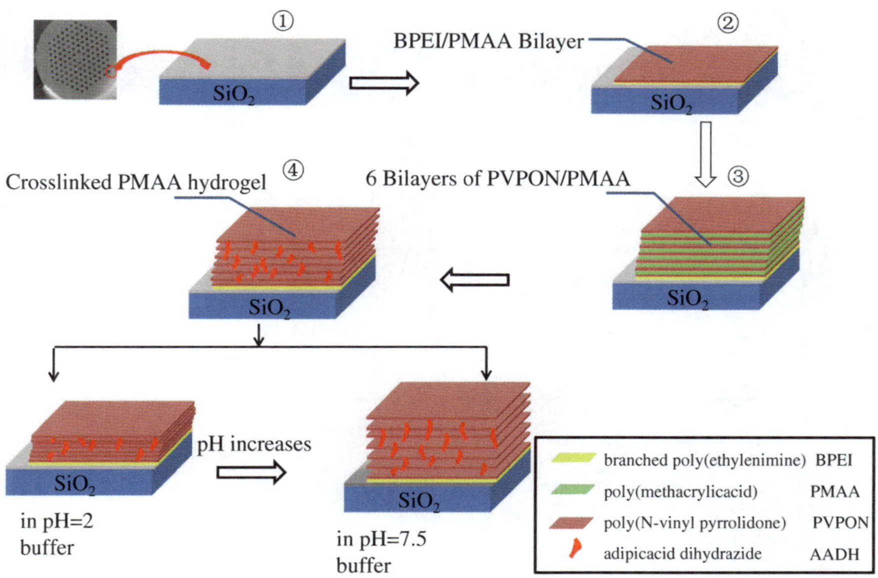

Fig. 15.10 Schematic of the procedures of LbL deposition of PVPON/PMAA and formation of PMAA hydrogel in PCF-LPG with BPEI/PMAA as adhesion bilayer. Reproduced with permission [50]

molecules were removed from the thin film by exposure to 0.01 M phosphate buffer solution with pH 7.5. The same procedures were performed using COF-LPG (2.49 cm in length and 710 μm in periodicity) in another set for comparison. Fluorescently labeled PMAA was used for the deposition in the last layer in order to examine the distribution of the thin film that's fabricated on the fiber under confocal microscopy. Shown in Fig. 15.11 is the confocal microscopy of the PMAA hydrogel on COF and in PCF, confirming that a PMAA hydrogel has been successfully deposited using both LPG platforms.

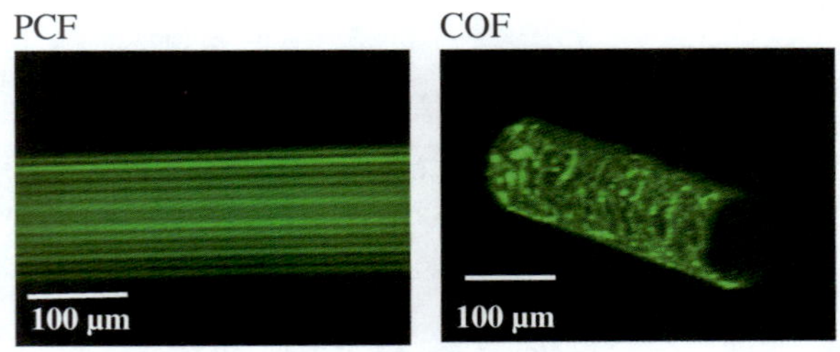

Fig. 15.11 Confocal microscopy of the PMAA hydrogel on *COF* and in *PCF*

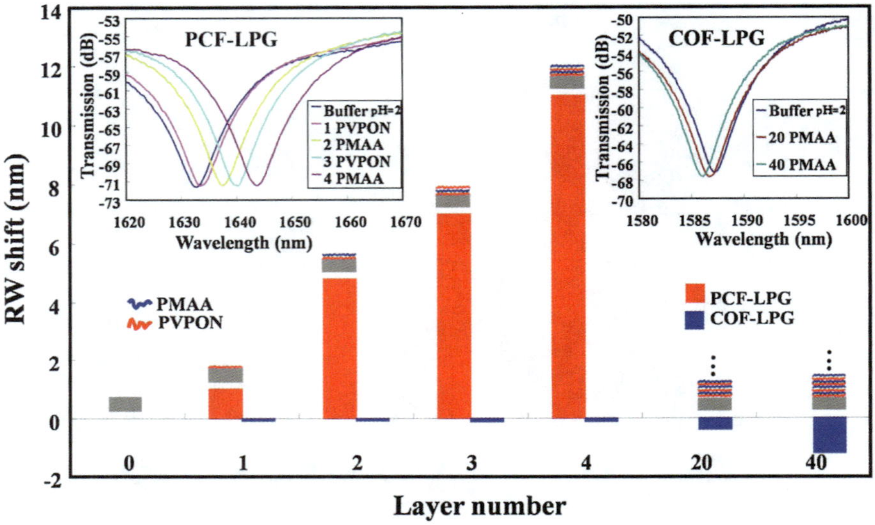

Fig. 15.12 Respective RW shifts in response to LbL assembly of PVPON and PMMA in PCF-LPG and on COF-LPG. Insets are the corresponding transmission spectra. reproduced with permission [50]

An optical spectrum analyzer was used to measure the transmission spectrum at each LbL step followed by buffer rinsing. The RW for PCF-LPG with buffer at approximately 1,678 nm arises from core mode to LP_{04} cladding mode coupling according to our numerical calculations and was used in our analysis. The RW of the PCF-LPG underwent a dramatic red shift (to longer wavelength) as the PVPON/PMAA LbL assembly was successively carried out in the cladding air channels. In contrast, a slight blue shift (to short wavelength) in RW took place for LbL processes on COF-LPG. Summarized in Fig. 15.12 are the respective shifts in RW as PVPON and PMAA layers were alternately deposited in PCF-LPG and on COF-LPG.

The red bars denote red shift in PCF-LPG and blue bars represent blue shift in COF-LPG. Six LbL bilayers in PCF-LPG produced a shift of $\sim$19.5 nm in RW (equivalent to $\sim$1.625 nm shift per LbL layer). In contrast, 20 LbL bilayers on

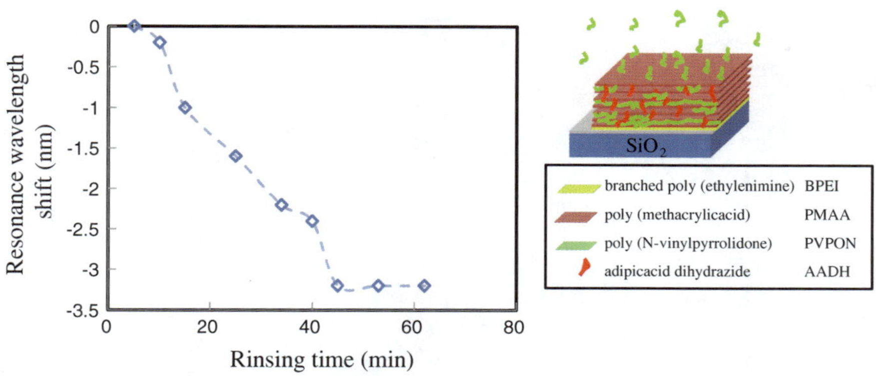

Fig. 15.13 Schematic of PVPON release from LbL film and time-dependent shift in RW as the process takes place in PCF-LPG. Reproduced with permission [50]

COF-LPG yielded only a shift of ~1 nm total (equivalent to ~0.025 nm/layer). The opposite direction in RW shift for PCF-LPG and COF-LPG accompanied with LbL deposition is a result of their drastically different dispersion characteristics of effective refractive indices of the cladding modes [51]. The ~65 fold increase in the sensitivity of PCF-LPG over COF-LPG can be attributed to higher evanescent field overlap with cladding air channels that also provide a far larger surface area for LbL growth. In contrast, COF-LPG only probes the LbL films on the outer surface of the all-solid fiber. The high sensitivity of PCF-LPG also enabled in situ monitoring of PVPON release from cross-linked PMAA layers in phosphate buffer at pH 7.5. Figure 15.13 shows blue shift in RW as PVPON was gradually released from the LbL film in PCF-LPG. The blue shift is expected per phase matching condition since the removal of PVPON results in a reduction in the effective index of refraction in the thin film. The stable RW after ~45 min suggests completion of the release process. Total release of PVPON transformed the LbL film to a pH-responsive PMAA hydrogel. This process afforded us the opportunity to test hydrogel-containing PCF-LPG as well as COF-LPG as potential pH sensors. Specifically, the RW shifts of the two systems were measured by exposing them to phosphate buffers (pH: 2–7.5). The pH-dependent transmission and RW shift are illustrated in Fig. 15.14.

There is a pronounced increase in the RW of the PCF-LPG (red bars) with decreasing pH, indicating higher refractive index of the thin film. In contrast, the RW shift for COF-PCF (blue bars) in response to pH variation is far less robust. It is well known that PMAA hydrogel tends to be increasingly deprotonated at higher pH, leading to stronger repulsion between the ionized carboxylic groups and swelling of the hydrogel. Increased swelling corresponds to decreased index of refraction of the hydrogel. The opposite shift in RW and the striking difference in sensitivity between PCF-LPG and COF-LPG can be similarly explained as for the results in Fig. 15.12. pH response behavior of the PMAA hydrogel is interpreted by the conformational change of PMAA as illustrated by Fig. 15.15.

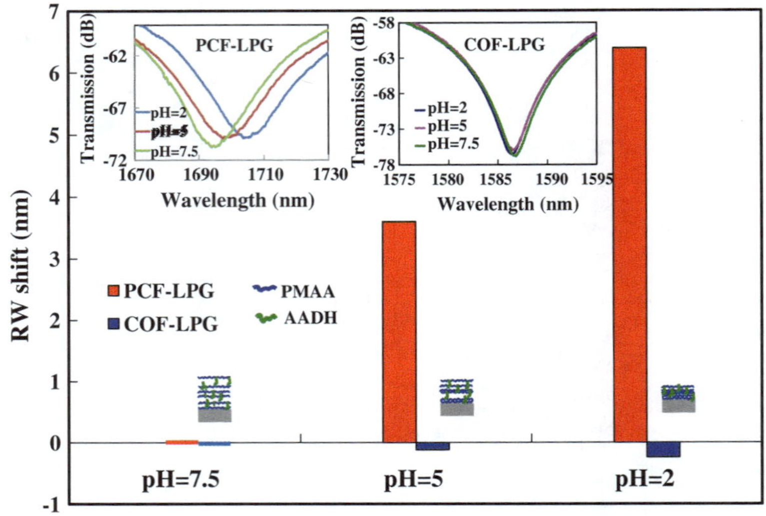

Fig. 15.14 pH-dependent transmission and RW shift for PCF-LPG and COF-LPG coated with PMAA hydrogels. Reproduced with permission [50]

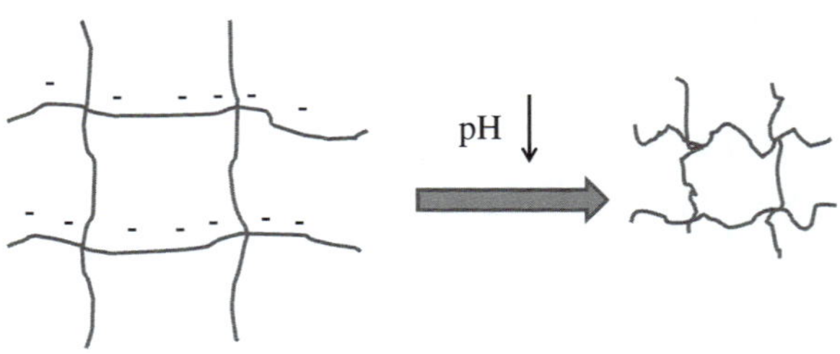

Fig. 15.15 Conformational change of PMAA thin film in buffer solution with different pH

The pH responsiveness of PMAA hydrogel has been demonstrated on flat geometry in [52]. The PMAA surface-attached hydrogels are highly swollen in water, and their swelling degree can be controlled between 1.2 and 3 by varying the degree of film crosslinking [52]. Most importantly, the swelling degree of these films is strongly pH dependent (Fig. 15.16).

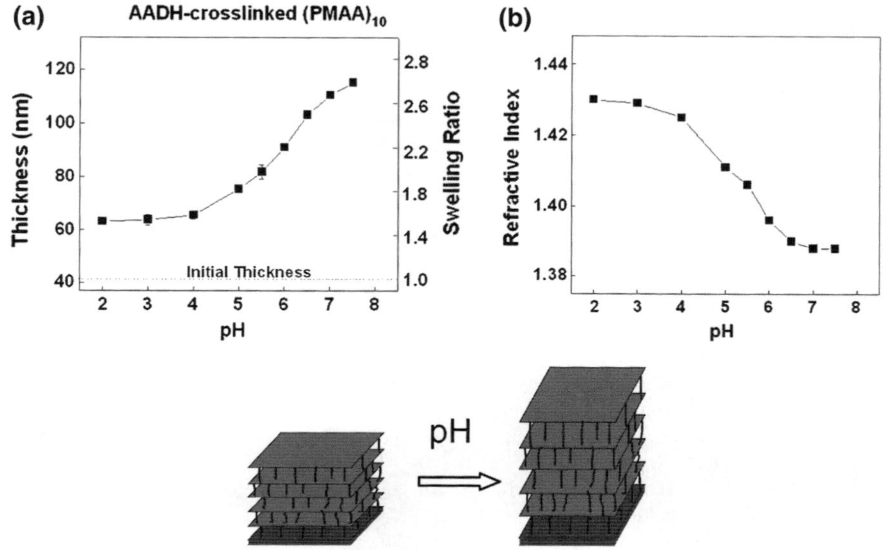

Fig. 15.16 Panels (**a**) and (**b**): swelling ratios and refractive indices of $(PMAA)_{10}$ films crosslinked with adipic acid dihydrozide (AADH) as a function of pH supported by 0.01 M phosphate buffer by in situ ellipsometry. The thickness of the dry $(PMAA)_{10}$ films was $\sim$41 nm. Reproduced with permission [52]

15.6 Concluding Remarks

The ability to design, fabricate, and functionalize PCF of seemingly unlimited microstructural features for vastly different optical characteristics affords us endless opportunities for research creativity and innovation. PCF as an inherent optofluidic platform allows the exploration and harvesting of intimate light-medium interactions in the air channels of ultra-large aspect ratio for chemical and biological synthesis, for process control, and for the development of novel sensing devices, all with real-time and in-situ capabilities. Significant progress has been made by the scientific community in advancing the frontier of PCF for related applications. PCF research in the context of lab-in-fiber optofluidics is still at its infancy. This field of research will undoubtedly undergo exponential growth in the coming decade, leading to new science, technology, and yet-to-be-foreseen applications.

Acknowledgments This chapter contains a significant amount of our prior and ongoing studies funded by the US National Science Foundation under grants ECCS-0404002, ECCS-0922175 and DMR-0906474. We thank the various colleagues of ours for their contributions to the work reviewed.

References

1. J.C. Knight, Photonic crystal fibres. Nature **424**, 847–851 (2003)
2. P.St.J. Russell, Photonic crystal fibers. Science **299**, 358–362 (2003)
3. J.C. Knight, T.A. Birks, P.S.J. *Russell, in Optics of Nanostructured Materials*, eds. by V.A. Markel, T.F. George (Wiley, New York, 2001), pp. 39–71
4. T.A. Birks, P.J. Roberts, P.J. Russell, D.M. Atkin, T.J. Shepherd, Full 2-D photonic bandgaps in silica/air structures. Electron. Lett. **31**, 1941–1943 (1995)
5. J.C. Knight, J. Broeng, T.A. Birks, P.J. Russell, Photonic band gap guidance in optical fibers. Science **282**, 1476–1478 (1998)
6. R.F. Cregan, B.J. Mangan, J.C. Knight, T.A. Birks, P.J. Russell, P.J. Roberts, D.C. Allan, Single-mode photonic band gap guidance of light in air. Science **285**, 1537–1539 (1999)
7. J.C. Knight, P.J. Russell, New ways to guide light. Science **296**, 276–277 (2002)
8. P.V. Kaiser, H.W. Astle, Low-loss single-material fibers made from pure fused silica. Bell Syst. Tech. J. **53**, 1021–1039 (1974)
9. A. Cerqueira S.Jr., F. Luan, C.M.B. Cordeiro, A.K. George, J.C. Knight, Hybrid photonic crystal fiber, Opt. Express **14**, 926–931 (2006)
10. J.C. Knight, T.A. Birks, P.J. Russell, D.M. Atkin, All-silica single-mode optical fiber with photonic crystal cladding. Opt. Lett. **21**, 1547–1549 (1996)
11. R. Bise, D.J. Trevor, Solgel-derived mircrostructured fibers: fabrication and characterization, in *Conference on Optical Fiber Communication, Technical Digest Series*, Anaheim, CA, vol. 3 pp. 269–271 (2005)
12. R. Bise, Manufacturing of microstructured optical fibers (2006). http://www.cns.cornell.edu/documents/RyanBiseOFSLaboratories.pdf
13. G.M. Whitesides, The origins and the future of microfluidics. Nature **442**, 368–373 (2006)
14. H.A. Stone, A.D. Stroock, A. Ajdari, Engineering flows in small devices: microfluidics toward a lab-on-a-chip. Ann. Rev. Fluid Mech. **36**, 381–411 (2004)
15. J. Hecht, *City of Light: The Story of Fiber Optics* (Oxford University Press, New York, 1999)
16. Y.Y. Huang, Y. Xu, A. Yariv, Fabrication of functional microstructured optical fibers through a selective-filling technique. Appl. Phys. Lett. **85**, 5182–5184 (2004)
17. Z. Huang, A review of progress in clinical photodynamic therapy. Technol. Cancer Res. Treat. **4**, 283–293 (2005)
18. N. Homann, Photochemical reactions as key steps in organic synthesis. Chem. Rev. **108**, 1052–1103 (2008)
19. F. Li, J. Zhuang, G. Jiang, H. Tang, A. Xia, L. Jaing, Y. Song, Y. Li, D. Zhu, A rewritable optical data storage material system by [2 + 2] photocycloreversion—photocycloaddition. Chem. Mater. **20**, 1194 (2008)
20. M. Grätzel, Solar energy conversion by dye-sensitized photovoltaic cells. Inorg. Chem. **44**, 6841–6851 (2005)
21. J.S.Y. Chen, T.G. Euser, N.J. Farrer, P.J. Sadler, M. Scharrer, P.J. Russell, Photochemistry in photonic crystal fiber nanoreactors. Chem.—Eur. J. **16**, 5607–5612 (2010)
22. J.S.Y. Chen, T.G. Euser, G.O. Williams, A.C. Jones, P.S.T. Russell, Photoswitching in Photonic Crystal Fiber, in *Conference on Optical Sensors, Karlsruhe Germany*, (Optical Society of America, 2010), paper SThB4
23. W.M. Sharman, C.M. Allen, J.E. van Lier, Photodynamic therapeutics: basic principles and clinical applications. Drug Discov. Today **4**, 507–517 (1999)
24. A.M. Cubillas, M. Schmidt, M. Scharrer, T.G. Euser, B.J.M. Etzold, N. Taccardi, P. Wasserscheid, P.S.J. Russell, Ultra-low concentration monitoring of catalytic reactions in photonic crystal fiber. Chem.—Eur. J. **18**, 1586–1590 (2012)
25. V.P. Minkovich, D. Monzon-Hernandez, J. Villatoro, G. Badenes, Microstructured optical fiber coated with thin films for gas and chemical sensing. Opt. Express **14**, 8413–8418 (2006)
26. M. Skorobogatiy, Microstructured and photonic bandgap fibers for applications in the resonant bio- and chemical sensors. J. Sens. **2009**, 524237 (2009)

27. T.M. Monro, S. Warren-Smith, E.P. Schartner et al., Sensing with suspended-core optical fibers. Opt. Fiber Technol. **16**, 343–356 (2010)
28. M. Fleischmann, P.J. Hendra, A.J. McQuillan, Raman spectra of pyridine adsorbed at a silver electrode. Chem. Phys. Lett. **26**, 163–166 (1974)
29. H. Yan, C. Gu, C. Yang, J. Liu, G. Jin, J. Zhang, L. Hou, Y. Yao, Hollow core photonic crystal fiber surface-enhanced Raman probe, Appl. Phys. Lett. **89**, Article ID 204101, (2006)
30. Y. Zhang, C. Shi; C. Gu, L. Seballos, J.Z. Zhang, Liquid core photonic crystal fiber sensor based on surface enhanced Raman scattering, Appl. Phys. Lett. **90**, Article ID 193504 (2007)
31. F.M. Cox, A. Argyros, M.C.J. Large, S. Kalluri, Surface enhanced Raman scattering in a hollow core microstructured optical fiber. Opt. Express **15**, 13675–13681 (2007)
32. A. Amezcua-Correa, J. Yang, C.E. Finlayson, A.C. Peacock, J.R. Hayes, P.J.A. Sazio, J.J. Baumberg, S.M. Howdle, Surface-enhanced Raman scattering using microstructured optical fiber substrates. Adv. Funct. Mater. **17**, 2024–2030 (2007)
33. Yan, J. Liu, C. Yang, G. Jin, C. Gu, and L. Hou, Novel index-guided photonic crystal fiber surface-enhanced Raman scattering probe, Opt. Express **16**, 8300–8305 (2008)
34. Y. Han, S. Tan, M.K. Khaing Oo, D. Pristinski, S. Sukhishvili, H. Du, Towards full-length accumulative surface-enhanced raman scattering-active photonic crystal fibers. Adv. Mater. **22**, 2647–2651 (2010)
35. R.G. Freeman, K.C. Grabar, K.J. Allison, R.M. Bright, J.A. Davis, A.P. Guthrie, M.B. Hommer, M.A. Jackson, P.C. Smith, D.G. Walter, M.J. Natan, Self-assembled metal colloid monolayers: an approach to SERS substrates. Science **267**, 1629–1632 (1995)
36. M. Erol, Y. Han, S.K. Stanley, C.M. Stafford, H. Du, S. Sukhishvili, SERS not to be taken for granted in the presence of oxygen. J. Am. Chem. Soc. **131**, 7480–7481 (2009)
37. M.K. Khaing Oo, Y. Han, R. Martini, S. Sukhishvili, H. Du, Forward-propagating surface-enhanced Raman scattering and intensity distribution in photonic crystal fiber with immobilized Ag nanoparticles, Opt. Lett. **34**, 968–970 (2009)
38. T.P. White, R.C. McPhedran, C.M. De Sterke, L.C. Botten, M.J. Steel, Confinement losses in microstructured optical fibers. Opt. Lett. **26**, 1660–1662 (2001)
39. Y. Zhu, H. Du, R. Bise, Design of solid-core microstructured optical fiber with steering-wheel air cladding for optimal evanescent-field sensing. Opt. Express **14**, 3541–3546 (2006)
40. M.K. Khaing Oo, Y. Han, J. Kanka, S. Sukhishvili, H. Du, Structure fits the purpose: Photonic crystal fibers for evanescent-field surface-enhanced Raman spectroscopy, Opt. Lett. **35**, 466–468 (2010)
41. H.W. Lee, Y. Liu, K.S. Chiang, Writing of long-period gratings in conventional and photonic-crystal polarization-maintaining fibers by CO_2-laser pulses. IEEE Photon. Technol. Lett. **20**, 132–134 (2008)
42. B.H. Kim, Y. Park, T.J. Ahn, B.H. Lee, Y. Chung, U.C. Paek, W.T. Han, Residual stress relaxation in the core of optical fiber by CO_2 laser radiation. Opt. Lett. **26**, 1657–1659 (2001)
43. D.D. Davis, T.K. Gaylord, E.N. Glytsis, S.C. Mettler, Very-high-temperature stable CO2 laser-induced longperiod fibre gratings. Electron. Lett. **35**, 740–742 (1999)
44. L. Rindorf, J.B. Jensen, M. Dufva, L.H. Pedersen, P.E. Høiby, O. Bang, Photonic crystal fiber long-period gratings for biochemical sensing. Opt. Express **14**, 8224–8231 (2006)
45. Z. He, F. Tian, Y. Zhu, N. Lavlinskaia, H. Du, Long-period gratings in photonic crystal fiber as an optofluidic label-free biosensor. Biosens. Bioelectron. **26**, 4774–4778 (2011)
46. G. Decher, Fuzzy nanoassemblies: toward layered polymeric multicomposites. Science **277**, 1232–1237 (1997)
47. S. Pavlukhina, S. Sukhishvili, Polymer assemblies for controlled delivery of bioactive molecules from surfaces. Adv. Drug Deliver. Rev. **63**, 822–836 (2011)
48. E. Kharlampieva, V. Kozlovskaya, S.A. Sukhishvili, Layer-by-layer hydrogen-bonded polymer films: from fundamentals to applications. Adv. Mater. **21**, 3053–3065 (2009)
49. Z. Tang, Y. Wang, P. Podsiadlo, N.A. Kotov, Biomedical applications of layer-by-layer assembly: from biomimetics to tissue engineering. Adv. Mater. **18**, 3203–3224 (2006)
50. F. Tian, J. Kanka, S. Sukhishvili, H. Du, Photonic crystal fiber for layer-by-layer assembly and measurements of polyelectrolyte thin films. Opt. Lett. **37**, 4299–4301 (2012)

51. Y. Zhu, Z. He, J. Kanka, H. Du, Numerical analysis of refractive index sensitivity of long-period gratings in photonic crystal fiber. Sens. Actuators, B **129**, 99–105 (2008)
52. S. Pavlukhina, Y. Lu, A. Patimetha, M. Libera, S. Sukhishvili, Polymer multilayers with pH-triggered release of antibacterial agents. Biomacromolecules **11**, 3448 (2010)

Chapter 16
Overview of Micro- and Nano-Structured Surface Plasmon Resonance Fiber Sensors

Byoungho Lee and Taerin Chung

Abstract We overview the micro- and nano-structured optical fiber sensors, directly associated with surface plasmon resonance (SPR) sensing. Fiber sensors combined with SPR technologies offer a new route to improving the sensing capability. Various approaches have been exploited and optimized. For example, D-shape, cladding-off, and tapered fiber structures as well as dielectric or metallic gratings have been introduced. These micro- and nano-structured SPR fiber sensors have received much attention due to their high sensitivity, sensor miniaturization, and the flexibility of optical fibers. Key issues for fiber SPR sensors are low loss and high overlap with sensing medium. These fiber sensors with the use of SPR have been continually studied for chemical and bio-sensing. In this chapter we will review the current status of these micro- and nano-structured optical SPR fiber sensors.

16.1 Introduction

An optical fiber sensor is a sensor that uses optical fibers as a sensing element and also, in many cases, as a means of relaying signals from a remote sensor to an electronic module called an interrogator that processes the signals. The fiber sensors have been used because of their small sensing element (sensor head) size and high sensitivity. Moreover, electrical power can be supplied to the remote

B. Lee (✉) · T. Chung (✉)
School of Electrical Engineering, Seoul National University,
1 Gwanak-Gu Gwanakro, Seoul 151-744, Korea
e-mail: byoungho@snu.ac.kr

T. Chung
e-mail: taerinc@gmail.com

A. Cusano et al. (eds.), *Lab-on-Fiber Technology*,
Springer Series in Surface Sciences 56, DOI: 10.1007/978-3-319-06998-2_16,

location of the signal processor and light source, not to the location of sensing head. Another advantage is that many sensors can be multiplexed along the length of a fiber by using different wavelengths of light for each sensor head or by measuring the time delay as light transmits along the fiber through each sensing element [1]. In addition to these properties, the fiber sensors have the advantages of light weight, structural versatility, immunity to electromagnetic interference, large bandwidth and environmental ruggedness [2]. To implement much higher sensitivity for some applications, fiber-optic surface plasmon resonance (SPR) sensors with micro- and nano-structures were introduced in 1983 as shown in Fig. 16.1. Since then, surface plasmons, which have strong evanescent field at the interface between metallic substrate and sensing medium, have been employed in a variety of fiber sensors [3–5].

In this chapter, we mainly describe the SPR fiber sensor as a sensing element which can measure the refractive index (RI) change of a sample medium. First, the fundamental principle of SPR is briefly explained and then various approaches to improve the sensitivity of conventional SPR fiber sensors are illustrated. For achieving sensitivity improvement in SPR fiber sensors, there are two key agendas—achieving high overlap of plasmonic waves with sample medium (to be measured) and achieving low loss. High overlap and low loss are necessary to attain high sensitivity and enough optical power of guided-light systems [6–8]. In an attempt to increase the overlapping area, the fiber cladding is modified through different approaches such as cladding-off, side-polishing, tapering, and angled fiber tip. Other approaches deal with the plasmonic structures such as metallic gratings, metallic nanoparticles and nanoholes or use gratings made in the core or cladding of optical fiber for efficient coupling of light with plasmonic waves.

Many developments on hybrid fiber SPR sensors can also be found in recent literatures [9–12]. Here we provide an integrated outlook of fiber-optic sensors advanced by micro- or nano-structure optical fibers with regard to physical characteristics, fabrication methodologies and sensing performance.

16.2 Fundamentals of Fiber-Optic Surface Plasmon Resonance Sensors

In this section, we discuss some basic principles of SPR in fiber based SPR sensors. An optical fiber is composed of a core and a cladding which has a slightly lower RI compared with that of the core. Light propagation is confined within the fiber due to total internal reflection (TIR). In typical SPR fiber sensors, the silica cladding is removed to reveal a small portion of the fiber core and a thin metallic layer is coated on it. The metallic layer is enclosed by a sensing layer, i.e., medium to be measured. When a light is launched in a given sensor, the evanescent field from the guided core mode excites surface plasmons at the interface of metal and sensing layers. The coupling efficiency of the core mode to the surface plasmons strongly depends on the wavelength, fiber parameters, sensor structure, and metal layer properties.

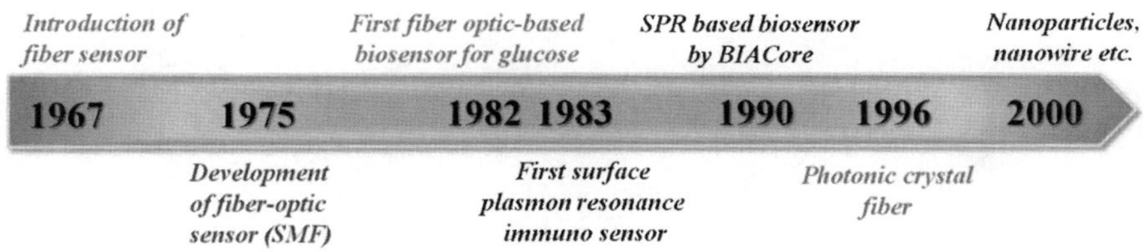

Fig. 16.1 The history of fiber-optic and fiber SPR sensors

16.2.1 Basic Principles of Surface Plasmon Resonance

SPR refers to the excitation of surface plasmon polaritons (SPPs) which are electromagnetic waves coupled with collective electron oscillations and trapped along metal-dielectric interfaces. SPPs are guided along the metal-dielectric interfaces in the same way that light is guided by an optical fiber, with the unique property of subwavelength-scale confinement perpendicular to the interface [13]. Fundamentally, the SPPs are excited by a longitudinal (TM- or *p*-polarized) wave that has a surface normal electric field component as shown in Fig. 16.2. The magnitude of electric and magnetic fields is at its maximum at the interface between the metal and dielectric layers. The propagation constant of the surface plasmon wave propagating along the metal-dielectric interface is expressed by

$$k_{sp}(\omega) = \mathrm{Re}\left(\frac{\omega}{c}\sqrt{\frac{\varepsilon_m \varepsilon_s}{\varepsilon_m + \varepsilon_s}}\right), \tag{16.1}$$

where ε_m and ε_s are the permittivity of the metal layer and dielectric (sample) medium, respectively, ω is the frequency of incident light, and c is the speed of light in vacuum.

16.2.2 Schematic and Sensing Principles of Fiber SPR Sensor

There are a few configurations that can excite surface plasmons. The general approach is the attenuated TIR method based on a prism coupler, called the Kretschmann method, as shown in Fig. 16.3a. Optical SPR fiber sensors are fundamentally analogous to Kretschmann's prism structure, where the prism is substituted by a fiber core, as represented in Fig. 16.3b. As shown in the schematic of SPR fiber sensor, the cladding has been etched off and the core is symmetrically coated with noble metal film.

In the configuration of Fig. 16.3a, the component of propagation wave-vector of light along dielectric medium should be equal to the propagation wave-vector of the surface plasmons in order to excite them. A prism with a high dielectric constant is required because the incident light directly from air or sample region

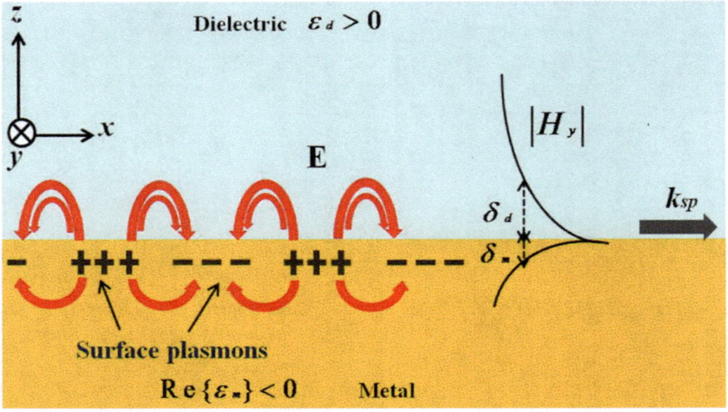

Fig. 16.2 Surface plasmon polaritons propagating along the interface between the dielectric and metal layers. The penetration depths to the dielectric and metal layers are denoted by δ_d and δ_m, respectively

cannot excite SPPs at the metal-air or metal-sample interface. When an angle of incident light beam is greater than the critical angle at the prism-air interface, the TIR of light beam can be observed. However, at a resonance angle the incident light excites the SPPs at the boundary of the metal layer and outside sample region. The evanescent tail of plasmonic waves as shown in Fig. 16.3a experiences the RI of the sample region. Hence, the effective propagation constant of the SPPs strongly depends on the RI of the sample region.

The interface-directional propagation constant of the incident wave at the interface is given by

$$k_i(\omega) = \frac{\omega}{c}\sqrt{\varepsilon_p}\sin\theta, \tag{16.2}$$

where ε_p indicates the dielectric constant of the prism material and θ is the angle of incident beam. The surface plasmon wave can be excited when the (16.1) and (16.2) are equal:

$$\mathrm{Re}\left(\sqrt{\frac{\varepsilon_m\varepsilon_s}{\varepsilon_m+\varepsilon_s}}\right) = n_p\sin\theta, \tag{16.3}$$

where $n_p(=\sqrt{\varepsilon_p})$ is the RI of the prism in Fig. 16.3a or RI of the fiber core in Fig. 16.3b. Hence, the SPP excitation condition sensitively responds to any change in the boundary condition. For example, the variation of RI in the sample medium, resulting from the biological interactions or reactions, leads to the change in the propagation constant of surface plasmon wave. Another key issue is how far the surface plasmon wave penetrates. As mentioned above, the surface plasmon wave has a maximum field at the interface and decays on both sides of metal and dielectric region. The depth of penetration (defined as the depth for which the field

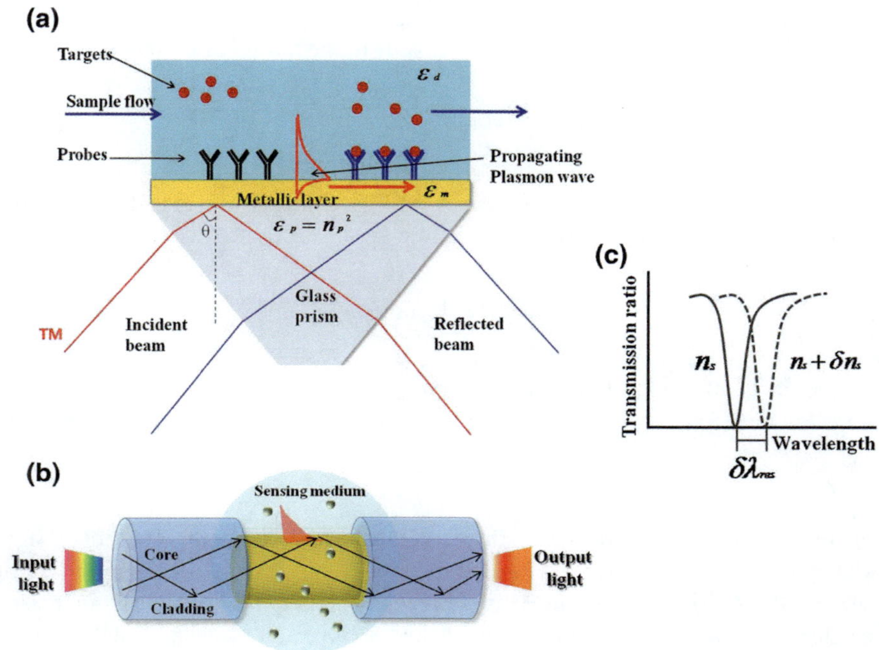

Fig. 16.3 Schematics of **a** Kretschmann configuration, **b** SPR fiber sensor and **c** sensing principle associated with SPR

is decreased by exp(−1) when compared with the field at the interface) through a dielectric (sample) medium can be given as

$$\delta_d = \left[\mathrm{Re}\left(j\frac{2\pi}{\lambda}\frac{\varepsilon_s}{\sqrt{\varepsilon_m + \varepsilon_s}}\right)\right]^{-1}. \tag{16.4}$$

If the normalized reflected intensity in Fig. 16.3a is measured as a function of incident angle θ by keeping other parameters and components unchanged, the sharp resonance dip is observed at the resonance angle θ_{res} due to an efficient transfer of energy to surface plasmons. When the wavelength interrogation method is used, the shift of resonance wavelength in response to the variation of RI in sample medium can be observed as in Fig. 16.3c.

To demonstrate the capability of SPR fiber sensor, sensitivity is very important for the evaluation of optical sensors. There are several ways to define the sensitivity. Here follows three examples: The first is to define it as the resonant angle shift with respect to the variation of surrounding RI as follows:

$$S_\theta = \frac{\partial\theta_{res}}{\partial n_s}. \tag{16.5}$$

Another definition of sensitivity in the spectral interrogation with wavelength modulation is the ratio of the wavelength shift to the variation of surrounding RI.

$$S_\lambda = \frac{\partial \lambda_{res}}{\partial n_s}. \tag{16.6}$$

When using a broadband light source, the figure of merit (FOM) defined as the ratio of (16.6) to the resonance width in the spectrum can be useful:

$$FOM = \frac{S_\lambda}{\Delta \lambda_{res}}. \tag{16.7}$$

The third definition of sensitivity is given by

$$S_R = \frac{\partial R}{\partial n_s}, \tag{16.8}$$

where R is the reflectivity of right at resonance wavelength. The input light wavelength is matched to the SPR wavelength when there is no change in RI of sample. If there is a change in RI of sample, the reflectivity of light is increased because the resonance wavelength is changed. Hence, for an incident light with given wavelength, the reflectivity increases with the change in RI of sample. When this is applied to optical SPR fiber sensor of Fig. 16.3b, the reflectivity becomes transmission coefficient through the optical fiber because the reflected light at the core-metal boundary propagates through the optical fiber. Hence, in that case, the following notation is more appropriate:

$$S_T = \frac{\partial T}{\partial n_s}, \tag{16.9}$$

where T denotes optical power transmission coefficient through the optical fiber. In some cases the sensitivity is represented as the minimum detectable change in RI. In fact this is the number the user of sensor is most interested.

16.3 Various Micro- and Nano-Structured SPR Fiber Sensors

16.3.1 Micro- and Nano-Structured SPR Fiber Sensors Based on Fiber Shaping

Optical fiber is a useful substitute to waveguide to easily induce the evanescent field and SPR. Compared with the prism SPR sensor structure, shortcomings in fiber SPR sensors including low sensitivity and detection limit are addressed. In addition, it is difficult to control the angle of the incident light, which allows the resonance curve of transmission spectrum broader. Hence, various approaches to overcome the drawbacks of SPR fiber sensors have been reported. There are mainly two categories to distinguish SPR fiber sensors as presented in Fig. 16.4a: One is fiber shaping such as tapering, D-shaping and side-polishing. Another is adding distinct structures featured by fiber gratings, subwavelength metallic gratings or the use of metallic layers in photonic crystal fibers. Reported sensitivity in various types of SPR fiber sensors is presented in Fig. 16.4b, showing the broadband wavelength range from visible to infrared regime. In attempt to resolve low sensitivity issue in conventional fiber sensors, micro- or nano-structured SPR fiber sensors have been continuously exploited.

To boost interactions between guided light and sensing materials, the fiber cladding can be taken off by polishing. Substantial overlapping area between sensing material and optical field can be achieved. It is capable of enhancing the sensitivity. In a tapered fiber, the power transmission is reduced in comparison with normal fiber. But, the transmission strongly depends on the outside material, which enables us to use it as a sensor. The small changes in the RI or the thickness of this overlay remarkably influence on the transmission properties in the multi-mode central region. In the tapered region, a critical factor to be addressed is taper ratio which determines the sensitivity in this sensor system. In Fig. 16.5, we can identify that the sensitivity increases with the taper ratio increments in a given tapered fiber geometry [14].

However, the decrease in signal light transmission with the increase of taper ratio would cause deterioration of signal-to-noise ratio in detection. The enlargement of the area of evanescent waves in contact with sensing regions is attributed to tapering fiber. At the metallic layer, incident wave vector and corresponding resonance wavelength shift rely on the taper ratio. As a consequence, sensitivity in the spectral interrogation method directly depends on the taper ratio. The modified fibers also involve side-polished fibers, D-shaped fibers and U-shaped fibers. Figure 16.6 describes the general and advanced configurations of SPR fiber sensors incorporating tapered fibers and the schematic of U-shaped fiber. A typical tapered fiber SPR sensing probe is shown in Fig. 16.6a [15, 16]. Changing the profile of tapered SPR sensing probe can also influence on the sensitivity of the sensor [15]. To improve the sensitivity, the SPR probe of uniform core with metallic coating sandwiched between two unclad tapered fiber regions was reported as shown in Fig. 16.6b. In this unique geometry, all guided rays can propagate up to the output end of the fiber. This is achieved by selecting the minimum value of the radius in the uniform core of the sensing region. Rays in the sensing region propagate close to the critical angle of the region. In Fig. 16.6c, an SPR fiber sensor with uniform metal coated U-shaped probe is adopted. The bending radius in U-shaped probe makes it possible to tune the sensitivity of the probe [17]. Furthermore, one-side metal coated fibers with and without remaining cladding, and structures with modified fiber tips such as flat or angled structures have been proposed [18, 19].

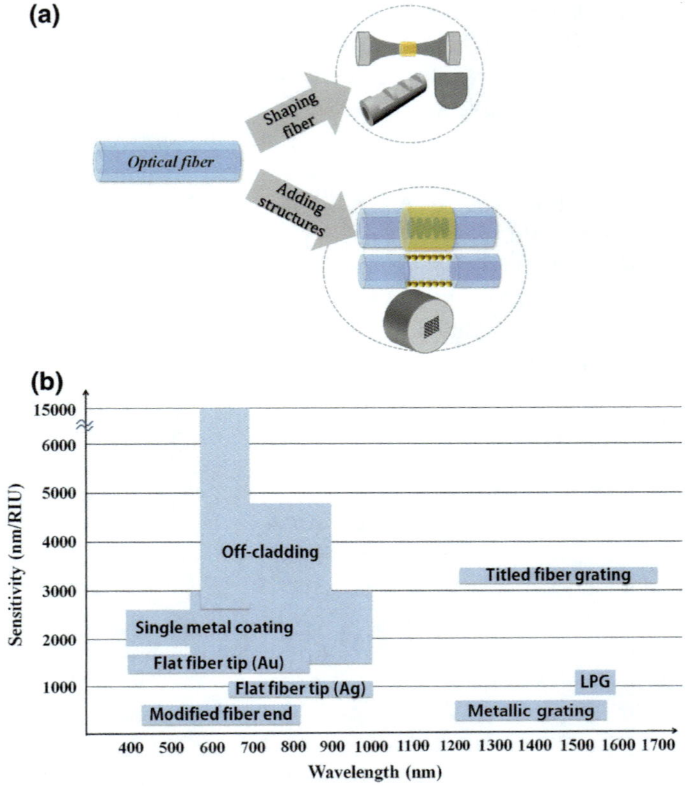

Fig. 16.4 **a** Various SPR fiber sensors and **b** reported sensitivity with respect to wavelength and sensing structure (*LPG* long-period grating, *RIU* refractive index unit)

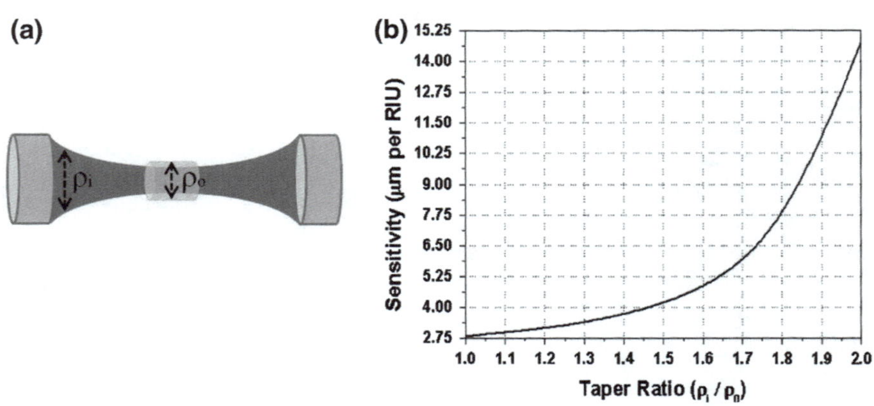

Fig. 16.5 **a** Tapered fiber and **b** sensitivity plot according to the taper ratio in SPR fiber sensor. Metal-coated length is a third of whole tapered length in step-index multimode fiber. Given fiber core diameter is 600 μm and gold layer thickness is 50 nm. ©IEEE [14]

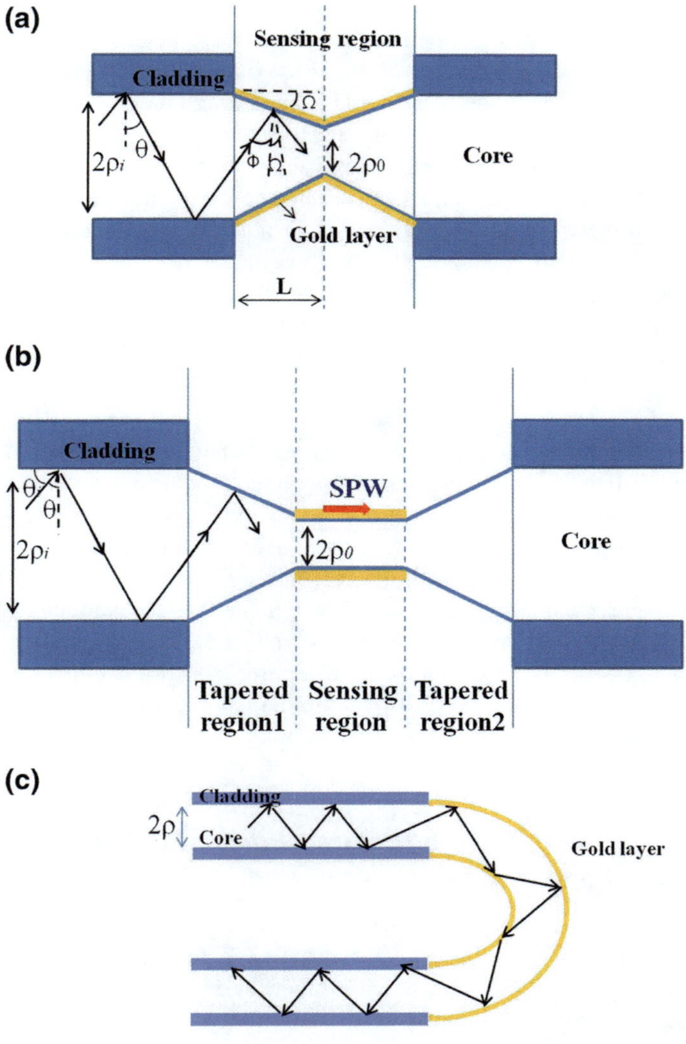

Fig. 16.6 SPR fiber sensors with **a** tapered probe, **b** uniform sensing probe sandwiched between two tapered fiber regions, and **c** U-shaped fiber sensing probe

The side-polished fibers including D-shaped fibers in Fig. 16.7 are usually fabricated by CO_2 laser machining or femtosecond laser. Traditionally, the plastic cladding of optical fiber was easily stripped to expose the fiber core. The operating sensing principle in side-polished fibers is based on attenuated TIR via multiple internal reflections along the fiber. The attenuated light intensity linearly responds to the increase of surrounding RI. The loss of light energy caused by the sensing portion of the fiber is detected by a sensor interrogation system [20–24]. The ability of multiple D-shaped zones in multi-mode fiber as a high sensitivity RI sensor was also demonstrated in Figs. 16.7a–b. The sensor resolution can be optimized with respect to the number of D-shaped zones in Fig. 16.7b [22]. When the number of D-shaped zones augments, the sensor resolution improves and then becomes worse after exhibiting the best value at five D-shaped zones in this work. In general, single-mode fiber SPR sensors are more sensitive than multi-mode fibers. Single mode tapered fiber with uniform waist with asymmetric metallic coating in Fig. 16.7c offers advanced immunity to deformation of optical fibers [25]. This desirable performance results from the use of polarization-maintaining optical fiber and thus enables more accurate SPR measurements. Side polished single-mode fiber SPR sensor with a thin metal overlayer in Fig. 16.7d was introduced as well. In this configuration, the main feature is single-polarization single-mode.

Aforementioned, many groups have reported the adjustment of resonance wavelength in the transmission spectrum and the improvement of sensitivity with regard to an overlayer in SPR fiber sensors. The overlayer is basically required to protect the metallic surface against oxidation. But, it needs an improvement to perform the behavior of sensor. For example, bimetallic layers with silver and gold are deposited on the optical fiber [26–28]. A double-clad fiber, characterizing a structure composed of three layers was proposed [29]. It is usually available in the high power fiber laser system with the core doping of rare earth elements and has the jacket made of a polymer which has lower index than the cladding. Thus, it can guide the incident light into the cladding as well as the core. The polymer can act as an outer cladding. These kinds of structures constitute either overlayer or multilayer structures. The overlayer or multilayer make it possible to tune the measurable range in SPR fiber sensor.

16.3.2 Micro- and Nano-Structured Fiber SPR Sensors Based on Gratings

In this section, SPR fiber sensors with various types of gratings are introduced. Gratings can couple light from the core mode into various cladding modes that might induce SPPs on metal-sample interface. In some cases, metallic gratings can compensate the phase mismatch between incident light and SPPs. In Fig. 16.8, fiber optic SPR sensors equipped with various gratings are illustrated. Long-period gratings (LPGs) in Fig. 16.8a typically have grating periods of hundreds of micrometers [30–32]. Although fiber Bragg gratings are widely used for optical sensors detecting strain and temperature, LPGs are more suitable to SPR fiber sensors. The LPG couples light from a guided mode into forward propagating cladding modes. The degree of coupling from the guided mode to cladding modes is wavelength dependent. LPGs have broad resonance spectrum to cladding modes. Meanwhile, other types of gratings advanced from the fundamental gratings have also been studied. Both tilted fiber gratings in Fig. 16.8b and metallic Bragg gratings exhibit multiple dips in the transmission spectrum of SPR fiber sensors [33, 34]. These gratings are capable of producing the sharp resonance dips

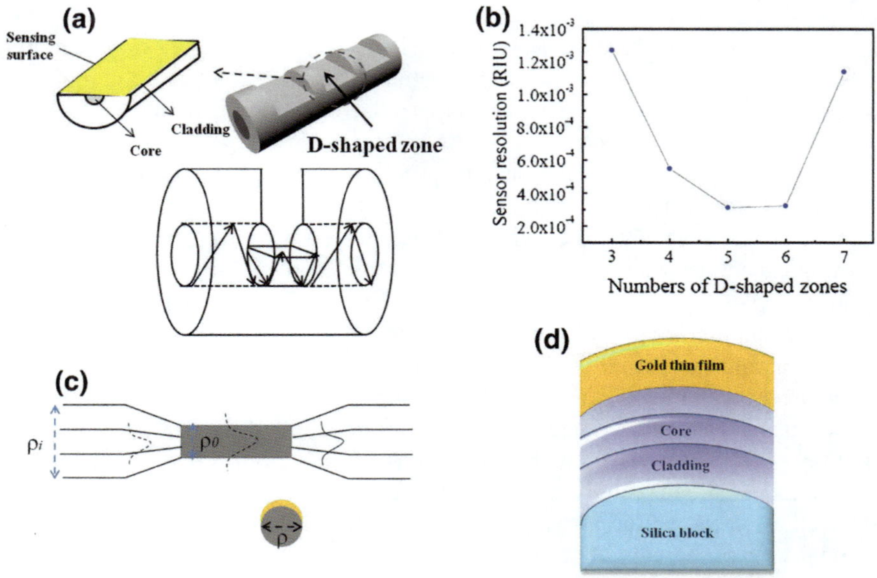

Fig. 16.7 Shaping fiber geometry with metal layer. **a** D-shaped fiber SPR sensor and **b** sensor resolution according to the number of D-shaped zones (Reproduced with permission from sensors; published by MDPI, 2011). **c** Single-mode tapered fiber with uniform waist made of high birefringence optical fiber. Asymmetric gold layer is deposited on it. **d** Side-polished single-mode fiber SPR sensors

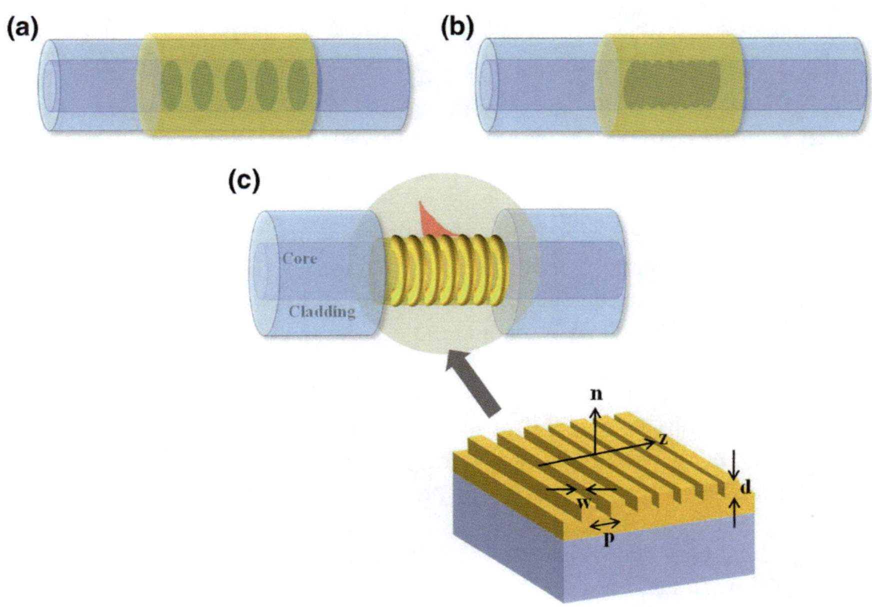

Fig. 16.8 Fiber SPR sensors using **a** long-period grating, **b** tilted grating, and **c** subwavelength metallic grating

and a superior value of signal to noise ratio (SNR), though the sensitivity is slightly deteriorated. Titled fiber gratings have a certain tilt angle between grating plane and fiber cross section, giving rise to the occurrence of more complicated mode coupling. It can be modeled as a reflection of core mode to backward propagating cladding modes.

In Fig. 16.8c, subwavelength metallic gratings are adopted into the optical fiber to enhance the sensitivity [35]. We numerically investigate the subwavelength metallic gratings in Fig. 16.8c via rigorous coupled wave analysis, to verify the sensing capability. The core is symmetrically enclosed by subwavelength metallic gratings whose dimensions are optimized. The structural parameters of given subwavelength gratings were set to the grating depth of 40 nm, the engraved part of 55 nm, and the period of 239 nm. Incident angle was set to 85° to match the resonance wavelength for the case of multi-mode fiber. Such a large incident angle allows subwavelength metallic gratings to be applied in SPR fiber sensor. TM-polarized plane waves with a fixed incident angle but different wavelengths were incident and reflected at the metallic grating layer. The calculated reflectance depending on the period and incident wavelength is shown in Fig. 16.9a. Strong resonance is observed within the wavelength range between 1,000 and 1,060 nm. As the period length in given subwavelength metallic gratings increases, effective dielectric constant depending on grating variables gets smaller. As a consequence, the resonance wavelength is slightly blue-shifted as the period length increases. For the purpose of examining the sensing capability in a given geometry, the reflectance spectra in response to the variation of surrounding RI from 1.3 to 1.36 are presented in Fig. 16.9b. The resonance wavelength is red-shifted in the near-infrared wavelength range with respect to the increase of corresponding RI. The calculated sensitivity in this subwavelength metallic grating structure shows approximately 3,583 nm/RIU.

To identify the electromagnetic field distribution arisen from this given structure, Fig. 16.10 provides the field distributions at the resonance wavelength of 1,000 nm. It implies that the field is coupled in the surface plasmon mode and produces strong evanescent wave in near-field regime. This evanescent field is in charge of the sensitivity in response to surrounding RI change.

In fabrication of metallic nanostructures onto the fiber end shown in Fig. 16.11, a standard optical communication fiber is used for this fiber-optic sensing probe. The diameter of fiber core is 9 μm and the cladding is 125 μm. Typically, gold is chosen as the deposited metal on the fiber end face, because it is sensitive to a change in the RI of the sensing layer and chemically stable. The fabrication procedure is as follows: the optical fiber was stripped of its polymer buffer layer and manually cut with a fiber cleaver for the purpose of realizing a flat and smooth fiber end surface. Next, the fiber end probe was coated with metal by e-beam evaporator. Before depositing metal, a titanium layer of 5 nm was first deposited on the cleaved fiber end. It plays as the adhesion layer between the fiber end face and metal layer. Then, gold is subsequently deposited over the titanium layer, approximately 850 nm thick. In patterning the nano-sized structure, focused ion

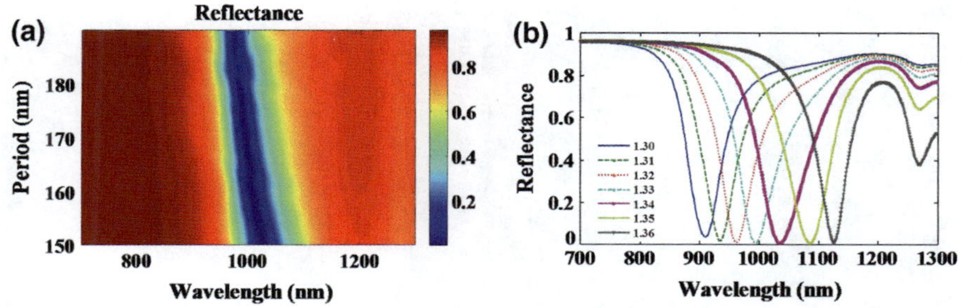

Fig. 16.9 **a** Reflectance according to the period of grating and incident wavelength, **b** reflectance spectra in response to the variation of surrounding refractive index from 1.3 to 1.36

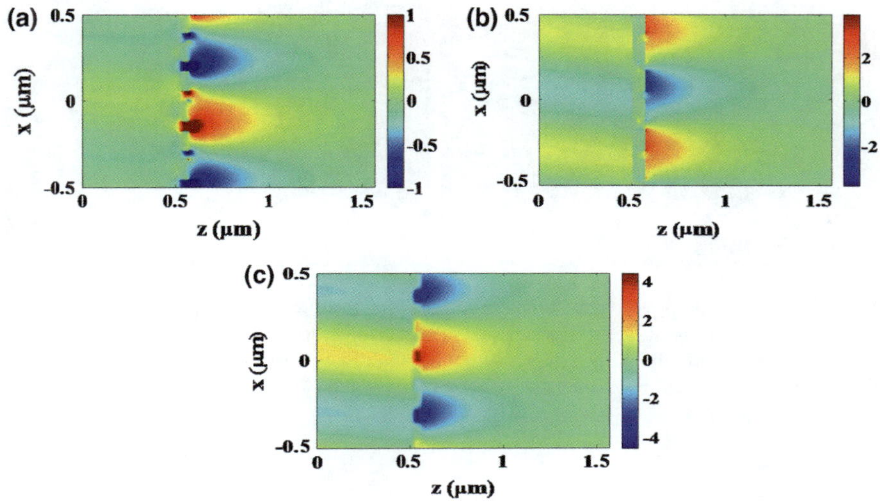

Fig. 16.10 Electromagnetic field distributions **a** x-component distribution of electric field, **b** z-component distribution of electric field, and **c** y-component distribution of magnetic field in x–z plane of given subwavelength metallic gratings when the incident wavelength is 1,000 nm and surrounding refractive index is 1.33

beam milling is used. By manipulating the shape and size of nanostructures in the metal layer, the location of the resonance wavelength can be adjusted.

Figure 16.11a–d present the SEM images of various periodic nanostructures mounted on the optical fiber tip. Figure 16.11a shows the overall fiber probe end which is patterned in the core region. The fabricated area is 13 μm by 13 μm square, which is enough to cover the fiber core. As mentioned above, these gratings mounted onto the fiber end are capable of enhancing the sensitivity [36]. In addition, the resonance wavelength can be tuned by engineering the dimension of metallic gratings.

Fig. 16.11 SEM images of various nanostructures fabricated on an optical fiber tip. **a** Fiber end tip with a 2D grating structure located in the center of cross section, **b** one-dimensional slit grating structures on a thick metal layer on a fiber end, **c** two-dimensional grating structures on the corresponding fiber end, and **d** split ring structures on the corresponding fiber end. The thickness of metal layer is 850 nm

16.3.3 Nano-Structured LSPR Fiber Sensors

Not only SPR but also localized SPR (LSPR) has been of interest for chemical and biological sensing applications. LSPR arisen from a metallic nanoparticle or nanostructure refers to collective oscillation of conduction electrons in a metallic nanoparticle or nanostructure at a resonant frequency which is determined by its size, shape, composition, and the RI of surrounding dielectrics. Sharma and Gupta proposed the use of metallic nanoparticle film for the purpose of enhancing the sensitivity in the optical fiber sensor shown in Fig. 16.12a [37]. Metallic nanoparticles are coated on the unclad portion of the fiber as presented in Fig. 16.12. On the other hand, the array of nanoapertures or nanoholes mounted on the tip of gold-coated optical fibers can also be employed as highly sensitive RI sensors as shown in Fig. 16.12b. As the localized RI of the medium in the vicinity of gold film possessing the array of nanoapertures is varied, a shift is detected in the

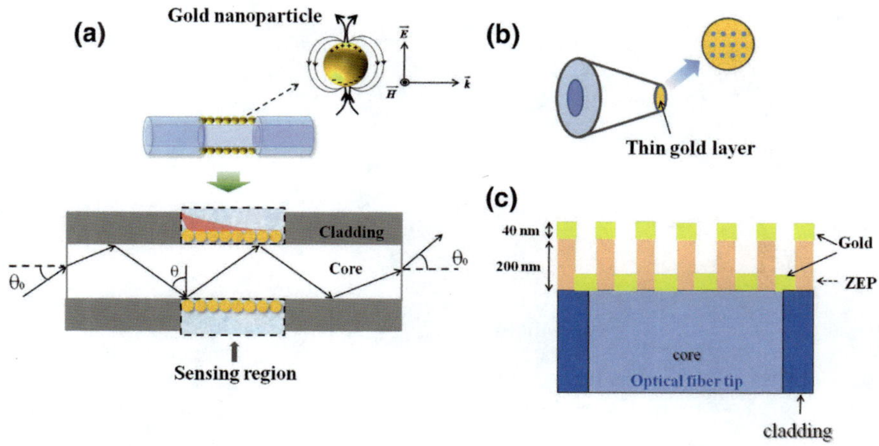

Fig. 16.12 Nano-structured LSPR fiber sensor **a** with metallic nanoparticle layer, **b** metallic nanohole array mounted on a fiber tip, and **c** hybrid metallo-dielectric nanostructures

transmission spectrum of metallic nanostructures. This sensor based on the arrays of nanoapertures is beneficial because several extraordinary transmission peaks in the transmission spectrum can be monitored, providing multiple reference points in response to the RI of the medium around the fiber.

Fundamentally, transmission of light through aperture smaller than the wavelength of incident light was discussed by Bethe [38]. According to him, the transmission coefficient of light at normal incidence to an aperture is proportional to the fourth power of the ratio of aperture radius to wavelength, which means that the transmission of light diminishes rapidly with the decrease of aperture radius. However, when the surface plasmon interaction is taken into account, considerably more light can be transmitted through the array of subwavelength apertures at the wavelength which depends on the periodicity of the array and material parameters. Therefore, the peak transmission and corresponding wavelength are mediated by the surrounding RI and can be made use of as the basis for sensors. Therefore, the binding of organic and biological molecules on the metallic surface could be monitored via the arrays of nanoholes in a gold film [39, 40]. In Fig. 16.12c, the sensing platform based on the integration onto the optical fiber tip of two-dimensional hybrid metallo-dielectric nanostructures supporting LSPR is shown and experimentally demonstrated [41]. This platform provides a sensitivity of ~ 125 nm/RIU for detecting the changes in bulk RI of different chemicals surrounding the fiber-tip device. For the sake of reliable fabrication of both dielectric and metallic nanostructures directly on the fiber tip, alternative approaches based on direct-write patterning of the fiber tip have been explored.

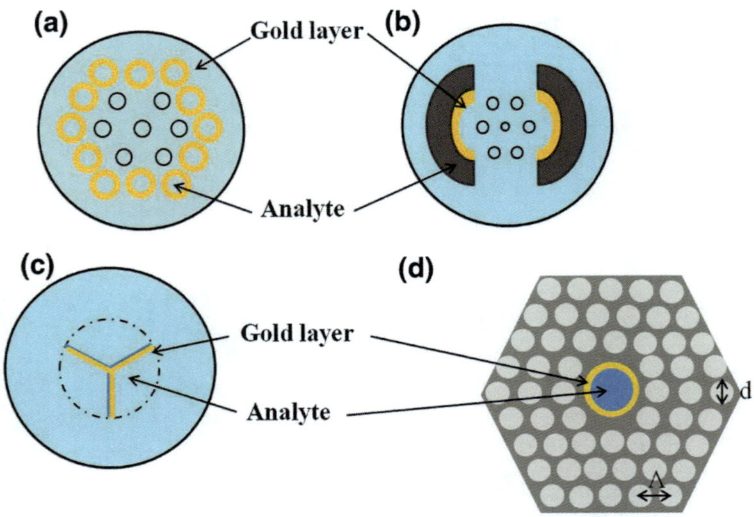

Fig. 16.13 Various PCF-based or microstructured SPR sensors: **a** micro-structured SPR fiber sensor with large metal-coated holes filled with analyte, **b** micro-structured SPR fiber sensor integrated with large semicircular metalized channes, **c** SPR-based three-hole micro-structured fiber sensor, and **d** PCF-SPR temperature sensor with liquid core

16.4 Other Structures of SPR Fiber Sensors

Besides the fiber-optic SPR sensors explained above, there are many other proposals and structures for micro- or nano-structured fiber-optic SPR sensors. Here let us take the examples of distinct photonic crystal fiber (PCF)-based or micro-structured SPR sensors. PCFs or holey fibers have multiple holes along the axial direction in cladding or core of optical fibers. There have been many researches on the use of holes for gas or liquid sensing because holes filled with gas or liquid affect the properties of guiding optical mode [42, 43]. Metallic layer coated in the holes of PCF or microstructured fiber allows the sensitivity to the RI of surrounding medium to increase due to SPR effect.

Figure 16.13 shows various exemplary structures relevant to PCF-based or micro-structured SPR sensors. The sensor in Fig. 16.13a consists of selectively metal-coated air holes containing analyte channels [44]. It helps to enhance the phase matching between the plasmonic mode and the core-guided more. Desirable sensitivity as high as 5,500 nm/RIU was reported in this sensor. In Fig. 16.13b, two large semicircular metalized channels are integrated into the fiber structure to raise microfluidic flow rate and the efficiency of plasmonic excitation [45]. SPR-based three-hole microstructured optical fiber was proposed and numerically demonstrated with 1×10^{-4} RI resolution for aqueous analytes in Fig. 16.13c [46]. Another PCF based on SPR sensor is found in Fig. 16.13d. PCF with the metal nanometer film coated on large central air core of photonic bandgap and infusion of

Table 16.1 Performance of micro- and nano-structured fiber SPR sensors

Structure modification		Property	Wavelength (nm)	RI range	Sensitivity (nm/RIU or RIU)	References
Fiber shaping	Cladding-off	Single metal coating	550–1,000	1.3335–1.4018	1,600–3,000	[48]
		With multi-layer	450–950	1–1.336	10^{-5} (RIU)	[49]
	Tapered fiber	Tapered probe with thin metal coating	600–900	1.33–1.34	2,700–4,900	[16]
		Sandwiched btw. two tapered fiber regions	580–700	1.33–1.34	2,750–15,250	[15]
	D-shaped fiber	Single-metal coating	750–910	1.35–1.42	5,000	[50]
		With over layer	750–950	1.329–1.353	5×10^{-7} (RIU)	[51]
	Modified fiber end	Flat fiber tip	400–840	1.333–1.3469	1,557	[52]
		Angled fiber tip	650	1.33–1.375	4,000 %/RIU (intensity)	[22]
Fiber gratings	LPG	3 μm core planar	~1,550	1.329–1.34	1,100	[30]
	Tilted grating	Angled FBG	1,220–1,700	1.3–1.38	3,365	[34]
	Metallic grating	Metal surface grating	1,200–1,500	1–1.41	500	[35]
Photonic crystal fiber	Three holes	Au	500–600	1.33–1.34	10^{-4}	[46]
	Hollow core Bragg fiber, honey comb PCF	Au			7×10^{-6}–5×10^{-5} RIU	[53]
Metallic Nanostructures	Nanoparticles	Nanoparticle film	400–800	1.342–1.357	1,600–1,900	[37]

the glycerin liquid has been proved to be temperature sensor with liquid core [47]. Hence, the performance of PCF-based SPR sensor can take the full advantage of adjusting the thickness of the metal layer, over-layer and hole diameter.

16.5 Conclusion and Prospect

In this chapter, we overviewed micro- and nano-structured fiber optic SPR sensors with regard to the types, characteristics, and geometry. Assorted fiber SPR sensor structures are briefly organized in Table 16.1. The collaboration of SPR technique with optical fiber technologies has brought considerable achievements in sensing applications. Novel technologies will be continuously tried to be combined with conventional fiber SPR sensors for further advancement. Although the fiber SPR sensors have not been competitive to commercial optical sensor yet, many researchers believe that they will be one of the promising sensor technologies, representing the unique properties of real-time and label-free detection.

Acknowledgments This work was supported by the National Research Foundation (NRF) of Korea grant funded by the Korea government (MSIP) through the Creative Research Initiatives Program (No. 2007-0054847).

References

1. O.S. Wolfbeis, Fiber-optic chemical sensors and biosensors. Anal. Chem. **78**, 3859–3873 (2006)
2. R. Slavik, J. Homola, J. Ctyroky, Miniaturization of fiber optic surface plasmon resonance sensor. Sens. Actuators B **51**, 311–315 (1998)
3. B. Lee, Review of the present status of optical fiber sensors. Opt. Fiber Technol. **9**, 57–79 (2003)
4. L.D. Maria, M. Martinelli, G. Vegetti, Fiber-optic sensor based on surface plasmon interrogation. Sens. Actuators B **12**, 221–223 (1993)
5. R.C. Jorgenson, S.S. Yee, A fiber-optic chemical sensor based on surface plasmon resonance. Sens. Actuators B **12**, 213–220 (1993)
6. J. Homola, R. Slavik, Fibre-optic sensor based on surface plasmon resonance. Electon. Lett. **32**, 480–482 (1996)
7. R. Slavik, J. Homola, J. Ctyroky, Single-mode optical fiber surface plasmon resonance sensor. Sens. Actuators B **54**, 74–79 (1999)
8. F. Bardin, I. Kasik, A. Trouillet, V. Matejec, H. Gagnaire, M. Chomat, Surface plasmon resonance sensor using an optical fiber with an inverted graded-index profile. Appl. Opt. **41**, 2514–2520 (2002)
9. D. Monzon-Hernandex, J. Villatoro, D. Talavera, D. Luna-Moreno, Optical-fiber surface-plasmon resonance sensor with multiple resonance peaks. Appl. Opt. **43**, 1216–1220 (2004)
10. B. Lee, S. Roh, J. Park, Current status of micro- and nano-structured optical fiber sensors. Opt. Fiber Technol. **15**, 209–221 (2009)
11. S. Roh, T. Chung, B. Lee, Overview of the characteristics of micro- and nano-structured surface plasmon resonance sensors. Sensors **11**, 1565–1588 (2011)

12. A. Sharma, R. Jha, B.D. Gupta, Fiber-optic sensors based on surface plasmon resonance: a comprehensive review. IEEE Sens. J. **7**, 1118–1129 (2007)
13. S.A. Maier, *Plasmonics—Fundamentals and Applications* (Springer, New York, 2007)
14. R.K. Verma, A.K. Sharma, B.D. Gupta, Modeling of tapered fiber-optic surface plasmon resonance sensor with enhanced sensitivity. IEEE Photon. Technol. Lett. **19**, 1786–1788 (2007)
15. R.K. Verma, A.K. Sharma, B.D. Gupta, Surface plasmon resonance based tapered fiber optic sensor with different taper profiles. Opt. Commun. **281**, 1486–1491 (2008)
16. Y.-C. Kim, W. Peng, S. Banerji, J.-F. Masson, K.S. Booksh, Tapered fiber optic surface plasmon resonance sensor for analyses of vapor and liquid phases. Opt. Lett. **30**, 2218–2220 (2005)
17. R.K. Verma, B.D. Gupta, Theoretical modeling of a bidimensional U-shaped surface plasmon resonance based fibre optic sensor for sensitivity enhancement. J. Phys. D **41**, Article ID 095106 (2008)
18. J. Zeng, D. Liang, Application of fiber optic surface plasmon resonance sensor for measuring liquid refractive index. J. Intell. Mater. Syst. Struct. **17**, 787–791 (2006)
19. Y.-J. Chang, Y.-C. Chen, H.-L. Kuo, P.-K. Wei, Nanofiber optic sensor based on the excitation of surface plasmon wave near fiber tip. J. Biomed. Opt. **11**, 014032-1-014032-5 (2006)
20. Y. Zhang, C Gu, A.M. Schwartzberg, J.Z. Zhang, Surface-enhanced Raman scattering sensor based on D-shaped fiber. Appl. Phys. Lett. **87**, 123105-1-123105-3 (2005)
21. S.-F. Wang, M.-H. Chiu, J.-C. Hsu, R.-S. Chang, F.-T. Wang, Theoretical analysis and experimental evaluation of D-type optical fiber sensor with a thin gold film. Opt. Commun. **253**, 283–289 (2005)
22. C.-H. Chen, T.-C. Tsao, J.-L. Tang, W.-T. Wu, A multi-D-shaped optical fiber for refractive index sensing. Sensors **10**, 4794–4804 (2010)
23. M.-H. Chiu, C.-H. Shih, Searching for optimal sensitivity of single-mode D-type optical fiber sensor in the phase measurement. Sens. Actuators B **131**, 1120–1124 (2008)
24. M.-H. Chiu, C.-H. Shih, M.-H. Chi, Optimum sensitivity of single-mode D-type optical fiber sensor in the intensity measurement. Sens. Actuators B **123**, 1120–1124 (2007)
25. M. Piliarik, J. Homola, Z. Manikova, J. Ctyroky, Surface plasmon resonance sensor based on a single-mode polarization-maintaining optical fiber. Sens. Actuators B **90**, 236–242 (2003)
26. A. Alvarez-Herrero, H. Guerrero, D. Levy, High-sensitivity sensor of low relative humidity based on overlay on side-polished fibers. IEEE Sens. J. **4**, 52–56 (2004)
27. A.K. Sharma, B.D. Gupta, On the sensitivity and signal-to-noise ratio of a step-index fiber-optic surface-plasmon resonance sensor with bimetallic layers. Opt. Commun. **245**, 159–169 (2005)
28. S.A. Zynio, A.V. Samoylov, E.R. Surovtseva, V.M. Mirsky, Y.M. Shirsov, Bimetallic layers increase sensitivity of affinity sensors based on surface plasmon resonance. Sensors **2**, 62–70 (2002)
29. T. Zhou, F. Pang, T. Wang, High temperature sensor properties of a specialty double cladding fiber. Proc. SPIE **8311**, Opt. Sens. Biophoton. **III**, 831100 (2011)
30. J.-L. Tang, S.-F. Cheng, W.-T. Hsu, T.-Y. Chiang, L.-K. Chau, Fiber-optic biochemical sensing with a colloidal gold-modified long period fiber grating. Sens. Actuators B **119**, 105–109 (2006)
31. Y.-J. He, Y.-L. Lo, J.-F. Huang, Optical-fiber surface-plasmon-resonance sensor employing long-period fiber gratings in multiplexing. J. Opt. Soc. Am. B **23**(5), 801–811 (2006)
32. G. Nemova, R. Kashyap, Theoretical model of a planar integrated refractive index sensor based on surface plasmon-polariton excitation with a long period grating. J. Opt. Soc. Am. B **24**, 2696–2701 (2007)
33. G. Nemova, R. Kashyap, Fiber-Bragg-grating-assisted surface plasmon polariton sensor. Opt. Lett. **31**, 2118–2120 (2006)

34. T. Allsop, R. Neal, S. Rehman, D.J. Webb, D. Mapps, I. Bennion, Characterization of infrared surface plasmon resonances generated from a fiber-optical sensor utilizing tilted Bragg gratings. J. Opt. Soc. Am. B **25**, 481–490 (2008)
35. S. Roh, H. Kim, B. Lee, Infrared surface plasmon resonance in a subwavelength metallic gratings under illumination at a large incidence angle. J. Opt. Soc. Am. B **28**, 1661–1667 (2011)
36. W. Ding, S.R. Andrews, T.A. Birks, S.A. Maier, Modal coupling in fiber tapers decorated with metallic surface gratings. Opt. Lett. **31**, 2556–2558 (2006)
37. A.K. Sharma, B.D. Gupta, Fiber optic sensor based on surface plasmon resonance with nanoparticle films. Photon. Nanotechnol. **17**, 124–131 (2006)
38. H.A. Bethe, Theory of diffraction by small holes. Phys. Rev. **66**, 163–182 (1944)
39. A. Dhawan, M.D. Gerhold, J.F. Muth, Plasmonic structures based on subwavelength apertures for chemical and biological sensing applications. IEEE Sens. J. **8**, 942–950 (2008)
40. Y. Lin, Y. Zou, R.G. Lindquist, A reflection-based localized surface plasmon resonance fiber-optic probe for biochemical sensing. Biomed. Opt. Express **2**, 478–484 (2011)
41. M. Consales, A. Ricciardi, A. Crescitelli, E. Esposito, A. Cutolo, A. Cusano, Lab-on-fiber technology: toward multifunctional optical nanoprobes. ACS Nano **6**, 3163–3170 (2012)
42. A.M.R. Pinto, M. Lopez-Amo, Photonic crystal fibers for sensing applications. J. Sens. **2012**, article ID 598178 (2012)
43. F.M. Cox, A. Argyros, M.C.J. Large, Liquid-filled hollow core microstructured polymer optical fiber. Opt. Express **14**, 4135–4140 (2006)
44. X. Yu, Y. Zhang, S. Pan, P. Shum, M. Yan, Y. Leviatan, C. Li, A selectively coated photonic crystal fiber based surface plasmon resonance sensor. J. Opt. **12**, 015005-1-015005-4 (2010)
45. A. Hassani, M. Skorobogatiy, Design criteria for microstrucutred-optical-fiber based surface-plasmon-resonances sensors. J. Opt. Soc. Am. B **24**, 1423–1429 (2007)
46. M. Hautakorpi, M. Mattinen, H. Ludvigsen, Surface-plasmon-resonance sensor based on three-hole microstructured optical fiber. Opt. Express **16**, 8427–8432 (2008)
47. P.B. Bing, Z.Y. Li, J.Q. Yao, Y. Lu, Z.G. Di, A photonic crystal fiber based on surface plasmon resonance temperature sensor with liquid core. Mod. Phys. Lett. B **26**, 1250082-1-1250082-9 (2012)
48. M. Kanso, S. Cuenot, G. Louarn, Sensitivity of optical fiber sensor based on surface plasmon resonance: modeling and experiments. Plasmonics **3**, 49–57 (2008)
49. A.K. Sharma, G.J. Mohr, Theoretical understanding of an alternating dielectric multilayer-based fiber optic SPR sensor and its application to gas sensing. New J. Phys. **10**, 023039 (2008)
50. J. Homola, R. Slavik, J. Ctyroky, Interaction between fiber modes and surface plasmon waves: Spectral properties. Opt. Lett. **22**, 1403–1405 (1997)
51. R. Slavik, J. Homola, J. Ctyroky, E. Brynda, Novel spectral fiber optic sensor based on surface plasmon resonance. Sens. Actuators B **74**, 106–111 (2001)
52. H. Suzuki, M. Sugimoto, Y. Matsui, J. Kondoh, Effects of gold film thickness on spectrum profile and sensitivity of a multimode-optical-fiber SPR sensor. Sens. Actuators B **132**, 26–33 (2008)
53. B. Gauvreau, A. Hassani, Majid Fassi Fehri, A. Kabashin, M. Skorobogatiy, Photonic bandgap fiber-based surface plasmon resonance sensors. Opt. Express **15**, 11413–11426 (2007)

Index

A. Cusano et al. (eds.), *Lab-on-Fiber Technology*,
Springer Series in Surface Sciences 56, DOI: 10.1007/978-3-319-06998-2,

Printed by Printforce, the Netherlands